DNA Damage and Repair: Methods and Applications

DNA Damage and Repair: Methods and Applications

Editor: Nas Wilson

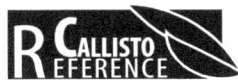

CALLISTO REFERENCE

www.callistoreference.com

Callisto Reference,
118-35 Queens Blvd., Suite 400,
Forest Hills, NY 11375, USA

Visit us on the World Wide Web at:
www.callistoreference.com

ISBN: 978-1-63239-984-7 (Hardback)

Trademark Notice: Registered trademark of products or corporate names are used only for explanation and identification without intent to infringe.

Cataloging-in-Publication Data

DNA damage and repair : methods and applications / edited by Nas Wilson.
    p. cm.
Includes bibliographical references and index.
ISBN 978-1-63239-984-7
1. DNA damage. 2. DNA repair. 3. Biochemical genetics. I. Wilson, Nas.
QH465.A1 D53 2018
616.042--dc23

Table of Contents

Preface..VII

Chapter 1 **Deoxycytidine Kinase Augments ATM-Mediated DNA Repair and Contributes to Radiation Resistance**... 1
Yuri L. Bunimovich, Evan Nair-Gill, Mireille Riedinger, Melissa N. McCracken, Donghui Cheng, Jami McLaughlin, Caius G. Radu, Owen N. Witte

Chapter 2 **Differential Radiosensitivity Phenotypes of DNA-PKcs Mutations Affecting NHEJ and HRR Systems following Irradiation with Gamma-Rays or Very Low Fluences of Alpha Particles**.. 16
Yu-Fen Lin, Hatsumi Nagasawa, John B. Little, Takamitsu A. Kato, Hung-Ying Shih, Xian-Jin Xie, Paul F. Wilson Jr., John R. Brogan, Akihiro Kurimasa, David J. Chen, Joel S. Bedford, Benjamin P. C. Chen

Chapter 3 **The Immune Strategy and Stress Response of the Mediterranean Species of the *Bemisia tabaci* Complex to an Orally Delivered Bacterial Pathogen**........................... 26
Chang-Rong Zhang, Shan Zhang, Jun Xia, Fang-Fang Li, Wen-Qiang Xia, Shu-Sheng Liu, Xiao-Wei Wang

Chapter 4 **Paraspeckle Protein 1 (PSPC1) is Involved in the Cisplatin Induced DNA Damage Response—Role in G1/S Checkpoint**... 35
Xiangjing Gao, Liya Kong, Xianghong Lu, Guanglin Zhang, Linfeng Chi, Ying Jiang, Yihua Wu, Chunlan Yan, Penelope Duerksen-Hughes, Xinqiang Zhu, Jun Yang

Chapter 5 **Exposure to Low-Dose Bisphenol A Impairs Meiosis in the Rat Seminiferous Tubule Culture Model: A Physiotoxicogenomic Approach**.. 45
Sazan Ali, Gérard Steinmetz, Guillaume Montillet, Marie-Hélène Perrard, Anderson Loundou, Philippe Duran, Marie-Roberte Guichaoua, Odette Prat

Chapter 6 **Characterization of DNA Repair Deficient Strains of *Chlamydomonas reinhardtii* Generated by Insertional Mutagenesis**... 60
Andrea Plecenikova, Miroslava Slaninova, Karel Riha

Chapter 7 **Ecotype Diversity and Conversion in *Photobacterium profundum* Strains**............................. 69
Federico M. Lauro, Emiley A. Eloe-Fadrosh, Taylor K. S. Richter, Nicola Vitulo, Steven Ferriera, Justin H. Johnson, Douglas H. Bartlett

Chapter 8 **Comparative DNA Damage and Repair in Echinoderm Coelomocytes Exposed to Genotoxicants**.. 79
Ameena H. El-Bibany, Andrea G. Bodnar, Helena C. Reinardy

Chapter 9 **The Progeroid Phenotype of Ku80 Deficiency is Dominant over DNA-PK$_{CS}$ Deficiency**... 88
Erwin Reiling, Martijn E. T. Dollé, Sameh A. Youssef, Moonsook Lee, Bhawani Nagarajah, Marianne Roodbergen, Piet de With, Alain de Bruin, Jan H. Hoeijmakers, Jan Vijg, Harry van Steeg, Paul Hasty

Chapter 10   **Dot1-Dependent Histone H3K79 Methylation Promotes the Formation of Meiotic Double-Strand Breaks in the Absence of Histone H3K4 Methylation in Budding Yeast**..97
Mohammad Bani Ismail, Miki Shinohara, Akira Shinohara

Chapter 11   **Analysis of the Role of Homology Arms in Gene-Targeting Vectors in Human Cells**...113
Ayako Ishii, Aya Kurosawa, Shinta Saito, Noritaka Adachi

Chapter 12   **Low-Dose Formaldehyde Delays DNA Damage Recognition and DNA Excision Repair in Human Cells**..122
Andreas Luch, Flurina C. Clement Frey, Regula Meier, Jia Fei, Hanspeter Naegeli

Chapter 13   **Sporadic Premature Aging in a Japanese Monkey: A Primate Model for Progeria**............................132
Takao Oishi, Hiroo Imai, Yasuhiro Go, Masanori Imamura, Hirohisa Hirai, Masahiko Takada

Chapter 14   **The Metagenomic Telescope**..141
Balázs Szalkai, Ildikó Scheer, Kinga Nagy, Beáta G. Vértessy, Vince Grolmusz

Chapter 15   **The Hypothetical Protein 'All4779', and not the Annotated 'Alr0088' and 'Alr7579' Proteins, is the Major Typical Single-Stranded DNA Binding Protein of the Cyanobacterium, *Anabaena* sp. PCC7120**......................................150
Anurag Kirti, Hema Rajaram, Shree Kumar Apte

Chapter 16   **DNA Binding Properties of the Actin-Related Protein Arp8 and its Role in DNA Repair**...161
Akihisa Osakabe, Yuichiro Takahashi, Hirokazu Murakami, Kenji Otawa, Hiroaki Tachiwana, Yukako Oma, Hitoshi Nishijima, Kei-ich Shibahara, Hitoshi Kurumizaka, Masahiko Harata

Chapter 17   **Maintenance of Sex-Related Genes and the Co-Occurrence of both Mating Types in *Verticillium dahliae***..173
Dylan P. G. Short, Suraj Gurung, Xiaoping Hu, Patrik Inderbitzin, Krishna V. Subbarao

Chapter 18   **Gemcitabine Induces Poly (ADP-Ribose) Polymerase-1 (PARP-1) Degradation through Autophagy in Pancreatic Cancer**...187
Yufeng Wang, Yasuhiro Kuramitsu, Kazuhiro Tokuda, Byron Baron, Takao Kitagawa, Junko Akada, Shin-ichiro Maehara, Yoshihiko Maehara, Kazuyuki Nakamura

Chapter 19   **Regulation of 53BP1 Protein Stability by RNF8 and RNF168 is Important for Efficient DNA Double-Strand Break Repair**...196
Yiheng Hu, Chao Wang, Kun Huang, Fen Xia, Jeffrey D. Parvin, Neelima Mondal

Permissions

List of Contributors

Index

Preface

DNA damage is termed as the damage that is caused to DNA by naturally occurring processes such as hydrolysis and metabolism and ultraviolet radiation. DNA repair is the process by which the body rebuilds damaged DNA. The symptoms of DNA damage is the development of lesions which cause breaks in the strands of DNA or remove base pairs from the DNA backbone. Aging, cancer, and apoptosis are long-term symptoms of DNA damage. This book provides significant information of this discipline to help develop a good understanding of DNA damage and repair. It aims to serve as a resource guide for students and experts alike and contribute to the growth of the discipline.

This book is a result of research of several months to collate the most relevant data in the field.

When I was approached with the idea of this book and the proposal to edit it, I was overwhelmed. It gave me an opportunity to reach out to all those who share a common interest with me in this field. I had 3 main parameters for editing this text:

1. Accuracy – The data and information provided in this book should be up-to-date and valuable to the readers.

2. Structure – The data must be presented in a structured format for easy understanding and better grasping of the readers.

3. Universal Approach – This book not only targets students but also experts and innovators in the field, thus my aim was to present topics which are of use to all.

Thus, it took me a couple of months to finish the editing of this book.

I would like to make a special mention of my publisher who considered me worthy of this opportunity and also supported me throughout the editing process. I would also like to thank the editing team at the back-end who extended their help whenever required.

Editor

Deoxycytidine Kinase Augments ATM-Mediated DNA Repair and Contributes to Radiation Resistance

Yuri L. Bunimovich[1,2]*, Evan Nair-Gill[1], Mireille Riedinger[1], Melissa N. McCracken[1], Donghui Cheng[4], Jami McLaughlin[6], Caius G. Radu[1,2,3], Owen N. Witte[1,4,5,6]*

1 Department of Molecular and Medical Pharmacology, University of California Los Angeles, Los Angeles, California, United States of America, 2 Crump Institute for Molecular Imaging, University of California Los Angeles, Los Angeles, California, United States of America, 3 Ahmanson Translational Imaging Division, David Geffen School of Medicine, University of California Los Angeles, Los Angeles, California, United States of America, 4 Howard Hughes Medical Institute, University of California Los Angeles, Los Angeles, California, United States of America, 5 Eli and Edythe Broad Center for Regenerative Medicine and Stem Cell Research, University of California Los Angeles, Los Angeles, California, United States of America, 6 Department of Microbiology, Immunology, and Molecular Genetics, David Geffen School of Medicine, University of California Los Angeles, Los Angeles, California, United States of America

Abstract

Efficient and adequate generation of deoxyribonucleotides is critical to successful DNA repair. We show that ataxia telangiectasia mutated (ATM) integrates the DNA damage response with DNA metabolism by regulating the salvage of deoxyribonucleosides. Specifically, ATM phosphorylates and activates deoxycytidine kinase (dCK) at serine 74 in response to ionizing radiation (IR). Activation of dCK shifts its substrate specificity toward deoxycytidine, increases intracellular dCTP pools post IR, and enhances the rate of DNA repair. Mutation of a single serine 74 residue has profound effects on murine T and B lymphocyte development, suggesting that post-translational regulation of dCK may be important in maintaining genomic stability during hematopoiesis. Using [^{18}F]-FAC, a dCK-specific positron emission tomography (PET) probe, we visualized and quantified dCK activation in tumor xenografts after IR, indicating that dCK activation could serve as a biomarker for ATM function and DNA damage response in vivo. In addition, dCK-deficient leukemia cell lines and murine embryonic fibroblasts exhibited increased sensitivity to IR, indicating that pharmacologic inhibition of dCK may be an effective radiosensitization strategy.

Editor: Alexander James Roy Bishop, University of Texas Health Science Center at San Antonio, United States of America

Funding: This work was supported by NIH grant R25T CA098010 [to YLB] and NIAID UCLA CMCR seed grant U19AI067769 [to YLB and ONW], NIH 5T32 AI06567 [to EN-G], CIRM TG2-01169 [to MNM]. ONW is an Investigator of the Howard Hughes Medical Institute and partially supported by the Eli and Edythe Broad Center of Regenerative Medicine and Stem Cell Research. The funders had no role in study design, data collection and analysis, decision to publish, or preparation of the manuscript.

Competing Interests: The authors have declared that no competing interests exist.

* Email: bunimovichyl@upmc.edu (YLB); OwenWitte@mednet.ucla.edu (ONW)

Introduction

Intracellular concentrations of deoxyribonucleotide triphosphates (dNTPs) are tightly regulated to avoid mutagenesis during DNA replication and repair [1]. Mammalian cells synthesize dNTPs by two mechanisms: 1) the *de novo* pathway converts glucose and amino acids to deoxyribonucleotides via ribonucleotide reductase (RNR); 2) the deoxyribonucleoside (dN) salvage pathway generates dNTPs through sequential phosphorylation of recycled deoxyribonucleosides [2]. Deoxycytidine kinase (dCK) is a rate-limiting enzyme in the dN salvage pathway, capable of phosphorylating deoxycytidine (dC), deoxyadenosine (dA) and deoxyguanosine (dG) [3,4]. Indirectly, dCK can also contribute to dTTP pools via the actions of deoxycytidylate deaminase and thymidylate synthase. Several studies have demonstrated increased dCK activity under various genotoxic conditions, including chemotherapy [5–7], ionizing [8–10] and UV [11] radiation, and inhibition of several protein kinases [12–14]. The potentiation of dCK activity was attributed to post-translational modifications that induced a conformational change of the enzyme [15–17]. Phosphorylation of serine 74 (Ser^{74}) was shown to be critical in regulating enzyme activity [18–20]. dCK can adopt an open state,

capable of substrate binding, or a closed, catalytically active, state [21,22]. Serine to glutamic acid (S74E) substitution mimicking Ser^{74} phosphorylation favors the open state and dramatically reduces phosphorylation of purines (dA and dG) but not pyrimidine dC [22].

Ataxia telangiectasia mutated (ATM) serine/threonine protein kinase is at the center of DNA double-strand break (DSB) repair [23]. ATM is a member of phosphoinositide 3-kinase (PI3K)-related protein kinase family, which also includes ataxia telangiectasia and Rad3-related protein (ATR) and catalytic subunit of DNA-dependent protein kinase (DNA-PKcs) [23]. ATM phosphorylates multiple substrates in the nucleus in response to DNA DSBs [24], and regulates several metabolic pathways which counteract oxidative stress and DNA damage [25–29]. In particular, ATM regulates NADPH and ribose-5-phosphate production via the pentose phosphate pathway by promoting phosphorylation of Hsp27, which binds and activates G6PD [25]. ATM also phosphorylates Ser^{72} in the RNR subunit p53R2, which stabilizes the enzyme against degradation and promotes DNA repair [26,27]. While there is much debate about the purpose of such regulatory mechanisms, it is likely that RNR regulation by ATM is needed to maintain dNTP pools and genomic stability [30].

Evidence from global proteomic analysis identified dCK as a target of ATM based on the phosphorylation of the $S^{74}Q$ motif of dCK after ionizing radiation (IR) [31], consistent with recent demonstration of the critical role of dN salvage in DSB repair [32]. While this manuscript was in preparation, Yang et al provided direct evidence for ATM phosphorylation of dCK at Ser^{74} [33]. Phosphorylated dCK was shown to interact with cyclin-dependent kinase 1 (Cdk1), thus inhibiting its activity and initiating the G2/M checkpoint. While Yang et al focused on dCK-dependent cell cycle regulation through protein-protein interaction, their work did not address whether ATM modulates the dN salvage pathway through dCK phosphorylation.

Here, we investigate how IR-induced activation of dCK modulates the metabolism of DNA precursors and the affect this has on DNA repair and radiation resistance. In a murine leukemia cell line, we confirm that ATM phosphorylates dCK after IR at Ser^{74}. dCK activation shifts its substrate specificity towards dC, resulting in higher rates of intracellular dC sequestration and dCTP production. dCK activation also augments DNA DSB repair, likely through homologous recombination (HR).

Our group has previously developed and characterized positron emission tomography (PET) probes specific for dCK which enable non-invasive measurements of enzyme activity [34,35]. We utilized one of these probes, $[^{18}F]$-FAC, to visualize dCK activation in subcutaneously implanted tumors following IR. We found that $[^{18}F]$-FAC PET is effective in measuring acute tumor responses to IR, indicating potential clinical utility of this imaging modality in assessing the ATM-mediated DNA damage response in vivo.

Given its emerging importance in DNA DSB repair, we hypothesized that dCK could be a target for new radiosensitizers. A murine leukemia cell line lacking dCK was more radiosensitive than isogenic cells with the restored levels of dCK. Mouse embryonic fibroblasts (MEFs) derived from dCK KO mice also demonstrated enhanced sensitivity to IR. Inhibitors of dCK may prove useful in sensitizing tumors to radiotherapy.

Finally, we hypothesized that endogenous cellular genotoxic stress could also trigger post-translational activation of dCK at Ser^{74}. dCK knockout (KO) mice exhibit specific partial blocks in the early stages of T and B lymphocyte development [36]. Furthermore, complete dCK inactivation in mice induces endogenous DNA damage in lymphoid and erythroid lineages [37]. We attempted to understand whether post-translational modification of dCK at Ser^{74} plays a role in the development of B and T lymphocytes. To that end, we performed bone marrow transplantations (BMT) using stem cells from dCK KO bone marrow expressing wild type (WT) dCK or Ser^{74} mutated isoform. Our results demonstrate that dCK Ser^{74} is critical for normal B and T cell development in murine bone marrow transplant model, suggesting that the regulation of dCK at that residue may occur during hematopoiesis in response to physiologic stress.

Materials and Methods

Ethics Statement

All animal studies were carried out according to the guidelines of the Department of Laboratory Animal Medicine (DLAM) and the Animal Research Committee at UCLA (Protocol number: ARC 2005-072). All surgery was performed under appropriate anesthesia, and all efforts were made to minimize suffering.

Cell Lines and Reagents

All reagents were purchased from Sigma-Aldrich unless stated otherwise. The murine leukemia line (L1210-10K) was a previous gift from Dr. Charles Dumontet (Universite Claude Bernard Lyon I, Lyon, France) [38]. The amphotrophic retrovirus packaging cell line 293T was used for the production of murine stem cell virus-based (MSCV) retroviruses [39] containing YFP or dsRed and human or mouse dCK (WT or Ser^{74} mutant). Cell sorting based on color marker using flow cytometry ensured a pure population of dCK-expressing cells that were matched to the same fluorescence intensity. All L1210-10K derived cell lines were cultured in RPMI medium 1640, supplemented with 5% FBS and 2 mM L-glutamine, at 5% CO_2 and 37 °C. ATR-defective (ATR-Seckel) cell line DK0064 (ATR^{A2101G}), wild-type human lympho-blastoid cell line CHOC6, and A-T cell lines ($AT224LA^{G170A/1402delAA}$ and $AT255LA^{IVS42+2delT/A9171T}$) were a gift from Dr. Richard Gatti (University of California, Los Angeles) [40].

Cloning and Mutagenesis of Deoxycytidine kinase

Total RNA was isolated from mouse thymus and cDNA was generated using RT-PCR (superscript III) with oligo dT priming. Human dCK triple mutant (A100V, R104M, D133A) was codon optimized and synthesized by DNA 2.0, Inc. Triple mutant was mutated back to wild type (5'-GGAGTTTTACTTTTCAAACC-TACGCTTGTCTGTCACGAATCAGAGCTCAACTGGCAA-GCCTC-3' and 5'-CTTTGAACGGTCTGTGTATAGTGAC-AGATACATTTTCGCTTCTAACCT-3') using a multi-site mutagenesis kit (Agilent). Serine 74 was mutated to glutamic acid (S74E, human 5'-GAGTTCGAAGAGCTGACAATGGAACA-GAAGAATGGAGGTAACGTC-3', mouse 5'-GGAATTTGA-GGAATTGACAACGGAGCAGAAGAGCGGTGGAAATGT-TC-3') or to alanine (S74A, human 5'- GTTCGAAGAGCTGA-CAATGGCTCAGAAGAATGGAGGTAAC-3', mouse 5'- TG-AGGAATTGACAACGGCTCAGAAGAGCGGTGG-3'). dCK was cut out with EcoRI and XhoI and ligated into MSCV-IRES-YFP, MSCV-IRES-DsRed and MSCV-6His-IRES-YFP.

Generation of Deoxycytidine Kinase Mouse Monoclonal Antibodies

Human dCK (6-His-tagged) was produced in bacteria, purified by Ni-NTA chromatography and used as an immunogen. Balb/c 6–8 week-old female mice were immunized by an intraperitoneal injection of 200 µg of 6-His-dCK in RIBI adjuvant (Sigma), followed by 4 monthly boosts of 100 µg of immunogen. Antibody titers were determined in the serum by ELISA. Spleens of the highest titer mice were excised and dissociated. Isolated splenocytes were fused to sp2/0 myeloma cells at a ratio of 5:1 splenocytes/myeloma using PEG1500 (Roche). Twenty percent of fusion was plated in HAT medium in 10x flat bottom 96-well plates at 200 µl/well. Fusion was cultured until 25–50% coverage of wells was achieved. Positive wells, determined by ELISA of the supernatant, were re-plated in 24-well plates in HT medium. Supernatants were screened by repeat ELISA and Western blot at 1:10 dilution in 5% milk/PBS-T. Positive wells were sub-cloned by limiting dilution, and the sub-clones were tested by ELISA and Western blot. Clones with the highest affinities were 3B1, 3E10, 6B9, 9D4, 10A1 and 10H10. Variable region of each monoclonal line was cloned and sequenced to identify individual clones. Clones 3E10, 6B9 and 9D4 have an identical sequence, while 3B1, 10A1 and 10H10 are unique clones. Preparative amounts of antibody were produced in CELLine-1000 flasks (Integra Bioscience) and purified by Prosep-G affinity chromatography (Millipore). Clone 9D4 is commercially available from Millipore.

Antibodies and Western Blotting

Protein content was measured with BCA assay (Pierce), 5–30 μg was subjected to SDS/PAGE on 4–20% acrylamide gel (Thermo Scientific) followed by transfer to nitrocellulose membrane. When detecting ATM, 7.5% Tris-HCl gel (Bio-Rad) was used, and protein was transferred to PVDF membrane (Millipore) overnight at 30 V. Membranes were blocked with 5% milk in TBS-T (Tween 0.1% wt/vol), and probed with the following antibodies: p53 pSer[15] (1:1000, Cell Signaling), p53 (1:1000, Cell Signaling), ATM pSer[1981] (1:1000, Cell Signaling), α-tubulin (1:2000, Santa Cruz), ERK2 (1:2000, Santa Cruz), dCK (Clone 9D4, 1:1000, Millipore). Polyclonal rabbit antibody against dCK pSer[74] was a gift from Dr. Francoise Bontemps (Universite Catholique de Louvain, Brussels, Belgium). ECL substrate (Millipore) was used for detection and development on GE/Amersham film. For separating nuclear and cytoplasmic lysates, NE-PER Extraction Reagents kit was used (Thermo Scientific).

In Vitro Uptake and Kinase Assay

In vitro uptake and kinase assay protocols were adapted from a previous report [41]. The following tritium-labeled compounds were purchased from Moravek Biochemicals: [5-^{3}H(N)]-2′-deoxy-cytidine (22 Ci/mmol), [2,8-^{3}H]-deoxyadenosine (8 Ci/mmol), [8-^{3}H]-2′-deoxyguanosine (6.6 Ci/mmol). Cells were irradiated using Cs-137 source at a rate of 7.16 Gy/min to a total dose of 3 Gy. Irradiated and untreated cells were placed in wells of a 96-well 0.22-μm MultiScreen filter bottom plates (Millipore) in triplicate, at 1×10^5 cells/well in 100 μl of 5% RPMI, and 1 μCi of [^{3}H]dC was added to each well. Cells were incubated at 37°C, 5% CO_2 for 1 hour, and washed five times with 200 μl of 5% RPMI using the Multiscreen vacuum manifold (Millipore). Plates were dried, wells were punched out into Wheaton scintillation Omni-Vials (Fisher), and Bio-Safe NA scintillation fluid was added (RPI Research Products). Radioactivity was determined with Beckman scintillation counter (LS 6500). Kinase assays were performed on either total cell lysates or Ni-NTA agarose (Qiagen) bound 6-His-dCK. Lysis buffer was prepared by adding fresh complete protease inhibitors (Roche) and Phosphatase Inhibitor Cocktails 2/3 (1:100, Sigma) to 50 mM Tris-HCl, 20% glycerol, 0.5% Nonidet P40, pH 7.6. Lysates were cleared by spinning at 16000×g, 4 °C for 20 min. One μl of 1 μg/μl total cell protein was added in triplicate to wells of a 96-well Microtest U-bottom plate (Becton Dickinson) containing 6 μl H2O and 2 μl of 5× kinase buffer (250 mM Tris-HCl, 5 mM ATP, 25 mM MgCl2, 10 mM DTT, 50 mM NaF, 5 mM thymidine, pH 7.6). One μCi of an appropriate tritium labeled deoxynucleoside was added to each well and plates were incubated at 37°C for 20 minutes. Reactions were quenched with 40 μl ice cold water, heated at 95°C for 2 minutes, and contents of each well were blotted on Whatman DE-81 filter discs (GE Healthcare). Dried disks were washed three times with 4 mM ammonium formate and twice with 95% ethanol. Radioactivity from dried disks was determined as above. In some cases, prior to irradiation cells were pretreated for 1 hour with ATM inhibitor Ku55933 (Tocris) or DNA-PK inhibitor Nu7441 (Tocris) diluted in DMSO.

His-Tagged dCK Purification

Cell lysates (650 μg total protein) were diluted 1:2 with Ni-NTA compatible buffer (40 mM NaH2PO4, 20% glycerol, 130 mM NaCl, 0.5% NP40, 10 mM imidazole) to which Phosphatase Inhibitor Cocktails 2/3 (1:100, Sigma) were added. Eighty μl of 50% Ni-NTA agarose beads (Qiagen) were washed with Ni-NTA buffer, added to the lysate and rotated overnight at 4°C. Beads were washed 4 times with Ni-NTA buffer. Equivalent of 50 μg of initial lysate was kept for Western blotting, and the remaining 600 μg equivalent was divided into 150 μg aliquots. dCK kinase assay was performed on each aliquot by adding 10 μl H2O2, 3 μl 5× kinase buffer and 1 μCi of tritium-labeled deoxyribonucleo-side, as described in the In Vitro Uptake and Kinase Assay section.

DNA Repair Assays

L1210-10K cells ± dCK (WT, S74A, S74E) were irradiated using a Cs-137 source at a rate of 7.16 Gy/min to a total dose of 3 Gy and kept in 15 ml Falcon tubes (BD Biosciences) at 37°C, 5% CO_2. At different time points after irradiation 2×10^5 cells were taken out and either cytospun onto a glass slide or fixed in 70% ice-cold ethanol overnight. Cytospun cells were fixed, and stained overnight at 4°C with anti-γH2A.X pSer[139] -Alexa647 mAb (1:50, Cell Signaling). Ethanol-fixed cells were washed 3 times in PBS and stained overnight at 4°C with anti-γH2A.X pSer[139] -Alexa647 mAb (1:100) in FACS buffer (PBS +3% FBS, 0.09% NaN3) with 0.1% Saponin. Cells were washed and resuspended in FACS buffer, and analyzed on FACSCanto flow cytometer (BD Biosciences). DNA recombination efficiency was measured using a transient transfection assay described previous [42,43]. Vector constructs pCMS-end and pCMS-hom to measure non-homologous end-joining (NHEJ) and homologous recombi-nation (HR), respectively, were a gift from Dr. Robert Schiestl (University of California, Los Angeles). Digestion of 500 μg plasmid DNA was carried out with XhoI/BamHI (pCMS-end) or XhoI (pCMS-hom) in 800 μl volume at 37°C for 3–4 hours. Digestion was extracted with 800 μl of phenol:chloroform:isoa-mylalcohol (25:24:1) and precipitated with sodium acetate (pH 5.2, 300 mM final) and 2 ml ethanol. Digested plasmids were pelleted at 16000×g, 4°C, washed once with 70% ethanol, dried briefly at 37°C, and re-suspended overnight in 120 μl of TE buffer at 4°C. L1210-10K cells ± dCK/dsRed were irradiated with 3 Gy, and incubated at 37°C and 5% CO_2 for 1 hour. Nucleofection was performed after combining 18 μg of digested plasmid with 5×10^5 cells in 100 μl Cell Line Nucleofector Solution V (Lonza), using program S018 on Nucleofector II (Amaxa). Immediately after nucleofection, cells were placed in pre-warmed 1.5 ml 5% RPMI media and incubated at 37°C, 5% CO_2 for 24 hours. Cells were analyzed on FACSCanto flow cytometer under YFP-GFP-RFP configuration.

Intracellular dNTP Pool Measurements

L1210-10K cells ± dCK (WT, S74A, S74E) were irradiated using a Cs-137 source at a rate of 7.16 Gy/min to a total dose of 3 Gy and kept in 15 ml Falcon tubes (BD Biosciences) at 37°C, 5% CO_2. Each hour after irradiation, 1×10^6 cells were taken out, washed once with PBS, pelleted and re-suspended in 500 μl of ice-cold 60% methanol. Cells were vortexed for 1 min and stored at − 20°C overnight. Next day lysates were exposed to 95°C for 3 minutes, and pelleted at 17,000×g for 15 minutes. Supernatants were evaporated for 4 hours in Speed VacPlus SC110A (Savant), and pellets were re-suspended in 100 μl of nuclease-free H2O. Previously described protocol was used to measure dNTP pools [44], with the following modifications. Reactions were carried out simultaneously in triplicate in 96-well U-bottom plates (Becton Dickinson) using 5 μl of lysate in 25 μl total volume per well. After incubation for 2 hours at 37°C, 20 μl from each well were transferred to 96-well DE81 Unifilter plates (Whatman). Plates were washed 4 times with Na2HPO4, once with diH2O and once with 95% ethanol using Multiscreen vacuum manifold (Millipore). Washed plates were dried, and 100 μl of scintillation fluid was added to each well. Radioactivity was measured with a BetaMax plate reader (PerkinElmer).

Bone Marrow Transplantation

Eight to twelve-week-old female dCK KO mice injected intraperitoneally with 150 mg/kg (200 µl) of 5-Fluorouracil (APP Pharmaceuticals) and four days later bone marrow was harvested. Bone marrow was cultured, stimulated with cytokines and infected with a retrovirus as previously described [45]. Ecotropic murine stem cell virus-based retrovirus containing either mouse WT or Ser[74] mutant dCK was generated in 293T cells with titers of $1-2 \times 10^7$/ml. B6.SJL 8–10 week-old female mice were irradiated using Co-60 source at a rate of 0.177 Gy/min to a total lethal dose of 950 rads, and 3 hours later 200 µl of HBSS containing $3-5 \times 10^5$ infected bone marrow cells were injected into their tail veins. Bone marrow was allowed to engraft for 5–8 weeks.

Thymus and Bone Marrow Analysis

Thymi, spleens and long bones were harvested from B6.SJL donors after 5–8 weeks of transplantation. Thymi and spleens were dissociated into single cells in 5% RPMI using frosted glass slides, and filtered through 40 µm sterile filters. Bone marrow single cell suspensions were obtained by flushing femur and tibia with 5% RPMI and filtering through 40 µm filters. Cells were washed in FACS buffer (PBS +3% FBS, 0.09% NaN₃) and counted. Dead cells were excluded by staining with DAPI (50 ng/ml). Antibodies were purchased from eBioscience, Inc. unless stated otherwise. Cell sorting and analysis was performed on FACSAria II flow cytometer (BD Biosciences). Single cell suspensions from spleens were stained with CD45.2-Alexa780 (donor marker, Clone 104) and CD45.1-PE-Cy5 (recipient marker, Clone A20) antibodies, and CD45.2$^+$YFP$^+$/CD45.2$^+$YFP$^-$ cells were sorted for Western blotting and dCK kinase assay. CD45.1$^+$ cells from a spleen of untreated B6.SJL mouse were sorted as control of dCK expression for Western blot. Thymocyte phenotyping was performed with the following antibodies: CD45.2-Alexa780 (Clone 104), CD45.1-PE-Cy5 (Clone A20), CD4-PE (Clone RM4-5), CD8-PE-Cy7 (Clone 53-6.7), CD25-Alexa700 (Clone PC61.5), CD44-PerCP-Cy5.5 (Clone IM7). Phenotyping of B cell development in the bone marrow was performed with the following antibodies: CD45.2-Alexa780 (Clone 104), CD45.1-PE-Cy5 (Clone A20), B220-PE-Cy7 (Clone RA3-6B2), IgM-biotin (Clone 11/41)+SA-eFluor710, CD43-PE (Clone S7, BDBiosciences), CD19-PerCP-Cy5.5 (Clone 1D3). Approximately $1-6 \times 10^6$ stained live thymocytes and bone marrow cells were pelleted, resuspended in 100 µl of Cytofix/Cytoperm buffer (BD Biosciences) and incubated on ice for 20 minutes. Cells were washed with 1 ml of $1 \times$ Perm/Wash buffer (BD Biosciences), re-suspended in 100 µl of Cytoperm Plus buffer (BD Biosciences) and incubated on ice for 10 minutes. Cells were washed, pelleted and stained with γH2A.X pS[139]-Alexa647 antibody (1:10, Clone 20E3, Cell Signaling) in 50 µl of Perm/Wash buffer overnight at 4°C. Cells were again stained with DAPI (4 µg/ml) in FACS buffer and analyzed.

In Vivo Treatment Model

L1210-10K ± dCK (WT, S74A, S74E) cells were resuspended in fresh RPMI media with no additives at a density of 4×10^7 cells/ml. Matrigel matrix (BD Biosciences) was added at a 1:1 volume ratio. NSG and CB17 SCID mice were anesthetized with an intraperitoneal injection of 1.2 mg ketamine/0.38 mg xylazine (Phoenix Pharmaceutical), and either the right or both front legs were injected s.c. with 2×10^6 cells in 100 µl volume. Tumors were allowed to develop for 10–15 days. Prior to irradiation, mice were anesthetized with an IP injection of ketamine/xylazine, placed on a platform and shielded with a cerrobend jig, except for a right front leg harboring a tumor. Exposed tumors were irradiated using a Gulmay RS320 x-ray unit filtered with 1.5 mm Cu and 3 mm Al

(Gulmay Medical) at a rate of 1.17 Gy/min, 300 kV and 10 mA. The x-rays were administered vertically with a focus-to-surface distance of 42.3 cm.

MicroPET/microCT Imaging and Image Analysis

All mouse imaging was conducted at UCLA Crump Institute for Molecular Imaging - Small Animal Imaging Center as previously described [46]. [^{18}F]-FDG was synthesized at the UCLA Ahmanson TranslationalImaging Division biomedical cyclotron facility as previously described [47]. [^{18}F]-FAC was synthesized at the Crump Preclinical Technology Center cyclotron facility as previously described [34]. Mice were kept warm under gas anesthesia (2% isoflorane). Approximately 200 µCi of either probe was injected into a tail vein 1 hour prior to the initiation of imaging. To look at the uptake of both probes after irradiation, probes were injected 1 hour after IR. Images were acquired with Siemens Preclinical Solutions microPET Focus 220 and Micro-CAT II CT. Image reconstruction and registration was performed as previously described [34]. AMIDE software was used for image analysis [48]. Three dimensional regions of interest (ROI) were drawn around tumors to quantify their volumes and accumulation of the probe as a mean percentage of injected radioactivity dose per weight (%ID/g).

Radiosensitivity Assays

Clonogenic survival assay was performed on MEFs from dCK KO and litter matched control dCK WT mice. Cells were plated at 300,000 cells/well at 37°C, 8% CO_2 in six well plates after Cs-137 irradiation. *In vivo* radiosensitivity assay was performed by implanting L1210-10K ± dCK cells into right front legs of 10-week-old CB17 SCID female mice and allowing the tumors to develop for 10 days. Tumors were measured daily with calipers, and volumes were estimated from the formula $(L \times W^2)/2$. Selective tumor irradiation with 3 Gy was performed twice daily (7 hours apart) for six days as described in *In Vivo Treatment Model* section. Tumors were then allowed to re-grow for 10 days. Imaging with [^{18}F]-FDG was performed as described in *MicroPET/MicroCT Imaging and Image Analysis* section.

Statistical Analyses

Data are presented as means ± SEM. Group comparisons were performed with one-sample t test function in column statistics of Prism 5 software (GraphPad Software). All P values are two-tailed, and P<0.05 was considered to be statistically significant. Graphs were generated with Prism 5 software.

Results

dCK activity increases after IR in an ATM-dependent manner and is correlated with Ser[74] phosphorylation

A large body of evidence has documented the activation of dCK after cell exposure to various genotoxic stresses [5–11]. IR is a quantifiable and reproducible exogenous source for generating DNA DSBs, and it is used to treat multiple types of cancer. We aimed to understand the mechanism of dCK regulation resulting from IR. To that end we chose to work with the gemcitabine-resistant L1210-10K cell line derived from a murine leukemia [38]. L1210-10K (10K) cells have a sequencing-confirmed mutation in the dCK gene which leads to a complete absence of the kinase [38]. By reintroducing dCK into 10K cells (10K+dCK) through retroviral transduction, we restored the dCK-dependent branch of the deoxyribonucleoside salvage pathway (Figure 1A). Two hours after irradiation of 10K+dCK cells with 3 Gy the rate of uptake of tritium-labeled dC (^{3}H-dC) from the media increased

1.5-fold (Figure 1A). Thirty minutes after irradiation the activity of dCK increased 3 to 4-fold, and remained elevated for 5 hours (Figure 1B). Ionizing radiation leads to oxidative stress through the formation of free radicals [49]. We treated 10K+dCK cells with hydrogen peroxide and observed a 2-fold increase in ^{3}H-dC uptake and 3-fold increase in dCK kinase activity (Figure 1A and B). Thus, both IR and oxidative stress induce the activation of dCK and stimulate the salvage of exogenous dC.

Protein from 10K+dCK cells was isolated at time-points after irradiation with 3 Gy. Increased dCK activity correlated with the phosphorylation of the enzyme at Ser^{74}, while the total amount of dCK remained unchanged (Figure 1C). Activation of the IR-induced DNA damage response was confirmed by measuring phosphorylation of p53 at Ser^{15} (Figure 1C), a target of DNA-PKcs, ATM and ATR [50]. Phosphorylation of p53 peaked at 30 minutes after IR and returned to baseline by 5 hours (Figure 1C). Our results confirm previous studies showing enhanced dCK activity after IR and demonstrate a direct correlation between IR-induced increase in enzyme activity and phosphorylation at Ser^{74}.

ATM is a serine/threonine protein kinase that is a dominant regulator of the cellular response to DNA damage and is critical to the repair of DSBs resulting from IR [23]. Using a large-scale proteomics screen, Matsuoka, et al implicated dCK as a substrate of ATM after IR-induced DNA damage response [31]. While this manuscript was in preparation, Yang et al demonstrated a direct phosphorylation of dCK by ATM at Ser^{74} in response to IR [33]. We also tested whether chemical inhibition or mutation of ATM

would block dCK activation by IR. Specific inhibition of ATM in 10K+dCK cells by Ku55933 [51] prevented IR-induced ATM autophosphorylation of Ser^{1981} (Figure 1D). Significantly, pharmacologic inhibition of ATM abrogated IR-induced dCK phosphorylation at Ser^{74} (Figure 1D). ATM inhibition also blocked the IR-induced increase in dCK activity in a dose-dependent manner (Figure 1F).

DNA-PKcs shares functional redundancy and overlapping substrates with ATM [23]. We tested whether dCK was activated after IR in the context of DNA-PKcs inhibition. Treatment of 10K+dCK cells with DNA-PK specific inhibitor Nu7441 [52] demonstrated partial reduction in phosphorylation of ATM at Ser^{1981} and dCK at Ser^{74} (Figure 1D), while the dCK activation after IR was completely preserved (Figure 1F).

ATM and ATR also share substrate specificity, but while ATM senses double-strand breaks, ATR senses single-strand DNA present during DSB processing or at stalled replication forks [53]. We tested IR-induced dCK activation in an ATR-defective (DK0064) human lymphoblastoid cell line (LCL) and compared it to the wild-type (WT) LCL CHOC6. Significant reduction in the level of ATR protein in ATR-defective human LCL compared to WT LCL was confirmed by western blot (Figure 1E). IR-induced phosphorylation of dCK at Ser^{74} was observed in the ATR-defective and WT LCLs (Figure 1E), corresponding to a 2.25 fold and 1.6 fold increase in dCK activity, respectively (Figure 1F). Selective chemical inhibition of ATM with Ku55933 in these LCLs completely abrogated IR-induced ATM autophosphory-

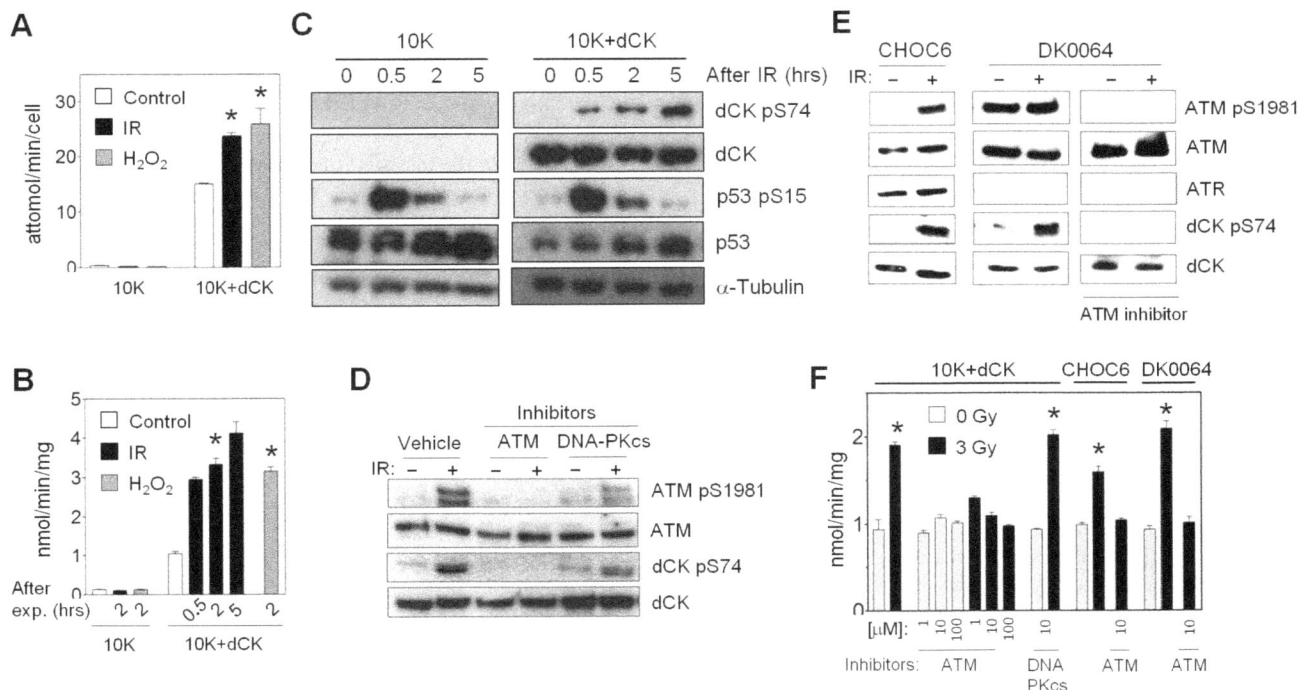

Figure 1. dCK is activated in an ATM-dependent manner after IR. (A) In vitro cell uptake assay of [^{3}H]-dC 2 hours after exposure of L1210-10K ± dCK (WT) to 3 Gy (*, P = 0.0002; N = 3) or 100 µM H$_2$O$_2$ (*, P = 0.0197; N = 3). (B) In vitro dCK kinase assay using L1210-10K ± dCK (WT) cell lysates, [^{3}H]-dC as substrate and performed at indicated times after exposure to 3 Gy (*, P = 0.0002; N = 3) or 100 µM H$_2$O$_2$ (*, P<0.0001; N = 3). (C) Western blot of L1210-10K ± dCK (WT) cell lysates obtained at indicated times after 3 Gy irradiation. (D) Western blot of lysates from 10K+dCK (WT) cells treated for 1 hour with either DMSO vehicle, 10 µM ATM inhibitor (Ku55933) or 10 µM DNA-PKcs inhibitor (Nu7441) and irradiated with 3 Gy. (E) Western blot of lysates from CHOC6 (WT LCL) and DK0064 (ATR-defective) cell lines before IR or 2 hours after 3 Gy exposure. Lysates from DK0064 cells pretreated with 10 µM ATM inhibitor Ku55933 are shown in the right panel. (F) In vitro dCK kinase assay using 10K+dCK (WT), CHOC6 and DK0064 cell lysates, [^{3}H]-dC as substrate and performed where indicated after 1 hour pretreatment with inhibitors of ATM and DNA-PKcs, and 2 hours after exposure to 3 Gy (*, 10K+dCK: P<0.0001; CHOC6: P = 0.001; DK0064: P = 0.0003; N = 3).

lation of Ser[1981], dCK phosphorylation at Ser[74] (Figure 1E) and dCK activation (Figure 1F). Interestingly, increased endogenous ATM phosphorylated at Ser[1981] was observed in ATR-Seckel cells compared to WT LCL (Figure 1E).

We have utilized two human A-T cell lines (AT224LA and AT255LA) to confirm the dependence of IR-induced dCK activation on ATM. These cell lines exhibit either a complete absence or a significant reduction of detectable ATM protein, respectively (Figure S1A). Two hours after irradiation the uptake of ^{3}H-dC was unchanged in A-T cells (Figure S1B), and the activation of dCK was significantly reduced compared to WT and ATR-defective LCLs (Figure S1C). In sum, these results utilizing a murine leukemia cell line model and human A-T LCLs confirm that the activation of dCK in response to IR is ATM-dependent, and demonstrate that dCK activation is likely not carried out by functionally related serine/threonine kinases.

Ser[74] is critical for dCK activation by IR

Ser[74] of dCK is phosphorylated in response to IR [31] (Figure 1). Mutation of Ser[74] was previously shown to dramatically affect the baseline activity of dCK [19]. The serine to alanine (S74A) mutation prevents Ser[74] phosphorylation and results in reduced dCK activity [19]. In contrast, the serine to glutamic acid (S74E) mutation mimics phosphorylation and results in enhanced catalytic activity [19]. We measured the effect of these positive and negative mutations at Ser[74] on the activity of dCK after IR. We first tested whether S74A substitution would lead to a loss of ATM-dependent activation of dCK after IR. The mutated kinase was expressed in L1210-10K cells through retroviral transduction followed by FACS sorting of YFP[+] cells with fluorescence intensity matched to that of the 10K+dCK WT cells. Similar expression levels of dCK WT and Ser[74] mutants were confirmed by Western blot (Figure 2A). Compared to un-irradiated cells containing WT dCK, and in agreement with previous studies, expression of S74A-dCK dramatically decreased the rate of ^{3}H-dC uptake and demonstrated a 3-fold reduction in enzyme activity in cell extracts (Figure 2B and C). Importantly, S74A is not a kinase-dead mutation. dCK activity measured in 10K+S74A-dCK cell extracts is significantly greater than that in 10K extracts. This mutation also did not affect the level of expression of the enzyme (Figure 2A). Moreover, previous enzyme kinetic analysis has shown that the k_{cat} for dC phosphorylation by the purified S74A-dCK is the same as that for the WT dCK [21]. It is possible that unlike the phosphorylated Ser[74] residue, the S74A mutation favors a closed dCK state, preventing substrate binding and increasing K_M for dC. Two hours after 3 Gy IR, ^{3}H-dC uptake and S74A-dCK activity remained unchanged in marked contrast to the WT-dCK (Figure 2B and C).

We next tested the effect of S74E substitution on the IR-induced dCK activation. We found that this mutation tripled dCK activity compared to un-irradiated WT (Figure 2C). Un-irradiated S74E mutants also exhibited an increased rate of ^{3}H-dC uptake compared to the WT dCK cells (Figure 2B). IR did not affect the ^{3}H-dC uptake or the kinase activity of the S74E mutant (Figure 2B and C). Based on our results analyzing Ser[74] mutations with positive and negative effects on the baseline dCK activity, we conclude that dCK is activated primarily by ATM-dependent phosphorylation of Ser[74] in response to IR-induced DNA damage.

Quantitative PET imaging of metabolic response to DNA damage

Our results indicate that ATM enhances dCK activity through post-translational modification in response to IR-induced DNA damage. Measuring dCK activity after IR *in vivo* may provide

information about cellular DNA damage response. PET probes specific for dCK [34] allow non-invasive prediction of tumor response to gemcitabine, a prodrug substrate of dCK [41]. Measuring the DNA damage response non-invasively via dCK activation could be a useful clinical tool for stratifying patients into responders versus non-responders during treatment with IR. This imaging strategy could also provide a useful biomarker for assessing the extent of irradiation of the target field, or predict a potential synergism of nucleoside analogs with radiotherapy [54]. We tested whether IR-induced ATM-dependent activation of dN salvage would result in accumulation of the dCK-specific PET probe, [18F]-FAC, in irradiated tumors. L1210-10K tumors expressing WT dCK or Ser[74] mutants were grown bilaterally on the front legs of NSG mice, and the right front leg was selectively irradiated with 3 Gy followed by tail-vein probe injection one hour later. The presence of WT dCK caused the irradiated tumor to retain significantly more [18F]-FAC compared to un-irradiated tumor (Figure 2D and E, top panel; Figure S2A). Taking the [18F]-FAC accumulation in 10K tumor as a background signal, IR induced a forty percent increase in [18F]-FAC accumulation in tumors expressing WT dCK (Figure 2E, top panel). Neither the 10K tumors nor Ser[74] mutants exhibited increased probe uptake after IR (Figure 2D and E, top panel).

We repeated this experiment with a different set of animals, which were imaged with [18F]-FDG. PET imaging with [18F]-FDG measures glycolysis and tumor viability, and is routinely performed clinically to diagnose and stage malignancy, and measure response to therapy [55]. Accumulation of [18F]-FDG by 10K tumors with WT or mutated dCK remained unchanged acutely after IR (Figure 2D and E, bottom panel; Figure S2B). Our results demonstrate that a new class of PET probes specific for dCK may be used to visualize and quantify the immediate metabolic response to genotoxic stress induced by IR.

Catalytic activity of dCK for deoxycytidine is selectively increased after IR

Deoxycytidine kinase is able to phosphorylate pyrimidine (dC) as well as purines (dA, dG) [56]. Previous studies demonstrated that phosphomimetic S74E mutation increases catalytic activity of dCK for dC, but not for dA and dG [21,22]. We hypothesized that ATM-dependent phosphorylation of dCK at Ser[74] following IR serves to preferentially increase the production of dCMP over dGMP and dAMP. Because dG and dA are not exclusively phosphorylated by dCK, we generated histidine-tagged constructs of WT dCK and Ser[74] mutants. Catalytic activities of affinity-purified His-tagged WT and mutated kinases for each dCK substrate were compared (Figure 3A). The rate of dC phosphorylation by the purified WT dCK doubled after IR, while the rates of dA and dG phosphorylation remained unchanged (Figure 3B). Confirming previous results, alanine substitution of Ser[74] decreased dCK activity for dC, but had little effect on dA and dG phosphorylation. S74E-dCK tripled the rate of dC phosphorylation, while decreasing dG and dA phosphorylation rates by one half (Figure 3B). Both mutants exhibited no enzyme activity change for dG and dA after IR (Figure 3B). We conclude that IR-induced ATM-dependent phosphorylation of dCK at Ser[74] increases the flux of deoxycytidine through the salvage pathway.

Activated dCK increases dCTP production and the rate of DNA double-strand break repair

Intracellular pools of dNTPs must be tightly regulated to avoid genomic instability, while maintaining adequate supply for DNA replication and repair. Having determined that IR-induced dCK

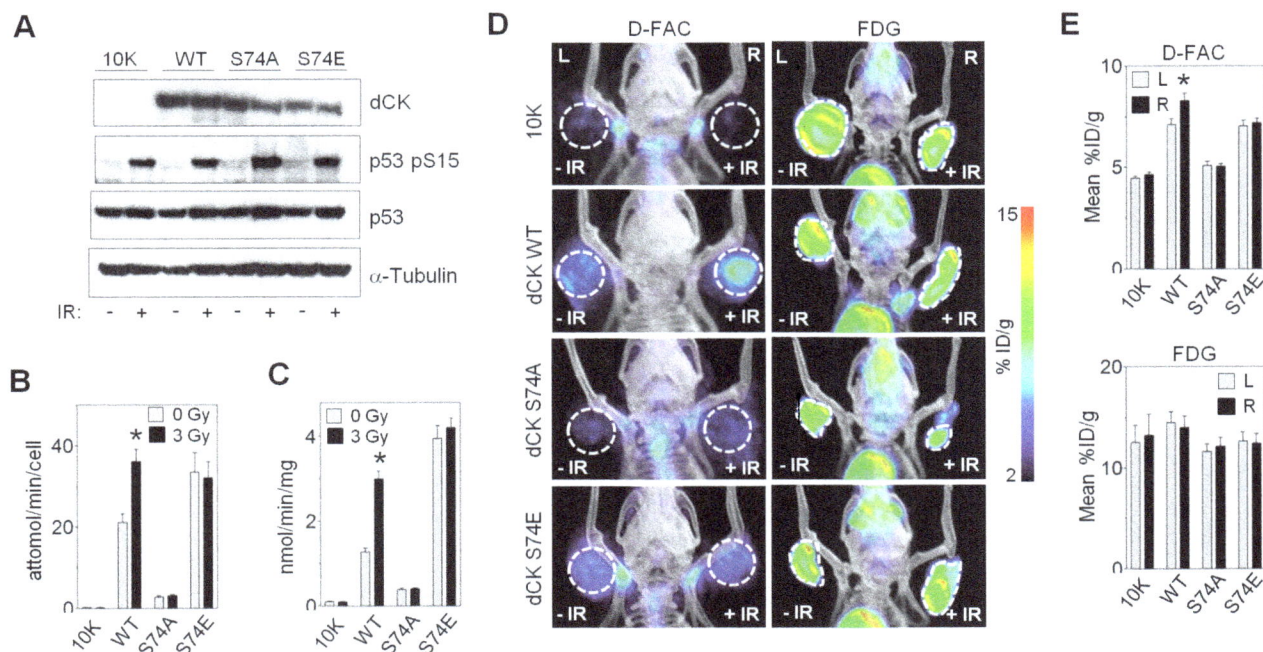

Figure 2. PET imaging of metabolic regulation during DNA damage response. (A) Western blot of 10K ± dCK (WT, S74A, S74E) before or 2 hours after exposure to 3 Gy. (B) *In vitro* cell uptake assay of [^{3}H]-dC after 3 Gy exposure of L1210-10K ± dCK WT (*, P = 0.0004; N = 14) or dCK Ser^{74} mutants. (C) *In vitro* dCK kinase assay using [^{3}H]-dC and cell lysates from10K ± dCK WT (*, P<0.0001; N = 9), dCK S74A or dCK S74E, 2 hours after 3 Gy irradiation. (D) [^{18}F]-FAC and [^{18}F]-FDG microPET/CT scans of NOD-SCID mice with bilateral 10K ± dCK(WT, S74A, S74E) tumors after 3 Gy irradiation of right tumor. (E) Averaged ROI values of [^{18}F]-FAC (top panel) or [^{18}F]-FDG (bottom panel) uptake in irradiated (R, right) and untreated (L, left) tumors (*, P = 0.0267; N = 4 mice per group).

regulation by ATM enhances phosphorylation of dC, we measured the dNTP pools before and after radiation exposure. The concentration of dCTP at baseline was significantly higher when WT dCK or S74E dCK were expressed in L1210-10K cells compared to 10K cells (Figure 3C). Following IR the concentration of dCTP sharply rose in cells with WT dCK, reaching the maximum by two hours, and then falling below initial value six hours after irradiation. Neither the 10K cells nor the Ser^{74} dCK mutants exhibited a similar surge in dCTP. S74A substitution did not affect dCTP pools at baseline or after radiation compared to 10K cells (Figure 3C). An S74E substitution did not elevate dCTP concentrations as much as the WT-dCK (Figure 3C). Perhaps this reflects a different steady state of intracellular dNTP pools in cells expressing constitutively active dCK. We conclude that at least one function of ATM regulation of dCK may be to provide a source of dCTP for DNA repair.

Most types of DNA repair require dNTPs, which may be salvaged from intracellular or exogenous sources, or synthesized *de novo*. It is possible that disabling dN salvage may slow DNA repair by requiring cells to rely solely on the *de novo* dNTP synthesis. Having determined that phosphorylation of dCK enhances dCTP production and pool size, we determined dCK-dependent kinetics of DNA DSB repair by measuring H2A.X phosphorylated at Ser^{139}, a marker of DSB-induced DNA damage response. Resolution of pSer^{139} H2A.X foci (γH2A.X) corresponds to successful repair of DNA DSBs [23]. Before radiation, 10K cells contained 8 fold more γH2A.X foci then 10K+dCK cells (Figure 4A). 10K and 10K+dCK cells showed similar numbers of γH2A.X foci per nucleus 30 min after IR. At 5 hours post irradiation there were significantly fewer positive foci per nucleus in the cells expressing dCK (Figure 4A and B). We utilized flow

cytometry to confirm these results and measure the effect of Ser^{74} mutation on DNA DSB repair rate. Again, the resolution of γH2A.X signal was significantly faster in the cells with WT dCK (Figure 4C and D). S74A substitution slowed DNA DSB repair rate, while S74E mutants were just as efficient at repairing DNA as the cells with WT dCK (Figure 4C and D).

Eukaryotic cells repair DNA DSB through either non-homologous end-joining (NHEJ) or homologous recombination (HR) [57]. Unlike NHEJ, HR requires extensive DNA synthesis, and is, therefore, dependent on the presence of sufficient dNTPs [58]. Blocking *de novo* dNTP synthesis via RNR inhibition suppresses HR [59]. Matsuoka, et al. observed that depletion of dCK in osteosarcoma U2OS cells by small interfering RNA increased baseline γH2A.X signal and compromised HR [31]. We tested the functionality of HR and NHEJ in 10K cells by utilizing a previously reported linearized plasmid transient transfection assay [42,43]. The plasmid can be introduced into the cells immediately after IR, avoiding unpredictable IR-induced damage to the plasmid. Cells were irradiated one hour prior to plasmid transfection to induce DNA DSBs and DNA damage response, and recombination efficiency (RE) after IR was calculated and normalized to the recombination efficiency before IR utilizing flow cytometry data. Cells reconstituted with WT dCK exhibited a higher efficiency of plasmid recombination via HR compared to 10K cells (Figure 4E). Non-homologous end-joining, being relatively insensitive to dNTP levels, was not affected by the non-functional dN salvage (Figure 4E). We conclude that, consistent with previously published results [31], ATM-dependent phosphorylation of dCK increases the rate of DNA DSB repair by supplying dCTP for HR.

A

B

C

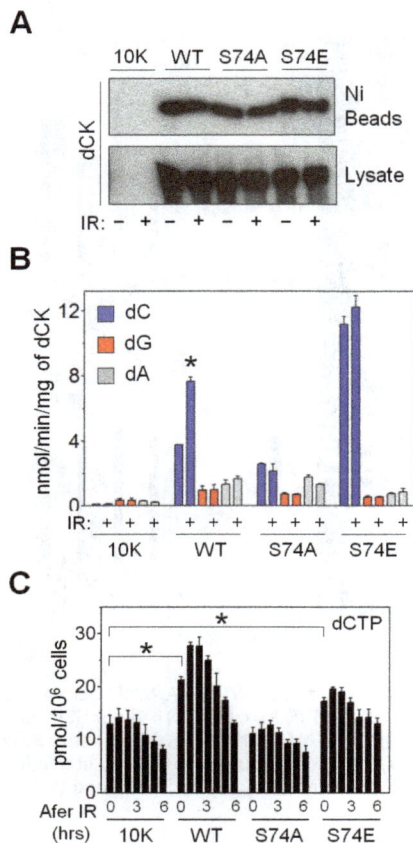

Figure 3. Changes in dCK substrate specificity and dCTP pool after IR. (A) Western blot of purified 6-His-tagged dCK (WT or Ser[74] mutants) from 10K cells, and corresponding total cell lysates. (B) *In vitro* kinase assay on purified 6-His-tagged dCK (WT, S74A, S74E) using either [3H]-dC, [3H]-dG or [3H]-dA as a substrate before or after 3 Gy exposure (*, P = 0.0025; N = 2). (C) Intracellular dCTP pools before and hourly after 3 Gy exposure of 10K ± dCK (WT, S74A, S74E) cells (*, 10K vs. WT P = 0.0006; 10K vs. S74E P = 0.032; N = 6 for each time point).

Deoxycytidine salvage modulates radiosensitivity

Radiation therapy and a large number of chemotherapy drugs act by damaging cellular DNA. Interference with a cancer cell's attempts to repair DNA further sensitizes it to therapy. The scope of strategies targeting DNA damage response in cancer is extremely broad [60]. Blockade of dNTP supply needed for replication and DNA repair is one such strategy. Inhibition of thymidylate synthase [61] and ribonucleotide reductase [62,63], two critical enzymes of the *de novo* dNTP synthesis, has been utilized for tumor radiosensitization. The importance of dN salvage in IR-induced DNA repair makes this pathway a possible target for novel sensitizers to genotoxic therapy. Recently, Yang et al demonstrated that stable knock down of dCK in HeLa cells leads to a significant increase in IR radiosensitivity [33]. We also hypothesized that the absence of dCK will increase cancer radiosensitivity. Fibroblasts derived from dCK KO mice were significantly more radiosensitive to 3 Gy irradiation compared to MEFs with WT dCK (Figure 5A). We carried out an *in vivo* IR treatment study, implanting 10K cell grafts into front right legs of CB17 SCID mice. Noting that the rates of DSB repair were significantly different between 10K and 10K+dCK cells, we performed hyperfractionated radiation delivery utilized in clinical

radiotherapy [64], delivering two 3 Gy doses daily. We hypothesized that 10K cells will be less efficient in repairing DNA between IR fractions, leading to accumulation of DNA damage and apoptosis. After a ten day engraftment, 36 Gy were delivered to the tumors in twelve fractions over six days (Figure 5B), followed by tumor re-growth. No differences in growth rates were observed between un-irradiated 10K and 10K+dCK tumors (Figure 5C). Tumors lacking dCK exhibited a greater reduction in volume in response to IR treatment compared to those expressing WT kinase (Figure 5C). In addition, PET imaging with [18F]-FDG on day 4 of IR treatment detected a significantly reduced viability and FDG uptake by tumors lacking dCK (Figure 5D and E). After completion of radiotherapy, the cancer relapsed as demonstrated by FDG-PET (Figure 5D and E), but 10K tumors remained smaller than those with WT dCK (Figure 5C and D). We conclude that the absence of dCK contributes to the enhanced radiosensitivity of L1210-10K murine leukemia cells and MEFs derived from dCK KO mice.

Mutating Ser[74] residue of dCK affects lymphocyte development

Thus far we described the response of dN salvage to the exogenous stress. Cells under endogenous, physiologic stress, resulting from metabolic production of reactive oxygen species (ROS) or DNA damage caused by rapid genomic replication [65,66], may also activate dN salvage through dCK regulation. Thus, maintenance of adequate dNTP pools for DNA replication and/or repair during the development of rapidly dividing cells such as B and T cell precursors may be one important function of deoxyribonucleoside salvage. dCK KO mice accumulate DNA damage in lymphoid and erythroid cells undergoing replicative stress, as indicated by increased γH2A.X signal [37]. We hypothesized that the regulation of dCK through Ser[74] phosphorylation observed *in vitro* may be important during lymphocyte development in mice. Bone marrow enriched for CD45.2[+] hematopoietic stem cells (HSCs) was harvested from dCK KO mice [36], infected with retrovirus carrying YFP and either WT or Ser[74]-mutated dCK, and transplanted into lethally irradiated B6.SJL recipients (CD45.1[+]). Five to eight weeks after BMT, CD45.2[+]YFP[+] lymphocytes from thymus and bone marrow were analyzed and compared to CD45.2[+]YFP[−] dCK KO lymphocytes. We compared dCK expression in retrovirally transduced bone marrow to intrinsic dCK levels in WT bone marrow. Splenic donor CD45.2[+]YFP[+] leukocytes expressed approximately one half the amount of WT or mutant dCK compared to CD45.1[+] WT B6.SJL splenic leukocytes (Figure 6A). The activities of WT dCK and Ser[74] mutants measured using total lysates from donor leukocytes were similar to those observed *in vitro* in 10K cell line (Figure 6B).

T cell development in the thymus proceeds through specific stages, which can be identified through distinct surface markers [67]. Immature thymocytes transition through CD4[−]CD8[−] double negative (DN) stage (sub-categorized into DN1, DN2, DN3 and DN4) to CD4[+]CD8[+] double positive (DP) stage to single positive (either CD4[+] or CD8[+]) phenotype [67]. Thymi of dCK KO mice are severely hypocellular and contain a partial block within DN developmental stage, at DN3 to DN4 transition [36]. Upon successful completion of TCR β gene rearrangement and β-selection check point in DN3a, thymocytes undergo extensive proliferation in DN3b, DN4 and DP stages [68]. We analyzed thymic development of CD45.2[+] dCK KO bone marrow reconstituted with WT or Ser[74] mutant dCK and YFP marker. Re-expression of WT dCK in KO HSCs rescued the impaired thymic development as evidenced by normalized cellularity

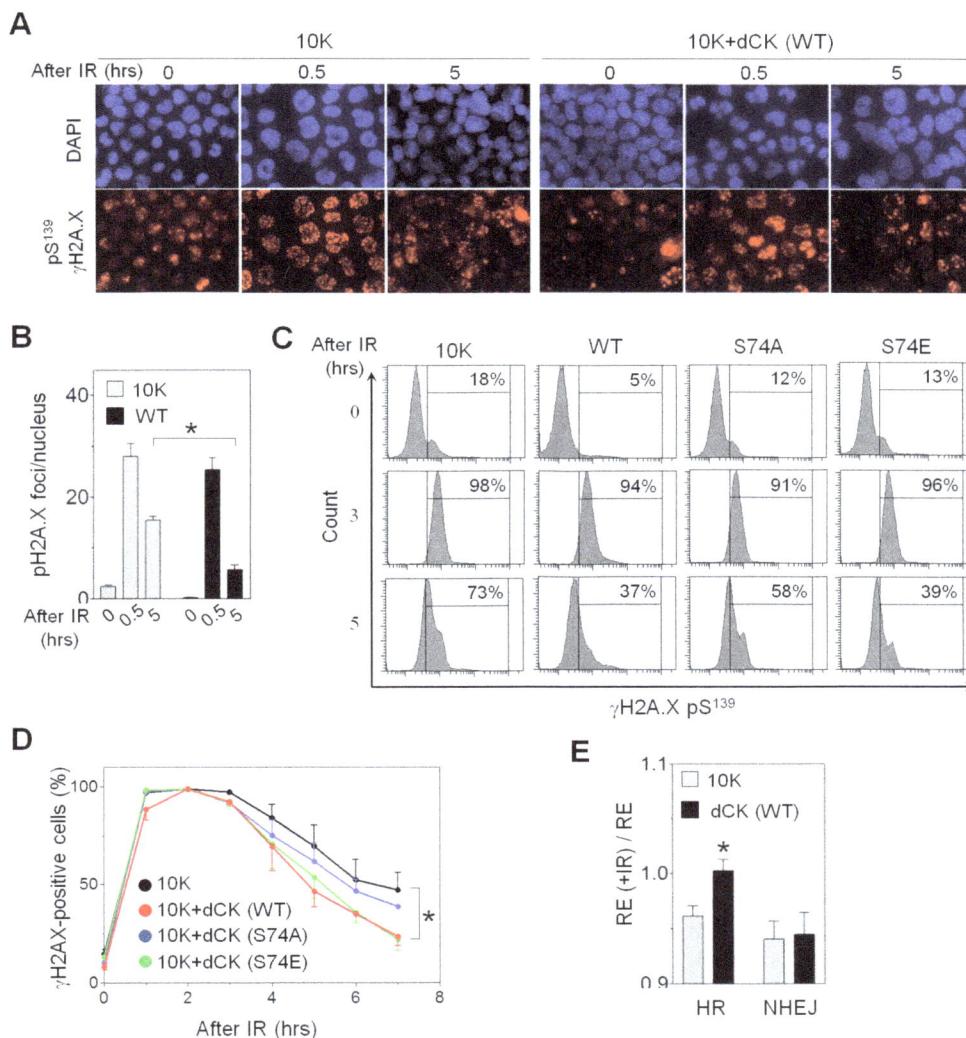

Figure 4. dCK influences the rate of IR-induced DNA repair. (A) Fluorescent images of DAPI and anti-pS[139] γH2A.X stained 10K ± dCK (WT) cells before or 0.5 and 5 hours after exposure to 3 Gy. (B) Number of γH2A.X foci per cell nucleus (*, P<0.0001; N = 10) calculated from (A). (C) FACS analysis of pS[139] γH2A.X positive 10K cells (indicated as % of total cell number) with or without dCK (WT, S74A, S74E) after 3 Gy irradiation. (D) Percentage of γH2A.X positive cells remaining over time after 3 Gy exposure (*, P = 0.01; N = 3 per time point), obtained as shown in (C). (E) Recombination efficiency (RE) of digested plasmid via homologous recombination (HR) and non-homologous end joining (NHEJ) 24 hours after nucleofection of 3 Gy irradiated 10K ± dCK (WT) cells divided by the RE of the untreated cells (*, P = 0.0078, N = 10).

(Figure 6C), overall predominance of DP cells in the thymus, and DN stage consisting almost exclusively of DN4 cells (Figure 6D–F). Presence of S74E mutant similarly led to the reversal of abnormal dCK KO thymic development (Figure 6C–F). The development of S74A-dCK expressing thymocytes phenocopied dCK KO cells, but the impairment was less pronounced (Figure 6C–F). Cellularity of S74A thymi was an order of magnitude lower compared to the WT (Figure 6C). The partial block in DN3 to DN4 transition was not as severe with S74A-dCK as in KO mice (Figure 6D–F). Therefore, the activity of dCK, dictated by the Ser[74] residue, influences T-cell development during bone marrow transplant reconstitution.

By staining DN3 thymocytes for pSer[139] γH2A.X, we also measured whether altering dCK activity through Ser[74] mutation affects the accumulation of DNA DSBs during T cell development. Accumulation of γH2A.X observed in DN3 thymocytes of KO and S74A-dCK animals was 8.5-fold higher compared to WT and

3.7 fold higher compared to S74E-dCK (Figure 6G), consistent with previous demonstration of enhanced DNA damage in dCK KO, but not WT, DN3 lymphocytes [37]. We conclude that hypofunctional, non-activatable S74A dCK leads to the accumulation of DNA double-strand breaks during thymocyte development.

B cells at various stages of development may also be phenotyped based on the expression of specific surface markers [69,70]. dCK KO mice have a partial block at the pro-B (CD43+CD19+) to pre-B cell (CD43−CD19+) transition [36]. At this stage, immature B cells are involved in immunoglobulin (Ig) gene rearrangement followed by clonal expansion [70]. Bone marrow reconstituted with the S74A-dCK mutant again phenocopied B-cell impairment in dCK KO mice (Figure 7A and B). Wild type dCK and the S74E mutant both rescued the dCK KO B-cell phenotype. Analysis of Hardy fractions [69] B–C (B220+IgM−CD19+CD43+), which include pro-B to pre-B transition, revealed significant

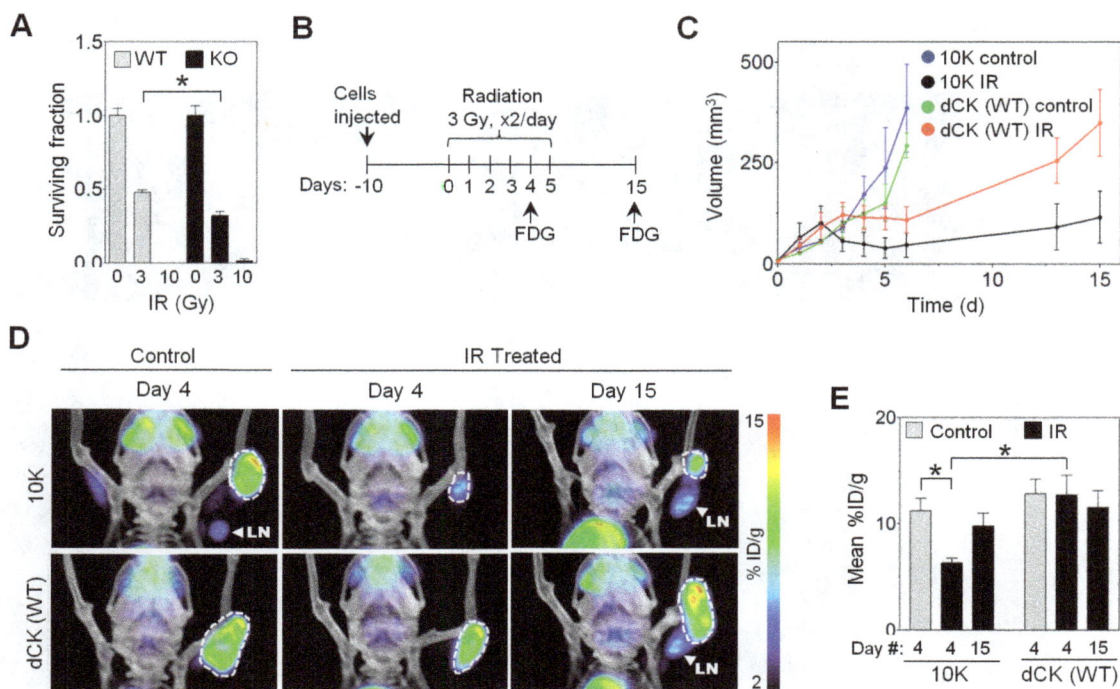

Figure 5. dCK absence sensitizes 10K cells and MEFs to IR. (A) Clonogenic survival assay of MEFs from dCK KO and dCK WT mice (*, P = 0.0383, N = 2). (B) Schematic of *in vivo* IR treatment schedule (FDG refers to PET imaging). (C) Volume measurements of *in vivo* untreated control and irradiated tumor model based on (L×W^2)/2 formula (N = 4 mice per group). (D) [^{18}F]-FDG microPET/CT scans of CB17 SCID mice treated as shown in (B) or untreated control. LN – lymph node. (E) Quantification of [^{18}F]-FDG uptake in irradiated and untreated 10K ± dCK *in vivo* tumor model (*, 10K ± IR at day 4: P = 0.0181; irradiated 10K vs. 10K+dCK at day 4: P = 0.0272; N = 4 mice per group).

accumulation of γH2A.X in the KO and S74A-dCK pro-B lymphocytes, but not in the pro-B cells with WT or S74E dCK (Figure 7A and C). Consistent with described DNA damage in developing dCK KO B cells [37], we demonstrate that dCK mutation at a single Ser^{74} residue is sufficient to impair B-cell development and cause accumulation of DNA DSBs in bone marrow reconstitution.

Discussion

The functional role of deoxycytidine kinase in DNA repair

Our results show that IR induces significant changes in deoxycytidine metabolism by activating dCK and increasing the salvage of deoxycytidine from the extracellular space in L1210-10K murine leukemia cell line. We show that ATM, but not ATR or DNA-PKcs, is critical for dCK activation by phosphorylating dCK at Ser^{74}. The importance of ATM in dCK activation that we observe agrees with a recent study by Yang et al who also show that ATM phosphorylates dCK at Ser^{74} after IR [33]. The authors of the study conclude that the significance of this post-translational modification is to enable dCK to participate in an inhibitory protein complex with Cdk1 in the context of DNA damage, thereby establishing the G2M checkpoint. This result posits an interesting role for dCK that is independent of its function as a metabolic enzyme. Here, we do not address the role of dCK in protein complexes. It is possible that activated dCK has a dual role after IR to ensure effective DNA repair: (1) to increase production of nucleotide precursors for DNA repair machinery, and (2) to stop cell-cycle progression and allow time for DNA repair to take place. Further investigations using kinase-dead dCK mutants or specific inhibitors of dCK enzymatic activity are necessary to dissect the

role of dCK in protein complexes from its metabolic function during DNA damage repair.

Regulation of nucleotide precursor metabolism in response to DNA damage

ATM has been implicated in affecting various cytoplasmic processes in addition to its more established role in the nucleus [71]. One particular emerging role of ATM in the cytoplasm appears to involve the regulation of cellular metabolic pathways [71]. We observed that dCK localizes in the cytoplasm of CHOC6 and L1210 cell lines (Figure S3). After IR, dCK was not observed in the nucleus, and its enhanced activity was only evident in the cytoplasm (Figure S3). ATM was detected in the cytoplasm of CHOC6 cells only after IR, consistent with accumulating evidence of the extranuclear shuttling of ATM [71]. We believe, therefore, that our data suggests yet another example of ATM function outside of the nucleus.

Maintenance of adequate and balanced dNTP pools is an important task that cells execute to avoid genomic instability during DNA damage response. DSB repair through HR, in particular, requires the presence of sufficient dNTPs to synthesize long stretches of DNA after 5'- end resection [58]. To maintain dNTP pools under endogenous or exogenous stress, cells must co-regulate the *de novo* and salvage dNTP production. While the *de novo* dNTP synthesis undoubtedly plays an important role in DNA repair, the contribution of deoxyribonucleoside salvage to DNA repair has been less well explored [1]. However, increasing evidence of co-regulation and feedback between these two pathways is emerging. For example, RNR subunit p53R2 suppresses MEK-ERK activity [72], while inhibition of MAPK pathway activates dCK [13]. RNR is allosterically regulated by

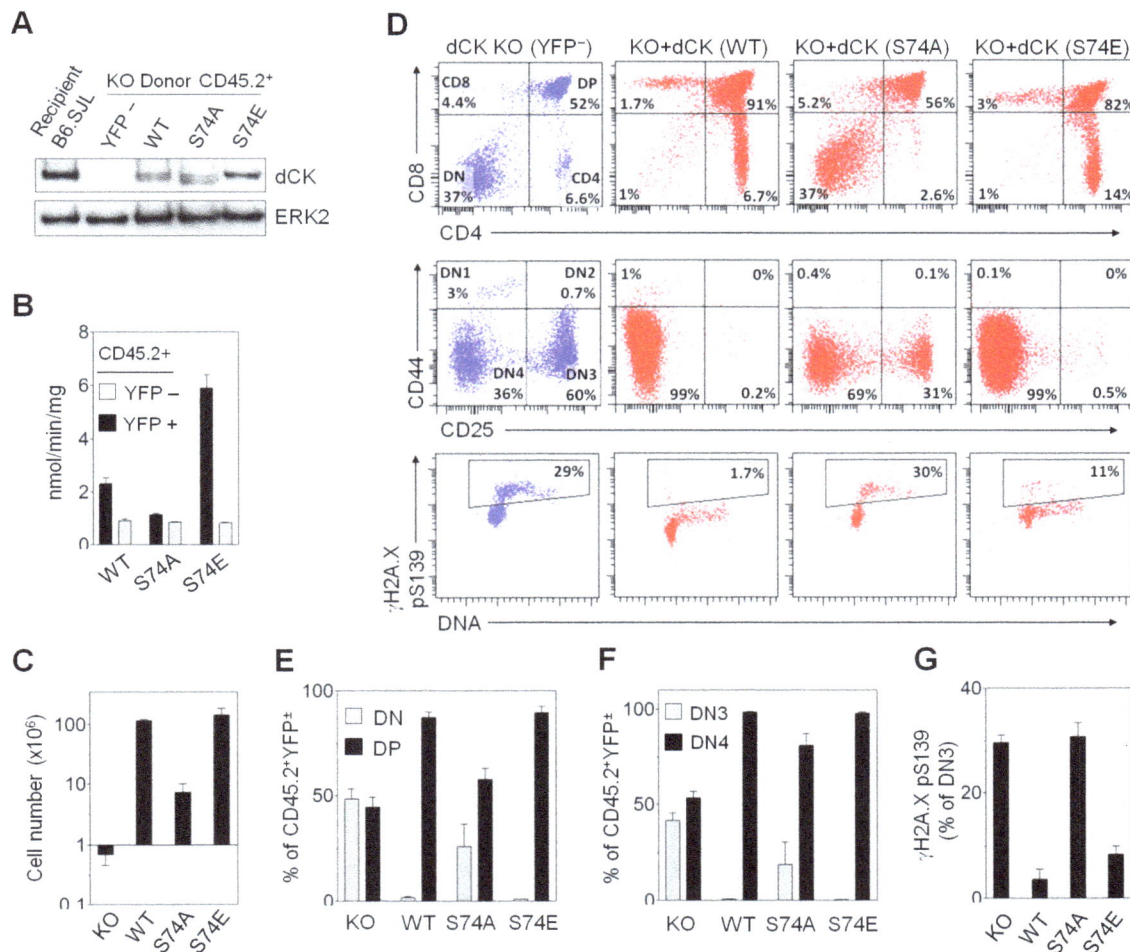

Figure 6. DCK Ser[74] influences T cell development. (A) Western blot of CD45.2[+] dCK KO cells (YFP[−]) carrying one of three dCK isoforms (WT, S74A, S74E; YFP[+]). Cells were isolated from recipient spleens and dCK expression level compared to endogenous dCK of CD45.1[+] B6.SJL recipient. (B) *In vitro* dCK kinase assay using [³H]-dC and lysates of dCK KO donor CD45.2[+] (dCK[+]YFP[+] or dCK[−]YFP[−]) cells FACS purified from spleens of B6.SJL recipients (N = 3 mice per group). (C) Cellularity of B6.SJL thymi repopulated with dCK KO donor cells carrying different dCK isoforms (CD45.2[+]YFP[+]; N = 3). (D) FACS analysis of dCK KO ± dCK (WT, S74A, S74E) thymocytes. Bottom panels show pSer[139] γH2A.X stained DN3 thymocytes. (E-F) Averaged percentages of single live donor CD45.2[+] thymocytes (± dCK/YFP) at different stages of development (N = 3). (G) Averaged percentages of DN3 thymocytes positive for pSer[139] γH2A.X (N = 3).

dTTP produced by the salvage pathway, shifting the specificity from pyrimidine to purine nucleotide reduction [73]. Finally, ATM regulates *de novo* dNTP synthesis through Ser[72] phosphorylation of RNR p53R2 subunit [26,27]. Mutation of that residue to alanine reduces basal and UV-induced dNTP levels [27]. While long-term dNTP balance may be achieved by increasing the expression of rate-limiting enzymes of dNTP synthesis, the majority of DNA DSB damage is repaired within the first 8 hours following the genotoxic insult. It is not surprising then that ATM regulates both *de novo* and salvage dNTP syntheses by post-translational modification of the rate limiting enzymes of these metabolic pathways, leading to a rapid increase of dNTP pools. Furthermore, each pathway may be responsible for supplying specific dNTPs for DNA repair. We show that the substrate specificity of phosphorylated dCK after IR shifts toward dC. Others demonstrated that the blockade of RNR by hydroxyurea causes selective depletion of purines (dATP and dGTP), but not pyrimidines [30]. ATM appears to coordinate the supply of dNTPs for DNA repair, and further work should aim at

uncovering additional details of this regulatory mechanism and specific *de novo* and salvage pathway contributions.

Building on the previous evidence, we demonstrated that in L1210 murine leukemia cell line the rate limiting enzyme in dN salvage, dCK, is activated via phosphorylation at Ser[74] by ATM in response to oxidative stress and IR-induced DNA damage. It is not clear whether oxidative stress can lead to dCK activation without inducing DNA damage. ATM is recruited to DNA double-strand breaks through association with Mre11-Rad50-Nbs1 (MRN) complex [74]. While MRN-dependent ATM autophosphorylation and monomerization has been one established mechanism of ATM activation, a separate mechanism involving direct oxidation and formation of an ATM dimer has been described [28]. Separating ROS production from DNA damage is difficult, and additional experiments are needed to further understand what activated form of ATM is responsible for regulating dN salvage pathway. We have observed that increased endogenous level of phosphorylated ATM at Ser[1981] in ATR-Seckel cells compared to WT LCL, as also reported by others [75], does not result in dCK

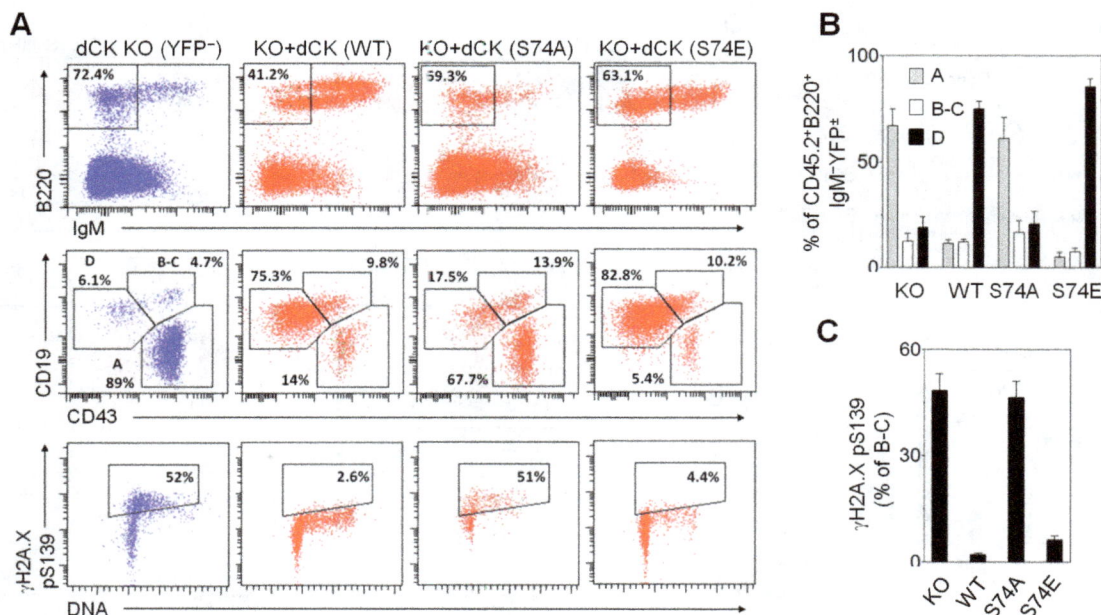

Figure 7. DCK Ser[74] influences B cell development. (A) FACS analysis of CD45.2+ donor dCK KO ± YFP/dCK (WT, S74A, S74E) B cell development in the bone marrow of B6.SJL recipients. FACS plots are representative examples. Bottom panels show γH2A.X stained B–C Hardy fractions. (B) Averaged percentages of single live donor CD45.2+B220+IgM− (± dCK/YFP) B cells at different stages of development (N = 3). (C) Averaged percentages of Hardy B–C fraction B cells positive for pSer[139] γH2A.X (N = 3).

activation in the absence of IR (Figure 1F). This suggests that phos-Ser[1981] modification of ATM is not sufficient for dCK activation. Furthermore, given the multitude of genotoxic agents which cause dCK activation, it remains unclear whether non-DSB forms of DNA damage such as single-strand DNA breaks lead to dN salvage activation, and whether this mechanism of activation is also ATM-dependent.

Role of deoxyribonucleoside salvage pathway in endogenous stress response

The Ser[74] residue of dCK is highly conserved across many species [19]. We hypothesized that ATM regulation of dCK is important in overcoming endogenous cellular stresses such as those resulting from oxidative species or DNA damage. Developing lymphocytes are particularly sensitive to the deletion of dCK [36]. Partial blocks in T and B lymphocyte development of dCK KO mice occur at the stages of VDJ and Ig gene recombination, respectively, and rapid proliferative expansion [67,70,76]. Reduced supply of dNTPs may slow programmed DNA repair and induce replication stress. However, gene rearrangement in developing lymphocytes occurs via non-homologous end joining pathway, which is relatively insensitive to dNTP levels [77]. Anabolic demands of major proliferative expansion result in increased metabolic activity, accumulation of reactive oxygen species and DNA damage [78]. It remains unclear whether dCK activation above baseline is needed to meet the demands of DNA replication and/or repair in the developing lymphocytes. ATM also plays an important role in VDJ recombination by activating cell-cycle checkpoints and directly stabilizing DNA DSB complexes [79]. Impairment of lymphocyte development in dCK KO mice is different from Atm−/− mice. Defective processing of VDJ recombination in Atm−/− thymocytes results in reduced production of mature CD4 and CD8 T cells and accumulation of double positive (DP) cells in the thymus [80]. Partially blocked transition

from DN3 to DN4 developmental stage seen in dCK KO thymi [36] is not observed in Atm−/− mice [80]. However, the regulation of dCK by ATM in vitro raises a possibility of similar mechanism occurring in vivo. The phenotype observed in T and B cell development with a mutation of a single Ser[74] dCK residue suggests that regulation of dCK through Ser[74] may play a role during hematopoiesis. Accumulation of γH2A.X foci in developing lymphocytes with KO dCK [37] and S74A-dCK mutants suggests that dCK activation based on post-translational modification may reduce endogenous genotoxic stress during clonal expansion. However, the S74A mutation compromises basal dCK activity. Mutating other residues critical for enzyme function which are not post-translationally modified should help determine whether reduced dCK activity exclusively accounts for the observed phenotype. Absence of a partial block at the double negative stage in Atm−/− thymocytes may be due to a functional redundancy in dN regulation. For example, other members of PIKK family, ATR and DNA-PKcs may regulate dN salvage in response to single-strand DNA present at stalled replication forks or generated by processed DSBs.

Exploiting the deoxyribonucleoside salvage pathway for cancer diagnosis and treatment

A tumor's response to genotoxic stress is an important clinical diagnostic and prognostic factor. Efficient DNA damage repair in cancer cells may indicate a resistance to chemo- or radiation therapy. For this reason, clinical tools that can measure tumor responses to DNA damaging agents are necessary to give insight into therapeutic efficacy. Non-invasive quantitative imaging of the DNA damage response would be useful for stratification of patients and making necessary adjustments to the therapy. Using a dCK specific PET probe, we visualized and quantified IR-induced ATM-dependent activation of the dN salvage pathway, manifested as an increased probe accumulation in the tumor 1 hour after tumor irradiation. This is a unique demonstration of a clinically-relevant quantitative imaging read-out of the acute tumor metabolic response to genotoxic therapy. Higher PET signal after localized irradiation indicates enhanced dCK activation and a potential increase in DNA repair rate. These tumor responses also indicate that dN analog prodrugs like gemcitabine and clofarabine, which are phosphorylated into their active forms by dCK, could be used in synergy with radiotherapy [54]. Finally, this technology may delineate areas of a single tumor lesion with variable sensitivities to the therapy, as well as provide whole-body information in the cases of total body/bone marrow irradiation for cancer therapy and bone marrow transplantation. Future work will be aimed at determining specific tumor types which exhibit robust dCK activation and [^{18}F]-FAC accumulation after IR.

In this study, we show that inhibition of deoxyribonucleoside salvage through the genetic deletion of dCK leads to enhanced radiosensitivity of a murine leukemia cell line and mouse embryonic fibroblasts. Yang et al showed enhanced radiosensitivity to IR in HeLa cells after stable dCK knockdown [33], suggesting that our result may be generalizable to other cell lines and types of human cancer expressing dCK. In principle, blockade of dCK by a small molecule inhibitor [81] may sensitize cancer to a wide range of genotoxic agents, including IR. Inhibitors of dCK Ser^{74} phosphorylation, in particular, may be useful in treating malignancy in conjunction with local radiotherapy. dCK is a promising target for cancer treatment and PET probes targeting dCK activity [35] will be useful in predicting and following clinical response to such therapy.

Supporting Information

Figure S1   IR-induced dCK activation is reduced in A-T cells. (A) Western blot of CHOC6 (WT LCL) and A-T cell lines (AT224LA, AT255LA). (B) *In vitro* cell uptake assay of [^{3}H]-dC by CHOC6, AT255LA and DK0064 cells before and 2 hours after exposure to 3 Gy (*, CHOC6: P = 0.034; DK0064: P = 0.023; N = 3). (C) Fold change in dCK activity of CHOC6, A-T cells (AT224LA, AT255LA) and DK0064 2 hours after 3 Gy (*, P< 0.0001; N = 9).

Figure S2   PET imaging of IR-induced ATM-dependent dCK activation. (A) [^{18}F]-FAC microPET/CT scans of four NOD-SCID mice with bilateral 10K+dCK (WT) tumors after 3 Gy irradiation of right tumor. Top two rows are coronal and transverse cross-sectional images, respectively. Bottom row: volume rendered images. (B) [^{18}F]-FAC and [^{18}F]-FDG micro-PET/CT scans of NOD-SCID mouse with bilateral 10 K tumors after 3 Gy irradiation of right tumor. Top row: coronal cross-section; bottom row: transverse cross-sectional images.

Figure S3   dCK is localized and activated after IR in the cytoplasm. (A) Western blot of nuclear (N) and cytoplasmic (C) fractions of CHOC6 (WT LCL) before and 2 hours after 3 Gy exposure. (B) *In vitro* dCK kinase assay using CHOC6 nuclear and cytoplasmic fraction lysates, [^{3}H]-dC as substrate and performed 2 hours after exposure to 3 Gy (*, P = 0.0049, N = 3). (C) Western blot of nuclear (N) and cytoplasmic (C) fractions of L1210 cell line before and 2 hours after 3 Gy exposure. (D) *In vitro* dCK kinase assay using L1210 nuclear and cytoplasmic fraction lysates, [^{3}H]-dC as substrate and performed 2 hours after exposure to 3 Gy (*, P = 0.0008, N = 3).

Acknowledgments

We thank Tanya Stoyanova for assistance with experiments; Wayne Austin for assistance with FACS and dCK KO mice; Christina Brown for assistance with A-T cells and Western blots; Dorthe Schaue for assistance with tumor irradiation; Robert Schiestl and Zorica Scuric for providing plasmids and assistance with plasmid recombination assay; Francoise Bontemps for providing pSer^{74} dCK antibody; Richard Gatti and William McBride for insightful discussion; David Stout and Waldemar Ladno for microPET/CT imaging; and the cyclotron group for the production of PET probes.

Author Contributions

Conceived and designed the experiments: YLB ENG CGR ONW. Performed the experiments: YLB ENG MR MNM DC JM. Analyzed the data: YLB ENG MNM DC. Contributed reagents/materials/analysis tools: CGR. Wrote the paper: YLB ENG CGR ONW.

References

1. Mathews CK (2006) DNA precursor metabolism and genomic stability. Faseb J 20: 1300–1314.
2. Reichard P (1988) Interactions between deoxyribonucleotide and DNA synthesis. Annu Rev Biochem 57: 349–374.
3. Chottiner EG, Shewach DS, Datta NS, Ashcraft E, Gribbin D, et al. (1991) Cloning and expression of human deoxycytidine kinase cDNA. Proc Natl Acad Sci U S A 88: 1531–1535.
4. Sabini E, Ort S, Monnerjahn C, Konrad M, Lavie A (2003) Structure of human dCK suggests strategies to improve anticancer and antiviral therapy. Nat Struct Biol 10: 513–519.
5. Csapo Z, Keszler G, Sasvari-Szekely M, Smid K, Noordhuis P, et al. (1998) Similar changes were induced by cladribine and by gemcitabine, in the deoxypyrimidine salvage, during short term treatments. Adv Exp Med Biol 431: 525–529.
6. Spasokoukotskaja T, Sasvari-Szekely M, Keszler G, Albertioni F, Eriksson S, et al. (1999) Treatment of normal and malignant cells with nucleoside analogues and etoposide enhances deoxycytidine kinase activity. Eur J Cancer 35: 1862–1867.
7. Cardoen S, Van den Neste E, Smal C, Rosier JF, Delacauw A, et al. (2001) Resistance to 2-chloro-2′-deoxyadenosine of the human B-cell leukemia cell line EHEB. Clin Cancer Res 7: 3559–3566.
8. Wei S, Ageron-Blanc A, Petridis F, Baumatin J, Bonnet S, et al. (1999) Radiation-induced changes in nucleotide metabolism of two colon cancer cell lines with different radiosensitivities. Int J Radiat Biol 75: 1005–1013.
9. Csapo Z, Keszler G, Safrany G, Spasokoukotskaja T, Talianidis L, et al. (2003) Activation of deoxycytidine kinase by gamma-irradiation and inactivation by hyperosmotic shock in human lymphocytes. Biochem Pharmacol 65: 2031–2039.
10. Haveman J, Sigmond J, Van Bree C, Franken NA, Koedooder C, et al. (2006) Time course of enhanced activity of deoxycytidine kinase and thymidine kinase 1 and 2 in cultured human squamous lung carcinoma cells, SW-1573, induced by gamma-irradiation. Oncol Rep 16: 901–905.
11. Van den Neste E, Smal C, Cardoen S, Delacauw A, Frankard J, et al. (2003) Activation of deoxycytidine kinase by UV-C-irradiation in chronic lymphocytic leukemia B-lymphocytes. Biochem Pharmacol 65: 573–580.
12. Staub M, Keszler G, Spasokukotskaja T, Sasvari-Szekely M, Peters F, et al. (1998) Potentiation of deoxycytidine kinase activity by inhibition of tyrosine kinase or DNA synthesis in different cells. Ann Oncol 9: 113–113.
13. Smal C, Cardoen S, Bertrand L, Delacauw A, Ferrant A, et al. (2004) Activation of deoxycytidine kinase by protein kinase inhibitors and okadaic acid in leukemic cells. Biochem Pharmacol 68: 95–103.
14. Smal C, Lisart S, Maerevoet M, Ferrant A, Bontemps F, et al. (2007) Pharmacological inhibition of the MAPK/ERK pathway increases sensitivity to 2-chloro-2′-deoxyadenosine (CdA) in the B-cell leukemia cell line EHEB. Biochem Pharmacol 73: 351–358.
15. Keszler G, Spasokoukotskaja T, Csapo Z, Talianidis I, Eriksson S, et al. (2004) Activation of deoxycytidine kinase in lymphocytes is calcium dependent and involves a conformational change detectable by native immunostaining. Biochem Pharmacol 67: 947–955.
16. Smal C, Bertrand L, Van Den Neste E, Cardoen S, Veiga-Da-Cunha M, et al. (2004) New evidences for a regulation of deoxycytidine kinase activity by reversible phosphorylation. Nucleos Nucleot Nucl 23: 1363–1365.
17. Keszler G, Spasokoukotskaja T, Sasvari-Szekely M, Eriksson S, Staub M (2006) Deoxycytidine kinase is reversibly phosphorylated in normal human lymphocytes. Nucleos Nucleot Nucl 25: 1147–1151.

18. Smal C, Vertommen D, Bertrand L, Rider MH, Van den Neste E, et al. (2006) Identification of phosphorylation sites on human deoxycytidine kinase after overexpression in eucaryotic cells. Nucleos Nucleot Nucl 25: 114.–1146.

19. Smal C, Vertommen D, Bertrand L, Ntamashimikiro S, Rider MH, et al. (2006) Identification of in vivo phosphorylation sites on human deoxycytidine kinase - Role of Ser-74 in the control of enzyme activity. J Biol Chem 281: 4887–4893.

20. Smal C, Van Den Neste E, Maerevoet M, Poire X, Theate I, et al. (2007) Positive regulation of deoxycytidine kinase activity by phosphorylation of Ser-74 in B-cell chronic lymphocytic leukaemia lymphocytes. Cancer Lett 253: 68–73.

21. McSorley T, Ort S, Hazra S, Lavie A, Konrad M (2008) Mimicking phosphorylation of Ser-74 on human deoxycytidine kinase selectively increases catalytic activity for dC and dC analogues. Febs Lett 582: 720–724.

22. Hazra S, Szewczak A, Ort S, Konrad M, Lavie A (2011) Post-translational phosphorylation of serine 74 of human deoxycytidine kinase favors the enzyme adopting the open conformation making it competent for nucleoside binding and release. Biochemistry 50: 2870–2880.

23. Lavin MF (2008) Ataxia-telangiectasia: from a rare disorder to a paradigm for cell signalling and cancer. Nat Rev Mol Cell Bio 9: 759–769.

24. Bensimon A, Aebersold R, Shiloh Y (2011) Beyond ATM: The protein kinase landscape of the DNA damage response. Febs Lett 585: 1625–1639.

25. Cosentino C, Grieco D, Costanzo V (2011) ATM activates the pentose phosphate pathway promoting anti-oxidant defence and DNA repair. Embo J 30: 546–555.

26. Eaton JS, Lin ZP, Sartorelli AC, Bonawitz ND, Shadel GS (2007) Ataxia-telangiectasia mutated kinase regulates ribonucleotide reductase and mitochondrial homeostasis. J Clin Invest 117: 2723–2734.

27. Chang LF, Zhou BS, Hul S, Guo R, Liu XY, et al. (2008) ATM-mediated serine 72 phosphorylation stabilizes ribonucleotide reductase small subunit p53R2 protein against MDM2 to DNA damage. Proc Natl Acad Sc U S A 105: 18519–18524.

28. Guo Z, Kozlov S, Lavin MF, Person MD, Paull TT (2010) ATM activation by oxidative stress. Science 330: 517–521.

29. Cheema AK, Timofeeva O, Varghese R, Dimtchev A, Shiekh K, et al. (2011) Integrated analysis of ATM mediated gene and protein expression impacting cellular metabolism. J Proteome Res 10: 2651–2657.

30. Hakansson P, Hofer A, Thelander L (2006) Regulation of mammalian ribonucleotide reduction and dNTP pools after DNA damage and in resting cells. J Biol Chem 281: 7834–7841.

31. Matsuoka S, Ballif BA, Smogorzewska A, McDonald ER, Hurov KE, et al. (2007) ATM and ATR substrate analysis reveals extensive protein networks responsive to DNA damage. Science 316: 1160–1166.

32. Kunos CA, Ferris G, Pyatka N, Pink J, Radivoyevitch T (2011) Deoxynucleoside salvage facilitates DNA repair during ribonucleotide reductase blockade in human cervical cancers. Radiat Res 176: 425–433.

33. Yang C, Lee M, Hao J, Cui X, Guo X, et al. (2012) Deoxycytidine kinase regulates the G2/M checkpoint through interaction with cyclin-dependent kinase 1 in response to DNA damage. Nucleic Acids Res 40: 962.–9632.

34. Radu CG, Shu CJ, Nair-Gill E, Shelly SM, Barrio JR, et al. (2008) Molecular imaging of lymphoid organs and immune activation by positron emission tomography with a new [(18)F]-labeled 2′-deoxycytidine analog. Nat Med 14: 783–788.

35. Shu CJ, Campbell DO, Lee JT, Tran AQ, Wengrod JC, et al. (2010) Novel PET probes specific for deoxycytidine kinase. J Nucl Med 51: 1092–1098.

36. Toy G, Austin WR, Liao HI, Cheng DH, Singh A, et al. (2010) Requirement for deoxycytidine kinase in T and B lymphocyte development. Proc Natl Acad Sci U S A 107: 5551–5556.

37. Austin WR, Armijo AL, Campbell DO, Singh AS, Hsieh T, et al. (2012) Nucleoside salvage pathway kinases regulate hematopoiesis by linking nucleotide metabolism with replication stress. J Exp Med 209: 2215–2228.

38. Jordheim LP, Cros E, Gouy MH, Galmarini CM, Peyrottes S, et al. (2004) Characterization of a gemcitabine-resistant murine leukemic cell line: Reversion of in vitro resistance by a mononucleotide prodrug. Clin Cancer Res 10: 5614–5621.

39. Hawley RG, Lieu FHL, Fong AZC, Hawley TS (1994) Versatile retroviral vectors for potential use in gene therapy. Gene Ther 1: 136–138.

40. O'Driscoll M, Ruiz-Perez VL, Woods CG, Jeggo PA, Goodship JA (2003) A splicing mutation affecting expression of ataxia-telangiectasia and Rad3-related protein (ATR) results in Seckel syndrome. Nat Genet 33: 497–501.

41. Laing RE, Walter MA, Campbell DO, Herschman HR, Satyamurthy N, et al. (2009) Noninvasive prediction of tumor responses to gemcitabine using positron emission tomography. Proc Natl Acad Sci U S A 106: 2847–2852.

42. Secretan MB, Scuric Z, Oshima J, Bishop AJR, Howlett NG, et al. (2004) Effect of Ku86 and DNA-PKcs deficiency on non-homologous end-joining and homologous recombination using a transient transfection assay. Mutat Res-Fund Mol M 554: 351–364.

43. Luo LZ, Gopalakrishna-Pillai S, Nay SL, Park SW, Bates SE, et al. (2012) DNA repair in human pluripotent stem cells is distinct from that in non-pluripotent human cells. PLoS One 7: e30541.

44. Mathews CK, Wheeler LJ (2009) Measuring DNA precursor pools in mitochondria. Methods Mol Biol 554: 371–381.

45. Wong S, McLaughlin J, Cheng D, Zhang C, Shokat KM, et al. (2004) Sole BCR-ABL inhibition is insufficient to eliminate all myeloproliferative disorder cell populations. Proc Natl Acad Sci U S A 101: 17456–17461.

46. Stout DB, Chatziioannou AF, Lawson TP, Silverman RW, Gambhir SS, et al. (2005) Small animal imaging center design: The facility at the UCLA Crump Institute for Molecular Imaging. Mol Imaging Biol 7: 393–402.

47. Hamacher K, Coenen HH, Stocklin G (1986) Efficient stereospecific synthesis of no-carrier-added 2-[F-18]-fluoro-2-deoxy-D-glucose using aminopolyether supported nucleophilic substitution. J Nucl Med 27: 235–238.

48. Loening AM, Gambhir SS (2001) AMIDE: A completely free system for medical imaging data analysis. J Nucl Med 42: 192P–192P.

49. Spitz DR, Azzam EI, Li JJ, Gius D (2004) Metabolic oxidation/reduction reactions and cellular responses to ionizing radiation: A unifying concept in stress response biology. Cancer Metastasis Rev 23: 311–322.

50. Shieh SY, Ikeda M, Taya Y, Prives C (1997) DNA damage-induced phosphorylation of p53 alleviates inhibition by MDM2. Cell 91: 325–334.

51. Hickson I, Yan Z, Richardson CJ, Green SJ, Martin NMB, et al. (2004) Identification and characterization of a novel and specific inhibitor of the ataxia-telangiectasia mutated kinase ATM. Cancer Res 64: 9152–9159.

52. Leahy JJJ, Golding BT, Griffin RJ, Hardcastle IR, Richardson C, et al. (2004) Identification of a highly potent and selective DNA-dependent protein kinase (DNA-PK) inhibitor (NU7441) by screening of chromenone libraries. Bioorg Med Chem Lett 14: 6083–6087.

53. Cimprich KA, Cortez D (2008) ATR: an essential regulator of genome integrity. Nat Rev Mol Cell Bio 9: 616–627.

54. Lee MW, Parker WB, Xu B (2013) New insights into the synergism of nucleoside analogs with radiotherapy. Radiat Oncol 8: 223–232.

55. Young H, Baum R, Cremerius U, Herholz K, Hoekstra O, et al. (1999) Measurement of clinical and subclinical tumour response using [F-18]-fluorodeoxyglucose and positron emission tomography: Review and 1999 EORTC recommendations. Eur J Cancer 35: 1773–1782.

56. Griffith DA, Jarvis SM (1996) Nucleoside and nucleobase transport systems of mammalian cells. Biochim Biophys Acta-Rev Biomembr 1286: 153–181.

57. Sancar A, Lindsey-Boltz LA, Unsal-Kacmaz K, Linn S (2004) Molecular mechanisms of mammalian DNA repair and the DNA damage checkpoints. Annu Rev Biochem 73: 39–85.

58. Filippo JS, Sung P, Klein H (2008) Mechanism of eukaryotic homologous recombination. Annu Rev Biochem 77: 229–257.

59. Burkhalter MD, Roberts SA, Havener JM, Ramsden DA (2009) Activity of ribonucleotide reductase helps determine how cells repair DNA double strand breaks. DNA Repair 8: 1258–1263.

60. Ljungman M (2009) Targeting the DNA damage response in cancer. Chem Rev 109: 2929–2950.

61. Kim SH, Brown SL, Kim JH (1998) The potentiation of radiation response in human colon carcinoma cells in vitro and murine lymphoma in vivo by AG337 (Thymitaq (TM)), a novel thymidylate synthase inhibitor. Int J Radiat Oncol Biol Phys 42: 789–793.

62. Shewach DS, Lawrence TS (1996) Gemcitabine and radiosensitization in human tumor cells. Invest New Drugs 14: 257–263.

63. Kunos CA, Radivoyevitch T, Pink J, Chiu SM, Stefan T, et al. (2010) Ribonucleotide reductase inhibition enhances chemoradiosensitivity of human cervical cancers. Radiat Res 174: 574–581.

64. Stuschke M, Thames HD (1997) Hyperfractionated radiotherapy of human tumors: Overview of the randomized clinical trials. Int J Radiat Oncol Biol Phys 37: 259–267.

65. Warren LA, Rossi DJ (2009) Stem cells and aging in the hematopoietic system. Mech Ageing Dev 130: 46–53.

66. Aude-Garcia C, Villiers C, Candeias SM, Garrel C, Bertrand C, et al. (2011) Enhanced susceptibility of T lymphocytes to oxidative stress in the absence of the cellular prion protein. Cell Mol Life Sci 68: 687–696.

67. Ciofani M, Zuniga-Pflucker JC (2010) Determining gamma delta versus alpha beta T cell development. Nat Rev Immunol 10: 657–663.

68. Taghon T, Yui MA, Pant R, Diamond RA, Rothenberg EV (2006) Developmental and molecular characterization of emerging beta- and gamma delta-selected pre-T cells in the adult mouse thymus. Immunity 24: 53–64.

69. Hardy RR, Carmack CE, Shinton SA, Kemp JD, Hayakawa K (1991) Resolution and characterization of pro-B and pre-pro-B cell stages in normal mouse bone marrow. J Exp Med 173: 1213–1225.

70. Hardy RR, Kincade PW, Dorshkind K (2007) The protean nature of cells in the B lymphocyte lineage. Immunity 26: 703–714.

71. Shiloh Y, Ziv Y (2013) The ATM protein kinase: regulating the cellular response to genotoxic stress, and more. Nat Rev Mol Cell Biol 14: 197–210.

72. Piao C, Jin M, Kim HB, Lee SM, Amatya PN, et al. (2009) Ribonucleotide reductase small subunit p53R2 suppresses MEK-ERK activity by binding to ERK kinase 2. Oncogene 28: 2173–2184.

73. Munch-Petersen B (2010) Enzymatic regulation of cytosolic thymidine kinase 1 and mitochondrial thymidine kinase 2: A mini review. Nucleos Nucleot Nucl 29: 363–369.

74. Uziel T, Lerenthal Y, Moyal L, Andegeko Y, Mittelman L, et al. (2003) Requirement of the MRN complex for ATM activation by DNA damage. Embo J 22: 5612–5621.

75. Stiff T, Walker SA, Cerosaletti K, Goodarzi AA, Petermann E, et al. (2006) ATR-dependent phosphorylation and activation of ATM in response to UV treatment or replication fork stalling. Embo J 25: 5775–5782.

76. Falk I, Biro J, Kohler H, Eichmann K (1996) Proliferation kinetics associated with T cell receptor-beta chain selection of fetal murine thymocytes. J Exp Med 184: 2327–2339.

77. Bassing CH, Alt FW (2004) The cellular response to general and programmed DNA double strand breaks. DNA Repair 3: 781–796.

78. Hesse JE, Faulkner MF, Durdik JM (2009) Increase in double-stranded DNA break-related foci in early-stage thymocytes of aged mice. Exp Gerontol 44: 676–684.

79. Bredemeyer AL, Sharma GG, Huang CY, Helmink BA, Walker LM, et al. (2006) ATM stabilizes DNA double-strand-break complexes during V(D)J recombination. Nature 442: 466–470.

80. Barlow C, Hirotsune S, Paylor R, Liyanage M, Eckhaus M, et al. (1996) ATM-deficient mice: A paradigm of ataxia telangiectasia. Cell 86: 159–171.

81. Yu XC, Miranda M, Liu ZY, Patel S, Nguyen N, et al. (2010) Novel potent inhibitors of deoxycytidine kinase identified and compared by multiple assays. J Biomol Screen 15: 72–79.

Differential Radiosensitivity Phenotypes of DNA-PKcs Mutations Affecting NHEJ and HRR Systems following Irradiation with Gamma-Rays or Very Low Fluences of Alpha Particles

Yu-Fen Lin[1][9], Hatsumi Nagasawa[2][9], John B. Little[3], Takamitsu A. Kato[2], Hung-Ying Shih[1], Xian-Jin Xie[4], Paul F. Wilson Jr.[5], John R. Brogan[2], Akihiro Kurimasa[6], David J. Chen[1], Joel S. Bedford[2], Benjamin P. C. Chen[1]*

1 Department of Radiation Oncology, University of Texas Southwestern Medical Center at Dallas, Dallas, Texas, United States of America, 2 Department of Environmental and Radiological Health Sciences, Colorado State University, Fort Collins, Colorado, United States of America, 3 Department of Genetics and Complex Diseases, Harvard School of Public Health, Boston, Massachusetts, United States of America, 4 Department of Clinical Sciences, University of Texas Southwestern Medical Center at Dallas, Dallas, Texas, United States of America, 5 Department of Biosciences, Brookhaven National Laboratory, Upton, New York, United States of America, 6 Institute of Regenerative Medicine and Biofunction, Graduate School of Medical Science, Tottori University, Tottori, Japan

Abstract

We have examined cell-cycle dependence of chromosomal aberration induction and cell killing after high or low dose-rate γ irradiation in cells bearing DNA-PKcs mutations in the S2056 cluster, the T2609 cluster, or the kinase domain. We also compared sister chromatid exchanges (SCE) production by very low fluences of α-particles in DNA-PKcs mutant cells, and in homologous recombination repair (HRR) mutant cells including Rad51C, Rad51D, and Fancg/xrcc9. Generally, chromosomal aberrations and cell killing by γ-rays were similarly affected by mutations in DNA-PKcs, and these mutant cells were more sensitive in G_1 than in S/G_2 phase. In G_1-irradiated DNA-PKcs mutant cells, both chromosome- and chromatid-type breaks and exchanges were in excess than wild-type cells. For cells irradiated in late S/G_2 phase, mutant cells showed very high yields of chromatid breaks compared to wild-type cells. Few exchanges were seen in DNA-PKcs-null, Ku80-null, or DNA-PKcs kinase dead mutants, but exchanges in excess were detected in the S2506 or T2609 cluster mutants. SCE induction by very low doses of α-particles is resulted from bystander effects in cells not traversed by α-particles. SCE seen in wild-type cells was completely abolished in Rad51C- or Rad51D-deficient cells, but near normal in Fancg/xrcc9 cells. In marked contrast, very high levels of SCEs were observed in DNA-PKcs-null, DNA-PKcs kinase-dead and Ku80-null mutants. SCE induction was also abolished in T2609 cluster mutant cells, but was only slightly reduced in the S2056 cluster mutant cells. Since both non-homologous end-joining (NHEJ) and HRR systems utilize initial DNA lesions as a substrate, these results suggest the possibility of a competitive interference phenomenon operating between NHEJ and at least the Rad51C/D components of HRR; the level of interaction between damaged DNA and a particular DNA-PK component may determine the level of interaction of such DNA with a relevant HRR component.

Editor: Jian Jian Li, University of California Davis, United States of America

Funding: This work was supported by grants CA166677 from the National Cancer Institute/National Institutes of Health, DE-FG02-07ER64350 and DE-FG02-05ER64089 from the U.S. Department of Energy Low Dose Radiation Research Program, and RP110465-P1 from the Cancer Prevention Research Institute of Texas. The funders had no role in study design, data collection and analysis, decision to publish, or preparation of the manuscript.

Competing Interests: The authors have declared that no competing interests exist.

* E-mail: benjamin.chen@utsouthwestern.edu

[9] These authors contributed equally to this work.

Introduction

The catalytic subunit of DNA dependent protein kinase (DNA-PKcs) is the key regulator of non-homologous end-joining (NHEJ), the predominant DNA double-strand break (DSB) repair mechanism in mammals. DNA-PKcs is recruited to DSBs through the DNA-binding heterodimer Ku70/80, and together with these factors form the kinase active DNA-PK holoenzyme [1]. The biological significance of DNA-PKcs first became evident with the finding that mutation within the gene encoding DNA-PKcs led to severe combined immunodeficiency (SCID) in mice and other

animals [2,3]. The other major phenotypic trait conferred by DNA-PKcs mutations was severe hypersensitivity to ionizing radiation (IR) and radiomimetic chemicals [4]. Kurimasa et al. confirmed the requirement of DNA-PKcs kinase activity for DSB rejoining after irradiation [5]. DNA-PKcs activation upon IR or treatment with radiomimetic chemicals rapidly results in phosphorylation of DNA-PKcs in the S2056 and the T2069 phosphorylation cluster regions [6–9]. Studies of DNA-PKcs mutant cell lines indicate that these phosphorylations are required for full DSB repair capacity and normal cellular radiosensitivity.

DNA-PKcs and its downstream NHEJ components are active in all cell cycle phases. In contrast, homologous recombination repair (HRR), another major DSB repair mechanism, contributes to DSB repair and cellular survival only during S and G_2 phases [10,11]. To clarify the significance of DNA-PKcs activities in NHEJ-mediated DSB repair and in radiosensitivity, it is important to study synchronous cell populations at different phases throughout the cell cycle. We reported previously that cells expressing DNA-PKcs with mutations in the S2056 cluster, the T2609 cluster, or the PI3K kinase domain have clear differences in radiosensitivities when mutant cells were irradiated in the G_1 phase [12]. Expression of DNA-PKcs with mutations in the T2609 cluster (L-3) or in the PI3K kinase domain (L-8, L-9, L-10, and L-11) results in extreme radiosensitivity, similar to that of Ku70/80-deficient xrs-5 and xrs-6 cells; however, mutations in the S2056 cluster (L-12) result in intermediate radiosensitivity [12].

DNA-PKcs mutants, V3 (DNA-PKcs null) and irs-20 (extreme c-terminal motif mutant) cause extreme and moderate radiosensitivity, respectively. These radiosensitive mutant cell strains respond to radiation in a cell-cycle-dependent manner and display enhanced radiation-induced cell cycle delay. In plateau phase G_1 cells, a greatly reduced potentially lethal damage repair (PLDR), sub-lethal damage repair (SLDR), and a great reduction or absence of a dose-rate effect are observed [13–17]. Chinese Hamster Ku70/80-deficient xrs-5 and xrs-6 cell lines are more radiosensitive than wild-type cells and the radiosensitivity does not depend on cell cycle stage. In addition, these cells show no PLDR and no dose-rate effect. In these respects xrs-5 and xrs-6 are similar to ATM-deficient cell strains [18–25].

In connection with the DNA-PKcs phosphorylation-defective mutants described above, we have also reported other results indicating that HRR was required for the induction of SCEs by alpha particles [26,27]. We have further investigated this in more detail in the present study by comparing SCE induction after very low doses of α-particles in cells that express the mutations in DNA-PKcs described above. The doses were sufficiently low that the observed levels of induced SCE could be attributed to effects produced in unirradiated "bystander" cells. In the present study, we compared radiosensitivity phenotypes among cell lines that express mutant versions of components of NHEJ system (DNA-PKcs, Ku80) or components of HRR system. We also examined the cell cycle dependence of chromosomal aberration induction and cell killing after high and low dose-rate γ irradiation.

Materials and Methods

Cell lines and synchrony

For these studies, we employed the wild-type Chinese hamster cell lines CHO [28] and AA8 [29], NHEJ- deficient mutant lines xrs-5 [30] and V3 [31], HRR mutant lines irs-3 [32], CL-V4B [33], 51D1 [34], Fanconi anemia (FA) mutant (KO40) [35], and cell lines derived from DNA-PKcs-null V3 cells complemented with human DNA-PKcs cDNA containing amino acid substitutions at various positions [5–7,12,36] that are described in Tables 1 and 2. The cells were maintained at 37°C in a humidified 95% air/5% CO_2 atmosphere in Eagle's minimal essential medium (MEM) supplemented with 10% heat-inactivated fetal bovine serum (FBS), penicillin (50 mg/ml), and streptomycin (50 µg/ml). When the cultures approached 30% confluence in T-25 tissue culture flasks or Mylar-dishes, the normal growth medium was replaced twice at 24-hour intervals with isoleucine-deficient MEM containing 5% 3× dialyzed FBS to synchronize the cells in the G_1 phase [37]. G_1 synchronized cells were released in normal growth medium for 12 hours to achieve synchrony in late S/G_2 phase of

Table 1. Amino acid substitutions in the DNA-PKcs constructs.

Cell line/code	DNA-PKcs mutants	S2056 cluster[1]					T2609 cluster[1]						PI3K		Rad-Sens[4]
		S2023	S2029	S2041	S2051	S2056	T2609	S2612	S2620	S2624	T2638	T2647	Y3715	D3921	
L-1	wild type														N*
L-3	V3-6A (T2609 cluster to A)						A	A	A	A	A	A			SSS*
L-6	V3-T2609A						A								S
L-7	V3 (empty vector)														SSS*
L-10	V3-KC23 (kinase dead)[2]												D		SSS*
L-11	V3-KD51 (kinase dead)[3]													N	SSS*
L-12	V3-5A (S2056 cluster to A)	A	A	A	A	A									S*
L-14	V3-3A (T2609A/T2638A/T2647A)						A				A	A			SS

1. Serine (S) and threonine (T) were replaced with alanine (A) at S2056 and T2609 cluster sites.
2. V3-KC23 mutant carries a frame-shift at position of amino acid 3715 that results in truncation of the protein after 10 amino acids and loss of the entire PI3K kinase domain.
3. In V3-KD51 mutant aspartic acid (D) at 3921 is replaced with asparagine (N).
4. N: Normal; S: Slightly sensitive; SS: Sensitive; SSS: Very sensitive. Levels are from a previous report on survival responses from this laboratory; Sensitivities marked by the * were those confirmed independently in the present study.

Table 2. Induction of SCE with 0.7 mGy α-particle irradiation.

Cell line	Defective gene	Origin	MDT[1] (hrs)	% in S-phase	SCE per chromosome	
					0 mGy	0.7 mGy
Wild type						
CHO	None	Wild type	14	3.4	0.336±0.030	0.437±0.058
AA8	None	CHO	13	1.4	0.332±0.009	0.440±0.016
NHEJ mutants						
xrs-5	Ku80−/−	CHO	18	10.5	0.442±0.048	1.281±0.170[2]
V3	DNA-PKcs−/−	AA8	18	0.8	0.387±0.069	0.607±0.073[3]
HR/FA mutants						
Irs-3	Rad51C	V79	14	6.3	0.161±0.009	0.169±0.001
CL-V48	Rad51C	V79	15	3.4	0.156±0.008	0.148±0.006
51D1	Rad51D	AA8	24	7.9	0.326±0.015	0.302±0.026
KO40	Fancg/xrrc9	AA8	20	13.9	0.332±0.014	0.443±0.017[3]
DNA-PKcs mutants						
L-1	V3 (wild-type)	V3	18	0.8	0.303±0.098	0.430±0.110
L-3	V3-6A (T2609 cluster to A)	V3	14	7.7	0.347±0.013	0.357±0.028
L-6	V3-T2609A	V3	17	12.2	0.305±0.069	0.327±0.003
L-10	V3-KC23 (kinase dead)	V3	18	12.1	0.421±0.042	1.316±0.014
L-11	V3-KD51 (kinase dead)	V3	16	2	0.505±0.056	0.916±0.002
L-12	V3-5A (S2056 cluster to A)	V3	16	2	0.269±0.012	0.425±0.053
L-14	V3-3A (T2609A/T2638A/T2647A)	V3	ND	6.1	0.332±0.014	0.332±0.014

[1]. MDT: mean doubling time.
[2]. Cells were irradiated with 0.35 mGy.
[3]. Cells were irradiated with 0.18 mGy;

cell cycle. As shown in Table 2, G_1 cell populations were relatively pure as judged by quantification of bromo-2′-deoxyuridine (BrdU) following a 30 minute-pulse labeling; BrdU specifically labels S phase cells. Experiments shown were performed with G_1 populations containing 1–12% S phase cells.

Irradiation and colony formation

For acute high dose-rate exposures, cells were irradiated with a J. L. Shepherd and Associates irradiator that emitted ^{137}Cs γ-rays at a dose rate of 2.5 Gy/min. For the low dose rate ^{137}Cs γ-ray irradiations, a J. L. Shepherd and Associates Model 81-14 beam irradiator containing a single relatively low activity (nominal 28 Ci) ^{137}Cs source placed 100 cm above a water-jacketed CO_2 incubator that was maintained at 37°C. For the colony formation assay, different dose rates were achieved by placing cultures on shelves in the incubator located at different distances below the source during 8 days of continuous low-dose-rate irradiation [17].

For α-particle irradiation, cells were cultured on the Mylar dishes and were placed over a Mylar window in the exposure well of specially constructed irradiator that provides a uniform source of well-characterized 3.07 MeV α-particles. The source consisted of 296 MBq of ^{238}PuO$_2$ electrodeposited onto one side of a 100-mm diameter stainless steel disk. The cells were irradiated from below in a helium environment, and the α-particles traversed a reciprocating collimator before reaching the Mylar window. The target-to-source distance is 42 mm in helium gas, 6 mm in air, and 3 mm in Mylar. Dose was controlled by a timer and precision photographic shutter, which allowed accurate doses of irradiation [38].

Survival curves were obtained by measuring the colony-forming ability of irradiated cell populations. Cells were plated immediately after irradiation onto 100-mm plastic Petri dishes and incubated for 8–10 days. Colonies were fixed with 100% ethanol and stained with 0.1% crystal violet solution. A colony with more than 50 cells was scored as a survivor.

Chromosome analysis

Synchronized G_1-phase cells were subcultured into three T25 plastic flasks and cultured in fresh medium containing 10 μM BrdU for the first and second cycle cells after irradiation. At 4 hour intervals beginning 13 hours after subculture, colcemid was added to one of the three flasks to arrest cells in the first mitosis after subculture; total sampling time thus covered 12 hours. Most cells moved into the first and second rounds of mitosis after 13–18 hours and 25–35 hours post irradiation, respectively. The cells were fixed in methanol:acetic acid (3:1), and chromosomes were spread by air dry method [39]. The differential staining of cells in first and second rounds of mitosis were performed by the fluorescence plus Giemsa technique [40]. Chromosome aberration was analyzed at peak mitotic indices after irradiation. In brief, exponentially growing cells were irradiated with ^{137}Cs γ-rays at a dose rate of 2.5 Gy/min. Colcemid was added to a final concentration of 0.1 μg/ml at 30 min after irradiation, and the cells were harvested 4 hours later; under these conditions the mitotic cells collected would have been in late S/G_2 phase of cell cycle at the time of irradiation [41]. For sister chromatid exchange (SCE), irradiated cells were cultured in complete MEM containing 10 μM BrdU for two rounds of cell replication, and colcemid was added prior to the peak of second mitoses [42].

Statistical analysis

All statistical analyses were performed using Prism GraphPad (version 6.02) software. Statistical significance was diagnosed by t-test and defined as $p<0.05$, but values of $p<0.01$, $p<0.001$ and $p<0.0001$ are shown as well to indicate level of confidence.

Results

Our previous investigation with G_1-synchronized V3 cells and derivative cell lines expressing DNA-PKcs mutants revealed the contributions of the S2056 and T2609 clusters and the PI3K domain to NHEJ-dependent DSB repair and clonogenic survival [12]. In the current study, we examined the impact of DNA-PKcs domains on radiosensitivity for cell killing and chromosomal aberration induction following irradiation during different cell cycle phases. Radiosensitivities for cell killing of wild-type and various NHEJ-deficient mutant cells (Table 1) were investigated in synchronous cell populations in G_1 or late S/G_2 phases of the cell cycle. The synchronized cell populations were irradiated with γ-rays and reseeded to evaluate clonogenic survival (Fig. 1). In comparison to the wild-type Chinese Hamster cells (CHO and AA8), Ku70/80-deficient xrs-5 cells were extremely radiosensitive, and, as previously reported, survival was not affected by the stage of the cell cycle at the time of irradiation [24]. DNA-PKcs-deficient V3 cells were also highly radiosensitive, but cells irradiated in the late S/G_2 phase of cell cycle had enhanced survival compared to those irradiated in G_1 (Fig. 1A). Similarly, cells expressing DNA-PKcs mutants defective at the T2609 cluster (L-3, Fig. 1B), the PI3K kinase domain (L-10 and L-11, Fig. 1C), or the S2056 cluster (L-12, Fig. 1D) all exhibited higher cell survival when irradiated in late S/G_2 than in G_1, although this effect was not as prominent as that observed for the wild-type CHO cells (Table 3).

Wild-type, xrs-5 and DNA-PKcs mutant cells were examined for chromosome and chromatid-type aberrations induced by various doses of γ-rays (Fig. 2). Generally, DNA-PKcs mutant cells displayed more chromosomal aberrations per unit dose than wild-type CHO cells regardless whether irradiation occurred at G_1 or late S/G_2 phases. When the cells were irradiated in the G_1 phase, relative to wild-type cells significantly elevated frequencies of radiation-induced breaks and exchanges were observed; levels of these aberrations were similar in cells with DNA-PKcs protein defective at the T2609 cluster, the PI3K domain or the S2056 cluster and in Ku-deficient xrs-5 cells (Fig. 2A, 2B). As previously noted, relatively few chromatid-type aberrations were induced by G_1 irradiation of wild-type cells but were numerous in NHEJ-deficient mutant cells (Fig. 2A, 2B) [12,15]. When the cells were irradiated in the late S/G_2 phase of the cell cycle, frequencies of radiation-induced chromosome breaks or deletions were similar in all DNA-PKcs mutant cells and were significantly different from wild-type cells (Fig. 2C). There were obvious differences between xrs-5 cells and DNA-PKcs mutant cells: After irradiation with 0.5 Gy in late S/G2 phase, approximately 3 to 5 times more chromatid breaks were observed in xrs-5 cells than in DNA-PKcs mutant cells (Fig. 2C). It is notable that after 1 Gy irradiation during the late S/G_2 phase of the cell cycle, very few xrs-5 mitotic cells were collected due to a long delay at the G_2 checkpoint as previously reported [43]. Results indicated that perhaps DNA-PKcs mutant cells showed lower frequencies of chromatid type exchanges (triradials and quadriradials) when the cells were irradiated in the late S/G_2 phase than when cells were irradiated

during the G_1 phase (Figs. 2B and 2D); L-3 cells were an exception to this, as discussed further below.

In cells proficient in so-called sub-lethal damage repair, there are marked reductions in effect per unit dose occur when radiation was delivered at low dose rates [44,45]. For dose rates that are sufficiently low, repair proficient wild-type cells are able to form colonies during continuous irradiation at this or lower dose rates. Mutant cells with even relatively minor defects in critical repair systems are less able to cope with the continuous irradiation, and above a critical dose rate have reduced abilities to form colonies during irradiation [17]. Based on these earlier observations, low dose-rate assay was carried out for the DNA-PKcs mutant cell lines used in the present study. In brief, exponentially growing cells were irradiated for 8 days at dose rates ranging from 1.7 to 5.5 cGy/hour. As illustrated in Figure 3, where the ability to form colonies is plotted against the dose rate experienced during the 8 day incubation period for colony formation, none of the three wild-type cell lines evaluated exhibited any significant reduction in colony forming ability during continuous irradiation at any of the dose-rates tested relative to unirradiated cells. The cells that express the L-3, L-10, and L-11 DNA-PKcs mutants, however, appeared to be even more sensitive than the V3 mutant, while the L-12 mutant appeared as intermediate in sensitivity (Fig. 3).

The baseline dose-response for SCEs induced by α-particles in wild-type cells is shown from 0 to 1.4 mGy for CHO, and L-1 (V3 corrected) cells in Figure 4. At the extreme low end of the sensitivity for SCE induction HRR-deficient CL-V4B (Rad51C), irs-3 (Rad51C), and 51D1(Rad51D) cells (Fig. 4A) as well as expressing DNA-PKcs with mutations in the T2609 cluster region (L-3/L-6 cells) (Fig. 4B) showed little or no SCE induction after α-particle irradiation. At the other extreme, the radiation sensitive xrs-5, V3, and L-10/L-11 (PI3K domain mutant or kinase dead) cells displayed large increases in SCE after 0.35 and 0.7 mGy α-particle irradiation (Figs. 4A and 4B). This contrasts with the very low or absent induction of SCEs observed in DNA-PKcs mutants L-3 and L-6 cells. Although FA-deficient KO40 cells displayed only intermediate radiosensitivity (P. Wilson, unpublished data), a moderate increase in SCE was observed in KO40 cells similar to the dose response for wild-type cells over the dose range tested (Fig. 4A) and this also appeared to be the case for L-12 (multiple S2056 cluster) and L-14 cells (see table 1) (Fig. 4B). Thus, greatly increased induced SCE frequencies were found in extreme radiosensitive xrs-5 and some of the more radiosensitive DNA-PKcs mutant cell strains including V3, L-10, and L-11, but in other instances the induced SCE frequencies were either similar to that for wild-type cells or in some cases no SCEs were induced at all. (see Figure 4 and Table 2 for induced frequencies at 0.7 mGy or, on average, 1 track traversal per 250 cell nuclei).

Discussion

We previously reported that V3 cells that express DNA-PKcs with substitution mutations in involving various regions of the protein displayed differential radiosensitivities when irradiated during the G_1 phase of the cell cycle [12]. V3-derivative cells expressing DNA-PKcs mutants at the T2609 cluster (L-2 and L-3) and the PI3K domain (L-10 and L-11) were extremely radiosensitive for cell killing when irradiated in the G_1 phase. In the current study, we observed the general trend that these DNA-PKcs mutant cells show increased radioresistance relative to their G1 responses when irradiated after cells had progressed into late S/G_2 phase of cell cycle (Fig. 1A–D). Ku70/80-deficient xrs-5 cell survival, however, was independent of the cell cycle occupied at the time of irradiation. In this, the Ku70/80-deficient cells are similar to

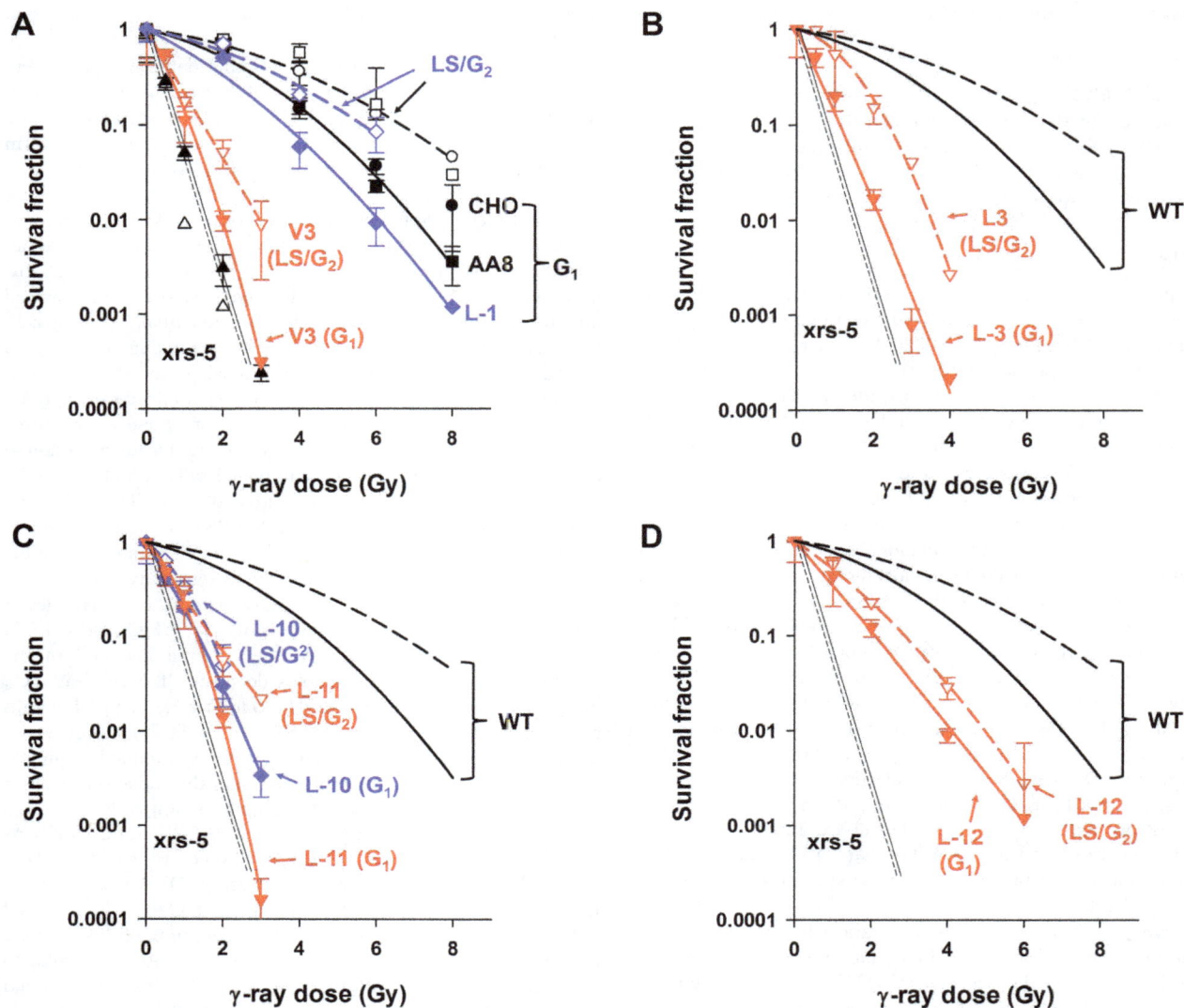

Figure 1. Effect of radiosensitivities on G$_1$- and late S/G$_2$-phase cells. G$_1$ (closed symbols) and late S/G$_2$ (LS/G$_2$, open symbols) synchronized cells [CHO and AA8 (WT), xrs-5, (A) V3, L-1, (B) L-3, (C) L-10, L-11, and (D) L-12] were irradiated by γ-rays and were reseeded immediately for analysis of colony formation. A colony with more than 50 cells was scored as a survivor. The results are means ± SEMs from more than three independent experiments with each cell line.

Table 3. Effect of cell cycle on radiosensitivity.

Cell line	D10 dose (Gy)[1]		Ratio of sensitivities of S/G$_2$ to G$_1$ cells[2]
	G$_1$ phase	S/G$_2$ phase	
CHO	4.1	6.4	1.6
AA8	4.6	7.7	1.7
xrs-5	0.8	0.8	1
V3	0.9	1.2	1.3
L-3	0.9	1.3	1.4
L-10	0.9	1.3	1.4
L-11	0.7	1.2	1.7
L-12	2.2	3.3	1.5

[1]. D10, radiation dose required to reduce survival to 10% in G$_1$ or S/G$_2$ synchronized cells.
[2]. Calculated by dividing D10 of S/G$_2$ phase by D10 of G$_1$ phase.

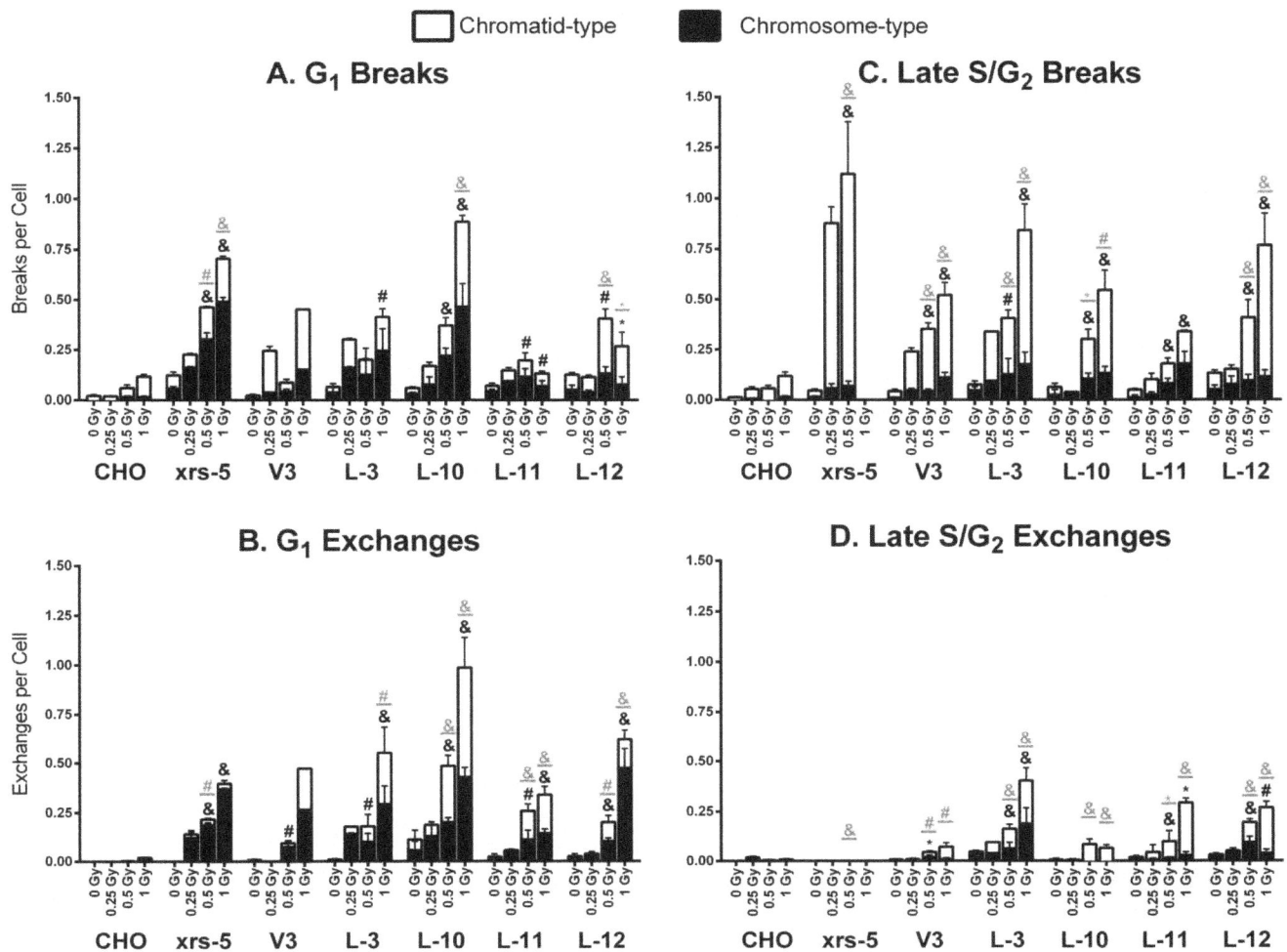

Figure 2. Gamma-ray induced chromosomal aberrations in G_1 and S/G_2 phases. Wild-type (CHO), NHEJ-deficient mutant lines (xrs-5, V3), and DNA-PKcs mutant strains were irradiated with doses of 0, 0.25, 0.5 or 1.0 Gy γ–rays during the G_1 or late S/G_2 phases of cell cycle. Chromatid-type ($\square$) and chromosome-type ($\blacksquare$) aberrations were analyzed and scored as breaks (A, C) or exchanges (B, D). The results are means $\pm$ SEMs from more than three independent experiments. Statistical analyses on the 0.5 Gy or 1 Gy-induced aberrations relative to CHO cells were performed using t-test. *, $p<0.05$; #, $p<0.01$; &, $p<0.001$; the black and gray symbols indicate the significant differences in chromosome-type and chromatid-type aberrations, respectively.

ATM-deficient cells, which are extremely sensitive to radiation, display no cell-cycle dependence of radiosensitivity, are deficient in SLD and PLD repair, and have a high frequency of radiation-induced chromosomal aberrations relative to wild-type cells [18–25,46].

The lack of a cell-cycle effect in ATM-deficient cells is likely due to the fact that ATM is involved in repair of DSB in both NHEJ and HRR pathways [47,48]. On the other hand, Ku protein is essential for DSB recognition and DNA-PKcs recruitment through only the NHEJ mechanism [1]. The lack of cell cycle effect on cell survival in xrs-5 cells could be explained by the fact that Ku protects DSB ends from non-specific processing [49], and thereby reduces the frequency of chromosomal exchange aberrations that might otherwise develop. The lack of Ku then might further reduce the proportion of break-pairs that are able to form exchanges, while at the same time resulting in more chromosomal breaks that fail to rejoin altogether resulting in deletion type aberrations. Such a blocking effect on DSB mis-rejoining may occur throughout all cell cycle phases as Ku exhibits similar kinetics for DSB rejoining regardless cell cycle status [50]. As we report here and previously Ku-deficient xrs-5 cells show dramatically increased induction of chromatid breaks but virtually no exchange aberrations when the cells are irradiated in late S/G_2 phase of cell cycle [41]. Interestingly, large increases in both chromosome breaks and exchanges were seen per unit dose in irs-5 cells when irradiated in G_1 (figure 2A and 2B). Additionally, Ku possesses 5′deoxyribose-5-phosphate (5′-dRP)/AP lyase activity that results in excision of abasic sites near DSBs *in vitro* [51], although it is not clear whether this activity contributes to cell cycle effects or aberration formation.

Low linear energy transfer (LET) irradiation induces base damage, single-strand breaks (SSBs), DSBs, and cross-links immediately after irradiation [41,45]. SSBs are rejoined in wild-type cells with a half-time of approximately 10 minutes. DSBs are more slowly rejoined with a half time of 1 to 2 hours [15,52,53]. Several reports have indicated that a two-component DSB rejoining system operates with a fast component ($t_{1/2}$ ~15 minutes) and slow component ($t_{1/2}$ ~1 to 2 hours) [15,52]. Mutations

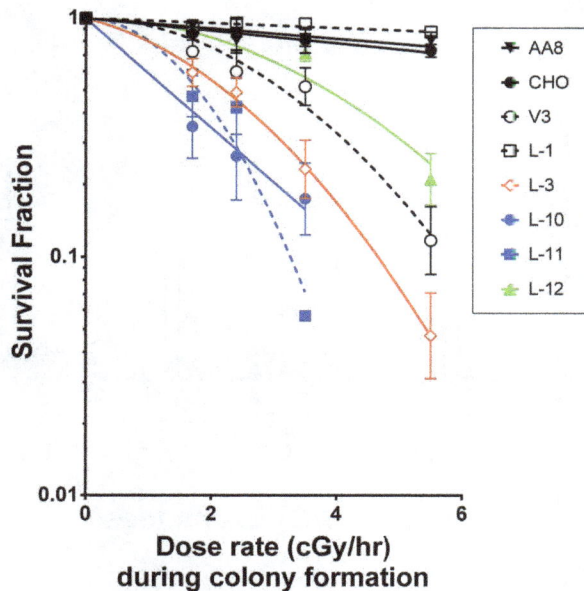

Figure 3. Effect of continuous low-dose-rate irradiation on colony formation. Asynchronous cells were continuously irradiated at a variety of low-dose rates for 8 days. A colony with more than 50 cells was scored as a survivor. Means ± SEMs from more than three independent experiments are shown.

leading to defects in dealing with dsbs strongly argue that prompt (or immediate) production of these dsbs by ionizing irradiation are the most important DNA lesions in cell killing, mutation, and tumor promotion for carcinogenesis by irradiation [15,45,54].

A close correlation between cell killing and the induction of chromosomal aberrations has also been reported in a number of investigations [46,52,55–61]. Our results agree with this conclusion. Cells with mutations in components of the NHEJ system were hypersensitive with respect to cell killing and chromosomal aberration induction by γ radiation. The degree of hypersensitivity varied depending on the nature of the mutation, but increases in numbers of chromosomal aberrations were correlated with increases in cell death in each of the cell lines.

In addition to the correlation between cell killing and chromosomal aberration induction, in most, although not all, radiosensitive mutant cells we observed high frequencies of chromosome-type aberrations after G_1 irradiation and also a very high frequency of chromatid-types aberrations. In wild-type cells, most aberrations were of chromosome types when cells were irradiated in G_1. This general observation has been reported for many radiosensitive DNA-PKcs mutant cells as well as for lymphocyte and fibroblast cells from ataxia-telangiectasia patients who have mutations in ATM [20,22,23,46,54,62–67]. Because virtually all base damaging agents that do not produce prompt DNA DSBs result in the production of only chromatid-type aberrations after treatment of G_1-phase cells, this suggests that ionizing radiation-sensitive mutant cells (e.g. NHEJ and ATM mutations) show both chromatid- and chromosome-type aberrations after G_1 irradiation because of a concomitant or partially overlapping deficiency or competitive inhibition between DSB and SSB rejoining systems. Late S/G_2-phase Ku80-deficient xrs-5 cells irradiated with doses of 25 and 50 cGy showed extremely high frequencies of total aberrations, and 80–90% were chromatid-type

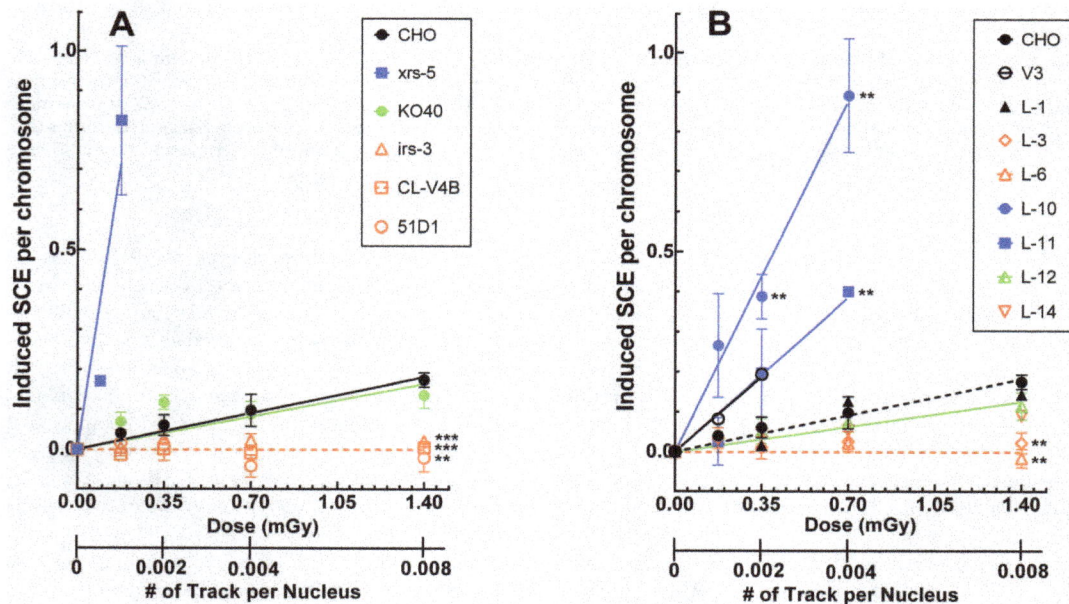

Figure 4. Induction of sister chromatid exchanges with extremely low dose α-irradiation. (A) Wild-type (CHO) and cells deficient in NHEJ (xrs-5), HRR (irs-3, CL-V48, 51D1), and FA (KO40) pathways and (B) V3 cells complemented with wild-type DNA-PKcs (L-1) or with mutants defective in T2609 cluster (L-3/6/14), S2056 cluster (L-12), or PI3K domain (L-10/11) were irradiated with extremely low doses of α-particles. SCEs were scored and normalized to the basal level of SCEs. Means ± SEMs from more than three independent experiments are given. Statistical analyses on the induced frequencies of SCE relative to CHO cells were performed using t-test. **, p<0.01; ***, p<0.001.

aberrations (Fig. 2). Although xrs-5 cells had an extremely long G_2 delay as compared with most other cell lines analyzed upon treatment with relatively low dose irradiation, these cells did progress from the G_2 checkpoint into mitosis [43]. Hashimoto and colleagues reported that the Ku 80-deficient cells apparently do not adequately repair DSBs before moving into mitosis [68]. They suggested that incompletely repaired DSBs result in SSBs that may appear as chromatid-type breaks in mitosis. This may be pertinent to the mechanism of G_2-chromosomal hypersensitivity originally suggested by Sanford and colleagues [69] and by Scott [70].

Three types of dose response curves for SCE induction were observed when DNA-PKcs mutant cells were irradiated by extremely low-dose α-particle irradiation (Fig. 4). Moderately radiosensitive L-12 (S2056 cluster mutant) cells showed similar (perhaps slightly lower) frequencies of SCEs relative to wild type (CHO and L-1) cells with up to 1.4 mGy α-particle irradiation (Fig. 4B). Only 0.8% of the nuclei were traversed by an α-particle by this dose (i.e., only one cell nucleus in 125 cells was traversed by an α-particle), so most of the surviving cells expressing SCEs occurred in unirradiated bystander cells. There were very few or no SCEs in L-3, L-6, and L-14 cells (T2609 cluster mutant) with up to 1.4 mGy α-particle irradiation (Fig. 4B). This suggests an overlap in the functional operation of the HRR system and the DNA-PKcs T2609 cluster mutant cells of the NHEJ system. The lack of SCE induction with extremely low dose α-particle fluences has been previously reported in CHO cell lines deficient in Rad51 paralogs (Rad51C, xrcc2, xrcc3) as well as another essential HRR protein Brca2 [26,27,71]. This was confirmed for other HRR mutant cells (irs-3, CL-V4B, and 51D1), which also lacked the ability to form SCEs in this study (Fig. 4A). We previously reported that mouse DNA-PKcs 3A knock-in mutant cells (identical design to the L-14 cell line used in this study) were defective in both HRR and FA repair pathways [72]. The lack of SCE induction in L-3, L-6, and perhaps L-14 cells suggests that each individual phosphorylatable residue in the T2609 cluster might contribute specifically and distinctively to the functional efficiency of the HRR process.

When cells were irradiated during the G_1 phase, the extent of cell killing depended on the number of amino acid replacements within the T2609 cluster. We previously reported that L-6 cells with a single residue replaced had near wild-type sensitivity, L-14 with three residues replaced was of intermediate sensitivity, and L-3 cells with six residues replaced was very radiosensitive [12]. The high radiosensitivity of the L-3 mutant was confirmed in the present study. Therefore, each phosphorylation in the T2609 cluster contributes to radiosensitivity after low LET irradiation based on their functionality underlying the NHEJ mechanism, and phosphorylation of these T2609 residues also influences the interaction with the HRR system to allow SCE induction after low dose α-particle irradiation.

In contrast to the lack of SCE induction in L-3, L-6, and L-14 cells (T2609 cluster mutants), a significant increase in SCEs

relative to wild-type cells were observed in L-10 and L-11 (PI3K mutant) cells. The L-10 and L-11 mutations led to approximately 3 and 10 times higher α-particle induced SCE frequencies, respectively, than observed in wild-type cells after treatment with 0.7 mGy (i.e., 0.4% of nuclei traversed by an α-particle) (Fig. 4B). Furthermore, relative to wild-type cells, Ku70/80-deficient xrs-5 cells showed enormously increased SCE frequencies at 0.13 or 0.17 mGy of α-particle irradiation, where on average less than 1 cell nucleus per 1000 cells was traversed by an α-particle. The reason for difference in sensitivity of one set of NHEJ mutants relative to the other with respect to SCE induction suggests the possibility that the wild-type NHEJ system has a minor modulating effect on the HRR system which is required for SCE formation after α-particle irradiation [26,27]. Severely reducing or eliminating the NHEJ function resulted in extreme hypersensitivity to cell killing and chromosomal aberration induction by low LET radiation but also removed any interference with the HRR system required for α-particle-induced SCE. The mutation in DNA-PKcs that affected phosphorylation of five residues in the S2056 cluster had only a modest effect on sensitivity of cells to γ-rays and no effect on HRR-dependent induction of SCE by α-particles.

In summary, the present study investigated the differential radiosensitivity phenotypes of cells expressing different DNA-PKcs mutants with comparison to cell lines defective in NHEJ or HRR components. Our new analyses are critical as we proceed with further mechanistic studies of DNA-PKcs mutations and their implication in carcinogenesis or other diseases. Significant alteration of DNA-PKcs expression has been correlated with cancer progression and resistance to radio- and/or chemotherapy treatment [73], although less emphasis has been put on mutation spectra analysis of DNA-PKcs probably due to the difficulty of analyzing the enormous DNA-PKcs-encoding PRKDC gene. Nonetheless, several point mutations in DNA-PKcs have been identified from breast tumor biopsies including a missense mutation that results in a Thr to Pro substitution at residue 2609 [74]. It is highly plausible that this substitution in the T2609 cluster is the driver for mutation accumulation, genome instability, and eventually carcinogenesis. With the advancement in deep sequencing techniques, it is foreseeable that DNA-PKcs mutations will be identified from tumor biopsies or other diseases. Our current analyses provide information necessary to delineate the molecular mechanism of phenotypic differences resulting from mutations in DNA-PKcs. Phenotypes and molecular mechanisms underlying them are not always predictable from genotypes.

Author Contributions

Conceived and designed the experiments: YFL HN BPC. Performed the experiments: YFL HN JRB. Analyzed the data: YFL HN TAK HYS XJX. Contributed reagents/materials/analysis tools: PFW AK DJC. Wrote the paper: YFL HN JBL JSB BPC.

References

1. Weterings E, Chen DJ (2008) The endless tale of non-homologous end-joining. Cell Res 18: 114–124.
2. Kirchgessner CU, Patil CK, Evans JW, Cuomo CA, Fried LM, et al. (1995) DNA-dependent kinase (p350) as a candidate gene for the murine SCID defect. Science 267: 1178–1183.
3. Perryman LE (2004) Molecular pathology of severe combined immunodeficiency in mice, horses, and dogs. Vet Pathol 41: 95–100.
4. Fulop GM, Phillips RA (1990) The scid mutation in mice causes a general defect in DNA repair. Nature 347: 479–482.
5. Kurimasa A, Kumano S, Boubnov NV, Story MD, Tung CS, et al. (1999) Requirement for the kinase activity of human DNA-dependent protein kinase catalytic subunit in DNA strand break rejoining. Mol Cell Biol 19: 3877–3884.
6. Chan DW, Chen BP, Prithivirajsingh S, Kurimasa A, Story MD, et al. (2002) Autophosphorylation of the DNA-dependent protein kinase catalytic subunit is required for rejoining of DNA double-strand breaks. Genes Dev 16: 2333–2338.
7. Chen BP, Chan DW, Kobayashi J, Burma S, Asaithamby A, et al. (2005) Cell cycle dependence of DNA-dependent protein kinase phosphorylation in response to DNA double strand breaks. J Biol Chem 280: 14709–14715.
8. Cui X, Yu Y, Gupta S, Cho YM, Lees-Miller SP, et al. (2005) Autophosphorylation of DNA-dependent protein kinase regulates DNA end processing and may also alter double-strand break repair pathway choice. Mol Cell Biol 25: 10842–10852.
9. Ding Q, Reddy YV, Wang W, Woods T, Douglas P, et al. (2003) Autophosphorylation of the catalytic subunit of the DNA-dependent protein

kinase is required for efficient end processing during DNA double-strand break repair. Mol Cell Biol 23: 5836–5848.

10. Rothkamm K, Kruger I, Thompson LH, Lobrich M (2003) Pathways of DNA double-strand break repair during the mammalian cell cycle. Mol Cell Biol 23: 5706–5715.

11. Jeggo PA, Lobrich M (2006) Contribution of DNA repair and cell cycle checkpoint arrest to the maintenance of genomic stability. DNA Repair (Amst) 5: 1192–1198.

12. Nagasawa H, Little JB, Lin YF, So S, Kurimasa A, et al. (2011) Differential role of DNA-PKcs phosphorylations and kinase activity in radiosensitivity and chromosomal instability. Radiat Res 175: 83–89.

13. Stackhouse MA, Bedford JS (1993) An ionizing radiation-sensitive mutant of CHO cells: irs-20. II. Dose-rate effects and cellular recovery processes. Radiat Res 136: 250–254.

14. Stackhouse MA, Bedford JS (1993) An ionizing radiation-sensitive mutant of CHO cells: irs-20. I. Isolation and initial characterization. Radiat Res 136: 241–249.

15. Stackhouse MA, Bedford JS (1994) An ionizing radiation-sensitive mutant of CHO cells: irs-20. III. Chromosome aberrations, DNA breaks and mitotic delay. Int J Radiat Biol 65: 571–582.

16. Lin JY, Muhlmann-Diaz MC, Stackhouse MA, Robinson JF, Taccioli GE, et al. (1997) An ionizing radiation-sensitive CHO mutant cell line: irs-20. IV. Genetic complementation, V(D)J recombination and the scid phenotype. Radiat Res 147: 166–171.

17. Priestley A, Beamish HJ, Gell D, Amatucci AG, Muhlmann-Diaz MC, et al. (1998) Molecular and biochemical characterisation of DNA-dependent protein kinase-defective rodent mutant irs-20. Nucleic Acids Res 26: 1965–1973.

18. Weichselbaum RR, Nove J, Little JB (1978) Deficient recovery from potentially lethal radiation damage in ataxia telengiectasia and xeroderma pigmentosum. Nature 271: 261–262.

19. Painter RB, Young BR (1980) Radiosensitivity in ataxia-telangiectasia: a new explanation. Proc Natl Acad Sci U S A 77: 7315–7317.

20. Nagasawa H, Little JB (1983) Comparison of kinetics of X-ray-induced cell killing in normal, ataxia telangiectasia and hereditary retinoblastoma fibroblasts. Mutat Res 109: 297–308.

21. Dritschilo A, Brennan T, Weichselbaum RR, Mossman KL (1984) Response of human fibroblasts to low dose rate gamma irradiation. Radiat Res 100: 387–395.

22. Little JB, Nagasawa H (1985) Effect of confluent holding on potentially lethal damage repair, cell cycle progression, and chromosomal aberrations in human normal and ataxia-telangiectasia fibroblasts. Radiat Res 101: 81–93.

23. Nagasawa H, Latt SA, Lalande ME, Little JB (1985) Effects of X-irradiation on cell-cycle progression, induction of chromosomal aberrations and cell killing in ataxia telangiectasia (AT) fibroblasts. Mutat Res 148: 71–82.

24. Nagasawa H, Chen DJ, Strniste GF (1989) Response of X-ray-sensitive CHO mutant cells to gamma radiation. I. Effects of low dose rates and the process of repair of potentially lethal damage in G1 phase. Radiat Res 118 559–567.

25. Nagasawa H, Little JB, Tsang NM, Saunders E, Tesmer J, et al. (1992) Effect of dose rate on the survival of irradiated human skin fibroblasts. Radiat Res 132: 375–379.

26. Nagasawa H, Wilson PF, Chen DJ, Thompson LH, Bedford JS, et al. (2008) Low doses of alpha particles do not induce sister chromatid exchanges in bystander Chinese hamster cells defective in homologous recombination. DNA Repair (Amst) 7: 515–522.

27. Nagasawa H, Peng Y, Wilson PF, Lio YC, Chen DJ, et al. (2005) Role of homologous recombination in the alpha-particle-induced bystander effect for sister chromatid exchanges and chromosomal aberrations. Radiat Res 164: 141–147.

28. Puck TT, Cieciura SJ, Robinson A (1958) Genetics of somatic mammalian cells. III. Long-term cultivation of euploid cells from human and animal subjects. J Exp Med 108: 945–956.

29. Thompson LH, Fong S, Brookman K (1980) Validation of conditions for efficient detection of HPRT and APRT mutations in suspension-cultured Chinese hamster ovary cells. Mutat Res 74: 21–36.

30. Jeggo PA, Kemp LM, Holliday R (1982) The application of the microbial "tooth-pick" technique to somatic cell genetics, and its use in the isolation of X-ray sensitive mutants of Chinese hamster ovary cells. Biochimie 64: 713–711.

31. Whitmore GF, Varghese AJ, Gulyas S (1989) Cell cycle responses of two X-ray sensitive mutants defective in DNA repair. Int J Radiat Biol 56: 657–665.

32. Jones NJ, Cox R, Thacker J (1987) Isolation and cross-sensitivity of X-ray-sensitive mutants of V79-4 hamster cells. Mutat Res 183: 279–286.

33. Zdzienicka MZ, Simons JW (1987) Mutagen-sensitive cell lines are obtained with a high frequency in V79 Chinese hamster cells. Mutat Res 178: 235–244.

34. Hinz JM, Tebbs RS, Wilson PF, Nham PB, Salazar EP, et al. (2006) Repression of mutagenesis by Rad51D-mediated homologous recombination. Nucleic Acids Res 34: 1358–1368.

35. Tebbs RS, Hinz JM, Yamada NA, Wilson JB, Salazar EP, et al. (2005) New insights into the Fanconi anemia pathway from an isogenic FancG hamster CHO mutant. DNA Repair (Amst) 4: 11–22.

36. Chen BP, Uematsu N, Kobayashi J, Lerenthal Y, Krempler A, et al. (2007) Ataxia telangiectasia mutated (ATM) is essential for DNA-PKcs phosphorylations at the Thr-2609 cluster upon DNA double strand break. J Biol Chem 282: 6582–6587.

37. Tobey RA, Ley KD (1971) Isoleucine-mediated regulation of genome repliction in various mammalian cell lines. Cancer Res 31: 46–51.

38. Metting NF, Koehler AM, Nagasawa H, Nelson JM, Little JB (1995) Design of a benchtop alpha particle irradiator. Health Phys 68: 710–715.

39. Hsu TC, Klatt O (1958) Mammalian chromosomes in vitro. IX. On genetic polymorphism in cell populations. J Natl Cancer Inst 21: 437–473.

40. Perry P, Wolff S (1974) New Giemsa method for the differential staining of sister chromatids. Nature 251: 156–158.

41. Nagasawa H, Brogan JR, Peng Y, Little JB, Bedford JS (2010) Some unsolved problems and unresolved issues in radiation cytogenetics: a review and new data on roles of homologous recombination and non-homologous end joining. Mutat Res 701: 12–22.

42. Nagasawa H, Little JB (1992) Induction of sister chromatid exchanges by extremely low doses of alpha-particles. Cancer Res 52: 6394–6396.

43. Nagasawa H, Keng P, Harley R, Dahlberg W, Little JB (1994) Relationship between gamma-ray-induced G2/M delay and cellular radiosensitivity. Int J Radiat Biol 66: 373–379.

44. Hall EJ, Bedford JS (1964) Dose Rate: Its Effect on the Survival of Hela Cells Irradiated with Gamma Rays. Radiat Res 22: 305–315.

45. Bedford JS, Dewey WC (2002) Radiation Research Society. 1952–2002. Historical and current highlights in radiation biology: has anything important been learned by irradiating cells? Radiat Res 158: 251–291.

46. Cornforth MN, Bedford JS (1987) A quantitative comparison of potentially lethal damage repair and the rejoining of interphase chromosome breaks in low passage normal human fibroblasts. Radiat Res 111: 385–405.

47. Thompson LH (2012) Recognition, signaling, and repair of DNA double-strand breaks produced by ionizing radiation in mammalian cells: the molecular choreography. Mutat Res 751: 158–246.

48. Shiloh Y, Ziv Y (2013) The ATM protein kinase: regulating the cellular response to genotoxic stress, and more. Nat Rev Mol Cell Biol 14: 197–210.

49. Sun J, Lee KJ, Davis AJ, Chen DJ (2012) Human Ku70/80 protein blocks exonuclease 1-mediated DNA resection in the presence of human Mre11 or Mre11/Rad50 protein complex. J Biol Chem 287: 4936–4945.

50. Shao Z, Davis AJ, Fattah KR, So S, Sun J, et al. (2012) Persistently bound Ku at DNA ends attenuates DNA end resection and homologous recombination. DNA Repair (Amst) 11: 310–316.

51. Roberts SA, Strande N, Burkhalter MD, Strom C, Havener JM, et al. (2010) Ku is a 5′-dRP/AP lyase that excises nucleotide damage near broken ends. Nature 464: 1214–1217.

52. Fornace AJ, Jr., Nagasawa H, Little JB (1980) Relationship of DNA repair to chromosome aberrations, sister-chromatid exchanges and survival during liquid-holding recovery in X-irradiated mammalian cells. Mutat Res 70: 323–336.

53. Iliakis GE, Metzger L, Denko N, Stamato TD (1991) Detection of DNA double-strand breaks in synchronous cultures of CHO cells by means of asymmetric field inversion gel electrophoresis. Int J Radiat Biol 59: 321–341.

54. Taylor AM, Metcalfe JA, Oxford JM, Harnden DG (1976) Is chromatid-type damage in ataxia telangiectasia after irradiation at G0 a consequence of defective repair? Nature 260: 441–443.

55. Dewey WC, Humphrey RM (1962) Relative radiosensitivity of different phases in the life cycle of L-P59 mouse fibroblasts and ascites tumor cells. Radiat Res 16: 503–530.

56. Dewey WC, Miller HH, Leeper DB (1971) Chromosomal aberrations and mortality of x-irradiated mammalian cells: emphasis on repair. Proc Natl Acad Sci U S A 68: 667–671.

57. Carrano AV (1973) Chromosome aberrations and radiation-induced cell death. I. Transmission and survival parameters of aberrations. Mutat Res 17: 341–353.

58. Bedford JS, Mitchell JB, Griggs HG, Bender MA (1978) Radiation-induced cellular reproductive death and chromosome aberrations. Radiat Res 76: 573–586.

59. Leenhouts HP, Chadwickt KH (1978) The crucial role of DNA double-strand breaks in cellular radiobiological effects. In: Lett JT, Adler H, editors. In Advances in Radiation Biology: Academic Press, New York. pp. 55–101.

60. Nagasawa H, Little JB (1981) Induction of chromosome aberrations and sister chromatid exchanges by X rays in density-inhibited cultures of mouse 10T1/2 cells. Radiat Res 87: 538–551.

61. Revell SH (1983) Relationship between chromosome damage and cell death. In: Ishihara I, Sasaki MS, editors. Radiation-induced chromosome damage in man New York: Alan R. Liss. pp. 215–233.

62. Higurashi M, Conen PE (1973) In vitro chromosomal radiosensitivity in "chromosomal breakage syndromes". Cancer 32: 380–383.

63. Taylor AM, Harnden DG, Arlett CF, Harcourt SA, Lehmann AR, et al. (1975) Ataxia telangiectasia: a human mutation with abnormal radiation sensitivity. Nature 258: 427–429.

64. Cox R, Hosking GP, Wilson J (1978) Ataxia telangiectasia. Evaluation of radiosensitivity in cultured skin fibroblasts as a diagnostic test. Arch Dis Child 53: 386–390.

65. Natarajan AT, Meyers M (1979) Chromosomal radiosensitivity of ataxia telangiectasia cells at different cell cycle stages. Hum Genet 52: 127–132.

66. Arlett CF, Harcourt SA (1980) Survey of radiosensitivity in a variety of human cell strains. Cancer Res 40: 926–932.

67. Zampetti-Bosseler F, Scott D (1981) Cell death, chromosome damage and mitotic delay in normal human, ataxia telangiectasia and retinoblastoma fibroblasts after x-irradiation. Int J Radiat Biol Relat Stud Phys Chem Med 39: 547–558.

68. Hashimoto M, Donald CD, Yannone SM, Chen DJ, Roy R, et al. (2001) A possible role of Ku in mediating sequential repair of closely opposed lesions. J Biol Chem 276: 12827–12831.

69. Parshad R, Sanford KK, Jones GM (1983) Chromatid damage after G2 phase x-irradiation of cells from cancer-prone individuals implicates deficiency in DNA repair. Proc Natl Acad Sci U S A 80: 5612–5616.

70. Scott D, Jones LA, Elyan SAG, Spreadborough A, Cowan R, et al. (1993) Identification of A-T heterozygotes. In: Gatt RA, Painter RB, editors. Ataxia-Telangiectasia: Springer Verlag, Berlin pp. 101–116.

71. Nagasawa H, Fornace D, Little JB (1983) Induction of sister-chromatid exchanges by DNA-damaging agents and 12-O-tetradecanoyl-phorbol-13-acetate (TPA) in synchronous Chinese hamster ovary (CHO) cells. Mutat Res 107: 315–327.

72. Zhang S, Yajima H, Huynh H, Zheng J, Callen E, et al. (2011) Congenital bone marrow failure in DNA-PKcs mutant mice associated with deficiencies in DNA repair. J Cell Biol 193: 295–305.

73. Hsu FM, Zhang S, Chen BP (2012) Role of DNA-dependent protein kinase catalytic subunit in cancer development and treatment. Transl Cancer Res 1: 22–34.

74. Wang X, Szabo C, Qian C, Amadio PG, Thibodeau SN, et al. (2008) Mutational analysis of thirty-two double-strand DNA break repair genes in breast and pancreatic cancers. Cancer Res 68: 971–975.

The Immune Strategy and Stress Response of the Mediterranean Species of the *Bemisia tabaci* Complex to an Orally Delivered Bacterial Pathogen

Chang-Rong Zhang[⑨], **Shan Zhang**[⑨¤], **Jun Xia, Fang-Fang Li, Wen-Qiang Xia, Shu-Sheng Liu, Xiao-Wei Wang***

Ministry of Agriculture Key Laboratory of Agricultural Entomology, Institute of Insect Sciences, Zhejiang University, Hangzhou, China

Abstract

Background: The whitefly, *Bemisia tabaci*, a notorious agricultural pest, has complex relationships with diverse microbes. The interactions of the whitefly with entomopathogens as well as its endosymbionts have received great attention, because of their potential importance in developing novel whitefly control technologies. To this end, a comprehensive understanding on the whitefly defense system is needed to further decipher those interactions.

Methodology/Principal Findings: We conducted a comprehensive investigation of the whitefly's defense responses to infection, via oral ingestion, of the pathogen, *Pseudomonas aeruginosa*, using RNA-seq technology. Compared to uninfected whiteflies, 6 and 24 hours post-infected whiteflies showed 1,348 and 1,888 differentially expressed genes, respectively. Functional analysis of the differentially expressed genes revealed that the mitogen associated protein kinase (MAPK) pathway was activated after *P. aeruginosa* infection. Three knottin-like antimicrobial peptide genes and several components of the humoral and cellular immune responses were also activated, indicating that key immune elements recognized in other insect species are also important for the response of *B. tabaci* to pathogens. Our data also suggest that intestinal stem cell mediated epithelium renewal might be an important component of the whitefly's defense against oral bacterial infection. In addition, we show stress responses to be an essential component of the defense system.

Conclusions/Significance: We identified for the first time the key immune-response elements utilized by *B. tabaci* against bacterial infection. This study provides a framework for future research into the complex interactions between whiteflies and microbes.

Editor: Kun Yan Zhu, Kansas State University, United States of America

Funding: Financial support for this study was provided by the Specialized Research Fund for the Doctoral Program of Higher Education (20120101110077) and Program for New Century Excellent Talents in University (NCET-12-0483). The funders had no role in study design, data collection and analysis, decision to publish, or preparation of the manuscript.

* E-mail: xwwang@zju.edu.cn

⑨ These authors contributed equally to this work.

¤ Current address: Department of Biological Sciences, National University of Singapore, Singapore, Singapore

Introduction

Insects interact and coexist with various types of microorganisms in many different ways and have evolved sophisticated strategies both to recognize and degrade entomopathogens, as well as to benefit from bacterial mutualists [1,2]. Over 20% of all insect species are known to be associated with endosymbionts [3]. Amongst these, phloem-sap-feeding insects such as whiteflies, psyllids and aphids have been shown to possess specialized bacteriocytes that harbor primary and secondary endosymbionts [2]. These endosymbionts provide essential amino acids, protect the host from pathogen infection and also help it to adapt to different environments [3,4,5]. How these insects protect themselves from bacteria pathogens, while retaining beneficial endosymbionts, however, remains to be discovered [6]. Elucidating the host defense mechanisms of these insects will not only shed light on this question, but will also facilitate the development of novel insect-pest control strategies.

The whitefly, *Bemisia tabaci* (Gennadius) (Hemiptera: Aleyrodidae), is a cryptic species complex composed of at least 36 morphologically indistinguishable species [7,8,9,10]. This species complex has members that rank as some of the most economically damaging insect pests [11]. *Bemisia tabaci* damages plants by direct sucking and by transmitting plant viruses [12]. In nature, whiteflies interact with various bacterial species, some of which are entomopathogenic and may serve as potential bio-control agents [13,14]. Previous pyrosequencing analysis has revealed a diverse range of bacteria present in *B. tabaci*, though latest report showed the diversity of bacterial communities in *B. tabaci* is relatively limited [15,16]. On the other hand, the primary (or obligatory) endosymbiont *Portiera* is considered to play a role in the synthesis of essential amino acids and carotenoids to the whitefly host [17,18], while secondary endosymbionts may regulate the life parameters in various ways [19,20]. These different interactions provide an excellent opportunity to study the immune system of an insect species that interacts simultaneously both with bacterial pathogens and endosymbionts. However, the molecular basis of the interactions between whitefly and those microbes remains largely unknown.

A previous study examined the transcriptome of whitefly exposed to the entomopathogenic fungus *Beauveria bassiana* and showed that only a limited number of canonical immune related genes were involved in the host's defense [21]. In addition, a genome-wide analysis and functional study showed that another hemipteran, the pea aphid, seems to lack many genes that are essential for the immune response in many other insects [22]. To investigate the immune system of Hemiptera further, we examined the gene expression profile of the whitefly *B. tabaci* under challenge from a well characterized Gram-negative bacterium *Pseudomonas aeruginosa* [23,24,25].

Most of the previous knowledge on insect immune response is based on host reactions after injection of bacteria into the insect's body cavity [1]. For most insects, however, the normal route of bacterial invasion is via oral ingestion [26,27]. More recent work has shifted from cavity injection to systematic intestinal immune responses under oral infection, which mimics the natural mode of bacterial invasion [28]. Here, we examined the whitefly's essential host-defense strategies after oral bacterial challenge. Functional analysis of the differentially expressed genes indicate that MAPK cascade, antimicrobial peptides (AMP) and gut epithelium renewal play critical roles in the whitefly defense system, whereas stress responses are also induced to improve host tolerance.

Materials and Methods

Plants, whiteflies and bacteria cultures

The Mediterranean (MED) species of the *B. tabaci* complex was used in all the experiments [7]. A culture of MED was maintained on cotton plants (*Gossypium hirsutum* L. cv. Zhemian 1793) in climate chambers at $27 \pm 1^{\circ}$C, 14 h light/10 h darkness with $70 \pm 10\%$ relative humidity (RH). The purity of the MED colony was checked every 5 generations by RAPDs and with sequencing of the mitochondrial cytochrome oxidase 1 gene [29],[30]. The *Pseudomonas aeruginosa* strain ATCC9027 was obtained from the Microorganisms Germplasm Bank of Guangzhou, China.

Bioassay

Bioassays were carried out at $27 \pm 1^{\circ}$C and $70 \pm 10\%$ RH. The infection solution was obtained from an overnight bacterial culture. The density of bacteria was adjusted to 1×10^8 CFU/ml in 10% sucrose solution. The same sucrose solution without *P. aeruginosa* was used as control. Feeding and body cavity injection are two main approaches to carry out infection. In this study, the former method was chosen to mimic a natural mode of bacterial infection as well as to avoid physical damages to the whitefly. Transparent plastics tubes (L10, ø4 cm) were prepared as the feeding chambers for whiteflies. One end of the chamber was covered with a sandwich of 2 layers of carefully stretched Parafilm membrane separated by a layer of 1 ml bacterial solution. Approximately 100 newly emerged adult whiteflies were fed for 60 hours in each tube. The dead whiteflies were counted and cleaned out of the container every 6 hours. All treatments were replicated four times.

Sample preparation for sequencing

Approximately 1,500 newly emerged adult whiteflies were collected for each treatment. At first, the control and 24 h treatment groups were fed, respectively, with sucrose and bacterial solution. After 18 hours, the 6 h treatment group was fed with bacteria. At 24 hours post-infection (hpi), approximately 1,000 whiteflies were collected from the control, 6 and 24 hpi treatments, respectively. This method ensured that the whiteflies were collected at the same developmental stage. In addition, because some whiteflies may not feed on the artificial diet at the beginning, the 6 hpi and 24 hpi whiteflies were fed with bacterial solution throughout the treatment to make sure that every whitefly individual can ingest enough bacteria. Samples were frozen immediately in liquid nitrogen and homogenated using the FastPrep system (MP Biomedicals). Total RNA was purified with SV total RNA isolation kit (Promega) according to the manufacturer's instructions. RNA quality was assessed by Nanodrop 2000 (Thermo Scientific) and 2100 Bioanalyzer (Agilent) as previously described [21,31]. For each treatment, two biological replicates were conducted and processed independently. One replicate was used in the digital gene expression (DGE) library preparation and the other was used for real time quantitative PCR (qPCR) analysis.

Digital gene expression (DGE) sequencing and tag annotation

The methodology for DGE sequencing was largely based on that described in previous studies [31,32]. In brief, the mRNA from each sample was purified with magnetic oligo (dT) and subjected to cDNA synthesis. The cDNA was subsequently digested with *Nla*III, which recognizes the CATG sites. Then adapter 1 was ligated to the site of *Nla*III cleavage. The purified cDNA fragments were digested using *Mme*I that cuts 17 bp downstream of the CATG site, thus producing tags with adapter 1. Then, the adapter 2 was ligated at the site of *Mme*I cleavage. After 15 cycles of PCR linear amplification, 6% TBE polyacrylamide gel electrophoresis was used to purify the tags. After digestion, single strand molecules were added to the Illumina sequencing flowcell and fixed. The purified tags were sequenced by using Illumina HiSeq 2000 platform at the Beijing Genomics Institute (Shenzhen, China). DGE library data sets obtained from this work are available at the NCBI Gene Expression Omnibus under the accession number of GSE52837.

Clean tags were generated after removing 3′ adaptor sequences, low quality sequences, empty reads and tags with a copy number of 1. A reference database containing all possible CATG+17 nucleotide tag sequences were created for the transcriptome of the MED whitefly (unpublished data, available upon request). Sequencing tags were mapped to the whitefly transcriptome reference database with no more than one nucleotide mismatch. The number of unambiguous clean tags for each gene was calculated for gene expression analysis and TPM (number of transcripts per million tags) was used to normalize the data.

Analysis of differentially expressed genes

The levels of gene expression were compared between: 1) the control library and the 6 hpi library; and 2) the control library and the 24 hpi library. False discovery rate (FDR)<0.05 and the absolute value of $\log_2$Ratio$\geq$1 were used as the threshold to judge the significance of gene expression difference [33,34]. Gene Ontology (GO) classification system provides a dynamic, controlled vocabulary for all eukaryotes [35] and was used to annotate the possible functions of differentially expressed genes (DEGs). Also, the Kyoto Encyclopedia of Genes and Genomes (KEGG) pathway analysis was used to depict the pattern of host response against bacterial challenge [36]. The number of DEGs in each GO term and KEGG pathway were also calculated. Using the MED transcriptome database as background, significantly enriched GO and KEGG pathway terms were determined using the hypergeometric test (P-value<0.05).

Real time quantitative PCR (qPCR) analysis

To confirm the results of the DGE analysis, the expression of 20 selected genes was measured using qPCR. cDNA was synthesized using the SYBR PrimeScript reverse transcription-PCR kit II (Takara). qPCRs were performed in 96-well plates using the ABI Prism 7500 fast real-time PCR system (Applied Biosystems) with SYBR green detection. Each gene was analyzed in triplicate, after which the average threshold cycle (C_T) was calculated per sample. The relative expression levels were calculated using the $2^{-\Delta\Delta C_t} = 2^{-[\Delta Ct\ (\text{treatment}) - \ \Delta Ct\ (\text{control})]}$ method. The reference gene actin was used to normalize the expression level of other genes [37]. All of the designed primers were synthesized at Boshang BioCompany (Table S1).

Results

Whitefly survival curve

This research focused on the response of whitefly to orally delivered bacteria; therefore it is important to find the important time points during bacterial infection. To achieve this, the survival rate of the whitefly was monitored for 60 h after *P. aeruginosa* ingestion. In this assay, a sharp decrease in the number of surviving whiteflies was observed during 12–36 hours post-infection (hpi) (Fig. 1). Therefore, 24 hpi was chosen as a sample collection time point. In addition, as sample collected earlier may reflect the whitefly's intestinal response to bacterial infection, the sample at 6 hpi was also collected for sequencing [26,38].

DGE library construction, sequencing and mapping

Three whitefly DGE libraries (Control, 6 hpi and 24 hpi) were sequenced and approximately 6 million raw tags were obtained for each library (Table S2). To determine the appropriate total tag sequencing number, a saturation analysis was carried out to check whether or not the number of detected genes kept increasing when the sequencing amount increased. Results for all three samples showed that when sequencing reached 5 million or higher, the increase in the number of detected genes was negligible (data not shown). After the removal of low quality reads, the numbers of

distinct tags were 128641, 138701 and 126216 in the libraries of the control, 6 hpi and 24 hpi, respectively (Table S2). The ratio of clean tag to total tag was about 97% in each library. For annotation, the short tags of these DGE libraries were mapped to the MED whitefly transcriptome reference database. Tags which can map to more than one gene were filtered out. About 80% of clean tags were mapped unambiguously to the transcriptome database, showing the high quality of the sequencing and reference database. As a result, each library generated ~17,000 tag-mapped genes, which also means that about 36% of genes in the whitefly transcriptome could be detected during DGE sequencing (Table S2).

Differentially expressed genes (DEGs) and qPCR validation

After annotation, we compared the control library with 6 hpi and 24 hpi bacterial challenged libraries to identify the DEGs that may play a central role in the host's defenses. Compared with the control, 948 genes were up-regulated and 400 genes were down-regulated at 6 hpi, while 758 genes were up-regulated and 1030 genes were down-regulated at 24 hpi (Fig. 2A, Table S3). Interestingly, at 6 hpi, the majority of DEGs were up-regulated. At this time point, the mortality of infected whiteflies started to increase, which indicated that the whiteflies had come into close interaction with the bacteria and had responded to pathogen infection. Therefore, 6 hpi reflects the early phase of whitefly defense against *P. aeruginosa*. At the 24 hpi, MED death increased rapidly. The bacteria had already clearly imposed substantial stress on the host. As a result, the sample at 6 hpi and 24 hpi represent different stages of MED's defense response.

The detected fold changes ($\log_2$ ratio) of gene expression ranged from −9.57 to 9.57, and the majority of genes were up- or down-regulated between 1.0- and 5.0-fold, respectively (Figure 2B). In previous studies, the DGE method has been proven to have high reliability [31,39,40]. Due to the relatively high expense of Illumina sequencing, only one sequencing run was performed for each sample. To validate the DGE data, 20 selected genes were quantified for their transcription levels with qPCR in the 6 hpi and 24 hpi treatments. In 34 out of 40 tests, qRT-PCR results were consistent with the DGE data, providing further evidence of the reliability of our sequencing results (Table S1).

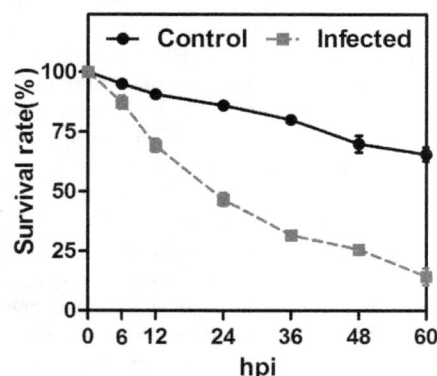

Figure 1. The survival curve of whitefly adults after infection of *P. aeruginosa* via oral ingestion. Error bars: ±SE of the mean.

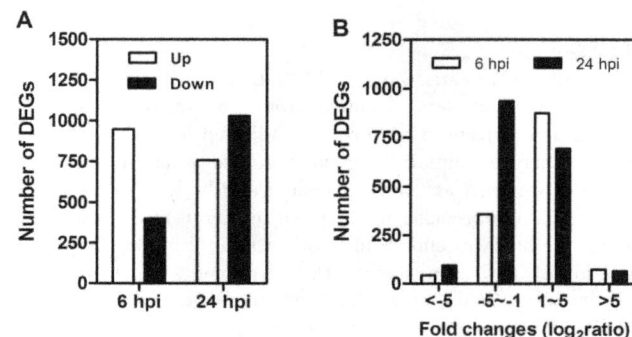

Figure 2. Analysis of differentially expressed genes. (A) An overview of differentially expressed genes (DEGs) between the whitefly libraries of 6 hpi and control, and of 24 hpi and control; the white and black bars indicate the up- and down- regulated genes, respectively. (B) The distribution of fold changes ($\log_2$ ratio) of the DEGs.

Table 1. Genes involved in immune related signaling[a].

Gene	Homologous function[b]	Accession	FC6[c]	FC24
c-type lectin				
comp29769_c0	C-type lectin like	XP_001989543.1	1.40	0.78
comp35820_c0	C-type lectin	EFA04178.1	0.83	1.28
MAPK				
comp38361_c0	TAK1-binding protein 1	XP_001640361.1	1.29	1.24
comp52776_c0	Afadin	XP_003736747.1	1.57	1.13
comp28150_c0	Insulin receptor	EGI60406.1	6.64	6.32
comp34866_c0	Ribosomal protein S6 kinase alpha-1	XP_001504130.1	2.39	2.09
comp31255_c0	MAPKKK13/LZK	XP_002068236.1	2.04	1.09
comp38783_c0	TRB2 protein	AAP04410.1	1.50	0.78
comp40004_c0	Raf serine/threonine-protein kinase	XP_001355538.2	1.06	0.94
comp22807_c0	Neurofibromin	EFZ16398.1	4.91	7.02
comp35826_c0	MAPKKK7/TAK1	ABY81296.1	0.89	1.48
comp34358_c0	HSP70	ADK94698.1	−1.04	−1.48
comp31282_c0	cheerio, isoform I	NP_001189238.1	−1.23	−1.51
comp21240_c0	HSP68	NP_001243928.1	−2.45	−4.70
comp31513_c0	Camp-dependent protein kinase 3	JAA55823.1	−1.04	−2.21
JAK-STAT				
comp39622_c0	CREB binding protein/P300	AAB53050.1	1.30	1.50
comp34743_c1	Phosphoinositide 3-kinase	XP_002001080.1	1.74	1.35

[a]The genes with fold change >2 fold ($|\log_2 ratio| > 1$) and FDR <0.05 are considered to be significant.
[b]Homologous function: the function of the homologous gene.
[c]FC: fold change ($\log_2 Ratio$) of gene expression, where ratio = TPM (6 or 24 hpi)/TPM (control). Underlined fold-change values represent significant changes to differentially expressed genes at given time point.

Ingestion of bacteria caused transcriptome reprogramming

Gene Ontology (GO) and Kyoto Encyclopedia of Genes and Genomes (KEGG) pathway analysis are widely used to examine the biological processes in a large group of genes [35,36]. First, different GO and KEGG Orthology (KO) terms were assigned to

DEGs. In the 6 hpi and 24 hpi data, 738 and 965 DEGs had GO annotations while 187 and 283 genes were annotated with KEGG terms. Enriched GO and KO terms in DEGs were then identified using hypergeometric analysis with the MED transcriptome database as background (Tables S4 and S5). In GO enrichment analysis, 124 and 136 GO terms under the category of biological process were over-presented in the 6 hpi and 24 hpi libraries,

Table 2. Genes involved in humoral immunity.

Gene	Homologous function	Accession	FC6	FC24
AMP				
comp28129_c0	Btk-1	ACT78451.1	1.62	0.47
comp33051_c0	Btk-2	ACT78451.1	1.61	0.27
comp31823_c0	Btk-3	ABC40571.1	1.90	0.67
Serine proteases				
comp39743_c0	Putative serine protease	EFR26537.1	1.34	1.35
comp20520_c0	Serine proteinase stubble	XP_001989504.1	−0.42	1.62
comp38353_c0	Putative serine protease	XP_002035636.1	−1.85	−1.83
comp42030_c0	Putative serine protease	EKC26449.1	−3.09	−7.41
comp32453_c0	Putative serine protease	XP_001653636.1	−0.53	−1.03
comp33254_c1	Putative serine protease	XP_002593630.1	0.18	−1.12
comp31974_c0	Putative serine protease	XP_783667.2	−0.93	−2.40
comp36335_c1	Putative serine protease	XP_002593630.1	−0.52	−2.94

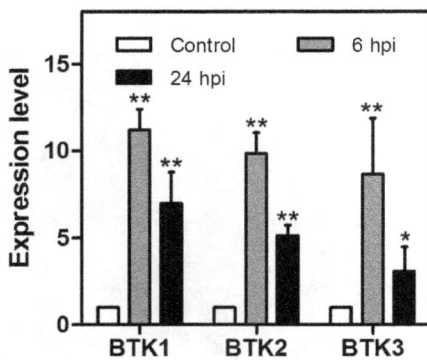

Figure 3. The relative expression level of BTK1, BTK2 and BTK3. The expression level of these genes in uninfected whiteflies was set to 1 (white). Grey and black represent 6 hpi and 24 hpi respectively. Statistical significance compared with Control of P≤0.05 (*) and P≤ 0.001 (**), Student's t test. Error bars: ±SE of the mean.

respectively (Table S4). At both time-points, cell cycle, cell proliferation, stress responses, genetic information processing and metabolic processes featured among the enriched GO terms. KEGG pathway enrichment analysis showed similar pattern and 21 and 17 pathways were enriched at 6 hpi and 24 hpi, respectively (Table S5). More importantly, the phagocytosis and melanogenesis related pathways were enriched in DEGs suggesting the critical role of the whitefly's cellular immune responses during bacterial infection. The involvement of multiple GO and KEGG pathways indicated that *P. aeruginosa* infection activated a number of cellular and molecular responses, which is discussed below.

Microbial recognition and signal transduction pathways

Pathogen recognition is the initial event of pathway activation and systematic responses. Host defense responses are initiated when microbial molecules such as lipopolysaccharides and peptidoglycans are detected by pattern-recognition receptors

(PRRs)[1]. Among those receptors, c-type lectins are sugar binding proteins specifically binding to polysaccharide chains on the pathogen's surface [41]. In infected whiteflies, two c-type lectins were up-regulated in both 6 hpi and 24 hpi treatment groups (Table 1). C-type lectins have been shown to enhance hemocyte encapsulation ability in cellular immunity and activate propheno-loxidase in humoral immunity [42]. Thus, MED c-type lectins are probably able to regulate cellular and humoral immunity upon bacterial infection.

Activation of signaling pathways follows pathogen recognition. The MAPK pathway, which comprises the ERK, JNK and p38 mediated kinase cascade, is a conserved insect host-defense repertoire [43]. Dysfunctions of JNK and p38 resulted in hypersensitivity toward infection and stress [44,45]. Here, several essential MAPK components were activated, including MAP3K7/ TAK1, its binding protein TAB1 and MAP3K13/LZK (Table 1). TAK1 is required for the activation of NF-κB and JNK pathways in *Drosophila*, while LZK is able to phosphorylate and activate JNK [46,47,48]. More interestingly, a putative peroxiredoxin was suppressed at 24 hpi and its homolog in *Drosophila* was identified as a negative regulator in the Tak1-JNK arm of the immune signaling system (Table S3) [49]. In addition, other genes associated with MAPK were activated, such as an insulin/growth factor receptor, a ribosomal protein S6 kinase and tribbles homolog 2. The findings provide additional evidence of the importance of the JNK pathways following *P. aeruginosa* infection. Moreover, the CREB binding protein/P300 and phosphoinositide 3-kinase, two players in JAK/STAT pathways, were also up-regulated significantly (Table 1).

Activation of AMPs and other effectors in immunity

Fast and massive production of AMPs has evolved in insects to be a central strategy of their immune system. Knottins are small proteins that have antimicrobial peptide activities and they are widely present in both plants and insects [50]. The DGE analysis revealed that three antimicrobial knottin genes (Btk 1, 2, and 3) were induced at 6 hpi and this was confirmed subsequently by the qPCR data (Table 2, Fig 3). In *Drosophila*, TAK1 and its binding protein are able to activate the NF-κB pathway and, ultimately,

Table 3. Genes involved in cell proliferation and related pathway.

Gene	Homologous function	Accession	FC6	FC24
Cell proliferation				
comp34792_c3	SMAD4	XP_003227329.1	1.52	1.14
comp28150_c0	Insulin receptor	EGI60406.1	6.64	6.32
comp35213_c0	Paired box protein Pax-6	ABS17534.1	3.25	1.58
comp34974_c0	Lysine-specific histone demethylase	XP_003814021.1	1.01	1.00
comp40004_c0	Raf serine/threonine-protein kinase	XP_001355538.2	1.06	0.94
comp38386_c0	SMAD1	NP_001259992.1	1.13	2.06
comp31937_c0	NADH dehydrogenase	XP_001977774.1	−1.06	−0.87
Wnt pathway				
comp35633_c0	wnt11	NP_001192557.1	1.44	0.20
Egfr pathway				
comp38516_c0	kekkon5, isoform A	NP_573382.1	3.32	3.17
comp37205_c0	fusilli-like	XP_002736133.1	1.38	1.05
comp35265_c0	Protein mago nash	EHJ65953.1	0.27	1.06
comp20819_c0	AP-2 complex subunit sigma	EGW03559.1	−0.34	−1.20

Table 4. Genes involved in protein folding and DNA repair.

Gene	Homologous function	Accession	FC6	FC24
Chaperones				
comp37167_c0	YLP motif-containing protein 1	EHB15431.1	1.05	1.16
comp40894_c0	HSP90 co-chaperone CPR7	XP_002105226.1	2.25	0.07
comp39790_c0	HSP90 co-chaperone CPR7	XP_002105226.1	1.29	0.51
comp38394_c4	Hsp90 co-chaperone/Cdc37	XP_002046556.1	1.05	0.76
comp37509_c0	DnaJ subfamily C member 14	XP_001978829.1	1.04	0.75
comp34358_c0	HSP70	ADK94698.1	−1.04	−1.48
comp21240_c0	HSP68	NP_001243928.1	−2.45	−4.70
comp40216_c0	beta-tubulin folding cofactor C	XP_002735209.1	−0.51	−1.19
comp13402_c0	DnaJ subfamily C member 19	XP_003488785.1	−0.75	−1.39
comp20054_c0	molecular chaperone DnaJ	XP_003248597.1	−0.45	−1.46
DNA repair				
comp30427_c0	G/T mismatch-specific thymine DNA glycosylase	XP_002011422.1	1.25	1.54
comp37437_c0	DNA repair protein XRCC2-like	XP_003134584.1	1.19	1.13
comp35539_c0	DNA-repair protein XRCC3	NP_001079887.1	1.38	0.66
comp32492_c0	Predicted methyltransferase	XP_001986788.1	1.04	0.04
comp39604_c0	DNA repair protein RAD18	EFZ21621.1	1.91	2.39
comp29666_c0	Nucleotide excision repair complex XPC-HR23B, subunit XPC/DPB11	CAA82262.1	0.93	1.45
comp31941_c0	DNA repair protein RAD50 isoform 1	XP_004042521.1	0.89	1.38
comp40044_c0	DNA repair protein REV1	NP_612047.1	0.63	1.12
comp38825_c1	DNA mismatch repair protein Msh2	XP_003473123.1	0.53	1.10
comp33285_c0	DNA damage-responsive repressor GIS1/RPH1	—	−2.03	−1.19

the production of AMPs [51,52]. Whether this signaling pathway is conserved in *B. tabaci*, however, remains to be discovered. Other than activation of AMPs, coagulation and melanization are also important for microbe sequestration and degradation. In insects, serine proteases and their inhibitors (serpin) are responsible for the activation and regulation of coagulation and melanization [53,54]. Several serine proteases are modulated in *P. aeruginosa* infected whiteflies. Interestingly though, 1 and 2 serine proteases are up- and down-regulated at 6 hpi, whereas 2 and 6 serine proteases are clearly up- and down-regulated at 24 hpi, respectively (Table 2).

Regulations of cell proliferation and epithelium renewal upon infection

The intestinal epithelium is the first protective barrier of the host from orally delivered microorganism infection. In *Drosophila*, stem cell division was activated upon infection to compensate for the cell damage due to the presence of bacteria and also to maintain gut homeostasis [38,55,56,57,58]. The enrichment of GO terms such as cell cycle (GO:0007049), positive regulation of cell proliferation (GO:0008284) and epithelium development (GO:0060429) indicated the involvement of cell proliferation and epithelium renewal in host defenses (Table S4). About 80% of genes were induced at both 6 hpi and 24 hpi and extensive genetic studies on the *Drosophila* gut have shown that this process is governed by JNK, JAK-STAT, Wingless and Epidermal growth factor receptor (Egfr) pathways [55,56,59,60], and all of these pathways were also regulated in our data. In addition, a Wingless like protein and several related genes in the Egfr pathway were induced (Table 3; Cordero et al, 2012; Jiang et al, 2011; Xu et al, 2011). Several other well-characterized genes in cell proliferation

were also activated, including SMAD1, SMAD4 and an insulin receptor (Table 3).

Stress response genes of whitefly

The high motility caused by *P. aeruginosa* clearly showed that a hyper-biotic stress had been imposed on the whitefly by 24 hpi. DEGs were consequently highly enriched in the GO term: regulation of response to stress (GO: 0080134) at both 6 hpi and 24 hpi (Table S4). This result is consistent with the activation of the MAPK, because their roles in the stress response are widely accepted [43].

Chaperones work in protein folding and quality control, which maintains cellular homeostasis and buffers environmental stress. Our analysis showed that several chaperones and cofactors were regulated. Five out of seven genes were induced at 6 hpi whereas five out of six genes were repressed at 24 hpi (Table 4). Induced genes include several cofactors of HSP90 and HSP40 proteins. In addition, a large group of genes involved in detoxification, such as Cytochrome P450, Glutathione S-transferase and Glutathione peroxidase were also regulated (Table 5). These genes are likely involved in the detoxification of reactive oxygen species, as well as other toxins produced by bacteria [61,62]. Interestingly, 15 out 16 Cytochrome P450 genes were suppressed at 24 hpi. Cytochrome P450 genes were also suppressed following pathogen infection [56,63]. It suggests there is a general trend of P450 suppression after pathogen infection in other insect species.

Another notable event in infected whitefly was the activation of DNA repair proteins. Five and seven DNA repair genes were induced in 6 hpi and 24 hpi whitefly, respectively (Table 4). The only repressed gene was a negative regulator in DNA damage

Table 5. Differentially expressed detoxification enzymes.

Gene	Homologous function	Accession	FC6	FC24
Glutathione S-transferase				
comp35665_c0	glutathione S-transferase	EFA01955.1	1.16	0.63
Glutathione peroxidase				
comp36914_c0	glutathione peroxidase	EFX89084.1	−0.29	−1.25
Cytochrome P450				
comp36389_c0	cytochrome P450	XP_001865029.1	1.77	0.85
comp33887_c0	Predicted similar to cytochrome P450	XP_966563.2	1.24	0.09
comp30063_c1	cytochrome P450	EFR21005.1	1.02	−0.96
comp402849_c0	cytochrome P450	XP_001987651.1	6.32	6.64
comp35292_c0	CYP6M1a	AFM08393.1	−1.09	−2.01
comp35528_c0	cytochrome P450	AEK21822.1	−1.12	−1.77
comp35166_c0	cytochrome P450 6BQ5	EFA02819.1	−1.17	−1.25
comp35935_c1	cytochrome P450 CYP6BK17	XP_969633.1	−1.47	−2.89
comp24705_c0	cytochrome P450	AEK21822.1	−1.49	−2.32
comp37361_c0	cytochrome P450	AEK21804.1	−1.69	−1.93
comp320941_c0	cytochrome P450 345C1	EFA12854.1	−1.26	−6.91
comp38461_c0	cytochrome P450	XP_966391.1	−0.21	−1.04
comp37745_c0	cytochrome P450	EFA02819.1	−0.05	−1.06
comp36384_c0	cytochrome P450	AFP49818.1	−0.74	−1.43
comp29930_c0	cytochrome P450 6a2-like isoform 2	XP_003248187.1	−0.15	−1.77
comp27263_c0	cytochrome P450 345C1	EHJ67475.1	−0.45	−1.78
comp37119_c0	cytochrome P450 protein	NP_001156683.2	−0.95	−1.95
comp37634_c1	cytochrome P450 6BQ13	EEZ99338.1	−0.25	−2.03
comp40428_c0	cytochrome P450	XP_001653674.1	−0.73	−2.73

repair (Table 4) [64]. DNA damage caused by pathogenesis might have triggered the DNA repair response, because several genes in mismatch repair and double-strand break response have been

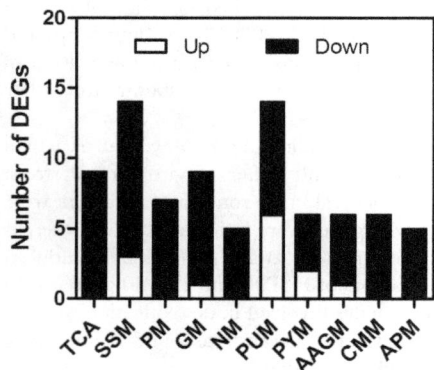

Figure 4. Modulation of basal metabolism-related genes in 24 hpi whiteflies. The numbers of up- and down-regulated genes are shown in white and black, respectively. The listed metabolism pathways are: TCA cycle (TCA, ko00020), Starch and sucrose metabolism (SSM, ko00500), Pyruvate metabolism (PM, ko00620), Galactose metabolism (GM, ko00052), Nitrogen metabolism (NM, ko00910), Purine metabolism (PUM, ko00230), Pyrimidine metabolism (PYM, ko00240), Alanine, aspartate and glutamate metabolism (AAGM, ko00250), Cysteine and methionine metabolism (CMM, ko00270), Arginine and proline metabolism (APM, ko00330).

reported to be highly activated to improve host defense [65]. An alternative explanation, however, is that these genes might also participate in cell-cycle checkpoint, as our data also show that cell proliferation is activated [66].

Modulation of basal metabolism

Orally-delivered infection of *P. aeruginosa* also modulated a large group of basal metabolism processes, especially in the late infection stage. Given that intestinal function is disturbed in pathogenesis, these regulations may act as a self-protective strategy to maintain and reallocate energy supply. Our data shows that the majority of metabolism-related genes were down-regulated at 24 hpi (Fig 4). Suspension of feeding has been discovered in other insects that encountered bacterial infection and thus the inhibition of metabolism might be a direct result of this [67]. Whitefly may have evolved this adaptive strategy to prevent further pathogen ingestion. In addition, this down-regulation might be due to the disruption of energy supply under bacterial infection.

Discussion

Despite its economic importance as a global pest, little is known about the immune response of the *B. tabaci* complex to bacterial infection. Recent developments in next-generation RNA-seq technology, however, allow a systematic study of the whitefly's host defense strategies upon intestinal bacterial infection. In whiteflies collected at both time points following bacterial infection, more than one thousand genes were modulated.

Functional analysis uncovered the complexity of the host's responses to pathogenic bacteria. Bacterial infection not only induced genes in immune signaling and several types of effectors, but also altered the expression pattern of several sets of genes in xenobiotics detoxification, protein folding, DNA repair and basal metabolism. Altogether, our analysis provides the first rough outline of the whitefly's defense mechanisms employed against pathogenic bacteria and raises further questions that deserve more investigation.

The major goal of our research was to decipher the immune strategies of *B. tabaci*. Humoral and cellular immunity are the two arms of the insect's defense system, and in the former response antimicrobial peptides (AMPs), enzymatic cascades and other soluble effectors are employed to degrade foreign invaders [1]. In our study, both DGE data and qPCR analysis showed the activation of AMP production, highlighting again its core status in the antimicrobial response. As with the cellular immune system, pathway analysis showed clearly the involvement of phagocytosis and melanogenesis, more research is needed to construct a comprehensive image. In addition, our data implicates the function of intestinal stem cell proliferation in gut homeostasis maintenance, which may be conserved in whitefly.

The signaling pathways governing immune processes were also revealed by our work. We observed significant activation of several arms of MAPK cascades. Raf and ribosomal protein S6 kinase are essential kinases in the ERK pathway, while TAK1, TAB1 and TRB2 can control both p38 and JNK pathways [43,68]. Studies in *Drosophila* showed TAK1 and TAB are located at the cross point of the JNK and Imd pathway in fruit fly [51,52]. Upon activation, TAK1 and TAB can regulate NF-κB and then activate AMP production. Ras/MAPK signaling can also mediate intestinal homeostasis and regeneration in *Drosophila*. Exploring downstream events controlled by MAPKs, therefore, may provide key answers to how those processes are regulated in whiteflies. Besides the MAPK pathway, only a few genes in other canonical pathways were identified in infected whiteflies. For instance, we failed to identify the canonical factors in the Imd pathway from whiteflies. Interestingly, the pea aphid, another hemipteran insect, also appears to lack the Imd pathway [22]. Nevertheless, the absence of these genes may be due to the limitation of our reference database, which only accounts for a part of the *B. tabaci* genome. Likewise, although several antimicrobial knottins were induced, their induction level was relatively modest compared to that of other insects. Surprisingly, several defensins were not up-regulated upon pathogen infection (data not shown). More analysis is needed to characterize the function of whitefly AMPs.

Our analyses also enable a close examination of the stress response strategy of *B. tabaci*. Though stress response genes are not involved directly in immunity, their importance in the host's defense systems is recognized. Activation of these genes can help the host maintain cellular homeostasis and increase its capacity to endure the infection [69]. The involvement of chaperones, detoxification enzymes and DNA damage repair is likely to help the whitefly build up a high tolerance towards infection. In fact, these genes also participate in environmental adaptation and the development of resistance to insecticides [70,71]. An understanding of the regulation of these gene sets in particular may help in the development of novel insecticides.

In summary, we report for the first time the results of an NGS investigation into the molecular interactions induced by the oral delivery of a bacterial pathogen, *P. aeruginosa*, to the whitefly *B. tabaci*. Functional analyses of DEGs indicated that at 6 hpi both humoral and cellular responses are involved in the whitefly's defense responses. Furthermore, MAPK cascade, AMP and gut epithelium renewal probably play critical roles in the defense system, whereas stress response genes are also induced to build stronger host tolerance. Of particular interest is that only a few genes in other canonical pathways were identified in infected whiteflies. Further research built upon these findings may present an opportunity for the development of a novel whitefly control technologies.

Supporting Information

Table S1 qRT-PCR primers and results.

Table S2 Overview of the DGE sequencing results.

Table S3 List of the differentially expressed genes.

Table S4 Results of the Gene Ontology enrichment analysis.

Table S5 Results of the KEGG pathway enrichment analysis.

Acknowledgments

We thank Professor John Colvin of the University of Greenwich, UK, for comments on an earlier version of the manuscript. The authors also wish to thank Dr. Jun-Bo Luan for advice on digital gene expression analyses.

Author Contributions

Conceived and designed the experiments: CRZ SZ JX SSL XWW. Performed the experiments: CRZ JX FFL. Analyzed the data: CRZ SZ WQX SSL XWW. Wrote the paper: CRZ SZ SSL XWW.

References

1. Lemaitre B, Hoffmann J (2007) The host defense of *Drosophila melanogaster*. Annu Rev Immunol 25: 697–743.
2. Baumann P (2005) Biology of bacteriocyte-associated endosymbionts of plant sap-sucking insects. Annu Rev Microbiol 59: 155–189.
3. Douglas A (1998) Nutritional interactions in insect-microbial symbioses: Aphids and their symbiotic bacteria *Buchnera*. Annu Rev Entomol 43: 17–37.
4. Montllor CB, Maxmen A, Purcell AH (2002) Facultative bacterial endosymbionts benefit pea aphid *Acyrthosiphon pisum* under heat stress. Ecol Entomol 27: 189–195.
5. Scarborough CL, Ferrari J, Godfray H (2005) Aphid protected from pathogen by endosymbiont. Science 310: 1781–1781.
6. Feldhaar H, Gross R (2008) Immune reactions of insects on bacterial pathogens and mutualists. Microbes Infect 10: 1082–1088.
7. De Barro PJ, Liu SS, Boykin LM, Dinsdale AB (2011) *Bemisia tabaci*: A statement of species status. Annu Rev Entomol 56: 1–19.
8. Liu SS, Colvin J, De Barro PJ (2012) Species concepts as applied to the whitefly *Bemisia tabaci* systematics: How many species are there? J Integr Agric 11: 176–186.
9. Firdaus S, Vosman B, Hidayati N, Supena J, Darmo E, et al. (2013) The *Bemisia tabaci* species complex: additions from different parts of the world. Insect Sci 20: 723–733.
10. Boykin LM, Bell CD, Evans G, Small I, De Barro PJ (2013) Is agriculture driving the diversification of the *Bemisia tabaci* species complex (Hemiptera: Sternorrhyncha: Aleyrodidae)? Dating, diversification and biogeographic evidence revealed. BMC Evol Biol 13: 228.
11. Dalton R (2006) Whitefly infestations: the Christmas invasion. Nature 443: 898–900.
12. Navas-Castillo J, Fiallo-Olivé E, Sánchez-Campos S (2011) Emerging virus diseases transmitted by whiteflies. Annu Rev Phytopathol 49: 219–248.

13. Ateyyat MA, Shatnawi M, Al-Mazra'awi MS (2009) Culturable whitefly associated bacteria and their potential as biological control agents. Jordan J Biol Sci 2: 139–144.

14. Davidson EW, Rosell RC, Hendrix DL (2000) Culturable bacteria associated with the whitefly, *Bemisia argentifolii* (Homoptera: Aleyrodidae). Fla Entomol 83: 159–171.

15. Xie W, Meng QS, Wu QJ, Wang SL, Yang X, et al. (2012) Pyrosequencing the *Bemisia tabaci* transcriptome reveals a highly diverse bacterial community and a robust system for insecticide resistance. PLoS ONE 7: e35181.

16. Jing X, Wong ACN, Chaston JM, Colvin J, McKenzie CL, et al. (2014) The bacterial communities in plant phloem-sap-feeding insects. Mol Ecol. doi: 10.1111/mec.12637

17. Thao ML, Baumann P (2004) Evolutionary relationships of primary prokaryotic endosymbionts of whiteflies and their hosts. Appl Environ Microbiol 70: 3401–3406.

18. Sloan DB, Moran NA (2012) Endosymbiotic bacteria as a source of carotenoids in whiteflies. Biol Lett 8: 986–989.

19. Himler AG, Adachi-Hagimori T, Bergen JE, Kozuch A, Kelly SE, et al. (2011) Rapid spread of a bacterial symbiont in an invasive whitefly is driven by fitness benefits and female bias. Science 332: 254–256.

20. Ruan YM, Xu J, Liu SS (2006) Effects of antibiotics on fitness of the B biotype and a non-B biotype of the whitefly *Bemisia tabaci*. Entomol Exp Appl 121: 159–166.

21. Xia J, Zhang CR, Zhang S, Li FF, Feng MG, et al. (2013) Analysis of whitefly transcriptional responses to *Beauveria bassiana* infection reveals new insights into insect-fungus interactions. PLOS ONE 8: e68185.

22. Gerardo NM, Altincicek B, Anselme C, Atamian H, Barribeau SM, et al. (2010) Immunity and other defenses in pea aphids, *Acyrthosiphon pisum*. Genome Biol 11: R21.

23. Apidianakis Y, Rahme LG (2009) *Drosophila melanogaster* as a model host for studying *Pseudomonas aeruginosa* infection. Nature Protocols 4: 1285–1294.

24. Limmer S, Haller S, Drenkard E, Lee J, Yu S, et al. (2011) *Pseudomonas aeruginosa* RhlR is required to neutralize the cellular immune response in a *Drosophila melanogaster* oral infection model. Proc Natl Acad Sci U S A 108: 17378–17383.

25. Tan MW, Ausubel FM (2000) *Caenorhabditis elegans*: a model genetic host to study *Pseudomonas aeruginosa* pathogenesis. Curr Opin Microbiol 3: 29–34.

26. Vodovar N, Vinals M, Liehl P, Basset A, Degrouard J, et al. (2005) *Drosophila* host defense after oral infection by an entomopathogenic *Pseudomonas* species. Proc Natl Acad Sci U S A 102: 11414–11419.

27. Jiravanichpaisal P, Lee BL, Söderhäll K (2006) Cell-mediated immunity in arthropods: hematopoiesis, coagulation, melanization and opsonization. Immunobiology 211: 213–236.

28. Buchon N, Broderick NA, Lemaitre B (2013) Gut homeostasis in a microbial world: insights from *Drosophila melanogaster*. Nat Rev Microbiol 11: 615–626.

29. Barro P, Driver F (1997) Use of RAPD PCR to distinguish the B biotype from other biotypes of *Bemisia tabaci* (Gennadius)(Hemiptera: Aleyrodidae). Aust J Entomol 36: 149–152.

30. Jiu M, Zhou XP, Tong L, Xu J, Yang X, et al. (2007) Vector-virus mutualism accelerates population increase of an invasive whitefly. PLOS ONE 2: e182.

31. Luan JB, Li JM, Varela N, Wang YL, Li FF, et al. (2011) Global analysis of the transcriptional response of whitefly to *Tomato yellow leaf curl China virus* reveals the relationship of coevolved adaptations. J Virol 85: 3330–3340.

32. Marioni JC, Mason CE, Mane SM, Stephens M, Gilad Y (2008) RNA-seq: an assessment of technical reproducibility and comparison with gene expression arrays. Genome Res 18: 1509–1517.

33. Veitch NJ, Johnson PC, Trivedi U, Terry S, Wildridge D, et al. (2010) Digital gene expression analysis of two life cycle stages of the human-infective parasite, *Trypanosoma brucei gambiense* reveals differentially expressed clusters of co-regulated genes. BMC Genomics 11: 124.

34. Voineagu I, Wang X, Johnston P, Lowe JK, Tian Y, et al. (2011) Transcriptomic analysis of autistic brain reveals convergent molecular pathology. Nature 474: 380–384.

35. Ashburner M, Ball CA, Blake JA, Botstein D, Butler H, et al. (2000) Gene Ontology: Tool for the unification of biology. Nat Genet 25: 25–29.

36. Ogata H, Goto S, Sato K, Fujibuchi W, Bono H, et al. (1999) KEGG: Kyoto encyclopedia of genes and genomes. Nucleic Acids Res 27: 29–34.

37. Su YL, He WB, Wang J, Li JM, Liu SS, et al. (2013) Selection of Endogenous Reference Genes for Gene Expression Analysis in the Mediterranean Species of the *Bemisia tabaci* (Hemiptera: Aleyrodidae) Complex. J Econ Entomol 106: 1446–1455.

38. Buchon N, Broderick NA, Chakrabarti S, Lemaitre B (2009) Invasive and indigenous microbiota impact intestinal stem cell activity through multiple pathways in *Drosophila*. Genes Dev 23: 2333–2344.

39. Wang XW, Luan JB, Li JM, Bao YY, Zhang CX, et al. (2010) *De novo* characterization of a whitefly transcriptome and analysis of its gene expression during development. BMC Genomics 11: 400.

40. AC Hoen P, Ariyurek Y, Thygesen HH, Vreugdenhil E, Vossen RH, et al. (2008) Deep sequencing-based expression analysis shows major advances in robustness, resolution and inter-lab portability over five microarray platforms. Nucleic Acids Res 36: e141–e141.

41. Tanji T, Ohashi-Kobayashi A, Natori S (2006) Participation of a galactose-specific C-type lectin in *Drosophila* immunity. Biochem J 396: 127–138.

42. Yu XQ, Gan H, R Kanost M (1999) Immulectin, an inducible C-type lectin from an insect, *Manduca sexta*, stimulates activation of plasma prophenol oxidase. Insect Biochem Mol Biol 29: 585–597.

43. Johnson GL, Lapadat R (2002) Mitogen-activated protein kinase pathways mediated by ERK, JNK, and p38 protein kinases. Science 298: 1911–1912.

44. Chen J, Xie C, Tian L, Hong L, Wu X, et al. (2010) Participation of the p38 pathway in *Drosophila* host defense against pathogenic bacteria and fungi. Proc Natl Acad Sci U S A 107: 20774–20779.

45. Rämet M, Lanot R, Zachary D, Manfruelli P (2002) JNK signaling pathway is required for efficient wound healing in *Drosophila*. Dev Biol 241: 145–156.

46. Boutros M, Agaisse H, Perrimon N (2002) Sequential activation of signaling pathways during Innate immune responses in *Drosophila*. Dev Cell 3: 711–722.

47. Ikeda A, Masaki M, Kozutsumi Y, Oka S, Kawasaki T (2001) Identification and characterization of functional domains in a mixed lineage kinase LZK. FEBS Lett 488: 190–195.

48. Park JM, Brady H, Ruocco MG, Sun H, Williams D, et al. (2004) Targeting of TAK1 by the NF-κB protein relish regulates the JNK-mediated immune response in *Drosophila*. Genes Dev 18: 584–594.

49. Radyuk SN, Michalak K, Klichko VI, Benes J, Orr WC (2010) Peroxiredoxin 5 modulates immune response in *Drosophila*. Biochimica et Biophysica Acta (BBA)-General Subjects 1800: 1153–1163.

50. Chiche L, Heitz A, Gelly JC, Gracy J, Chau PT, et al. (2004) Squash inhibitors: from structural motifs to macrocyclic knottins. Curr Protein Pept Sci 5: 341–349.

51. Silverman N, Zhou R, Erlich RL, Hunter M, Bernstein E, et al. (2003) Immune activation of NF-κB and JNK requires *Drosophila* TAK1. J Biol Chem 278: 48928–48934.

52. Sun L, Deng L, Ea CK, Xia ZP, Chen ZJ (2004) The TRAF6 ubiquitin ligase and TAK1 kinase mediate IKK activation by BCL10 and MALT1 in T lymphocytes. Mol Cell 14: 289–301.

53. Hoffmann JA, Kafatos FC, Janeway CA, Ezekowitz R (1999) Phylogenetic perspectives in innate immunity. Science 284: 1313–1318.

54. Cerenius L, Söderhäll K (2004) The prophenoloxidase-activating system in invertebrates. Immunol Rev 198: 116–126.

55. Buchon N, Broderick NA, Kuraishi T, Lemaitre B (2010) *Drosophila* EGFR pathway coordinates stem cell proliferation and gut remodeling following infection. BMC Biology 8: 152.

56. Buchon N, Broderick NA, Poidevin M, Pradervand S, Lemaitre B (2009) *Drosophila* intestinal response to bacterial infection: activation of host defense and stem cell proliferation. Cell Host & Microbe 5: 200–211.

57. Cronin SJ, Nehme NT, Limmer S, Liegeois S, Pospisilik JA, et al. (2009) Genome-wide RNAi screen identifies genes involved in intestinal pathogenic bacterial infection. Science Signaling 325: 340.

58. Jiang H, Patel PH, Kohlmaier A, Grenley MO, McEwen DG, et al. (2009) Cytokine/Jak/Stat signaling mediates regeneration and homeostasis in the *Drosophila* midgut. Cell 137: 1343–1355.

59. Jiang H, Grenley MO, Bravo MJ, Blumhagen RZ, Edgar BA (2011) EGFR/Ras/MAPK signaling mediates adult midgut epithelial homeostasis and regeneration in *Drosophila*. Cell Stem Cell 8: 84–95.

60. Xu N, Wang SQ, Tan D, Gao Y, Lin G, et al. (2011) EGFR, Wingless and JAK/STAT signaling cooperatively maintain *Drosophila* intestinal stem cells. Dev Biol 354: 31–43.

61. Gonzalez FJ (2005) Role of cytochromes P450 in chemical toxicity and oxidative stress: studies with *CYP2E1*. Mutat Res 569: 101–110.

62. Parkes TL, Hilliker AJ, Phillips JP (1993) Genetic and biochemical analysis of glutathione-S-transferase in the oxygen defense system of *Drosophila melanogaster*. Genome 36: 1007–1014.

63. Aronstein KA, Murray KD, Saldivar E (2010) Transcriptional responses in honey bee larvae infected with chalkbrood fungus. BMC Genomics 11: 391.

64. Jang YK, Wang L, Sancar GB (1999) RPH1 and GIS1 are damage-responsive repressors of PHR1. Mol Cell Biol 19: 7630–7638.

65. Toller IM, Neelsen KJ, Steger M, Hartung ML, Hottiger MO, et al. (2011) Carcinogenic bacterial pathogen *Helicobacter pylori* triggers DNA double-strand breaks and a DNA damage response in its host cells. Proc Natl Acad Sci U S A 108: 14944–14949.

66. Zhou B-BS, Elledge SJ (2000) The DNA damage response: putting checkpoints in perspective. Nature 408: 433–439.

67. Vallet-Gely I, Lemaitre B, Boccard F (2008) Bacterial strategies to overcome insect defences. Nat Rev Microbiol 6: 302–313.

68. Kiss-Toth E, Bagstaff SM, Sung HY, Jozsa V, Dempsey C, et al. (2004) Human tribbles, a protein family controlling mitogen-activated protein kinase cascades. J Biol Chem 279: 42703–42708.

69. Schneider DS, Ayres JS (2008) Two ways to survive infection: what resistance and tolerance can teach us about treating infectious diseases. Nat Rev Immunol 8: 889–895.

70. Karunker I, Benting J, Lueke B, Ponge T, Nauen R, et al. (2008) Over-expression of cytochrome P450 *CYP6CM1* is associated with high resistance to imidacloprid in the B and Q biotypes of *Bemisia tabaci* (Hemiptera: Aleyrodidae). Insect Biochem Mol Biol 38: 634–644.

71. Mahadav A, Kontsedalov S, Czosnek H, Ghanim M (2009) Thermotolerance and gene expression following heat stress in the whitefly *Bemisia tabaci* B and Q biotypes. Insect Biochem Mol Biol 39: 668–676.

Paraspeckle Protein 1 (PSPC1) Is Involved in the Cisplatin Induced DNA Damage Response—Role in G1/S Checkpoint

Xiangjing Gao[1,2⑨], Liya Kong[3⑨], Xianghong Lu[4], Guanglin Zhang[1,2], Linfeng Chi[1,2], Ying Jiang[5], Yihua Wu[1,2], Chunlan Yan[1,2], Penelope Duerksen-Hughes[6], Xinqiang Zhu[2]*, Jun Yang[1,7,8]*

1 Collaborative Innovation Center for Diagnosis and Treatment of Infectious Diseases, The First Affiliated Hospital, Zhejiang University, Hangzhou, Zhejiang, China, 2 Department of Toxicology, Zhejiang University School of Public Health, Hangzhou, Zhejiang, China, 3 Department of preventative medicine, Zhejiang Chinese Medical University, Hangzhou, China, 4 Lishui People's Hospital, Lishui, Zhejiang, China, 5 Center Testing International Corporation, Shenzhen, Guangdong, China, 6 Department of Basic Science, Loma Linda University School of Medicine, Loma Linda, Califorina, United States of America, 7 Department of Toxicology, Hangzhou Normal University School of Public Health, Hangzhou, Zhejiang, China, 8 Department of Biomedicine, College of Biotechnology, Zhejiang Agriculture and Forestry University, Hangzhou, China

Abstract

Paraspeckle protein 1 (PSPC1) was first identified as a structural protein of the subnuclear structure termed paraspeckle. However, the exact physiological functions of PSPC1 are still largely unknown. Previously, using a proteomic approach, we have shown that exposure to cisplatin can induce PSPC1 expression in HeLa cells, indicating the possible involvement for PSPC1 in the DNA damage response (DDR). In the current study, the role of PSPC1 in DDR was examined. First, it was found that cisplatin treatment could indeed induce the expression of PSPC1 protein. Abolishing PSPC1 expression by siRNA significantly inhibited cell growth, caused spontaneous cell death, and increased DNA damage. However, PSPC1 did not co-localize with γH2AX, 53BP1, or Rad51, indicating no direct involvement in DNA repair pathways mediated by these molecules. Interestingly, knockdown of PSPC1 disrupted the normal cell cycle distribution, with more cells entering the G2/M phase. Furthermore, while cisplatin induced G1/S arrest in HeLa cells, knockdown of PSPC1 caused cells to escape the G1/S checkpoint and enter mitosis, and resulted in more cell death. Taken together, these observations indicate a new role for PSPC1 in maintaining genome integrity during the DDR, particularly in the G1/S checkpoint.

Editor: Marco Muzi-Falconi, Universita' di Milano, Italy

Funding: National Natural Science Foundation of China (Nos. 81172692, 81302398 and 81202241; http://isisn.nsfc.gov.cn/egrantindex/funcindex/prjsearch-list); Postdoctoral Science Foundation of China (Nos. 2011M501020, 2012M511378, and 2013M530286; http://jj.chinapostdoctor.org.cn/V1/Program1/Info_Show. aspx?InfoID=9e750e10-db65-4ef5-96a0-55285269580d); Zhejiang Provincial Natural Science Foundation (No. LY12H2600; http://www.zjnsf. gov.cn/). The funders had no role in study design, data collection and analysis, decision to publish, or preparation of the manuscript.

Competing Interests: Although one author is employed by a commercial company "Center Testing International Corporation".

* E-mail: gastate@zju.edu.cn (JY); zhuxq@zju.edu.cn (XQZ)

⑨ These authors contributed equally to this work.

Introduction

Cells are continuously faced with exogenous and endogenous stress that can induce DNA damage, potentially leading to genomic instability and cell death [1]. To maintain genomic integrity, cells have evolved the DNA damage response (DDR), a complex network of interacting pathways. Usually, DNA damage is primarily detected by the MRE11–RAD50–NBS1 (MRN) complex, which is followed by the activation of the phosphatidylinositol 3-kinase-like protein kinase (PIKKs) family members: ataxia telangiectasia mutated protein (ATM), ataxia telangiectasia and Rad3-related protein (ATR) and DNA dependent protein kinase (DNA-PK) [2–4]. These kinases phosphorylate and activate a variety of substrates to execute various cellular functions such as DNA repair, cell cycle arrest and cell death. One substrate is the histone variant H2AX, which can be phosphorylated at Ser-139 (termed γH2AX) and is directly involved in DNA repair [5,6]. Phosphorylation of H2AX is required to recruit a number of DDR proteins including repair factors and chromatin remodeling complexes [7–9]. For this reason, γH2AX foci formation has been recognized as an effective indicator of DNA damage, even when only a few DNA double-strand breaks (DSBs) are elicited [10–12]. As a mediator/adaptor of DDR, 53BP1 can facilitate ATM-dependent phosphorylation events, including the efficient phosphorylation of checkpoint kinase 2 (CHK2), and is required for ATM-dependent repair of DSBs through the non-homologous end-joining (NHEJ) pathway [13–15]. Similarly, in the homologous recombination (HR) pathway, the Rad51 protein interacts with the ssDNA-binding protein (SSBs) and re-localizes with the nucleus to form distinct foci, which represent repair active sites [16]. It is well known that proteins involved in DNA repair usually, either bind directly to the DNA at a damaged site such as Ku and Rad52 proteins [17,18], or interact with other repair proteins as part of the repair complex at the damaged site (referred as the "repair foci") [19]. These proteins, together with many other DNA repair proteins, are important in maintaining genome stability. As would be expected, defective DNA damage repair is associated with various developmental, immunological, and neurological disorders, and is a major driver in cancer [20].

During DDR, cell cycle checkpoints, including the G1/S and G2/M checkpoints, can be activated before replication or mitosis ensues, respectively [21,22]. Cells can arrest the cell cycle temporarily to allow for: (i) cellular damage to be repaired; (ii) the dissipation of an exogenous cellular stress signal; or (iii) availability of essential growth factors, hormones or nutrients [23,24]. If the damage can be effectively repaired during cell cycle, cells can regain normal functions and resume the cell cycle. Alternatively, if cell cycle checkpoint fails and the damage cannot be successfully repaired, chronic DDR can trigger cell death through mechanisms such as apoptosis or cellular senescence [25,26]. The checkpoint response, which prevents cells from accumulating mutations through replication and possibly developing into cancer, is a critical part of the DDR [27,28].

Because of the importance of DDR in cell growth and survival, numerous studies have been conducted to identify the many proteins/molecules involved and to reveal the underlying mechanisms. High-throughput technologies, such as genomics and proteomics, can generate huge amounts of information, and data mining of this information can reveal previously unknown or unexpected associations. Therefore, such technologies are useful tools for identifying new molecules/pathways involved in cellular activities such as DDR. Previously, using such an approach, e.g., nuclear proteomics, we investigated the induction of DDR in HeLa cells by cisplatin, a first-line chemotherapeutic agent with DNA damaging properties. Interestingly, among the many proteins affected by cisplatin treatment, we found that the expression of paraspeckle protein 1 (PSPC1) could be induced by cisplatin, suggesting it as a newly-discovered participant in cisplatin-induced DDR [29].

PSPC1 was first identified as a structural protein of a specific type of nuclear body called the paraspeckle [30]. Paraspeckles are involved in transcriptional and post-transcriptional gene regulatory functions, such as controlling expression of hyper edited mRNAs, mRNA biogenesis, pre-mRNA 3′-end formation, cyclic AMP signaling, and nuclear receptor-dependent transcriptional regulation [31–33]. PSPC1 contains two copies of the RNA recognition motif (RRM), which is the most prevalent RNA-binding domain in eukaryotes and a prerequisite for the localization of PSPC1 to paraspeckles. Another two proteins, polypyrimidine tract-binding protein associated splicing factor (PSF) and 54 kDa nuclear RNA binding protein (p54nrb) contain two RRMs and together with PSPC1, comprise the protein core of paraspeckles in HeLa cells. In addition to their functional role in the paraspeckle, previous studies also showed their role in cell survival or proliferation. For example, it was shown that attenuating p54nrb expression in human colon cancer HCT-116 cells resulted in smaller colony size and lower plating efficiency [34], but knockdown of p54nrb had no effect on long-term survival in HeLa cells [35]. PSF knockdown severely inhibited cell proliferation in DLD-1 cells [36], and caused a more severe loss of cell viability in the Rad51D-deficient mouse embryonic fibroblast (MEF) cells than in the corresponding Rad51D-proficient cells [37]. Also, it has been shown that PSF and p54nrb form a stable complex *in vivo*, which is involved in the repair of DSBs *via* the HR pathway [34,38]. Furthermore, the PSF·p54nrb complex is involved in NHEJ in vertebrates [39,40].

In contrast, the functions of PSPC1 are largely unknown with the exception of its possible involvement in regulating either gene expression or RNA processing. For example, Myojin *et al* showed that PSPC1 has RNA-binding activity [41], and Fox *et al* reported

that PSPC1 might be involved in the regulation of mRNA splicing [42]. Other studies suggested that PSPC1 might regulate androgen receptor-mediated transcriptional activity [43]. Interestingly, one earlier study, which analyzed ATM and ATR substrates in an effort to reveal the extensive protein network activated in response to DNA damage, identified PSPC1 as a possible phosphorylation substrate of ATM/ATR [44]. Furthermore, Ha *et al* reported that PSF could promote the recruitment of PSPC1 to sites of DNA damage following knockdown of p54nrb [40]. Such information, combined with our observation that PSPC1 expression can be induced by cisplatin as well as evidence that the other two paraspeckle proteins, PSF and p54nrb, are involved in DNA repair, all lead to the hypothesis that PSPC1 is very likely a participant in the DDR. However, the precise role of PSPC1 in DDR has not yet been carefully investigated. To address this question, we carried out a series of analyses designed to reveal a possible role of PSPC1 in the DDR, and as reported here, we provide the first piece of evidence for the direct involvement of PSPC1 in DDR. Specifically, we provide evidence for its function at the G1/S checkpoint.

Methods

Cell culture and cell cycle synchronization

Human cervical carcinoma (HeLa) cells obtained from the ATCC were grown in Minimal Essential Medium (MEM) supplemented with 10% new born calf serum (NCS) with 5% CO_2 at 37°C. Cell cycle synchronization was carried out by double thymidine blockage at the G1/S boundary as described in [45]. Briefly, cells were grown in the presence of 2 mM thymidine (Sigma, St. Louis, MO) for 18 h, then washed with PBS, and grown in fresh medium without thymidine for 8 h. Thymidine was added again at 2 mM and incubated another 18 h to block cells at the G1/S boundary.

Chemicals and antibodies

Cisplatin was purchased from Sigma; PSF and p54nrb antibodies were purchased from Santa Cruz Biotechnology (Santa Cruz, CA), mouse monoclonal anti-β-actin antibody and the Annexin V-fluoresce isothiocyanate (FITC)/propidium iodide (PI) apoptosis detection kit were obtained from Multisciences Biotechnology (Hangzhou, China). γH2AX, Rad51 and 53BP1 antibodies were purchased from Millipore (Billerica, MA); Caspase-3 and PARP antibodies were supplied by Bioworld Technology (St. Louis Park, MN); and an affinity-purified peptide antibody against PSPC1 was generated in rabbits in our laboratory as described by Fox *et al* [42]. Alexa Fluor 488-conjugated and IR Dye-conjugated goat anti-mouse and goat anti-rabbit IgG were obtained from Life Technologies (Carlsbad, CA, USA).

Transfection of small interfering RNA (siRNA) and detection of PSPC1 expression

Two sets of siRNA oligo nucleotides for the human PSPC1 gene corresponding to nucleotides 1257—1275 (siPSPC1) and negative control siRNA were synthesized by Shanghai GenePharma Co., Ltd and used for transfection. siRNAs were transfected into HeLa cells using Lipofectamine2000 (Invitrogen, Carlsbad, CA), essentially as directed by the manufacturer and using a siRNA concentration of 40 nM. In short, cells were seeded into a 6-well cell culture plate, siRNA-Lipofectamine2000 complexes were added to each well after 24 h, and the medium was changed after 6 h incubation. After 18 h incubation, the attenuation of mRNA levels was detected by real-time reverse transcriptase PCR (RT-PCR). Total RNA was isolated using Trizol Reagent

(Invitrogen), and 2 μg of total RNA was used for first-strand cDNA synthesis with Super Script III Reverse Transcriptase (Invitrogen). RT-PCR was performed in 20 μl using the TaKaRa SYBR *Premix Ex Taq* Kit (TaKaRa Biotechnology, Dalian, China) and 100 ng of input cDNA template. β-actin was used as an internal standard. Primers for PSPC1 were 5′-AGACGCTTG-GAAGAACTCAGA-3′ and 5′-TTGGAGGAGGACCTTGGT-TAC-3′; primers for β-actin were 5′-TGCGTGACATTAAG-GAGAA-3′ and 5′-AAGGAAGGC TGGAAGAGT-3′.

Plasmid vectors and transfection

The pPSPC1 and pCON plasmids were constructed by Shanghai Genechem Co., Ltd (G006). Cells were transfected with 2 μg plasmid as well as the empty vector in Opti-MEM medium (Invitrogen) with X-tremeGENE HP DNA transfection reagent (Roche) according to the manufacturer's protocol.

Immunoblotting

Cells were lysed in RIPA lysis buffer (Beyotime, Nantong, China), and protein concentrations were determined using the bicinchoninic acid (BCA) Protein Assay Kit (Beyotime). Denatured protein extracts were loaded and separated on 15% or 8% SDS–polyacrylamide gels (Mini-Protean II, Bio-Rad) and transferred to an Immunoblot polyvinylidene fluoride (PVDF) Membrane (Millipore). After blocking with 3% non-fat milk in Tris-buffed saline with 0.1% (v/v) Tween-20 (TBST), membranes were incubated with primary antibodies at 4°C overnight, followed by incubation of IR Dye-conjugated secondary antibodies for 1 h at room temperature. After three washes, membrane-bound proteins of interest were detected using an Odyssey Infrared Imaging System (Li-Cor, USA).

Assessment of cell viability

Cell viability was determined using the Trypan blue exclusion assay as described previously [46]. In short, cells were treated with trypsin, removed from the plate and centrifuged for 5 min at 250 g. The pellet was suspended in MEM. Equal volumes of 0.4% Trypan blue and the cell suspension were mixed and 10 μl of the mixture was applied to a hemocytometer. The stained (non-viable) and unstained (viable) cells were counted under a microscope.

Analysis of apoptosis

The Annexin V-FITC/PI kit (Multiscience) was used to analyze the extent of apoptosis. Briefly, cells were collected by trypsinization and washed three times with phosphate-buffered saline (PBS), then resuspended in 500 μl binding buffer with 5 μl Annexin V-FITC and 10 μl PI. Cells were incubated for 5 min in the dark at room temperature. The cells were then analyzed using a FC500 MCL machine (Beckman Coulter) at 10,000 events/sample.

Immunofluorescence microscopy

For immunofluorescent staining, cells were fixed in 4% paraformaldehyde for 15 min, permeabilized with 0.5% triton and blocked with 3% BSA for 1 h at 37°C. The cells were incubated with primary antibodies overnight, washed three times in PBS, and then incubated with Alexa Fluor 488-conjugated secondary antibodies for 1 h. DNA was counterstained with 1 μg/ml DAPI for 15 min at 37°C. Cells mounted on cover slips were observed with a Leica DMI 4000 immunofluorescent microscope or a Zeiss confocal laser scanning microscope.

Cell cycle analysis

For flow cytometry measurements of the cell cycle, 36 h-post transfection cells were trypsinized, centrifuged at 300 g for 5 min and fixed overnight in 70% cold ethanol at −20°C. After washing twice with PBS, the cells were resuspended in 500 μl of fresh PBS containing 50 μl of 2 mg/ml RNaseA and 10 μl of 1 mg/ml PI (Sigma). Cells were incubated for 15 min at 37°C. The cells were then analyzed immediately using a FC500 MCL machine (Beckman Coulter) at 10,000 events/sample.

Statistical analysis

Statistical analysis was performed using the Student's t-test or one-way ANOVA. Each experiment was conducted at least three times independently. Data were presented as mean ± SD and a probability level of $P < 0.05$ was considered significant.

Results

PSPC1 expression in HeLa cells is induced by cisplatin

Previously, we had employed nuclear proteome analysis to demonstrate that PSPC1 could be induced by cisplatin in HeLa cells [29]. To further validate this observation, HeLa cells were treated with different doses of cisplatin for 12 h, and the expression of PSPC1 was examined by Western blot. As shown in Figure 1, the level of PSPC1 was indeed increased by cisplatin treatment. Cisplatin concentrations at 10 μM or higher were not examined as significant loss of cell viability was induced (data not shown). Therefore, all the following experiments using cisplatin were conducted at concentrations of either 2.5 or 5 μM.

Knockdown of PSPC1 reduces cell survival

To explore the possible biological functions of PSPC1, we first examined the effects of PSPC1 siRNA knockdown on cell growth and cell death. Transfection with PSPC1 siRNA consistently reduced mRNA and protein expression by about 95% compared with control siRNA, as assessed by both RT-PCR and Western blot (Figure 2A). Trypan blue exclusion assay results showed that PSPC1 knockdown significantly inhibited cell growth (Figure 2B, left panel). Furthermore, although there was a slight increase at

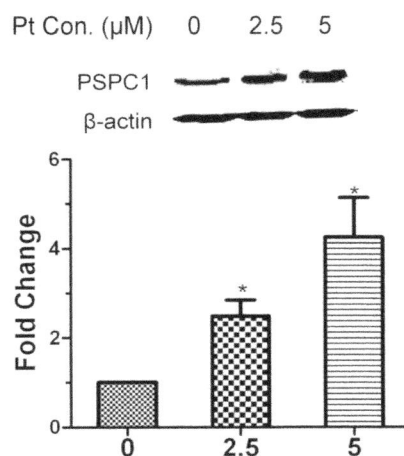

Figure 1. PSPC1 is induced by cisplatin. HeLa cells were treated with 2.5 or 5 μM of cisplatin (Pt) for 12 h, and expression of PSPC1 was detected by Western blot. The results are shown as the mean ±SD of three independent experiments. *P<0.05, compared with the control group.

early hours (up to 36 h), the number of live cells then gradually decreased, eventually dropping to less than the originally seeded number of cells by 72 h in the siPSPC1 group (Figure 2B, right panel). This observation implies an important role for PSPC1 in maintaining cell viability. Therefore, we further evaluated the effects of PSPC1 on cell death. As shown in Figure 3A, about 10% of the cells were Annexin V and PI-positive in the control group, in contrast, after PSPC1 knockdown, the percentage of dual-positive cells was 15%, a slight but significant increase. In addition, we also assessed the level of cleaved Caspase-3 and cleaved PARP by Western blot, which are considered markers of apoptosis. As shown in Figure 3B, cleaved Caspase-3 and cleaved PARP were significantly up-regulated after knockdown of PSPC1 in HeLa cells, suggesting that some of the PSPC1-knockdown cells undergo apoptosis by caspase and/or PARP-dependent mechanisms.

Alteration of PSPC1 expression influences the formation of γH2AX foci

As our interest was the possible role of PSPC1 in DDR, we then measured the extent of cisplatin-induced DNA damage in the presence or absence of PSPC1 using γH2AX foci formation as a sensitive indicator. Interestingly, Western blot data showed that PSPC1 knockdown resulted in a marked increase in the level of γH2AX in cells even without cisplatin exposure (Figure 4A). Cisplatin treatment induced a dose-dependent increase in γH2AX protein levels, and the level of this increase was much stronger in each siPSPC1 group as compared with the corresponding siControl group (Figure 4A). Flow cytometry and immunofluorescence results demonstrated the same trend (Figure 4B and 4C).

To further verify whether PSPC1 expression can influence cisplatin-induced DNA damage, HeLa cells were transfected with an overexpression plasmid of PSPC1. As shown in Figure 4D,

overexpression of PSPC1 in HeLa cells significantly inhibited the increase of γH2AX protein level compared to control cells, implying less severe DNA damage. Together, these findings suggested that PSPC1 is important in maintaining DNA stability and minimizing genomic insults in cells.

PSPC1 does not form distinct foci with γH2AX, 53BP1 nor Rad51

As noted above, cisplatin can induce increased expression of PSPC1 (Figure 1), and the loss of PSPC1 results in increased DNA damage (Figure 3). Therefore, it is reasonable to predict that PSPC1 might play a role in DNA repair and in this way protect cells from cisplatin-induced damage. To investigate this possibility, we examined the distribution of PSPC1, as well as its relationship with several key factors involved in DNA repair, including γH2AX, 53BP1, and Rad51. The results (Figure 5A) showed that there were no significant changes in the relatively diffuse distribution pattern of PSPC1 in the nucleus in both control and cisplatin treated cells. In contrast, cisplatin induced the formation of distinct Rad51, 53BP1 and γH2AX foci as compared with their respective controls. In addition, upon close examination, PSPC1 did not co-localize with Rad51, 53BP1, or γH2AX to form distinct foci after cisplatin treatment (Figure 5A). Taken together, these results fail to support the idea that PSPC1 participates in the specific DNA repair events mediated by Rad51, 53BP1 and γH2AX.

Studies of the DNA repair function of p54nrb showed that knockdown of p54nrb could lead to a delay in the repair of DNA damage [34]. This suggested an alternate mechanism for PSPC1 action, and to further examine the possible DNA repair activity of PSPC1, we measured the level of γH2AX during a 48 h period as an indicator of DNA repair in the presence and absence of PSPC1.

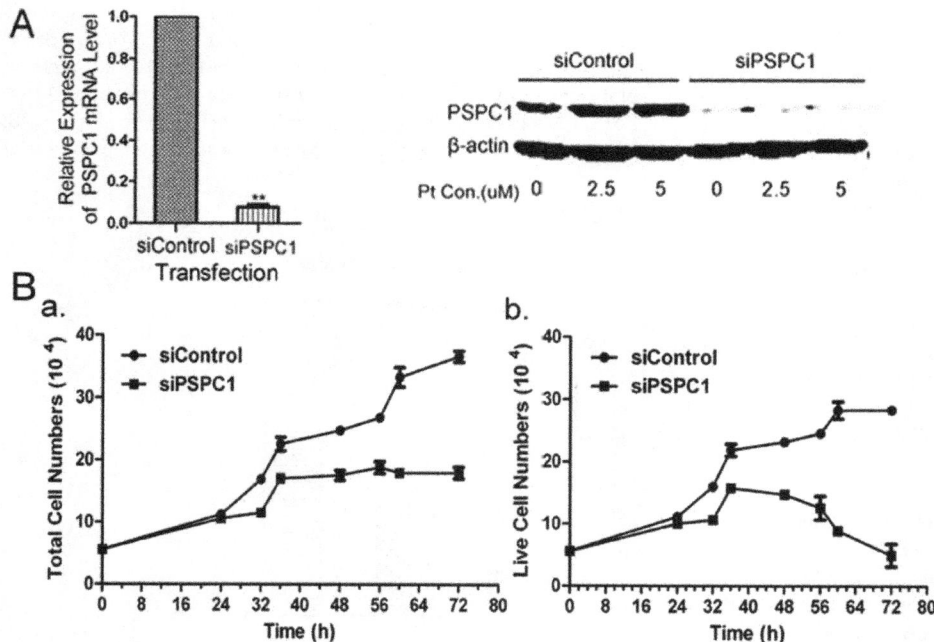

Figure 2. Attenuation of PSPC1 expression inhibits cell proliferation. (A) HeLa cells were transfected with 40 nM PSPC1 siRNAs (siPSPC1) or control siRNA (siControl) ('Materials and Methods' section). 24 h later, expression of PSPC1 was analyzed using quantitative real-time PCR (left histogram) and Western blot (right panels). β-actin was used as the loading control. (B) Cell proliferation of HeLa cells transfected with siPSPC1 or siControl was measured by the Trypan blue exclusion assay. Left, total cell number; Right, viable cell number. Data represents the average of three independent experiments with six replicate measurements (mean ± SD).

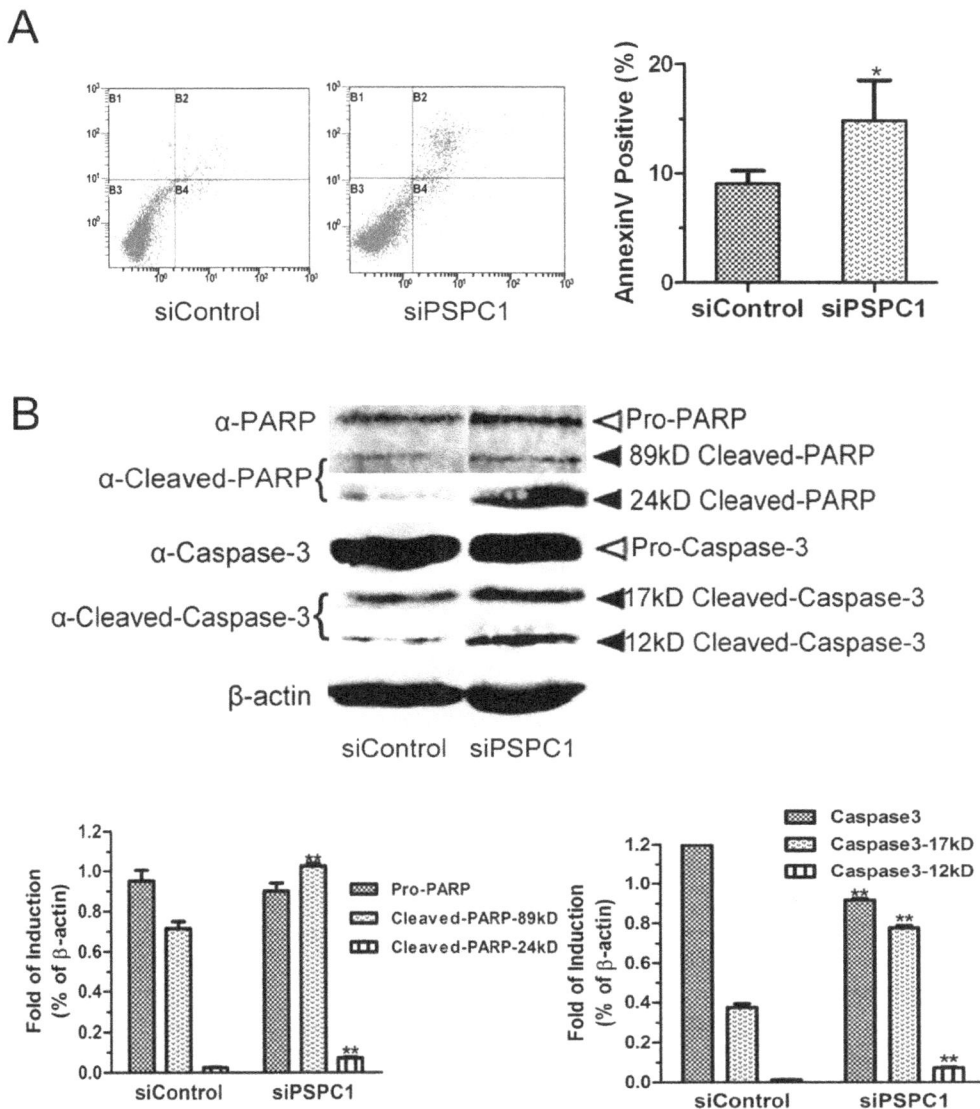

Figure 3. Knockdown of PSPC1 induces cell death. (A) HeLa cells harvested at 24 h post-transfection were analyzed by dual-parameter flow cytometry utilizing Annexin V-FITC and PI. Representative dot plot data from three independent experiments are shown in the left panel, and the histogram graph at right represents the percentage of dual-parameter positive cells pooled from three independent experiments. (B) HeLa cells harvested at 24 h post-transfection were analyzed by Western blotting to evaluate the expression of Caspase-3 and PARP. Densitometric data of three independent experiments are presented below the immunoblot, and β-actin was used as an internal standard. Data are presented as mean ± SD. *$P<$ 0.05, **$P<$ 0.01, compared with control group.

The results showed that in control siRNA cells, the γH2AX foci level remained low, as expected. In contrast, knockdown of PSPC1 with siRNA led to a burst of γH2AX formation at 16 h. These lesions were repaired rapidly, and the level of γH2AX decreased to a level slightly higher than that of control cells after 20 h (Figure 5B, top panel). Following cisplatin treatment, the increase in γH2AX foci appeared earlier in PSPC1 knockdown cells than in the control cells. These cells also showed a burst in γH2AX formation at about 16 h, followed by the rapid repair, although γH2AX level remained higher than in control cells (Figure 5B, lower panel). Therefore, although the repair kinetic curve is quite different in the presence and absence of PSPC1, there is no clear delay of repair in PSPC1-knockdown cells as compared with control cells.

Loss of PSPC1 causes cells to enter G2/M phase

Upon DNA damage, mammalian cells may activate cell-cycle arrest to stop or delay cell division to allow the damage to be repaired [47]. As the above results did not support a direct role for PSPC1 in DNA repair, we asked whether PSPC1 might function in cell cycle progression. siPSPC1 or siControl-transfected HeLa cells were first synchronized at the S phase, then allowed to grow in fresh medium for 24 h, and subjected to cell cycle analysis. The results showed that for control siRNA transfected cells, 48% of the cells were in G1, 35% in S, and 17% in the G2/M phase; however, for siPSPC1 cells, the ratio was: 35% in G1, 27% in S, and 38% in the G2/M, a more than 2-fold increase in the number of cells entering G2/M (Figure 6A).

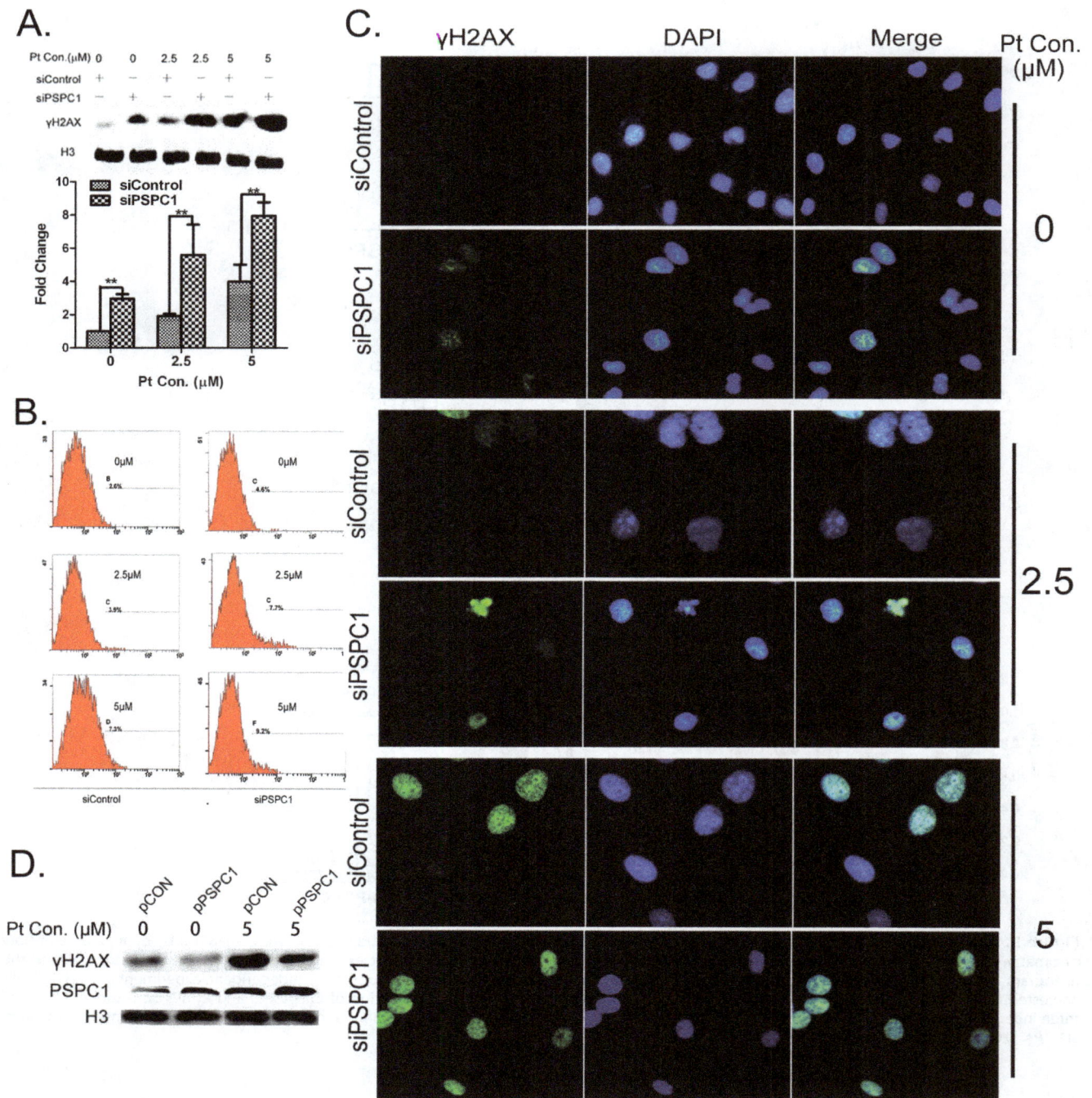

Figure 4. Alteration of PSPC1 expression influences the formation of γH2AX foci. HeLa cells were transfected with siPSPC1 or siControl. 24 h post-transfection, cells were treated with 2.5 or 5 μM of cisplatin for 12 h, and the expression of γH2AX was examined by Western blot (A), flow cytometry (B), and immunofluorescence microscopy (C). (D) HeLa cells were transfected with either pPSPC1 or pCON to overexpress PSPC1. 24 h post-transfection, cells were treated with 5 μM of cisplatin for 12 h, and the expression of γH2AX or PSPC1 was examined by Western blot. *$P < 0.05$, compared with control.

To confirm whether these cells were indeed entering the G2/M phase, the expression levels of phospho-histone H3, cyclinB and Cdc2, known regulatory proteins of the G2/M phase [48], were measured by Western blot. As shown in Figure 6B, PSPC1 knockdown markedly increased the level of phospho-histone H3. Similarly, the levels of cyclinB and Cdc2 were also increased

significantly after attenuation of PSPC1 expression (Figure 6B). Therefore, these data pointed out a possible involvement for PSPC1 in regulating the cell cycle.

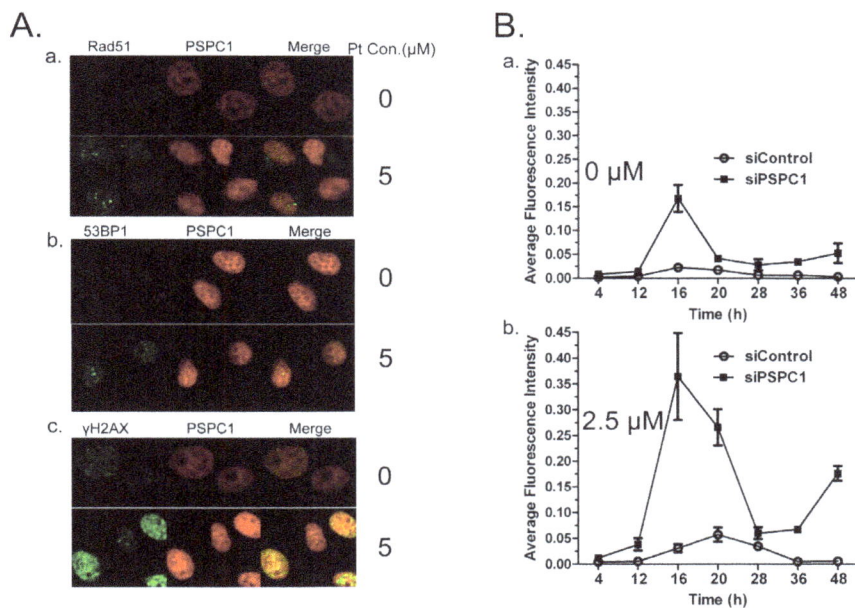

Figure 5. PSPC1 may not participate in DNA repair. (A) PSPC1 does not form distinct foci with γH2AX, 53BP1 or Rad51. Representative confocal laser scanning images of HeLa cells were analyzed 12 h after 5 μM cisplatin treatment. (B) DNA repair kinetic curve in siControl and siPSPC1 cells as calculated by the intensity of γH2AX measured by immunofluorescence microscopy. Quantitative analysis of average density (Fluorescence intensity per unit area) was determined by Image-Pro Plus 6.0.

PSPC1 is involved inG1/S phase arrest induced by cisplatin

To further clarify the function of PSPC1 in cell cycle regulation, we examined the cell cycle distribution of HeLa cells upon cisplatin exposure. Cisplatin treatment is known to induce S arrest in cells [49], and our results showed that after 24 h of cisplatin treatment, siControl exhibited a clear S phase arrest, with about 57% cells in S phase (Figure 7A, compared to 35% in Figure 6A) and 8% in the G2/M phase. On the other hand, knockdown of PSPC1 by siRNA attenuated the cisplatin-induced S phase-arrest, with only 42% in S phase but about 30% of cells in the G2/M phase, almost a 4-fold increase compared with the siControl cells (Figure 7A).

In addition, cisplatin-induced cell death was also measured. As shown in Figure 7B, compared with siControl, the percentage of dead cells was significantly increased following PSPC1 siRNA knockdown (~2-fold). Taken together, these data suggest that PSPC1 might play a role in regulation of the G1/S checkpoint, whereas disruption of its function could lead to cells escaping the G1/S arrest and entering G2/M phase. These events have the potential to increase DNA damage and to cause more cell death.

Discussion

PSPC1 is found in paraspeckles in transcriptionally active cells as well as perinucleolar caps in cells that are not actively transcribing Pol II genes [42]. It belongs to the Drosophila Behavior Human Splicing (DBHS) family, which is composed of two classical RRMs followed by a proline-rich coiled-coil motif [32]. Together with other two DBHS family members, PSF and p54nrb, PSPC1 forms the protein core of paraspeckles. To date, many studies have been conducted to investigate the functions of

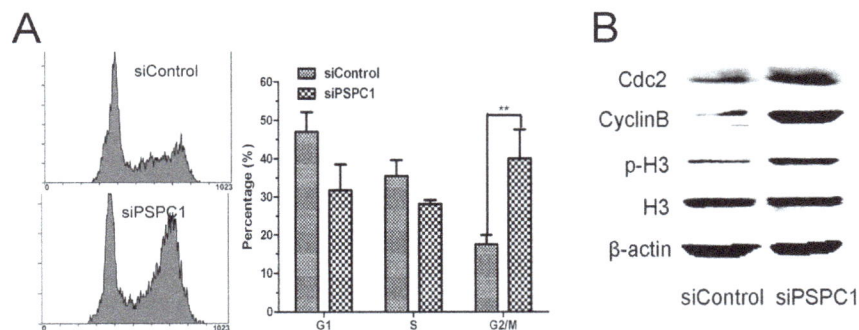

Figure 6. Loss of PSPC1 causes cells to enter G2/M phase. (A) HeLa cells were cultured in the presence of 2 mM thymidine for 18 h, washed with PBS, and transfected with siRNAs using Lipofectamine2000. 8 h after transfection, thymidine was added again to 2 mM to block cells at the G1/S boundary. After another 18 h, cells were transferred to fresh medium for 24 h, then harvested and analyzed by flow cytometry. (B) Using the same cells, the levels of representative G2/M phase proteins (phospho-histone H3 (Ser10), Cdc2 and cyclinB [55]) were examined by Western blot.

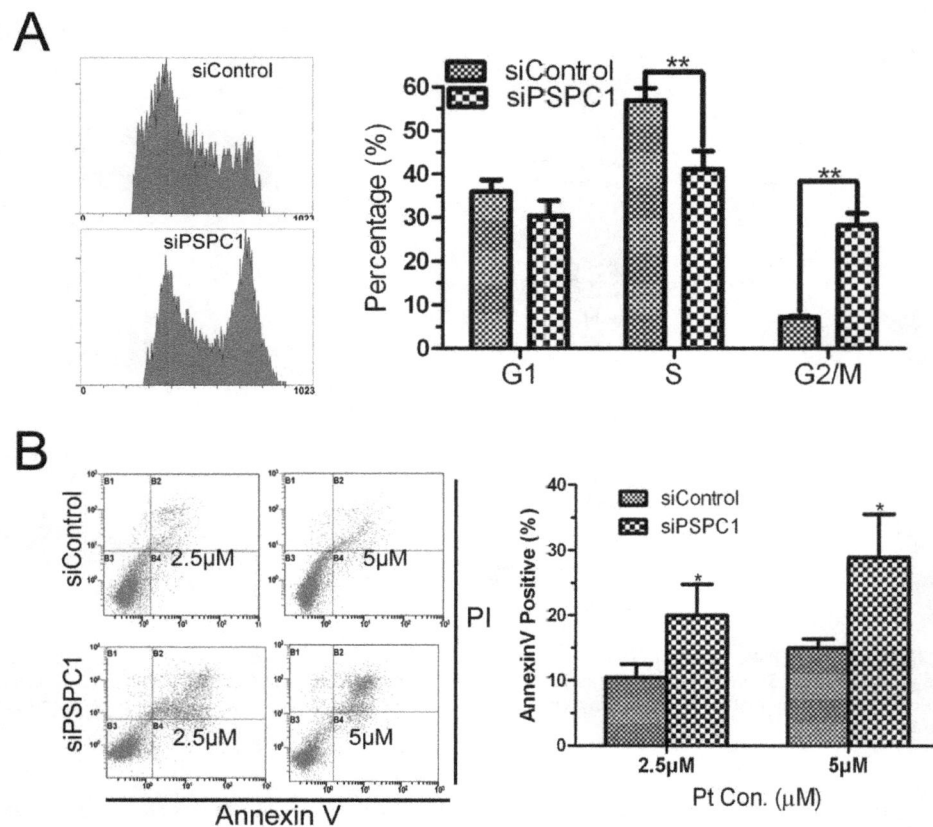

Figure 7. Knockdown of PSPC1 causes cells to escape cisplatin-induced G1/S arrest. (A) HeLa cells were synchronized by double thymidine blockage at the G1/S boundary as described previously, released into fresh medium with 5 μM cisplatin for 24 h, and then harvested and detected by flow cytometry for cell cycle distribution. (B) The same synchronized HeLa cells were treated with 2.5 or 5 μM of cisplatin for 24 h, then apoptosis was analyzed by dual-parameter flow cytometry with Annexin V-FITC and PI staining. Representative dot plot data from three independent experiments are shown left, and the right histogram represents the percentage of non-viable cells pooled from three independent experiments. *$P < 0.05$, **$P < 0.01$, compared with control group.

PSF and p54nrb, and the results pointed out their important roles in DNA repair. In contrast, PSPC1 is more selectively expressed, acting as a coactivator of transcription [43]. It is also known that PSPC1 can dimerize with p54nrb through the coiled-coil domain to regulate pre-RNA processing, but not with PSF [50]. Nevertheless, until now, little is known about its other functions, especially in DDR.

Previously, as part of an effort to investigate the DDR, we conducted a nuclear proteomics screen for DDR-related proteins. This screen identified PSPC1 as a novel molecule possibly participating in cisplatin-induced DDR [29]. Combined with previous reports stating that (i) PSPC1 could be phosphorylated by ATM/ATR [44], (ii) p54nrb and PSF are involved in DSB repair, and (iii) PSF could promote the recruitment of PSPC1 to sites of DNA damage after p54nrb knockdown [40], the involvement of PSPC1 in DDR seemed a reasonable possibility. To test this hypothesis, we first showed that PSPC1 could indeed be induced by cisplatin (Figure 1). This phenomenon is characteristic for proteins participating in the DDR, for example, p53 and proline-rich acidic protein 1 (PRAP1), key regulators of DDR, can be induced under conditions of DNA damage [51,52]. Thus, this is the first piece of evidence linking PSPC1 to DDR. To clarify the physiological function of PSPC1, we then inhibited the expression of PSPC1 by siRNA, and examined the effects of this knockdown on cell growth and survival. These results showed that depletion of

PSPC1 significantly inhibited cell proliferation (Figure 2). The effects of knocking down either PSF or p54nrb on cell survival or proliferation have been previously investigated by others. Those studies indicate that the effects of these proteins on cell proliferation are likely to be cell-type specific due to different genetic backgrounds. Nonetheless, our data indicated an important role for PSPC1 in maintaining normal cell growth, at least in HeLa cells.

Additionally, our results showed that after loss of PSPC1, the number of live cells was dramatically reduced (Figure 2), indicating the occurrence of cell death. The activation of Caspase-3 and PARP further demonstrated that knockdown of PSPC1 indeed can cause apoptosis (Figure 3). Similarly, PSF knockdown also induced Caspase-3 mediated apoptosis in DLD-1 cells [36], suggesting that PSPC1 and PSF might share certain common functions. However, it should be noted that the loss of PSPC1 increased the number of apoptotic cells only to a small extent, while the number of live cells decreased rather dramatically. Thus, it is believed that other types of cell death, including necrosis, autophagy, or necroptosis may also be occurring, and is an area of ongoing and future study.

As our focus is the relationship between PSPC1 and DDR, we next evaluated the effects of PSPC1 knockdown on cisplatin-induced DNA damage. Our results showed that depletion of PSPC1 sensitized cells to cisplatin-induced DNA damage, as assessed by the appearance of γH2AX foci (Figure 4). This is

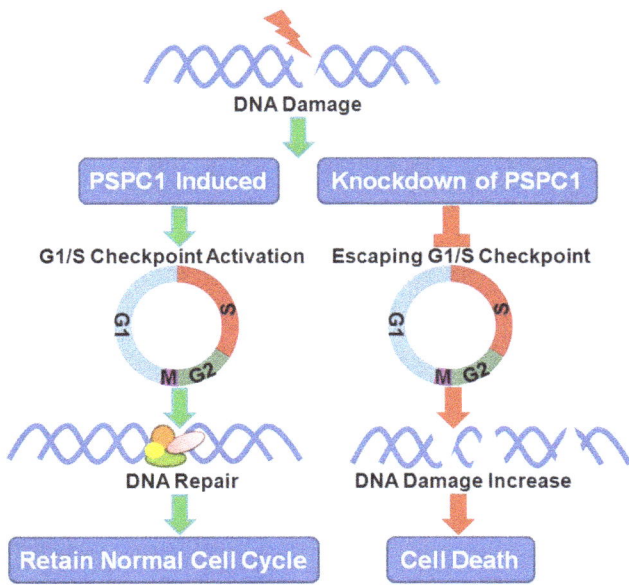

Figure 8. A schematic model for the proposed role of PSPC1 in DDR.

consistent with previous reports indicating that knocking down p54nrb sensitized cells to radiation [34]. Such similarity provides another piece of evidence for the involvement of PSPC1 in DDR.

As PSF and p54nrb play important roles in DNA repair during DDR, we were wondering whether PSPC1 also had a similar function. Previously, PSF and p54nrb have been shown to bind directly to DNA, and can interact with other repair proteins such as Rad51 [35,53]. For this reason, we used confocal microscopy to ask whether PSPC1 could form foci upon cisplatin exposure, and whether it interacted with other repair proteins. Unexpectedly, we did not observe the formation of distinct PSPC1 foci following cisplatin treatment. Furthermore, the fluorescent image did not support a co-localization of PSPC1 with any of the three DDR tested proteins (Figure 5A), thus casting doubt on the idea that PSPC1 participates in DNA repair through direct interactions with these proteins. We then reasoned that if a protein is involved in DNA repair, disruption of its function would be expected to lead to the delay of DNA repair, as in the case for p54nrb [34,35]. Thus, we monitored DNA repair in PSPC1 knockdown cells using the disappearance of γH2AX as an indicator. As shown in Figure 5B, although the extent of DNA damage was much severe

in PSPC1 knockdown cells, the rate of repair was almost the same as in control cells. Together, these data implied that PSPC1 might not function in DNA repair, a situation that is quite different from that seen for PSF and p54nrb.

If PSPC1 were not involved in DNA repair, then what was its role in the DDR? One thing we noticed is that in the DNA repair curve, the burst of γH2AX occurred starting at about 12 h (Figure 5B), at which time cells might be entering the S phase. Such information pointed to a possible relationship between PSPC1 and the cell cycle. The following cell cycle analysis indeed revealed that knockdown of PSPC1 disrupted normal cell cycle distribution, with decreased cell numbers in the G1 and S phases, but significantly increased number of cells in the G2/M phase (Figure 6). These results provide evidence that PSPC1 plays a role in cell cycle regulation. Furthermore, cisplatin is known to induce G1/S arrest, during which time damaged DNA can be repaired [49]. However, in PSPC1 knockdown cells, cisplatin-induced G1/S arrest was abolished, and cells continued the cell cycle and entered G2/M (Figure 7). This observation led to the speculation that PSPC1 might be involved in regulation of the G1/S checkpoint.

Based on the above results, the following hypothesis is proposed (Figure 8). In normal cells, PSPC1 is required to regulate the G1/S transition. Upon cisplatin exposure, PSPC1 is induced, and coupled with other proteins of the G1/S checkpoint machinery, G1/S arrest is induced, thereby allowing the repair of DNA damage. However, when PSPC1 is knocked down, cisplatin-induced DNA damage cannot activate the appropriate G1/S checkpoint machinery, and cells "slip" through and enter G2/M. As a consequence, cells with unrepaired DNA damage entered mitosis prematurely but cannot complete mitosis, eventually leading to cell death [54]. This could explain the significant cell death observed in PSPC1 knockdown cells.

Acknowledgments

Jun Yang is a recipient of the Zhejiang Provincial Program for the Cultivation of High-level Innovative Health Talents.

Author Contributions

Conceived and designed the experiments: GXJ KLY LXH ZGL CLF JY WYH YCL DHP ZXQ YJ. Performed the experiments: GXJ KLY. Analyzed the data: GXJ KLY LXH ZGL CLF JY WYH YCL DHP ZXQ YJ. Contributed reagents/materials/analysis tools: GXJ KLY LXH ZGL CLF JY WYH YCL DHP ZXQ YJ. Wrote the paper: GXJ KLY. Revising the article critically for important intellectual content: GXJ KLY LXH ZGL CLF JY WYH YCL DHP ZXQ YJ. Final approval of the version to be published: GXJ KLY LXH ZGL CLF JY WYH YCL DHP ZXQ YJ.

References

1. Yu Y, Zhu W, Diao H, Zhou C, Chen FF, et al. (2006) A comparative study of using comet assay and gammaH2AX foci formation in the detection of N-methyl-N'-nitro-N-nitrosoguanidine-induced DNA damage. Toxicol In Vitro 20: 959–965.

2. Ward IM, Chen J (2001) Histone H2AX is phosphorylated in an ATR-dependent manner in response to replicational stress. J Biol Chem 276: 47759–47762.

3. Stiff T, O'Driscoll M, Rief N, Iwabuchi K, Lobrich M, et al. (2004) ATM and DNA-PK function redundantly to phosphorylate H2AX after exposure to ionizing radiation. Cancer Res 64: 2390–2396.

4. Huen MS, Chen J (2010) Assembly of checkpoint and repair machineries at DNA damage sites. Trends in biochemical sciences 35: 101–108.

5. Rogakou EP, Boon C, Redon C, Bonner WM (1999) Megabase chromatin domains involved in DNA double-strand breaks in vivo. J Cell Biol 146: 905–916.

6. Rogakou EP, Pilch DR, Orr AH, Ivanova VS, Bonner WM (1998) DNA double-stranded breaks induce histone H2AX phosphorylation on serine 139. J Biol Chem 273: 5858–5868.

7. Chowdhury D, Keogh MC, Ishii H, Peterson CL, Buratowski S, et al. (2005) gamma-H2AX dephosphorylation by protein phosphatase 2A facilitates DNA double-strand break repair. Mol Cell 20: 801–809.

8. Kusch T, Florens L, Macdonald WH, Swanson SK, Glaser RL, et al. (2004) Acetylation by Tip60 is required for selective histone variant exchange at DNA lesions. Science 306: 2084–2087.

9. Paull TT, Rogakou EP, Yamazaki V, Kirchgessner CU, Gellert M, et al. (2000) A critical role for histone H2AX in recruitment of repair factors to nuclear foci after DNA damage. Curr Biol 10: 886–895.

10. Rao VA, Agama K, Holbeck S, Pommier Y (2007) Batracylin (NSC 320846), a dual inhibitor of DNA topoisomerases I and II induces histone gamma-H2AX as a biomarker of DNA damage. Cancer Res 67: 9971–9979.

11. Redon CE, Dickey JS, Bonner WM, Sedelnikova OA (2009) gamma-H2AX as a biomarker of DNA damage induced by ionizing radiation in human peripheral blood lymphocytes and artificial skin. Adv Space Res 43: 1171–1178.

12. Rothkamm K, Lobrich M (2003) Evidence for a lack of DNA double-strand break repair in human cells exposed to very low x-ray doses. Proc Natl Acad Sci U S A 100: 5057–5062.

13. Schultz LB, Chehab NH, Malikzay A, Halazonetis TD (2000) p53 binding protein 1 (53BP1) is an early participant in the cellular response to DNA double-strand breaks. J Cell Biol 151: 1381–1390.

14. Rappold I, Iwabuchi K, Date T, Chen J (2001) Tumor suppressor p53 binding protein 1 (53BP1) is involved in DNA damage-signaling pathways. J Cell Biol 153: 613–620.

15. Anderson L, Henderson C, Adachi Y (2001) Phosphorylation and rapid relocalization of 53BP1 to nuclear foci upon DNA damage. Mo Cell Biol 21: 1719–1729.

16. Tarsounas M, Davies AA, West SC (2004) RAD51 localization and activation following DNA damage. Philosophical transactions of the Royal Society of London Series B, Biological sciences 359: 87–93.

17. Pierce AJ, Hu P, Han M, Ellis N, Jasin M (2001) Ku DNA end-binding protein modulates homologous repair of double-strand breaks in mammalian cells. Genes Dev 15: 3237–3242.

18. Van Dyck E, Stasiak AZ, Stasiak A, West SC (1999) Binding of double-strand breaks in DNA by human Rad52 protein. Nature 398: 728–731.

19. Bekker-Jensen S, Mailand N (2010) Assembly and function of DNA double-strand break repair foci in mammalian cells. DNA Repair (Amst) 9: 1219–1228.

20. Murga M, Fernandez-Capetillo O (2007) Genomic instability: on the birth and death of cancer. Clin Transl Oncol 9: 216–220.

21. Ciccia A, Elledge SJ (2010) The DNA damage response: making it safe to play with knives. Mol Cell 40: 179–204.

22. Chapman JR, Taylor MR, Boulton SJ (2012) Playing the end game: DNA double-strand break repair pathway choice. Mol Cell 47: 497–510.

23. Gartel AL, Tyner AL (2002) The role of the cyclin-dependent kinase inhibitor p21 in apoptosis. Molecular cancer therapeutics 1: 639–649.

24. Bakkenist CJ, Kastan MB (2003) DNA damage activates ATM through intermolecular autophosphorylation and dimer dissociation. Nature 421: 499–506.

25. Campisi J, d'Adda di Fagagna F (2007) Cellular senescence: when bad things happen to good cells. Nat Rev Mol Cell Biol 8: 729–740.

26. Halazonetis TD, Gorgoulis VG, Bartek J (2008) An oncogene-induced DNA damage model for cancer development. Science 319: 1352–1355

27. Bartkova J, Horejsi Z, Koed K, Kramer A, Tort F, et al. (2005) DNA damage response as a candidate anti-cancer barrier in early human tumorigenesis. Nature 434: 864–870.

28. Gorgoulis VG, Vassiliou LV, Karakaidos P, Zacharatos P, Kotsinas A, et al. (2005) Activation of the DNA damage checkpoint and genomic instability in human precancerous lesions. Nature 434: 907–913.

29. Wu W, Yan C, Gan T, Chen Z, Lu X, et al. (2010) Nuclear proteome analysis of cisplatin-treated HeLa cells. Mutat Res 691: 1–8.

30. Andersen JS, Lyon CE, Fox AH, Leung AK, Lam YW, et al. (2002) Directed proteomic analysis of the human nucleolus. Curr Biol 12: 1–11.

31. Prasanth KV, Prasanth SG, Xuan Z, Hearn S, Freier SM, et al. (2005) Regulating gene expression through RNA nuclear retention. Cell 123: 249–263.

32. Bond CS, Fox AH (2009) Paraspeckles: nuclear bodies built on long noncoding RNA. J Cell Biol 186: 637–644.

33. Amelio AL, Miraglia LJ, Conkright JJ, Mercer BA, Batalov S, et al. (2007) A coactivator trap identifies NONO (p54nrb) as a component of the cAMP-signaling pathway. Proc Natl Acad Sci U S A 104: 20314–20319.

34. Li S, Kuhne WW, Kulharya A, Hudson FZ, Ha K, et al. (2009) Involvement of p54(nrb), a PSF partner protein, in DNA double-strand break repair and radioresistance. Nucleic Acids Res 37: 6746–6753.

35. Krietsch J, Caron MC, Gagne JP, Ethier C, Vignard J, et al. (2012) PARP activation regulates the RNA-binding protein NONO in the DNA damage response to DNA double-strand breaks. Nucleic Acids Res 40: 10287–10301.

36. Tsukahara T, Haniu H, Matsuda Y (2013) PTB-associated splicing factor (PSF) is a PPARgamma-binding protein and growth regulator of colon cancer cells. PLoS One 8: e58749.

37. Rajesh C, Baker DK, Pierce AJ, Pittman DL (2011) The splicing-factor related protein SFPQ/PSF interacts with RAD51D and is necessary for homology-directed repair and sister chromatid cohesion. Nucleic Acids Res 39: 132–145.

38. Salton M, Lerenthal Y, Wang SY, Chen DJ, Shiloh Y (2010) Involvement of Matrin 3 and SFPQ/NONO in the DNA damage response. Cell Cycle 9: 1568–1576.

39. Bladen CL, Udayakumar D, Takeda Y, Dynan WS (2005) Identification of the polypyrimidine tract binding protein-associated splicing factor.p54(nrb) complex as a candidate DNA double-strand break rejoining factor. J Biol Chem 280: 5205–5210.

40. Ha K, Takeda Y, Dynan WS (2011) Sequences in PSF/SFPQ mediate radioresistance and recruitment of PSF/SFPQ-containing complexes to DNA damage sites in human cells. DNA Repair (Amst) 10: 252–259.

41. Myojin R, Kuwahara S, Yasaki T, Matsunaga T, Sakurai T, et al. (2004) Expression and functional significance of mouse paraspeckle protein 1 on spermatogenesis. Biol Reprod 71: 926–932.

42. Fox AH, Lam YW, Leung AK, Lyon CE, Andersen J, et al. (2002) Paraspeckles: a novel nuclear domain. Curr Biol 12: 13–25.

43. Kuwahara S, Ikei A, Taguchi Y, Tabuchi Y, Fujimoto N, et al. (2006) PSPC1, NONO, and SFPQ are expressed in mouse Sertoli cells and may function as coregulators of androgen receptor-mediated transcription. Biol Reprod 75: 352–359.

44. Matsuoka S, Ballif BA, Smogorzewska A, McDonald ER 3rd, Hurov KE, et al. (2007) ATM and ATR substrate analysis reveals extensive protein networks responsive to DNA damage. Science 316: 1160–1166.

45. Fang G, Yu H, Kirschner MW (1998) Direct binding of CDC20 protein family members activates the anaphase-promoting complex in mitosis and G1. Mol Cell 2: 163–171.

46. Zhou C, Li Z, Diao H, Yu Y, Zhu W, et al. (2006) DNA damage evaluated by gammaH2AX foci formation by a selective group of chemical/physical stressors. Mutat Res 604: 8–18.

47. Hoeijmakers JH (2001) Genome maintenance mechanisms for preventing cancer. Nature 411: 366–374.

48. Juan G, Traganos F, James WM, Ray JM, Roberge M, et al. (1998) Histone H3 phosphorylation and expression of cyclins A and B1 measured in individual cells during their progression through G2 and mitosis. Cytometry 32: 71–77.

49. Wagner JM, Karnitz LM (2009) Cisplatin-induced DNA damage activates replication checkpoint signaling components that differentially affect tumor cell survival. Molecular pharmacology 76: 208–214.

50. Fox AH, Bond CS, Lamond AI (2005) P54nrb forms a heterodimer with PSP1 that localizes to paraspeckles in an RNA-dependent manner. Mol Biol Cell 16: 5304–5315.

51. Huang BH, Zhuo JL, Leung CH, Lu GD, Liu JJ, et al. (2012) PRAP1 is a novel executor of p53-dependent mechanisms in cell survival after DNA damage. Cell death & disease 3: e442.

52. Laptenko O, Prives C (2006) Transcriptional regulation by p53: one protein, many possibilities. Cell Death Differ 13: 951–961.

53. Morozumi Y, Takizawa Y, Takaku M, Kurumizaka H (2009) Human PSF binds to RAD51 and modulates its homologous-pairing and strand-exchange activities. Nucleic Acids Res 37: 4296–4307.

54. Vitale I, Galluzzi L, Castedo M, Kroemer G (2011) Mitotic catastrophe: a mechanism for avoiding genomic instability. Nat Rev Mol Cell Biol 12: 385–392.

55. Zhang B, Huang B, Guan H, Zhang SM, Xu QZ, et al. (2011) Proteomic profiling revealed the functional networks associated with mitotic catastrophe of HepG2 hepatoma cells induced by 6-bromine-5-hydroxy-4-methoxybenzalde-hyde. Toxicol Appl Pharmacol 252: 307–317.

Exposure to Low-Dose Bisphenol A Impairs Meiosis in the Rat Seminiferous Tubule Culture Model: A Physiotoxicogenomic Approach

Sazan Ali[1], Gérard Steinmetz[2], Guillaume Montillet[3], Marie-Hélène Perrard[3], Anderson Loundou[4], Philippe Durand[3¤], Marie-Roberte Guichaoua[1⑨], Odette Prat[2*⑨]

1 Institut Méditerranéen de Biodiversité et d'Ecologie marine et continentale (IMBE), Centre National de la Recherche Scientifique (CNRS) UMR 7263/ Institut de Recherche pour le Développement (IRD) 237, Faculté de Médecine, Aix-Marseille Université (AMU), Marseille, France, 2 Institute of Environmental Biology and Biotechnology (IBEB), Life Science division, French Alternative Energy and Atomic Energy Commission (CEA), Marcoule, Bagnols-sur-Cèze, France, 3 Institut de Génomique Fonctionnelle de Lyon (IGFL), Centre National de la Recherche Scientifique (CNRS) UMR 5242/ Institut National de la Recherche Agronomique (INRA), Ecole Normale Supérieure de Lyon (ENS), Lyon, France, 4 Unité d'Aide Méthodologique à la Recherche clinique, Faculté de Médecine, Aix-Marseille Université (AMU), Marseille, France

Abstract

Background: Bisphenol A (BPA) is one of the most widespread chemicals in the world and is suspected of being responsible for male reproductive impairments. Nevertheless, its molecular mode of action on spermatogenesis is unclear. This work combines physiology and toxicogenomics to identify mechanisms by which BPA affects the timing of meiosis and induces germ-cell abnormalities.

Methods: We used a rat seminiferous tubule culture model mimicking the *in vivo* adult rat situation. BPA (1 nM and 10 nM) was added to the culture medium. Transcriptomic and meiotic studies were performed on the same cultures at the same exposure times (days 8, 14, and 21). Transcriptomics was performed using pangenomic rat microarrays. Immunocyto-chemistry was conducted with an anti-SCP3 antibody.

Results: The gene expression analysis showed that the total number of differentially expressed transcripts was time but not dose dependent. We focused on 120 genes directly involved in the first meiotic prophase, sustaining immunocytochemistry. Sixty-two genes were directly involved in pairing and recombination, some of them with high fold changes. Immunocytochemistry indicated alteration of meiotic progression in the presence of BPA, with increased leptotene and decreased diplotene spermatocyte percentages and partial meiotic arrest at the pachytene checkpoint. Morphological abnormalities were observed at all stages of the meiotic prophase. The prevalent abnormalities were total asynapsis and apoptosis. Transcriptomic analysis sustained immunocytological observations.

Conclusion: We showed that low doses of BPA alter numerous genes expression, especially those involved in the reproductive system, and severely impair crucial events of the meiotic prophase leading to partial arrest of meiosis in rat seminiferous tubule cultures.

Editor: Xuejiang Guo, Nanjing Medical University, China

Funding: The Research Consortium ECCOREV n° 3098 (Ecosystemes Continentaux et Risques Environnementaux) CNRS/Aix-Marseille Université funded this study (AOI 2010, grant number 7). The funders had no role in study design, data collection and analysis, decision to publish, or preparation of the manuscript.

Competing Interests: P. Durand is currently affiliated with Kallistem but his contribution to this study was made when he was an employee of the IGFL, UMR 5242 CNRS INRA Ecole Normale Supérieure de Lyon 1, F-69342 Lyon France. The authors can affirm that no conflicting interest exists in that case.

* Email: odette.prat@cea.fr

⑨ These authors contributed equally to this work.

¤ Current address: Kallistem SAS, ENS, Lyon, France

Introduction

Bisphenol A (4, 4′-isopropylidenediphenol), or BPA, is one of the world's most highly produced chemicals, used to manufacture epoxy resins and polycarbonate plastics. According to physiologically based pharmacokinetic studies, BPA is found in human serum, urine, milk and fat, with plasma levels ranging from 0.2 to 20 ng/mL (or 1 to 100 nM) [1,2,3,4,5,6]. This substance is mainly absorbed by the digestive tract. BPA is an endocrine-disrupting chemical (EDC), which could interact with both α- and β-estrogen receptors [7] and bind to androgen receptors [8]. Emerging evidence suggests that this molecule may influence multiple endocrine-related pathways [9]. Several controversies have divided scientific opinion regarding the adverse effects of BPA in the testis and reproductive organs [10,11]. Indeed, previous studies indicate that there are no reproductive effects of BPA [12,13,14,15]. Nevertheless, numerous findings suggest that BPA adversely affects the male reproductive system. Tohei et al [16] showed that bisphenol A inhibits testicular function in adult male rats. In mice, testicular hypotrophy and decreased daily sperm production were observed in the presence of BPA [17,18]. In humans, BPA also reduced sperm concentration, motility and morphology [19]. BPA exposure may also induce apoptosis in rat germ cells *in vivo* [20] and in cultured rat Sertoli cells [21], and has the potential to redistribute several known Sertoli cell junctional proteins [22,23]. Subsequent studies also demonstrated that BPA is genotoxic. The accumulation of DNA damage in germ cells was induced by BPA exposure via oxidative stress [24]. BPA causes meiotic abnormalities in oocytes [25,26,27,28,29,30] and in male germ cells of the adult rat [31]. Despite these numerous studies of the effects of BPA on sperm quality, few investigations have been conducted on the crucial meiotic step of spermatogenesis [24,32]. Thus, the molecular action of BPA on spermatogenesis remains largely unknown.

We conducted a fine analysis of the first meiotic prophase with low doses of BPA (1 nM and 10 nM), approximating levels in biological fluids [33]. Decreased efficiency of sperm production in mice appeared at a dose of 20 µg/kg/day (20 ng/g body weight/day) [34]. This study was performed using a validated rat seminiferous tubule culture model [35], able to reproduce spermatogenesis *ex vivo*. This model allows the analysis of cellular responses induced by exposure to low doses of toxic substances for three weeks. This period of time corresponds, in the rat, to the development of spermatogenesis and mimics puberty, a critical period of life with regard to endocrine disruptors [36,37].

It appears that a model "sensitive" to the possible adverse effects of chemicals is indeed of the highest importance for toxicological studies. These models must be able to respond to very low concentrations of toxicants. It must also be underlined that, in order to prevent "false-positive" results, particular attention must be paid to toxicant test concentrations, which must be realistic. Our intention was to investigate whether BPA, at the selected doses for three weeks, could alter the chronology of meiosis and induce morphological abnormalities, and to apprehend its mechanisms of action, combining toxicogenomic and physiological approaches.

Microarray-based transcriptional profiling is a powerful and ultrasensitive tool for monitoring altered cellular functions and pathways under the action of toxicants, providing a wealth of information for sketching the mode of action of toxic substances [38,39,40] or for finding new toxicity bioindicators [41]. However, to date very few transcriptome analyses have been conducted to comprehend the molecular action of BPA on spermatogenesis [42,43,44]. Using transcriptomics, we were able to detect changes in gene expression at biological doses, in the nanomolar range. We studied whether specific patterns of gene modulation could be associated with cytological changes of the first meiotic prophase, observed by immunostaining of the synaptonemal complexes (SC) with an anti-SCP3 antibody. This model allowed for immunocytochemistry and transcriptomic experiments using the same cultures at the same exposure times.

Material and Methods

Animals

The entire study was performed *ex vivo* using cultures of seminiferous tubules. For these cultures, Male 23-day-old Sprague Dawley rats from Charles River France Inc. (supplier: Janvier, France), having undergone no treatment, were used. Animals were housed 3/ cage, at temperature $21 \pm 3°C$, light cycle 12–12 (6 pm-6 am), diet made of SDS VRF1(from Special Diets Services), water filtered 0.1 µm in bottle system, sawdust bedding, autoclaved. Rats were anesthetized with chloroform then decapitated. At the age of 22–23 days the most advanced germ cells are late spermatocytes [45] allowing to study the whole meiotic phase under our culture conditions [35].

In order to counterbalance interanimal variations, testes from eight rats were pooled in every culture, and used immediately, as previously described [35].

The same population was seeded for control cultures and cultures exposed to toxicant. Analyses were performed at days 8 (D8), 14 (D14) and 21 (D21) of the cultures. All procedures were approved by the Scientific Research Agency (approval number 69306) and conducted in accordance with the guidelines for care and use of laboratory animals. The experimental protocol was designed in compliance with recommendations of the European Economic Community (EEC) (86/ 609/EEC) for the care and use of laboratory animals.

Preparation and culture of seminiferous tubules

The technique of seminiferous tubule culture has been described previously [35]. Cultures were performed with and without BPA. When required, BPA was added beginning from day 2, at 1 nM or 10 nM (Sigma-Aldrich Corporation, St. Louis, USA) in the basal compartment of the bicameral culture chamber. BPA concentrations were selected on the basis of those found in human and rat plasma, i.e. 0.2 to 20 ng/mL, meaning 1 to 100 nM [1,2,3,4,6]. 0.3% DMSO was used as the BPA dilution vehicle; the same solvent concentration was introduced in control cultures.

RNA extraction, labelling and microarray experiments

Two different pools of seminiferous tubules were exposed to two concentrations of BPA (1 nM and 10 nM) or to complete medium with vehicle (control cells) for 8, 14, and 21 days. Total RNA was extracted using the RNeasy Mini kit (Qiagen). RNAs were quantified with the Nanodrop 1000 spectrophotometer; their qualities were assessed with the Agilent 2100 Bioanalyzer. RNA samples were amplified and labeled with the cyanine-3 fluorophore using a Low Input QuickAmp Labeling Kit (Agilent). Hybridization was performed using Agilent Oligo Microarrays (Rat V3 4×44K). Fluorescence was scanned and signal data were extracted with Feature Extraction Software (Agilent).

Cytological methods

Samples treatment. Spreading, and immunocytological localization of SC axial and lateral elements, were performed according to [46]. After spreading by cytocentrifugation at 30 g, slides were fixed in 2% paraformaldehyde (Merk Darmstad, Germany). A rabbit polyclonal anti-SCP3 antibody (Abcam, Cambridge, UK Ab 15093) was used at a 1:100 dilution, to reveal axial elements and lateral elements of the SC. Detection was performed with an FITC-conjugated anti-goat immunoglobulin G (Abcam, Cambridge, UK) at a dilution of 1:100. Slides were mounted in antifade medium (Vectashield, Vector Laboratories, Burlingame, USA).

Microscope analysis. A Zeiss Axioplan 2 Fluorescence Photomicroscope (Carl Zeiss, Oberkochen, Germany) was used to observe the spermatocyte nuclei. Primary spermatocytes, stained with the anti-SCP3 antibody, were selected to evaluate the respective percentages of leptotene, zygotene, pachytene and diplotene stages: 100 to 200 nuclei were analyzed for each culture, for control cultures, and for each time and BPA dose condition. We evaluated the percentages of the three pachytene substages, P1, P2 and P3, corresponding to early, mid and late pachytene substages, in the rat [37]. These substages were defined according to the condensation degree of the sex bivalent during the pachytene stage; 50 nuclei were analyzed for each condition. The pachytene index (PI) was evaluated for each culture and for each time and BPA dose. We defined the PI in rat by the ratio P3/P1+P2+P3 [37]. The percentages of nuclei showing SC abnormalities were quantified at each time point, in both control cultures and cultures exposed to 1 nM and 10 nM BPA. For each stage, and for each abnormality, we researched a possible dose-and-time variation.

Statistical analysis

Transcriptomic analysis. In this experimental design, six independent analyses were conducted versus each specific control for the considered time point: a) 1 nM BPA-exposed cells for 8 days, b) 1 nM BPA-exposed cells for 14 days, c) 1 nM BPA-exposed cells for 21 days, d) 10 nM BPA-exposed cells for 8 days, e) 10 nM BPA-exposed cells for 14 days, f) 10 nM BPA-exposed cells for 21 days. For each analysis, eight raw fluorescence data files (four controls and four tests) were submitted to GeneSpring Software GX11 (Agilent Technologies) using a widely used method for determining the significance change of gene expression [38,47]. The fold change cutoff between control and exposed samples was set to 1.5. Genes significantly up- or downregulated were determined by an unpaired t-test, with a p-value <0.05 and a Benjamini-Hochberg false discovery rate correction. We thus obtained probe sets that were significantly induced or repressed after exposure to BPA.

Immunocytochemistry (ICC). Statistical analysis was performed using PASW Statistics Version 17.0.2 (IBM SPSS Inc., Chicago, IL, USA). Continuous variables are expressed as means±SD. Comparisons of means between two groups were performed using a Student's t-test. All tests were two-sided. The statistical significance was defined as p<0.05. Three biological replicates were analyzed for D8 and D14, and two replicates for D21. Each experiment included controls (vehicle only) and tests (BPA). The total number of nuclei analyzed was 4630, combining all doses and time points.

Biological analysis

Lists of genes significantly induced or repressed after exposure to BPA were uploaded into Ingenuity Pathway Analysis Software (IPA, Ingenuity Systems, www.ingenuity.com) for biological analysis by comparison with the Ingenuity Knowledge Database. These lists of altered genes were then processed to investigate the functional distribution of these genes, as defined by Gene Ontology. Datasets and known canonical pathways associations were measured by IPA by using a ratio (R) of the number of genes from a dataset that map to a specific pathway divided by the total number of genes that map to this canonical pathway. A Fisher's exact test was used to determine a p-value representing the significance of these associations.

Quantitative RT-PCR

Total RNA was isolated according to the manufacturer's instructions using the RNeasy Kit (Qiagen), and treated with DNase. RNA purity and concentration were determined by UV on a Nanodrop Spectrophotometer and integrity was assessed on an Agilent 2100 Bioanalyzer (Agilent Technologies). All the samples used in this study showed 28S/18S ratio signing intact and pure RNA. Differential analysis of RNA from cells exposed to

NPs and from unexposed cells was performed by qRT-PCR with the Sybr Green PCR Master Mix (Finzyme) Kit according to the manufacturer's instructions, on Opticon II (Biorad). Primer (Sigma) sequences were, for *Stra8*: 5' CAGCCTCAAAGTGG-CAGGTA 3' (forward) and 5' GGGAGAGGAGTGGGACA-GAT 3' (reverse); for *Mlh1*: 5' CGCCATGCTGGCCTTA-GATA 3' (forward) and 5' CCTCCAAAGGCGGCACATA 3' (reverse); for *Prdm9*: 5' AGAATGAGAAAGCCAACAGCA 3' (forward) and 5' AGACTCCTTAGAAGTTTTAGCAGA 3' (reverse); for *Sycp1*: 5' GAGAGAAGACCGTTGGGCA 3' (forward) and 5' TCCATTGCAAGTAAAAGCAACA 3' (reverse); for *Fpr3*: 5' ACTGTGAGCCTGGCTAGGAA 3' (forward) and 5' CTCGTGAAGCACGGCTAGAA 3' (reverse); for *Dmc1*: 5' CTTTCCGTCCAGATCGCCTT 3' (forward) and 5' AAAATGCCGGCTTCTTCGTG 3' (reverse); for *Card11*: 5' CTCAGGCCCAGTTTCTCCAG 3' (forward) and 5' CTGTTGAGCTCTGTGGAGGG 3' (reverse); for *Nfkb1*: 5' GGAGATGGCCCACTGCTATC 3' (forward) and 5' TTCGGAAGGCCTCGAATGAC 3' (reverse). For *Stra8*, *Mlh1*, *Prdm9*, *Sycp1*, *Fpr3*, *Dmc1*, *Card11*, *Nfkb1*, the amplicon sizes were 347, 200, 100, 286, 313, 160, 92, and 223 bp, respectively. The measurements were the means of six individual results and normalization was based on the total RNA mass quantified on the Nanodrop. Expression ratios were calculated according to Pfaffl et al. [48], where the relative expression ratio (R) of a target gene was calculated using PCR efficiency (E) and the CT (number of cycles at threshold) deviation of an unknown sample versus a control. The target gene fold change was then expressed as follows: $E^{(CT\ mean\ control\ -\ CT\ mean\ treatment)}$. Statistical significance was tested by Pair Wise Fixed Reallocation Randomization Test (REST software) where a p-value less than 0.05 was considered significant.

Results

Transcriptome analysis

Figure 1 indicates the number of significantly differentially expressed genes for each dose and time point. These figures encompass up- and downregulated transcripts. The number of genes affected by BPA increased markedly over the exposure time. At 1 nM BPA, this modification was time dependent. At 10 nM, there was, curiously, a decrease in the number of modulated genes at D14, but this number increased again at D 21. The entire list of significantly up- and downregulated genes for each dose and time point (FC>1.5 with p-value <0.05) is provided as supplementary material (Table S1).

Altered physiological functions and canonical pathways. We analyzed the distribution of altered genes per function, as defined in Gene Ontology (Figure 2), using Ingenuity Pathway Analysis. Radar plots helped to apprehend the complexity of toxicity, both in terms of amplitude and effect. This resulted in a specific pattern representing the toxicity for each individual dose and time point. Graphic overlay of all time points allows a visual comparison of the extension of adverse effects throughout the exposure time. The distribution of functions altered by BPA on the radar plot delineated similar patterns from D8 to D21, meaning that adverse effects were amplified with time but not with dose (Fig. 2A and 2B for 1 nM and 10 nM, respectively). Whatever the dose and time point the top three altered functions were cancer, cell death and cellular development, as shown in Fig. 2. For instance, at D21/1 nM BPA, the numbers of genes related to cancer, cell death and cellular development were 2927, 2422 and 1833, respectively. For reproductive system disease and DNA replication and repair, the numbers of genes were 746 and

Time and dose response relationship

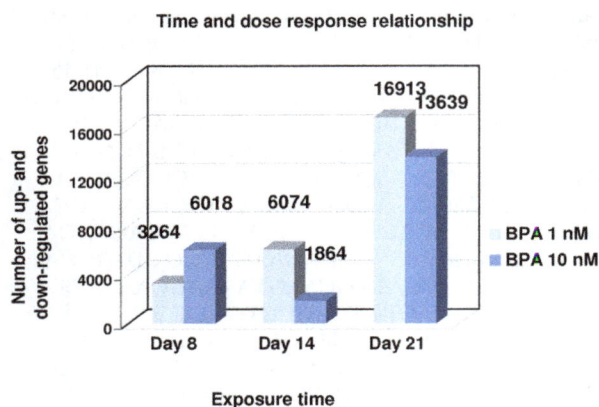

Figure 1. Total number of genes differentially up- or down-regulated by 1 nM and 10 nM BPA after 8, 14 and 21 days of exposure. Genes were selected with a fold change cutoff ≥1.5 (p-value <0.05). The global expression change compared with control cells was time but not dose dependent.

511, respectively. For other dose and time points, the numbers of genes per altered function are indicated in Figure 2.

Figure 3 shows the common canonical pathways disturbed by BPA (1 nM at D8, D14, and D21). Each canonical pathway is

constituted of a finite number of genes. For each time point, we calculated a ratio indicating the percentage of altered genes in our dataset belonging to a given canonical pathway (for precise calculation, see Material and Methods). We selected nine canonical pathways on the basis of the most significant p-values. These nine main canonical pathways were altered by BPA in a time-dependent manner. The following scores are given for D21/ 1 nM BPA, but all were altered early, at D8.

– *DNA double-strand break repair by homologous recombination,* $R = 0.56$, p-value 3.52×10^{-3}
– *LXR/RXR activation,* $R = 0.68$, p-value 4.26×10^{-3}
– *Aryl hydrocarbon receptor signaling,* $R = 0.69$, p-value 2.75×10^{-4}
– *NF-κB signaling,* $R = 0.66$, p-value 4.78×10^{-3}
– *NRF2-mediated oxidative stress response,* $R = 0.74$, p-value 1.29×10^{-8}
– *Xenobiotic metabolism signaling,* $R = 0.66$, p-value 6.76×10^{-4}
– *Apoptosis signaling,* $R = 0.72$, p-value 2.95×10^{-4}
– *Role of BRCA1 in DNA damage response,* $R = 0.63$, p-value 9.77×10^{-3}
– *Androgen signaling,* $R = 0.56$, p-value 2.59×10^{-2}

Transcription changes of genes expressed in meiotic and premeiotic cells. Of the 746 deregulated genes of the

A) BPA 1 nM

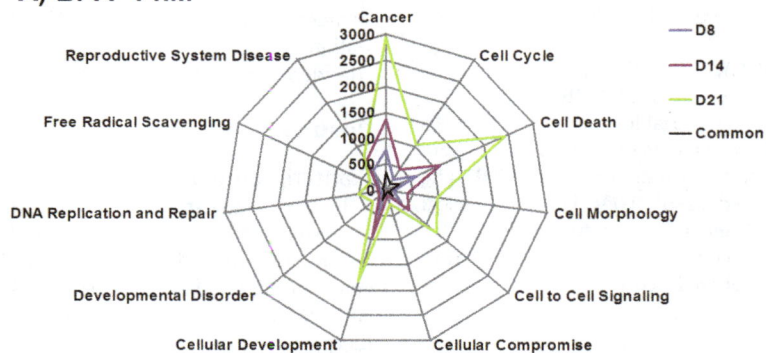

B) BPA 10 nM

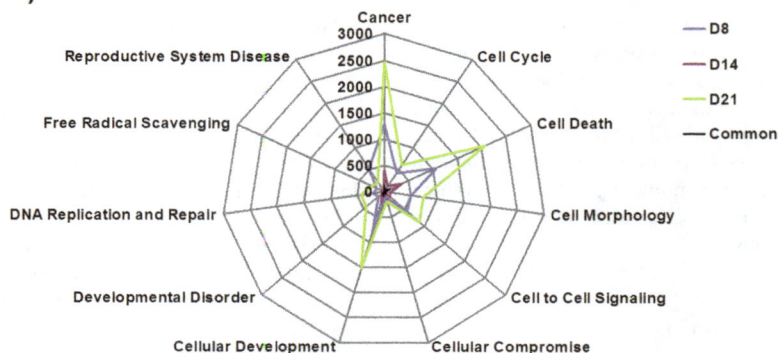

Figure 2. Distribution of differentially expressed genes per altered function. Genes significantly up- or downregulated after cell exposure to BPA for 8 days (blue), 14 days (red), or 21 days (green), and genes common to all time points (black) were examined and classified per function according to Gene Ontology. The results comprise a specific pattern of BPA toxicity, similar for all time points. The amplitude of the toxicity is given by the scale representing the number of modulated genes for each type of altered function. **A** – 1 nM BPA. **B** – 10 nM BPA.

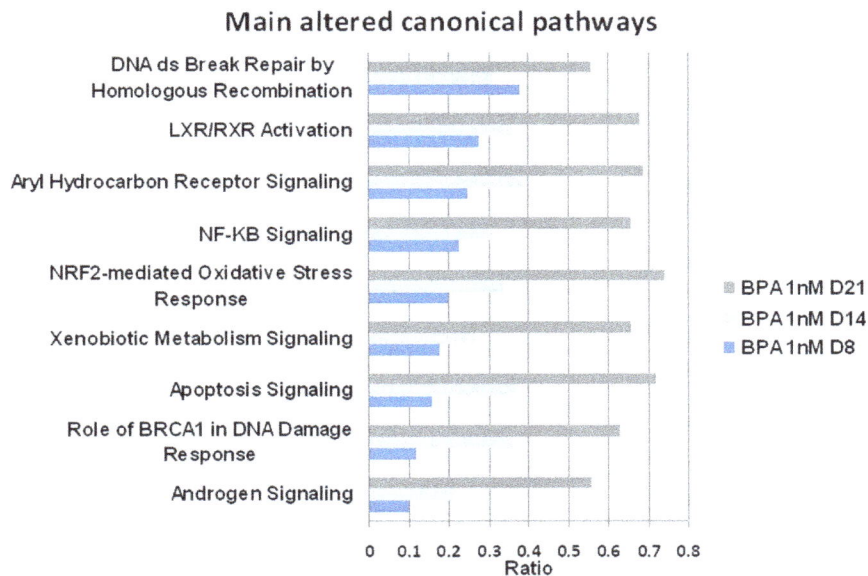

Figure 3. Comparative analysis of canonical pathways significantly altered by 1 nM BPA at D8, D14 and D21. The x-axis depicts gene ratios within a dataset mapping to the considered pathway (see Methods for calculation). A Fisher's exact test was used to determine a p-value representing the significance of these associations (p<0.01). In all cases, the ratios increased with exposure time.

reproductive system, we focused on 120 genes known to be involved in premeiotic steps or in the first meiotic prophase (Table S2). As for the total set of genes (Fig. 1), the number of BPA-affected genes increased markedly over the time of exposure, except for a decrease at D14/10 nM BPA. The highest fold changes were observed at D21/1 nM. Among these 120 genes, the number of downregulated genes (62.2%) widely exceeded the number of upregulated genes. The genes that had the greatest fold change were downregulated, except for Nos2, which was upregulated. The greatest fold change was observed for Stra8 (−37.83) which was deregulated in all conditions. Figure 4 shows that BPA deregulates genes involved in all the important processes of premeiotic steps and first meiotic prophase. The genes affected with the greatest fold change were mainly involved in meiotic initiation and recombination. Indeed, of the 120 genes, 62 were directly involved in these two functions. Some modified genes were involved in functions other than meiotic events but nevertheless essential to meiosis, such as transcriptional regulation, cell cycle, chromatin organization, protein stability, stress-induced responses and repression of retrotransposable elements. Most of this study's meiotic genes coded for nuclear proteins, some for cytoplasmic proteins, and rarely for plasma membrane proteins (Table S2). All of these genes are represented in Fig. 4. They are classified according to their respective functions in meiotic initiation, recombination and pairing. All have been shown to be interconnected in a network (Fig. 5) obtained by IPA.

qRT-PCR validation of the microarray data focused on meiosis. Quantitative RT-PCR was performed using the same batches of RNA as those evaluated by microarrays. Validation of the microarray data was investigated on five genes belonging to BPA-1 nM/D21, and on 7 genes belonging to BPA-10 nM/D21. Table 1 reports the compared fold changes obtained with microarray and qRT-PCR (p-value <0.05).

Immunocytological analysis of the meiotic prophase

Effect of BPA on the percentage of SCP3-stained meiotic stages. Each stage's normal morphological aspect of the first meiotic prophase – leptotene, zygotene, pachytene and diplotene – has been described previously [37]. The percentages of these four stages in the control cultures were of the same order of magnitude as those obtained previously. These percentages were evaluated from 100 to 200 nuclei for each culture, for every BPA dose and time condition, and for control cultures. We showed in the present study that BPA disrupted the progression of meiotic prophase in the cultures analyzed, for the two doses at the three time points.

The most obvious changes were observed at the leptotene and diplotene stages. In the treated cultures, the percentage of leptotene stage (Fig. 6A) increased for all days and concentrations compared with control cultures. This increase was at the limit of significance for 1 nM at D8 (11.4±2.8 versus 8.1±0.6, p = 0.06). The increase in leptotene stage was significant (p<0.05) for 10 nM at D8 (11.6±1.5 versus 8.1±0.6 in control), for 1 and 10 nM at D14 and D21 (D14: 13.8±1.1 and 11.7±2.3, respectively, versus 6.6±0.5 in control; D21: 13.1±0.2 and 17.6±3.8, respectively, versus 6.3±1.9 in control). In the same cultures, diplotene stage decreased for all days and concentrations compared with control cultures (Fig. 6B). The decrease in diplotene stage was not significant for 1 nM at D8. This decrease was significant (p< 0.05) for 10 nM at D8 (2.4±0.2 versus 5.8±1.5 in control), for 1 and 10 nM at D14 and D21 (D14: 3.2±1.5 and 4.2±1.5, respectively, versus 14.5±2.3 in control; D21: 2.1±1.2 and 3.2±1.5 versus 16.0±1.8 in control). Nevertheless, these changes of leptotene and diplotene stages were independent of the BPA concentration and of the exposure time.

Zygotene stage slightly decreased in the BPA-treated cultures, whereas pachytene stage slightly increased, but these variations were not significant, whatever the doses and time points.

Effect of BPA on the pachytene index. Although the percentage of pachytene stage did not vary significantly in this study for the majority of doses and time points, we observed a

DNA replication

PapolA PapolB PapolG Pola1

Enter in meiosis

Cyp26b1 Prdm9 Sohlh1
Stra8 wee1wee2

Pairing

condensins

Smc2 Smc4

Synaptonemal complex proteins

Fkbp6 Scp2L Syce1 Syce2 Sycp1
Sycp2 Sycp3 Stag3 Tex12

*Other functions in pairing,
desynapsis*

AurkB Cyp26B1 HspA2
Terc Ubr2

Checkpoints/Meiotic control

Btrc Chk1 Chk2 Fpr3
Ubr2

Pachytene checkpoint

Chk1 Atr Atm Dmc1 Hus1 Rad1
Mlh1 Msh5 Rad51 Rad54B Wee1

Apoptosis

Bax Bcl2,Bcl2A1,13,14
Card6,9,10,11 Casp1,2,3,4,6,7,8,9,14
Ccna1 Mapk14 Nos2 Ppp1Cc

Sister chromatid cohesion

Pds5A Pds5B Rad21 Rad21L Rec8
Smc1A Smc 2 Smc3 Smc5 Syn1

Genetic recombination

Atm Atr Chk1 Chk2
Hus1Spo11 MaelMei1Mei4
Msh2 Sumo2

Chk1 Chk2 Dmc1 Exo1Mre11
Msh4 Msh5 Rad51 Rad51C
Rad52 Trip13 Hormad 1,2

Brca1 Brca2 Cdk2 Dmc1 Fanca
FancD2 Gen1 Mei1 Mlh1Mlh3 Msh2
Msh3 Pms1 Rad1 Rad51 Rad51B
Rad51C Rad54B Rpa1 Rpa2 Sumo2
Sumo3 Tex11 Tex12 Tex15 Top2A
Top2B Top3A Top3B Ube2B

Other functions

Asz1 Atrx Boll Daz2 Dazap1,2 Ccna2 Ccne2
Ewsr1 Figla Hist1h2ba Hist1h2bb Nos2 Sumo1
Stmn1 Tex13B Tex 14 Tex 19.1 Tex101
Top3B

Double Strand Breaks

L

*Homology search/DNA strand exchange/
Stable Holliday Junctions*

L/Z

*DSB repair/CO formation/dissociation of
Holliday Junctions*

P

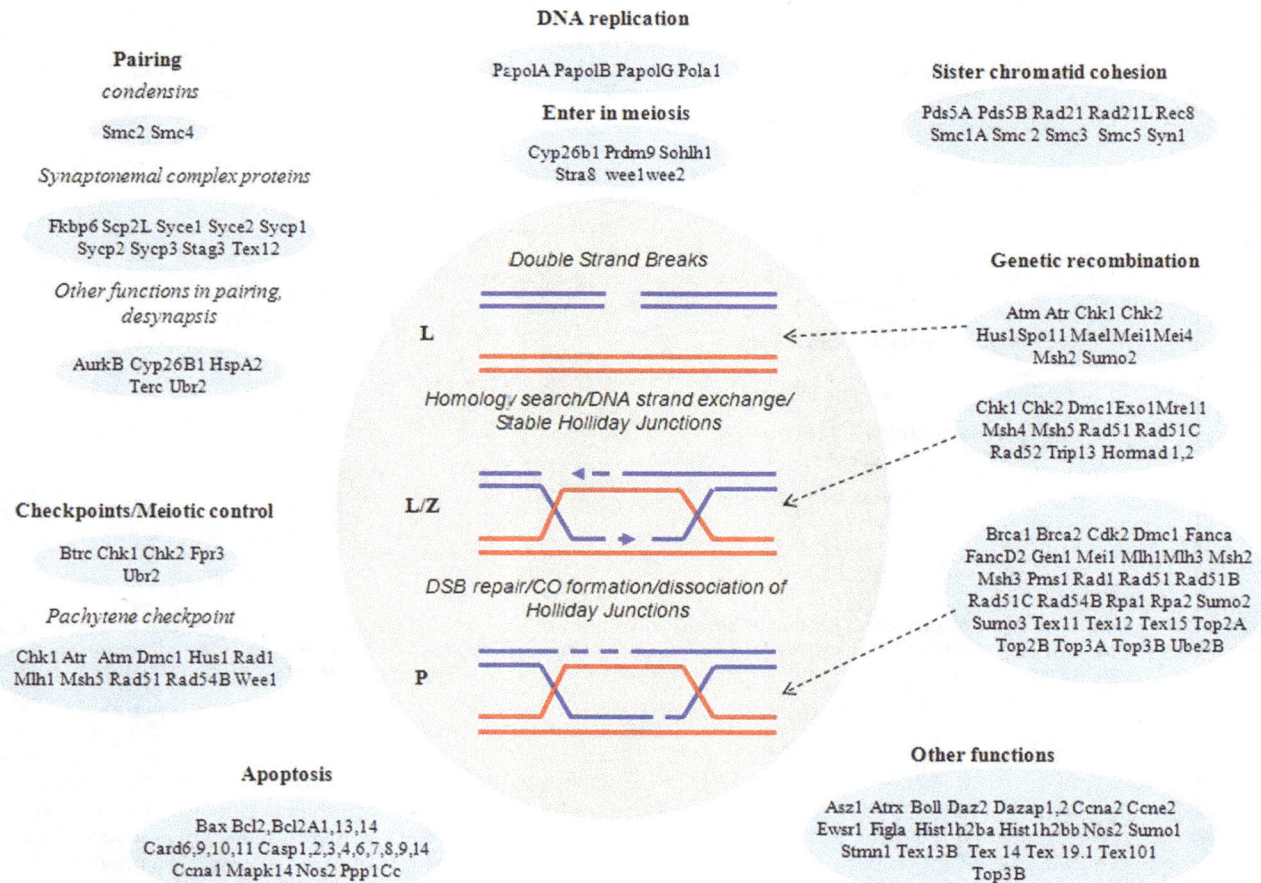

Figure 4. Diagram of the main stages of meiotic recombination and the corresponding stages of the meiotic prophase (L = leptotene, Z = zygotene and P = pachytene). The 120 genes showing a fold change ≥ 1.5 (p value ≤ 0.05) in BPA-treated cultures compared with controls, and involved in events of the meiotic prophase, were classified according to their function. This figure shows that the main functions of the first meiotic prophase are altered by BPA. Genes having several functions appear several times in this figure.

decrease of the P3 substage at all dose and time points (p<0.05). For 1 nM and 10 nM, at D8: 13.0±2.6 and 10.4±2.3, respectively, versus 22.0±2.6 in control cultures; at D14: 9.3±3.1 and 7.3±2.3, respectively, versus 24.7±3.1 in control cultures; at D21: 8.3±0.2 and 8.3±1.7, respectively, versus 29.1±2.3 in control cultures. Consequently, the PI also decreased at all doses and time points (p<0.05) (Fig. 7). For 1 nM and 10 nM, at D8: 0.13±0.03 and 0.10±0.03, respectively, versus 0.22±0.05 in control cultures; at D14: 0.06±0.03 and 0.05±0.02, respectively, versus 0.25±0.03 in control cultures; at D21: 0.08±0.007 and 0.08±0.02, respectively, versus 0.29±0.03 in control cultures.

BPA-induced axial element and SC abnormalities. SYCP3 revealed BPA-induced abnormalities of axial elements and SC at the leptotene, pachytene and diplotene stages.

– At *leptotene*, in the absence of BPA, thin and discontinuous axial cores held the nucleus area of the leptotene nuclei [36,37]. With BPA, abnormally long stretches of axial cores without indication of polarization appeared in these nuclei at both BPA concentrations and at three culture time points (Fig. 8A). For 1 nM and 10 nM at D8: 25.9±1.2 and 30.2±1.3, respectively; at D14: 22.4±2.1 and 34.9±2.1, respectively; at D21: 23.0±6.7 and 26.5±6.1, respectively.

– *At pachytene*, the prevalent abnormality observed in the presence of BPA was asynapsis, especially total asynapsis (Fig. 8B). The percentage of asynapsis increased significantly (p<0.05) for all doses and time points with no dose or time dependency. For 1 and 10 nM, at D8: 24.0±3.2 and 26.4±2.8, respectively, versus 5.2±2.2 in control cultures. At D14: 27.2±2.9 and 24.6±3.7, respectively, versus 9.1±3.7 in control cultures. At D21: 44.8±9.4 and 52.8±7.8, respectively, versus 9.5±1.4 in control cultures (Fig. 9A).

– The pulverized SC nuclei (Fig. 8C), proving apoptosis [49], significantly increased (p<0.05) for all doses and time points (Fig. 9B). For 1 nM and 10 nM at D8: 6.2±0.8 and 8.5±3.0, respectively, versus 1.9±0.5 in control cultures; D14: 11.3±1.9 and 10.0±1.0, respectively, versus 3.2±0.6, in control cultures; D21: 42.5±7.5 and 52.1±11.5 versus 20.4±1.6 for 10 nM).

– *At diplotene*, all spermatocytes contained univalents and fragmented lateral elements of SC (Fig. 8D).

Discussion

BPA effects were investigated on male meiosis, using a validated and reproducible seminiferous tubule culture model [35]. Under the present experimental ex-vivo conditions, testosterone, pro-

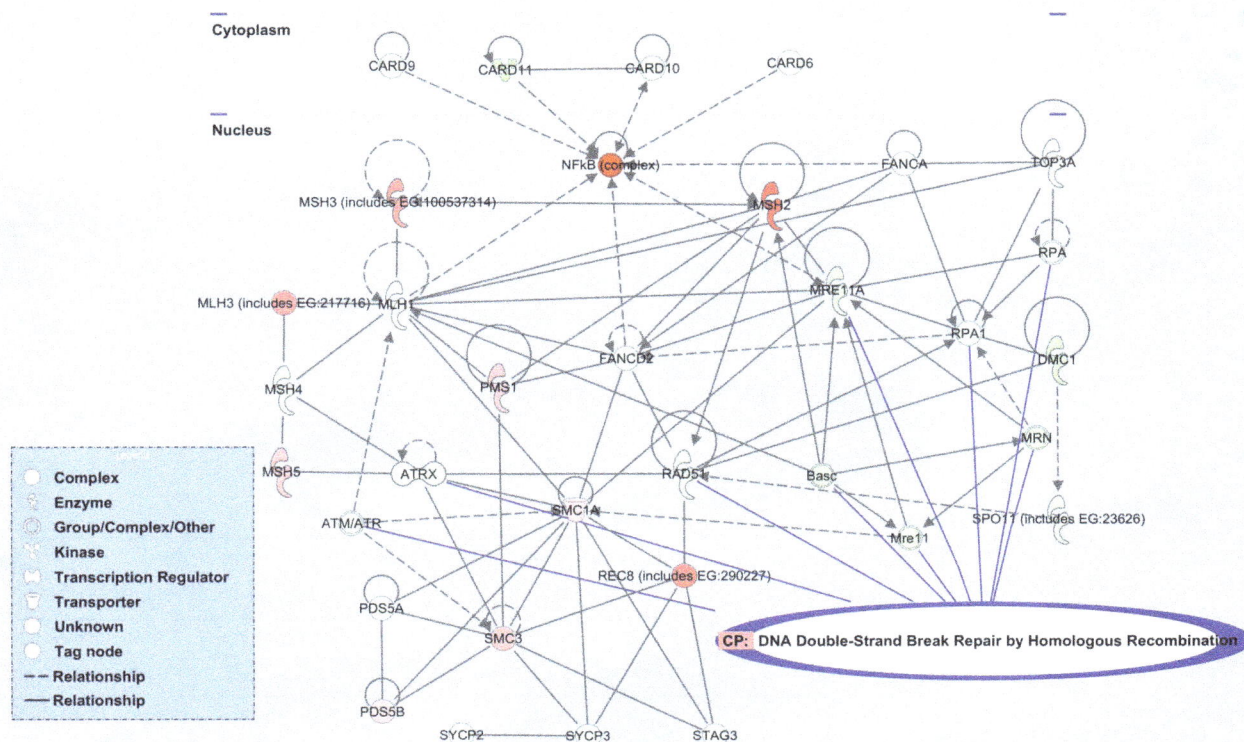

Figure 5. The top-ranked Ingenuity network identified within the group of 120 genes preceding meiotic divisions, involved in the first meiotic prophase, or essential to meiosis that were found to be differentially expressed in our datasets. This literature-based network shows the high level of connections between genes linked to DNA DSB repair in our datasets, considering all BPA doses and time points. The indicated values of fold change (induction or repression) are the maximal values found in the time course with 1 nM and 10 nM BPA. The red nodes are upregulated, the green nodes are downregulated.

duced in vivo by the Leydig cells and acting on the Sertoli cells, is added directly to the culture medium as described in Staub et al. (2002). As for the relationship between the Sertoli cells and the germ cells, we have shown previously that, under our experimental conditions, cellular junctions between the Sertoli cells and the germ cells, which are most important for germ cell differentiation, are maintained [50,51]. This model allows the analysis of induced responses by germ cell exposure to low doses of toxic substances for three weeks. This period of time corresponds, in the rat, to the development of spermatogenesis and mimics puberty, a critical period of life sensitive to endocrine disruptors. Indeed, a tiny concentration of endocrine disruptors can produce long-term adverse effects on the reproductive system [52]. Seminiferous tubules from 23-day-old Sprague-Dawley rats were used. At this age there are no round spermatids in the rat testes. Thus, we are sure that the round spermatids originate from the meiotic divisions which occurred in vitro [35]. Moreover, we previously showed similarities between the meiotic processes in vivo and ex vivo [36,37]. We performed transcriptomic analyses in BPA-treated cultures versus controls without any a priori concerning the results, varying the doses and time points. We performed germ cell ICC analyses on the same cultures, at the same doses and time points. We show that the transcriptomic results and morphological observations are consistent. The percentages of the four populations of spermatocytes under the control conditions, as well as the pachytene substages, were found in the present study to be very

similar to those described in our previous publications [36,37]. BPA concentrations were selected on the basis of the concentrations found in human and rat sera, i.e. 0.2 to 20 ng/mL, meaning 1 to 100 nM [1,2,3,4,6,33,53].

BPA alters important biological functions and canonical pathways in the seminiferous tubule cultures

The overall number of altered genes is a very good indicator of the level of cellular disturbance induced by a toxic compound [43,54,55]. Here, we observed no dose dependency in terms of number of significantly differentially expressed genes, but a time dependency (Fig. 1).

Radar plots showed that BPA alters important biological functions. We analyzed, for all doses and time points, the distribution of altered genes per function, as defined in Gene Ontology. The same functions were altered at each dose and time point, but the number of genes increased only with the exposure time (Fig. 2). Altered genes involved in cancer, cell death and cellular development were predominant. Notably, genes involved in disease of the reproductive system and in DNA replication and repair were also altered.

Although the functions described provide valuable information on the action of the involved genes, the canonical pathways help in understanding, in a faster and more drastic manner, the interactions between these genes themselves and the cellular mechanisms to which they belong. As shown in Fig. 3, the nine

Table 1. Microarray gene expression validation by qRT-PCR.

1 nM BPA, D21

Gene ID	Primers (5'-3')	Microarray fold change BPA/Ctrl	qRT-PCR fold change BPA/Ctrl
Stra8	F: CAGCCTCAAAGTGGCAGGTA	−37.8	−6.1
	R: GGGAGAGGAGTGGGACAGAT		
Mlh1	F: CGCCATGCTGGCCTTAGATA	−3.1	−1.1
	R: CCTCCAAAGGCGGCACATA		
Prdm9	F: AGAATGAGAAAGCCAACAGCA	−25.4	−5.6
	R: AGACTCCTTAGAAGTTTTAGCAGA		
Sycp1	F: GAGAGAAGACCGTTGGGCA	−9.1	−3.9
	R: TCCATTGCAAGTAAAAGCAACA		
Fpr3	F: ACTGTGAGCCTGGCTAGGAA	−25.1	−3.5
	R: CTCGTGAAGCACGGCTAGAA		

10 nM BPA, D21

Gene ID	Primers (5'-3')	Microarray fold-change BPA/Ctrl	qRT-PCR fold change BPA/Ctrl
Stra8	F: CAGCCTCAAAGTGGCAGGTA	−4.6	−6.9
	R: GGGAGAGGAGTGGGACAGAT		
Mlh1	F: CGCCATGCTGGCCTTAGATA	−2.3	−1.6
	R: CCTCCAAAGGCGGCACATA		
Prdm9	F: AGAATGAGAAAGCCAACAGCA	−3.9	−2.2
	R: AGACTCCTTAGAAGTTTTAGCAGA		
Frp3	F: ACTGTGAGCCTGGCTAGGAA	−2.6	−2.7
	R: CTCGTGAAGCACGGCTAGAA		
Dmc1	F: CTTTCCGTCCAGATCGCCTT	3.9	2.5
	R: AAAATGCCGGCTTCTTCGTG		
Card11	F: CTCAGGCCCAGTTTCTCCAG	−3.9	−11
	R: CTGTTGAGCTCTGTGGAGG3		
Nfkb1	F: GGAGATGGCCCACTGCTATC	3	5.9
	R: TTCGGAAGGCCTCGAATGAC		

Changes in the mRNA expression measured by transcriptomics and by quantitative real-time PCR. The fold changes in cells treated with BPA at 1 nM and 10 nM are expressed versus untreated cells, at D21. For qRT-PCR, the measurements were the means of six measurements (triplicates of two independent experiments) and normalization was based on the total RNA mass quantified on Nanodrop spectrophotometer. All expression levels in treated cells were significantly different from controls ($p < 0.05$). F = forward; R = reverse.

main disturbed canonical pathways in our cultures support the literature findings. The disturbance of these canonical pathways was time dependent. Apart from these observations, analyzing the direction of expression (induction or repression) of each specific transcript is difficult in the context of current knowledge. Since the mRNAs are produced in an oscillatory manner (in bursts) [56], their production is likely to obey extremely finely tuned processes, depending on many features, and should not be overinterpreted at this stage.

The *aryl hydrocarbon receptor signaling* pathway was altered as expected, as BPA is an aromatic substance. The *xenobiotic metabolism signaling* pathway by cytochrome P450 was highlighted, as expected, including P450-family genes. Tumor necrosis factor, Tnf, was strongly increased, up to 39 times at D14/1 nM BPA. Several genes of the Nfkb family were upregulated at all

doses and time points. Seventy genes encoding heat-shock proteins were mobilized. Their role is to prevent protein aggregation under stress conditions.

The strong activation of *the NRF2-mediated oxidative stress response* pathway induced the overexpression of genes encoding Phase I and II metabolizing enzymes, as well as antioxidant proteins.

The *LXR/RXR activation* pathway was highly disturbed, meaning an effect of BPA on lipid metabolism, inflammation and cholesterol metabolism through the low-density lipoprotein receptor (LdlR). Genes regulated by LXR included ATP-binding cassette transporter A1 (Abca1), which was upregulated. This observation confirms previous descriptions of the implication of BPA in lipid metabolism [57,58].

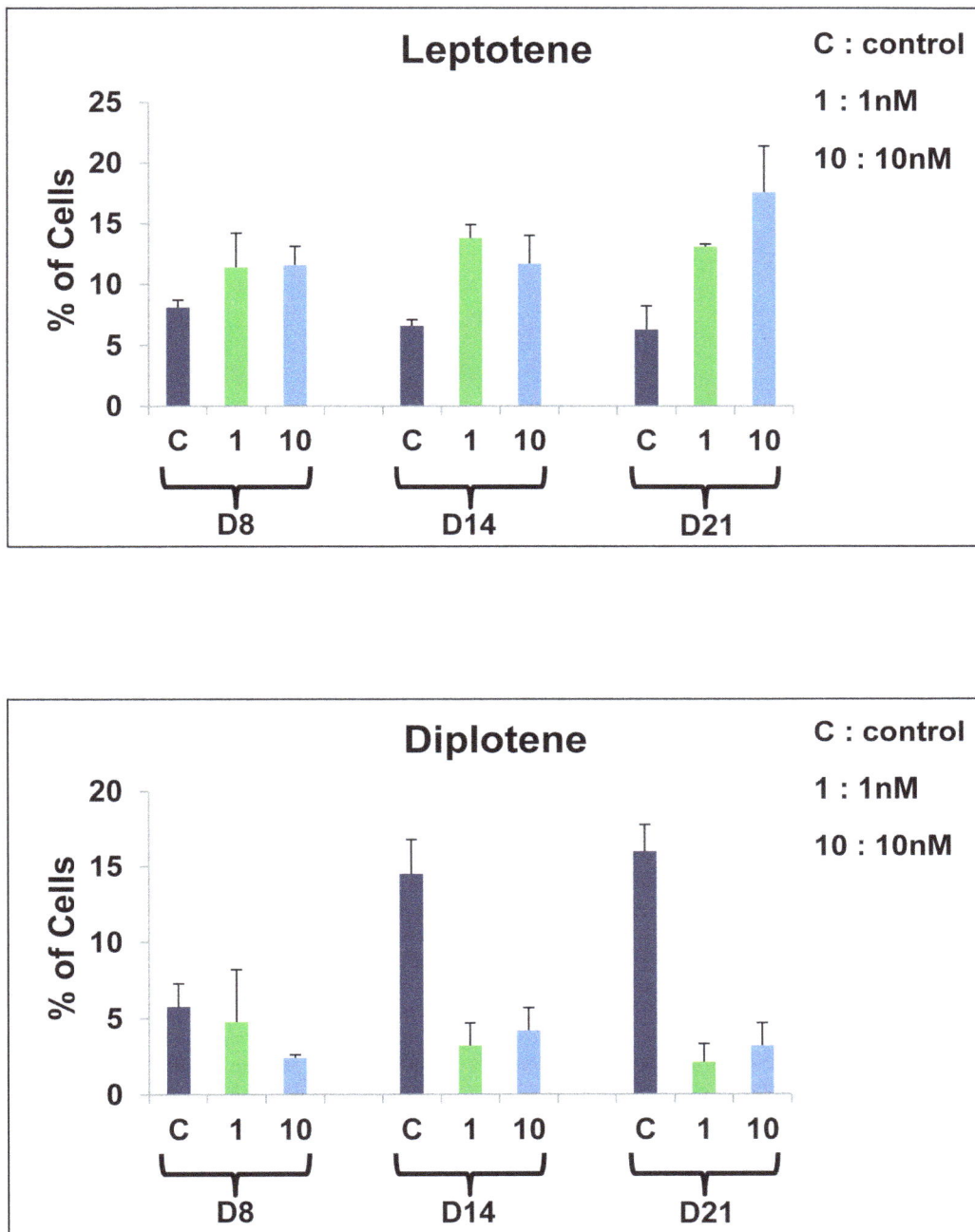

Figure 6. Effects of BPA on the percentages of leptotene and diplotene stages. A – percentage of leptotene stage increased at all doses and time points. The increase was at the limit of significance for 1 nM at D8 (p = 0.06) and significant for D14 and D21 at the two doses (p<0.05). **B** – diplotene stage percentage decreased at all doses and time points. The increase was not significant for 1 nM at D8 and was significant for D14 and D21 at the two doses (p<0.05). Each bar represents the mean ±SD (n = 3 cultures for D8 and D14, and 2 cultures for D21) versus control. C = control culture, 1 and 10 = 1 nM and 10 nM BPA, respectively.

An interesting finding was the alteration of the *Androgen signaling* pathway as proof of the role of BPA as an endocrine disruptor. At D21/1 nM BPA, the androgen receptor (AR) was downregulated (−2.7), and the androgen signaling pathway included 76 altered genes. BPA has also been described as an estrogen-like substance [9,11]. Indeed, in our hands, BPA leaves the marks of endocrine disruption. Numerous prolactin family genes were repressed, especially at 21 days of culture and more intensely at 1 nM than 10 nM. BPA altered the expression of tens of GPCR. At D21/1 nM BPA, the estrogen receptors, Esr1 and Esr2, were also significantly downregulated (−15.3 and −14.3, respectively). Members of the aldo-keto reductase family (Akr1c), and especially Akr1c12/C13 and Akr1c3, were downregulated. This late gene encodes an enzyme essential for testosterone synthesis and could explain the decreased testosterone synthesis already described with BPA *in vivo* [52].

Pachytene (PI)

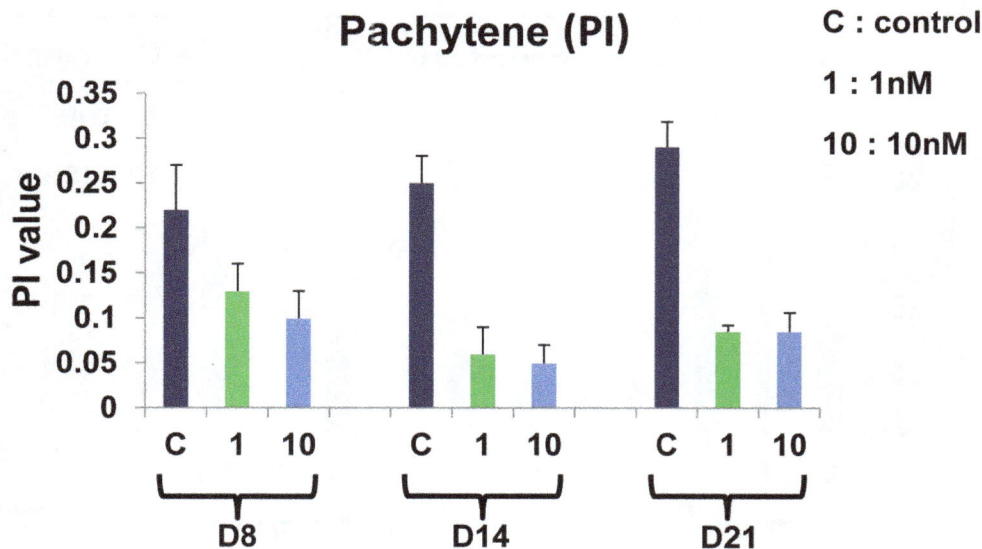

C : control

1 : 1nM

10 : 10nM

Figure 7. Effect of BPA on the pachytene stage progression. The pachytene index (PI) decreased at both dose and time points (p<0.05). C = control culture, 1 and 10 = 1 nM and 10 nM BPA, respectively. Each bar represents the mean ±SD (n = 3 cultures for D8 and D14, and 2 cultures for D21) versus control.

One can, logically, wonder why these features are visible at 1 nM and not at 10 nM. A lack of dose dependency is often a mark of endocrine disruptors, which induce nonmonotonic dose-response curves [5]. An endocrine disruptor can be equally as potent as endogenous hormones in some systems, causing biological effects at levels as low as picomolar [59]. One of the possible explanations for the absence of dose dependency might be that, after the receptor is bound by BPA and transcription of target genes has occurred, the reaction eventually must reach a plateau until the bound receptor must be inactivated [60]. As an antiandrogen, BPA can also trigger other mechanisms of action that are more complex or even antagonist to the previous one [61].

Many genes were implicated in genetic recombination, in particular the process of *DNA double-strand break repair by homologous recombination*, and are discussed below. The *role of BRCA1 in DNA damage response* is also a major altered pathway and is consistent with the many genes playing a role in cancer functions. Atm, Atr and Chk2, regulators of the DNA damage response, were downregulated in several experimental conditions (Table S1). Thus, our transcriptomic analysis on cultured seminiferous tubules supports current knowledge regarding BPA action.

Transcriptomic analysis sustains immunocytological observations

We showed in this study that ICC analysis revealed concomitant modifications of meiotic prophase chronology, defects of chromosome pairing, and induction of germ-cell apoptosis in cultures exposed to BPA. Combined transcriptomic analysis showed that BPA deregulates numerous genes involved in premeiotic steps and in the first meiotic prophase: DNA replication and sister chromatid cohesion, initiation of meiosis, pairing and genetic recombination, meiotic control and checkpoints, germ-cell apoptosis, and genes implicated in several other cellular processes essential to the life of the germ cells (Fig. 4). In the scope of this article, to explain the abnormalities revealed by ICC, we chose to focus in particular on 120 genes, preceding meiotic divisions, involved in the first meiotic prophase, or essential to meiosis (Table

S2). This table shows that differentially expressed genes are almost the same at both doses of BPA for a same time of exposure. This observation and the fact that most of these genes/proteins interact with each other as shown in Figure 5 could explain why the phenotype observed by ICC is the same for 1 and 10 nM.

Quantitative analysis of meiotic prophase revealed that BPA increases the percentage of leptotene nuclei, one-third of these showing abnormally long stretches of axial core. Abnormalities of the leptotene stage could be consistent with an alteration of cells to progress towards the zygotene stage. According to Zickler and Kleckner [62], the leptotene/zygotene transition appears to be an unusually complex and critical transition. Indeed, during this transient period the polarization of meiotic chromosome telomeres (bouquet stage) occurs, which is essentially concomitant with the onset of synaptonemal complex (SC) formation. This stage is also concomitant with the progression between DSBs and stable strand-exchange intermediates (double Holliday junctions). These observations are consistent with the microarray analysis, showing transcriptional changes of the genes involved in the early steps of genetic recombination as a homology search, DNA strand exchange and stable Holliday junction formation (Fig. 4). Thus, it is impossible for most germ cells to pass through the leptotene-zygotene transition. The most deregulated gene, Stra8 (Table S2), participates as a fundamentally positive regulator in the commitment of spermatocytes to meiosis, and regulates progression through the early stages of meiotic prophase [63]. In the study of Mark et al. [63], a small percentage of *Stra8-/-* spermatocytes progressed into the later stages of meiosis and showed a prolonged bouquet stage configuration. Leptotene nuclei with abnormally long stretches of axial core were also reported in Dmc1-/- mice [64]. In our study, Dmc1 was strongly repressed at D21/1 nM BPA and upregulated at D21/10 nM. The regulation of some key genes of meiosis was validated by qRT-PCR (Table 1).

Morphological analysis of pachytene spermatocytes indicated the prevalence of asynapsis – almost exclusively extended asynapsis. Homologous pairing failure has also been described in pachytene oocytes of C.elegans and mice exposed to BPA [25,30,65]. This abnormality is obviously related to the changes

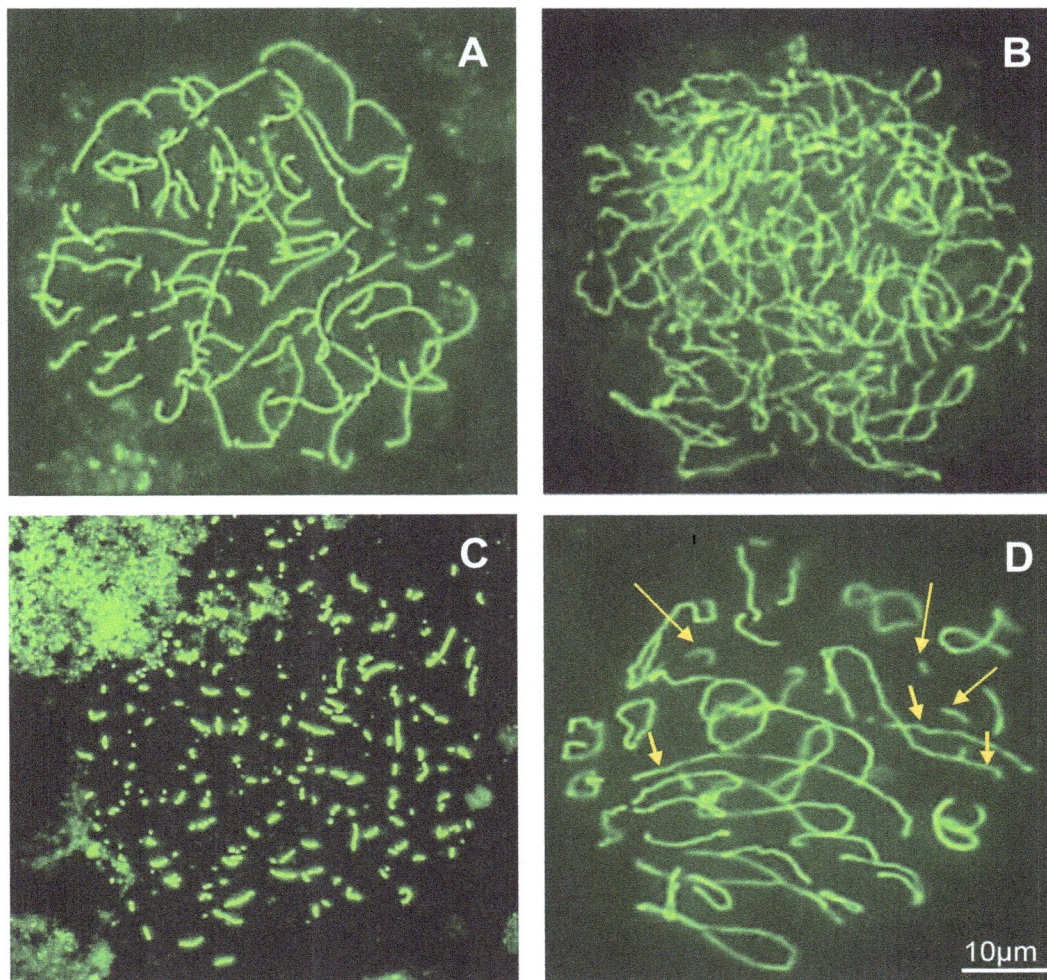

Figure 8. Pictures of abnormal spermatocyte nuclei cultured in the presence of BPA and stained with the anti-SCP3 antibody.
A – leptotene nucleus with abnormally long stretches of axial cores without indication of polarization; **B** – pachytene nucleus with total asynapsis; **C** – pachytene nucleus with pulverized synaptonemal complexes, proving apoptosis; **D** – diplotene nucleus with univalents (short arrows) and fragments (long arrows). Bar = 10 µm.

we observed in the transcription of many genes involved in homologous pairing and recombination. We previously reported [66] that high levels of extended asynapsis could arise from defects affecting these two crucial steps of meiosis. In the particular case of BPA, Allard and Colaiacovo [65] and Liu et al. [31] postulated that the DSB's repair machinery is impaired. In the present study, we showed that several genes involved in this process were deregulated by BPA (Table S2, Fig. 4). Several members of the Rec/Rad-51 family, which is known to play an important role in DNA repair by homologous recombination, were up- or down-regulated, with a high fold change for Rad51C. Genes involved in the DNA damage response, such as Brca1, Atm, Atr and Chk2, were deregulated. Nevertheless, other impaired processes may be involved in asynapsis formation. We showed that several genes involved in DSB generation (Spo11), in homology search and DNA strand exchange, and in the formation of Holliday junctions (Exo1, Hormad2, Msh4, Rad51C, Rad52, Tex15) were strongly deregulated. The gene, Fpr3, which functions in a checkpoint-like manner to ensure that chromosome synapsis is contingent on the initiation of recombination, was strongly underexpressed. Of the

proteins of genetic recombination, the SC proteins can be directly involved in the mechanisms of asynapsis. Indeed, the corresponding genes, Sycp1, Sycp2, Sycp2L and Sycp3, were clearly downregulated. It has also been reported that Stra8 plays a direct role in SC assembly [63]. The above discussion about asynapsed pachytene could explain the presence of univalents at the diplotene stage. Deregulation of genes which stabilize the Holliday junction (Msh4, Msh5) or maintain genomic integrity during DNA replication and recombination (Mlh1, Mlh3) could be involved in diplotene abnormalities. We emphasize that all these genes interact with each other. Their interconnections were visualized in a literature-based network showing the high level of connections between differentially expressed genes linked to DNA DSB repair in our datasets (Fig. 5). Thus, the phenotypes observed with ICC could result from the action of several of these interconnected genes and probably other genes expressed in germ cells and whose functions are not yet clearly defined.

Expression changes of genes revealed in our study could also explain some published data. We saw above that the increased number of Rad51 foci observed by Allard and Colaiacovo [65]

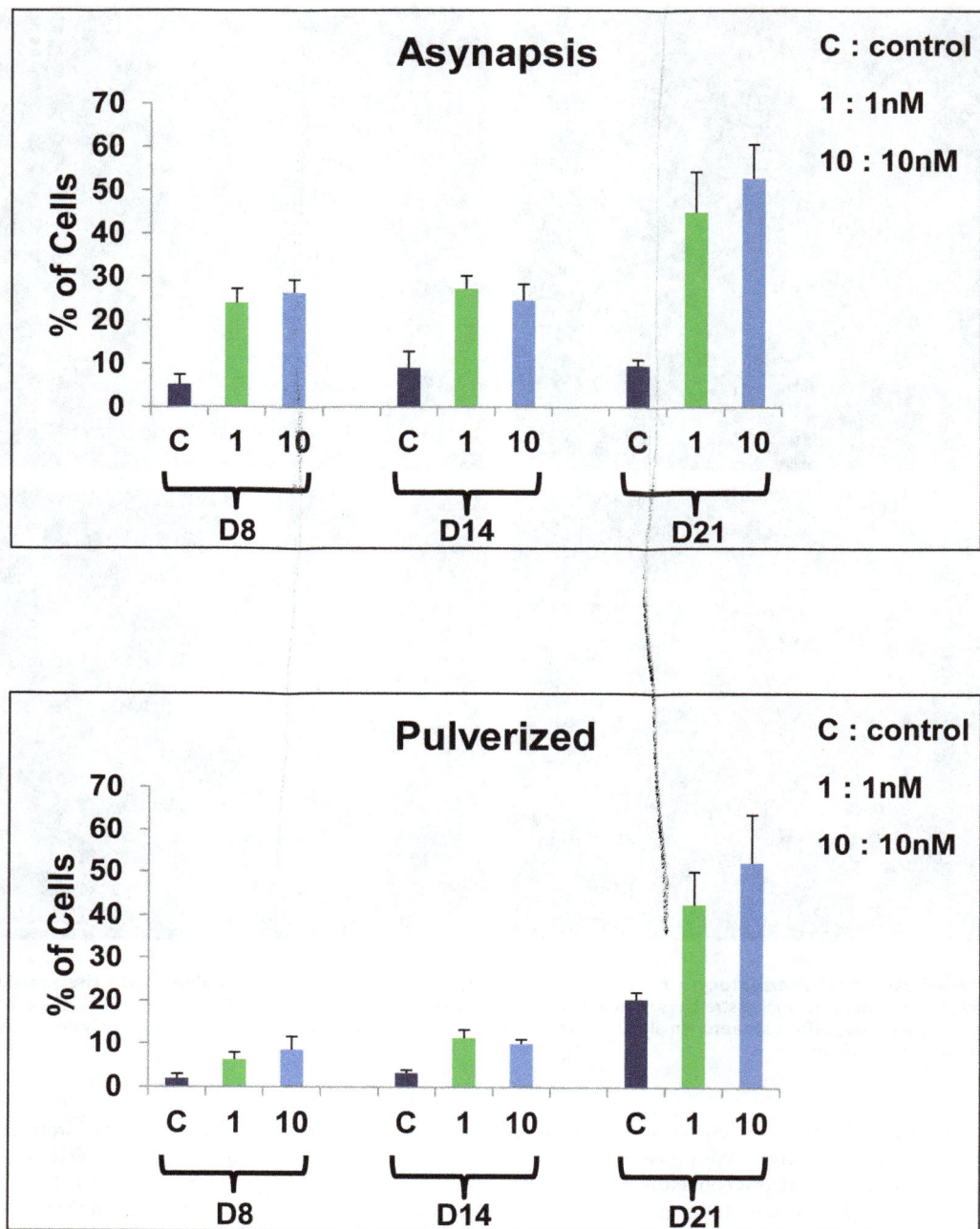

Figure 9. Changes of synaptonemal complex (SC) abnormality percentages during the pachytene stage in BPA-cultured spermatocytes. A – the percentage of nuclei with asynapsed SC increases at each dose and time point (p<0.05). **B** – the percentage of nuclei with pulverized SC (apoptotic nuclei) increases at each dose and time point (p<0.05). Each bar represents the mean ±SD (n = 3 cultures for D8 and D14, and 2 cultures for D21) versus control. **C** = control culture, 1 and 10 = 1 nM and 10 nM BPA, respectively.

might be consistent with transcriptional changes of the Rec/Rad51 family. The increased number and modified distribution of Mlh1 foci reported in pachytene oocytes following BPA exposure [25,30] might be consistent with the strong downregulation of Prdm9, a major player in hotspot specification [67].

Morphological analysis revealed the presence of apoptotic (pulverized) spermatocytes (Fig. 8C) whose percentage significantly increased at D21. Early features of apoptosis were also observed with the transcriptomic analysis. Radar plots (Fig. 2) show that cell

death is one of the most represented functions within the deregulated genes. Among the BPA-responsive apoptosis genes, we identified genes from the Card and Casp families. The antiapoptotic gene, Bcl2, was downregulated whereas the proapoptotic gene, Bax, was upregulated (Table S2). In addition Nfκb, which modulates the differentially expressed Card-family genes, was strongly induced.

Apoptosis might be induced in cells that fail to recombine and/or pair their homologous chromosomes [68]. Several checkpoints that monitor the progression of meiotic recombination are activated in response to the unrepaired meiotic DSBs. For example, Atm and Atr are checkpoint kinases that regulate meiotic DSB repair; both were downregulated in our study. Thus, the accumulation of apoptotic spermatocytes coupled with a decrease of the pachytene index is an indication of pachytene checkpoint activation by BPA. We have localized this checkpoint at the end of the P2 pachytene substage in humans [66] and rats [37]. The majority of asynapsed spermatocytes could, thus, be eliminated in this way. This could explain the decreasing percentage of diplotene nuclei in our BPA-exposed cultures.

These results demonstrate the interest of the analysis of SC in toxicological studies because they underline the specificities of each toxic substance. SC analysis is a highly sensitive indicator of potentially heritable effects of genotoxic agents [69]. All agents tested by Allen et al. [70,71] and Backer et al. [69] caused dose-dependent SC damage, which varied with the chemicals. We also previously showed with the same culture model that Cr (VI) treatment led to SC fragmentation whereas Cd treatment induced moth-eaten SC [36,37]. Presently, we show that BPA alters meiotic cell progression and increases asynapsis, without dose dependency.

Does BPA produce aneuploid gametes?

The pachytene checkpoint prevents chromosome missegregation by eliminating asynapsed pachytene spermatocytes that would lead to the production of aneuploid gametes [68]. Nevertheless, as shown in Fig. 4, BPA elicited expression changes of 11 genes implicated in pachytene checkpoint function, leading to a failure of checkpoint activation that could alleviate the meiotic arrest at this point.

Moreover, we previously showed that the pachytene checkpoint was not an absolute barrier [66], abnormal meiotic cells being able to complete spermatogenesis. In the present study, the presence of diplotene spermatocytes with univalents leads to the assumption that the existence of aneuploid metaphases II cannot be excluded. If so, low-level BPA exposure could induce errors in chromosome segregation and could produce aneuploid germ cells. Consistent with this hypothesis, studies of oocyte meiosis from female mice exposed to BPA indicate that BPA can affect chromosome segregation by disturbing synapsis and recombination [28,30]. A second mechanism involving the cell division machinery was suspected to explain a potential aneugenic effect of BPA [26]. It was demonstrated that BPA alters the centrosome dynamic and increases the number of mitotic and meiotic spindles with unaligned chromosomes. Nevertheless, according to Eichenlaub-Ritter et al. [27,72], low-level chronic BPA exposure does not appear to pose a risk for the induction of errors in chromosome segregation at first meiosis in mouse oocytes. These authors preferentially suggested that BPA induced meiotic arrest. However, we did not find in the literature any sperm chromosomal analyses of BPA-exposed male rodents to demonstrate the existence of aneuploidy.

According to Hunt et al. [73], studies in rodents allow predictions about humans to be made regarding reproductive effects of EDCs. Chalmel et al. [74] reported a cross-species expression profile between rodents and humans. Thus, our cytological and transcriptional results could be a predictor of the deleterious effects of low-dose BPA on human spermatogenesis.

Another point is that, although *ex vivo* models might be questionable for their lack of biotransformation and clearance compared to *in vivo* models, they do nevertheless represent a good alternative to animal testing regarding the necessary reproductive toxicity assays of thousands of chemicals.

Conclusion

This study provides arguments for the deleterious effects of BPA at low doses on male germ cells, by combining transcriptomic analyses and immunocytochemistry in an *ex vivo* rat seminiferous tubule culture model. Transcriptomic analyses showed that BPA altered the expression of genes involved in events preceding meiosis, its initiation and progression. Of the numerous genes differentially expressed by low-dose BPA exposure, we focused on 120 premeiotic and meiotic genes; some showed very elevated fold changes. Nevertheless we did not observe any dose dependency between 1 nM and 10 nM with both the techniques used. Only the gene expression analysis underlined a time dependency between D8, D14 and D21. Immunocytochemistry showed that low-dose BPA had deleterious effects on meiotic progression, and that the main alterations induced by BPA are asynapsis and apoptosis. These results bring additional arguments for the hypothesis that BPA alters pairing and recombination. Moreover, many differentially expressed genes were also involved in other important physiological functions, corroborating published findings, such as the triggering of xenobiotic metabolism, the disturbance of lipid metabolism, and endocrine disruption. Further analysis of our transcriptomic data could help to provide candidates for predictive biomarkers of meiotic abnormalities related to the toxic effects of chemicals on spermatogenesis.

Supporting Information

Table S1   List of genes up- or down-regulated at 1 nM and 10 nM BPA (D8, D14 and D21).

Table S2   List of the 120 genes differentially expressed under BPA exposure involved in the premeiotic and first meiotic prophase. Fold change values and cellular localization (N nuclear, C cytoplasm, PM plasma membrane, Un undertermined), are reported for the two BPA concentrations (1 and 10 nM) and at the three time-points (D8, D14 and D21). Red = up-regulated genes; green = down-regulated genes. The number of deregulated genes is time dependent but not dose-dependent.

Acknowledgments

We thank the Research Consortium ECCOREV n° 3098 (Ecosystemes Continentaux et Risques Environnementaux) CNRS / Aix-Marseille Université who funded this study (AOI 2010, grant number 7). We acknowledge Marc Fraterno for his help in creation of figure 8 and Thierry Orsière for fruitful discussions.

Author Contributions

Conceived and designed the experiments: OP MG MHP PD. Performed the experiments: SA GS GM. Analyzed the data: OP MG. Contributed reagents/materials/analysis tools: OP AL. Contributed to the writing of the manuscript: OP MG PD.

References

1. Ikezuki Y, Tsutsumi O, Takai Y, Kamei Y, Taketani Y (2002) Determination of bisphenol A concentrations in human biological fluids reveals significant early prenatal exposure. Hum Reprod 17: 2839–2841.

2. Takeuchi T, Tsutsumi O (2002) Serum bisphenol a concentrations showed gender differences, possibly linked to androgen levels. Biochem Biophys Res Commun 291: 76–78.

3. Takeuchi T, Tsutsumi O, Ikezuki Y, Takai Y, Taketani Y (2004) Positive relationship between androgen and the endocrine disruptor, bisphenol A, in normal women and women with ovarian dysfunction. Endocr J 51: 165–169.

4. Takeuchi T, Tsutsumi O, Nakamura N, Ikezuki Y, Takai Y, et al. (2004) Gender difference in serum bisphenol A levels may be caused by liver UDP-glucuronosyltransferase activity in rats. Biochem Biophys Res Commun 325: 549–554.

5. Vandenberg LN, Chahoud I, Heindel JJ, Padmanabhan V, Paumgartten FJ, et al. (2012) Urinary, circulating, and tissue biomonitoring studies indicate widespread exposure to bisphenol A. Cien Saude Colet 17: 407–434.

6. Zhang HQ, Zhang XF, Zhang LJ, Chao HH, Pan B, et al. (2012) Fetal exposure to bisphenol A affects the primordial follicle formation by inhibiting the meiotic progression of oocytes. MolBiolRep 39: 5651–5657.

7. Kuiper GG, Lemmen JG, Carlsson B, Corton JC, Safe SH, et al. (1998) Interaction of estrogenic chemicals and phytoestrogens with estrogen receptor beta. Endocrinology 139: 4252–4263.

8. Sohoni P, Sumpter JP (1998) Several environmental oestrogens are also anti-androgens. JEndocrinol 158: 327–339.

9. Rubin BS (2011) Bisphenol A: an endocrine disruptor with widespread exposure and multiple effects. J Steroid Biochem Mol Biol 127: 27–34.

10. Vandenberg LN, Hunt PA, Myers JP, Vom Saal FS (2013) Human exposures to bisphenol A: mismatches between data and assumptions. Rev Environ Health 28: 37–58.

11. Vandenberg LN, Maffini MV, Sonnenschein C, Rubin BS, Soto AM (2009) Bisphenol-A and the great divide: a review of controversies in the field of endocrine disruption. Endocr Rev 30: 75–95.

12. Ashby J, Tinwell H, Lefevre P, Joiner R, Haseman J (2003) The effect on sperm production in adult Sprague-Dawley rats exposed by gavage to bisphenol A between postnatal days 91-97. Toxicol Sci 2003 74(1): 129–138.

13. Howdeshell KL, Furr J, Lambright CR, Wilson VS, Ryan BC, et al. (2008) Gestational and lactational exposure to ethinyl estradiol, but not bisphenol A, decreases androgen-dependent reproductive organ weights and epididymal sperm abundance in the male long evans hooded rat. Toxicol Sci 102: 371–382.

14. LaRocca J, Boyajian A, Brown C, Smith SD, Hixon M (2011) Effects of in utero exposure to Bisphenol A or diethylstilbestrol on the adult male reproductive system. Birth Defects Res B Dev Reprod Toxicol 92: 526–533.

15. Tyl R, Myers C, Marr M, Thomas B, Keimowitz A, et al. (2002) Three-generation reproductive toxicity study of dietary bisphenol A in CD Sprague-Dawley rats. Toxicol Sci 68(1): 121–146.

16. Tohei A, Suda S, Taya K, Hashimoto T, Kogo H (2001) Bisphenol A inhibits testicular functions and increases luteinizing hormone secretion in adult male rats. Exp Biol Med (Maywood) 226: 216–221.

17. Aikawa H, Koyama S, Matsuda M, Nakahashi K, Akazome Y, et al. (2004) Relief effect of vitamin A on the decreased motility of sperm and the increased incidence of malformed sperm in mice exposed neonatally to bisphenol A. Cell Tissue Res 315: 119–124.

18. Al-Hiyasat AS, Darmani H, Elbetieha AM (2002) Effects of bisphenol A on adult male mouse fertility. EurJOral Sci 110: 163–167.

19. Meeker JD, Ehrlich S, Toth TL, Wright DL, Calafat AM, et al. (2010) Semen quality and sperm DNA damage in relation to urinary bisphenol A among men from an infertility clinic. Reprod Toxicol 30: 532–539.

20. Qiu LL, Wang X, Zhang XH, Zhang Z, Gu J, et al. (2013) Decreased androgen receptor expression may contribute to spermatogenesis failure in rats exposed to low concentration of bisphenol A. Toxicol Lett 219: 116–124.

21. Iida H, Maehara K, Doiguchi M, Mori T, Yamada F (2003) Bisphenol A-induced apoptosis of cultured rat Sertoli cells. Reprod Toxicol 17: 457–464.

22. Fiorini C, Tilloy-Ellul A, Chevalier S, Charuel C, Pointis G (2004) Sertoli cell junctional proteins as early targets for different classes of reproductive toxicants. ReprodToxicol 18: 413–421.

23. Li MW, Mruk DD, Lee WM, Cheng CY (2010) Connexin 43 is critical to maintain the homeostasis of the blood-testis barrier via its effects on tight junction reassembly. ProcNatlAcadSciUSA 107: 17998–18003.

24. Wu HJ, Liu C, Duan WX, Xu SC, He MD, et al. (2013) Melatonin ameliorates bisphenol A-induced DNA damage in the germ cells of adult male rats. Mutat Res 752: 57–67.

25. Brieno-Enriquez MA, Robles P, Camats-Tarruella N, Garcia-Cruz R, Roig I, et al. (2011) Human meiotic progression and recombination are affected by Bisphenol A exposure during in vitro human oocyte development. HumReprod 26: 2807–2818.

26. Can A, Semiz O, Cinar O (2005) Bisphenol-A induces cell cycle delay and alters centrosome and spindle microtubular organization in oocytes during meiosis. Molecular Human Reproduction 11: 389–396.

27. Eichenlaub-Ritter U, Vogt E, Cukurcam S, Sun F, Pacchierotti F, et al. (2008) Exposure of mouse oocytes to bisphenol A causes meiotic arrest but not aneuploidy. Mutat Res 651: 82–92.

28. Hunt PA, Koehler KE, Susiarjo M, Hodges CA, Ilagan A, et al. (2003) Bisphenol a exposure causes meiotic aneuploidy in the female mouse. CurrBiol 13: 546–553.

29. Lenie S, Cortvrindt R, Eichenlaub-Ritter U, Smitz J (2008) Continuous exposure to bisphenol A during in vitro follicular development induces meiotic abnormalities. Mutat Res 651: 71–81.

30. Susiarjo M, Hassold TJ, Freeman E, Hunt PA (2007) Bisphenol A exposure in utero disrupts early oogenesis in the mouse. PLoSGenet 3: e5.

31. Liu C, Duan W, Li R, Xu S, Zhang L, et al. (2013) Exposure to bisphenol A disrupts meiotic progression during spermatogenesis in adult rats through estrogen-like activity. Cell Death Dis 4: e676.

32. Liu C, Duan W, Zhang L, Xu S, Li R, et al. (2014) Bisphenol A exposure at an environmentally relevant dose induces meiotic abnormalities in adult male rats. Cell Tissue Res 355(1): 223–232.

33. Vandenberg LN, Chahoud I, Heindel JJ, Padmanabhan V, Paumgarten FJ, et al. (2010) Urinary, circulating, and tissue biomonitoring studies indicate widespread exposure to bisphenol A. Environ Health Perspect 118: 1055–1070.

34. vom Saal FS, Cooke PS, Buchanan DL, Palanza P, Thayer KA, et al. (1998) A physiologically based approach to the study of bisphenol A and other estrogenic chemicals on the size of reproductive organs, daily sperm production, and behavior. ToxicolIndHealth 14: 239–260.

35. Staub C, Hue D, Nicolle JC, Perrard-Sapori MH, Segretain D, et al. (2000) The whole meiotic process can occur in vitro in untransformed rat spermatogenic cells. ExpCell Res 260: 85–95.

36. Geoffroy-Siraudin C, Perrard MH, Chaspoul F, Lanteaume A, Gallice P, et al. (2010) Validation of a rat seminiferous tubule culture model as a suitable system for studying toxicant impact on meiosis effect of hexavalent chromium. ToxicolSci 116: 286–296.

37. Geoffroy-Siraudin C, Perrard MH, Ghalamoun-Slaimi R, Ali S, Chaspoul F, et al. (2012) Ex-vivo assessment of chronic toxicity of low levels of cadmium on testicular meiotic cells. ToxicolApplPharmacol 262: 238–246.

38. Fisichella M, Berenguer F, Steinmetz G, Auffan M, Rose J, et al. (2012) Intestinal toxicity evaluation of TiO2 degraded surface-treated nanoparticles: a combined physico-chemical and toxicogenomics approach in caco-2 cells. Part Fibre Toxicol 9: 1743–8977.

39. Guyton K, Kyle A, Aubrecht J, Cogliano V, Eastmond D, et al. (2009) Improving prediction of chemical carcinogenicity by considering multiple mechanisms and applying toxicogenomic approaches. Mutat Res 681(2–3): 230–240.

40. Hartung T, McBride M (2011) Food for Thought ... on mapping the human toxome. ALTEX 28(2): 83–93.

41. Prat O, Berenguer F, Steinmetz G, Ruat S, Sage N, et al. (2010) Alterations in gene expression in cultured human cells after acute exposure to uranium salt: Involvement of a mineralization regulator. Toxicol In Vitro 24: 160–168.

42. Lopez-Casas PP, Mizrak SC, Lopez-Fernandez LA, Paz M, de Rooij DG, et al. (2012) The effects of different endocrine disruptors defining compound-specific alterations of gene expression profiles in the developing testis. Reprod Toxicol 33: 106–115.

43. Naciff JM, Hess KA, Overmann GJ, Torontali SM, Carr GJ, et al. (2005) Gene expression changes induced in the testis by transplacental exposure to high and low doses of 17{alpha}-ethynyl estradiol, genistein, or bisphenol A. Toxicol Sci 86: 396–416.

44. Tainaka H, Takahashi H, Umezawa M, Tanaka H, Nishimune Y, et al. (2012) Evaluation of the testicular toxicity of prenatal exposure to bisphenol A based on microarray analysis combined with MeSH annotation. J Toxicol Sci 37: 539–548.

45. Clermont Y (1972) Kinetics of spermatogenesis in mammals: seminiferous epithelium cycle and spermatogonial renewal. Physiol Rev 52: 198–236.

46. Metzler-Guillemain C, Guichaoua MR (2000) A simple and reliable method for meiotic studies on testicular samples used for intracytoplasmic sperm injection. FertilSteril 74: 916–919.

47. Wright WR, Parzych K, Crawford D, Mein C, Mitchell JA, et al. (2012) Inflammatory transcriptome profiling of human monocytes exposed acutely to cigarette smoke. PLoS One 7: 17.

48. Pfaffl MW, Horgan GW, Dempfle L (2002) Relative expression software tool (REST) for group-wise comparison and statistical analysis of relative expression results in real-time PCR. Nucleic Acids Res 30: e36.

49. Longepied G, Saut N, Aknin-Seifer I, Levy R, Frances AM, et al. (2010) Complete deletion of the AZFb interval from the Y chromosome in an oligozoospermic man. Hum Reprod 25: 2655–2663.

50. Gilleron J, Carette D, Durand P, Pointis G, Segretain D (2009) Connexin 43 a potential regulator of cell proliferation and apoptosis within the seminiferous epithelium. Int J Biochem Cell Biol 41(6): 1381–1390.

51. Godet M, Sabido O, Gilleron J, Durand P (2008) Meiotic progression of rat spermatocytes requires mitogen-activated protein kinases of Sertoli cells and close contacts between the germ cells and the Sertoli cells. Dev Biol 315 (1): 173–188.

52. Della Seta D, Minder I, Belloni V, Aloisi A, Dessi-Fulgheri F, et al. (2006) Pubertal exposure to estrogenic chemicals affects behavior in juvenile and adult male rats. Horm Behav 50(2): 301–307.

53. Shin BS, Kim CH, Jun YS, Kim DH, Lee BM, et al. (2004) Physiologically based pharmacokinetics of bisphenol A. J Toxicol Environ Health A 67: 1971–1985.

54. Lobenhofer EK, Cui X, Bennett L, Cable PL, Merrick BA, et al. (2004) Exploration of low-dose estrogen effects: identification of No Observed Transcriptional Effect Level (NOTEL). Toxicol Pathol 32: 482–492.

55. Ludwig S, Tinwell H, Schorsch F, Cavaille C, Pallardy M, et al. (2011) A molecular and phenotypic integrative approach to identify a no-effect dose level for antiandrogen-induced testicular toxicity. Toxicol Sci 122: 52–63.

56. Raj A, Peskin CS, Tranchina D, Vargas DY, Tyagi S (2006) Stochastic mRNA synthesis in mammalian cells. PLoS Biol 4: e309.

57. Miyawaki J, Sakayama K, Kato H, Yamamoto H, Masuno H (2007) Perinatal and postnatal exposure to bisphenol a increases adipose tissue mass and serum cholesterol level in mice. J Atheroscler Thromb 14: 245–252.

58. Rubin BS, Soto AM (2009) Bisphenol A: Perinatal exposure and body weight. Mol Cell Endocrinol 304: 55–62.

59. Welshons WV, Thayer KA, Judy BM, Taylor JA, Curran EM, et al. (2003) Large effects from small exposures. I. Mechanisms for endocrine-disrupting chemicals with estrogenic activity. Environ Health Perspect 111: 994–1006.

60. Vandenberg LN, Colborn T, Hayes TB, Heindel JJ, Jacobs DR, Jr., et al. (2012) Hormones and endocrine-disrupting chemicals: low-dose effects and nonmonotonic dose responses. Endocr Rev 33: 378–455.

61. Kortenkamp A, Faust O, Evans R, McKinlay R, Orton F, et al. (2011) State of the Art Assessment of Endocrine Disrupters. European Commission document Final report.

62. Zickler D, Kleckner N (1998) The leptotene-zygotene transition of meiosis. AnnuRevGenet 32: 619–697.

63. Mark M, Jacobs H, Oulad-Abdelghani M, Dennefeld C, Feret B, et al. (2008) STRA8-deficient spermatocytes initiate, but fail to complete, meiosis and undergo premature chromosome condensation. J Cell Sci 121: 3233–3242.

64. Pittman DL, Cobb J, Schimenti KJ, Wilson LA, Cooper DM, et al. (1998) Meiotic prophase arrest with failure of chromosome synapsis in mice deficient for Dmc1, a germline-specific RecA homolog. MolCell 1: 697–705.

65. Allard P, Colaiacovo MP (2010) Bisphenol A impairs the double-strand break repair machinery in the germline and causes chromosome abnormalities. ProcNatlAcadSciUSA 107: 20405–20410.

66. Guichaoua MR, Perrin J, Metzler-Guillemain C, Saias-Magnan J, Giorgi R, et al. (2005) Meiotic anomalies in infertile men with severe spermatogenic defects. HumReprod 20: 1897–1902.

67. Baudat F, Buard J, Grey C, de MB (2010) Prdm9, a key control of mammalian recombination hotspots. MedSci(Paris) 26: 468–470.

68. Roeder GS, Bailis JM (2000) The pachytene checkpoint. Trends Genet 16: 395–403.

69. Backer L, Gibson J, Moses M, Allen J (1988) Synaptonemal complex damage in relation to meiotic chromosome aberrations after exposure of male mice to cyclophosphamide. Mutat Res 203(4): 317–330.

70. Allen J, Gibson J, Poorman P, Backer L, Moses M (1988) Synaptonemal complex damage induced by clastogenic and anti-mitotic chemicals: implications for non-disjunction and aneuploidy. Mutat Res 201(2): 313–324.

71. Allen JW, Poorman P, Backer L, Gibson J, Westbrook-Collins B, et al. (1988) Synaptonemal complex damage as a measure of genotoxicity at meiosis. Cell Biol Toxicol 4(4): 487–494.

72. Eichenlaub-Ritter U, Vogt E, Cukurcam S, Sun F, Pacchierotti F, et al. (2008) Evaluation of aneugenic effects of bisphenol A in somatic and germ cells of the mouse Exposure of mouse oocytes to bisphenol A causes meiotic arrest but not aneuploidy. MutatRes 651: 64–70.

73. Hunt PA, Susiarjo M, Rubio C, Hassold TJ (2009) The bisphenol A experience: a primer for the analysis of environmental effects on mammalian reproduction. BiolReprod 81: 807–813.

74. Chalmel F, Rolland AD, Niederhauser-Wiederkehr C, Chung SS, Demougin P, et al. (2007) The conserved transcriptome in human and rodent male gametogenesis. Proc Natl Acad Sci USA 104: 8346–8351.

6

Characterization of DNA Repair Deficient Strains of *Chlamydomonas reinhardtii* Generated by Insertional Mutagenesis

Andrea Plecenikova[1,2¤], Miroslava Slaninova[2], Karel Riha[1,3]*

1 Gregor Mendel Institute, Austrian Academy of Sciences, Vienna Biocenter (VBC), Vienna, Austria, **2** Department of Genetics, Faculty of Natural Sciences, Comenius University in Bratislava, Bratislava, Slovakia, **3** Central European Institute of Technology (CEITEC), Masaryk University, Brno, Czech Republic

Abstract

While the mechanisms governing DNA damage response and repair are fundamentally conserved, cross-kingdom comparisons indicate that they differ in many aspects due to differences in life-styles and developmental strategies. In photosynthetic organisms these differences have not been fully explored because gene-discovery approaches are mainly based on homology searches with known DDR/DNA repair proteins. Here we performed a forward genetic screen in the green algae *Chlamydomonas reinhardtii* to identify genes deficient in DDR/DNA repair. We isolated five insertional mutants that were sensitive to various genotoxic insults and two of them exhibited altered efficiency of transgene integration. To identify genomic regions disrupted in these mutants, we established a novel adaptor-ligation strategy for the efficient recovery of the insertion flanking sites. Four mutants harbored deletions that involved known DNA repair factors, DNA Pol zeta, DNA Pol theta, SAE2/COM1, and two neighbouring genes encoding ERCC1 and RAD17. Deletion in the last mutant spanned two *Chlamydomonas*-specific genes with unknown function, demonstrating the utility of this approach for discovering novel factors involved in genome maintenance.

Editor: John R. Battista, Louisiana State University and A & M College, United States of America

Funding: This work was funded by grants from the Slovak Agency for Research and Development (SK-AT-0023-10; APVV-0661-10; www.apvv.sk), Comenius University (UK/63/2012), Austrian Agency for International Cooperation (OeAD, SK12/2011; www.oead.at), and Austrian Science Fund (FWF; Y418-B03; www.fwf.ac.at). The funders had no role in study design, data collection and analysis, decision to publish, or preparation of the manuscript.

Competing Interests: The authors have declared that no competing interests exist.

* Email: karel.riha@gmi.oeaw.ac.at

¤ Current address: Institute of Botany, Slovak Academy of Sciences, Bratislava, Slovakia

Introduction

DNA double strand breaks (DSBs) pose a serious threat to genome integrity as their erroneous repair may lead to chromosomal rearrangements with potentially lethal consequences for an organism. The response to DSBs elicits a highly organized and complex cellular program, called the DNA damage response (DDR), which sets in motion processes that mitigate the adverse effects of DNA damage and facilitate DNA repair. Broken DNA is usually repaired by one of two mechanistically distinct pathways: homologous recombination (HR) and non-homologous end joining (NHEJ). While HR uses a homologous DNA strand as a template for error-free repair, NHEJ is inherently error-prone and does not rely on sequence homology. The preferred mode of repair and cellular consequences of DDR varies between organisms and is also dependant on cell type and cell cycle context [1,2]. For example, while HR is the preferred mode of repair in many unicellular organisms such as budding and fission yeast, NHEJ is the prevalent pathway in plants and animals.

In many aspects, plants seem to respond differently to DNA insults than do animals. The constant risk of tumor formation in animals has led to evolution of DDR that assures precise genome maintenance, often resulting in apoptotic death of damaged cells.

The lack of such a strong selective constraint presumably permitted evolution of a more relaxed DDR in plants, making them more tolerant to genome damage [3,4]. Furthermore, plant cells are exposed to high levels of genotoxic stress resulting from long-term exposure to solar ultraviolet (UV) irradiation, photosynthesis and extended periods of desiccation [5,6]. Thus, some features of plant DDR and DSB repair may deviate from models primarily established from studies in yeasts and mammals. Functional characterization of plant DDR and DSB repair is mainly limited to a few model organisms including *Arabidopsis thaliana*, rice, maize, and the moss *Physcomytrella patens* [6,7,8]. The key experimental strategy for DDR gene discovery in plants is based on homology searches for conserved DNA repair proteins known from yeast and animals [9,10]. Forward genetic screens have a much higher potential to identify unknown genes with novel DDR and DSB repair functions, and they have been successfully applied for dissecting the mechanism of meiotic recombination [11,12]. However, only a few screens were performed with the aim to identify novel components of somatic DSB repair [13,14,15]. This is partially due to the complicated nature of these screens, as the scored phenotype is often represented by sensitivity to a genotoxic treatment and hence the death of the plant that carries the desired mutation.

In this study we exploited the potential of the unicellular green algae *Chlamydomonas reinhardtii* in discovery of genes related to DDR and DSB repair in the plant kingdom. *C. reinhardtii* is an established model that is traditionally used for studying photosynthesis and cell motility. Interest in this organism has also been sparked by prospects of using algae as source of hydrogen or lipids for biofuel production. In addition, sharing a common ancestor 1.1 billion years ago, *C. reinhardtii* provides an excellent complementary model to higher plants for comparative gene analyses within the plant kingdom [16]. Whereas unicellular *C. reinhardtii* appears to be a relatively simple organism, in regards to genome complexity and mechanisms that govern genome maintenance and expression it is comparable with higher eukaryotes. The draft genome sequence revealed more than 15,000 genes that contain a large number of introns and numerous repetitive DNA sequences and transposable elements [17]. The presence of DNA methylation, siRNA and microRNAs indicates that epigenetic and post-transcriptional regulation of gene expression is more complex than in many other unicellular models [18,19]. In addition, *C. reinhardtii* displays an extremely low efficiency of HR, which is unusual for a unicellular organism [20] and similar to higher plants. *C. reinhardtii* therefore offers the unique opportunity to study some of the cellular processes that are featured in higher eukaryotes while providing the advantages of a unicellular experimental system.

Here, we performed a forward genetic screen of a population of *C. reinhardtii* mutagenized by insertional mutagenesis for sensitivity to genotoxic stress and describe an efficient strategy for identification of disrupted loci. We characterized five mutants and determined that they encompass deletions in known DNA repair genes, which validated our approach, as well as in genes with unknown functions, demonstrating the potential of this approach in identification of novel DDR related genes.

Materials and Methods

Strain and growth media

The *C. reinhardtii* strain *cw15-302 arg2* [21] was obtained from the laboratory of Dr. Christoph Beck (University of Freiburg, Germany). Cells were grown under constant light at 22°C in Tris-acetate phosphate media (TAP) supplemented with L-arginine (100 mg/L) and with 1% agar when solid medium was required [22].

Genetic screen

Plasmid pHyg3 [23] was digested with *Hind*III and the linear fragment containing the chimeric *aph7"* selection marker gene was used as the insertional cassette (Figure 1A). Insertional mutagenesis of *C. reinhardtii* cells was performed by the glass-bead transformation method [24]. Immediately after transformation, cells were spread onto selective TAP plates containing 10 mg/L of hygromycin B (Calbiochem) and grown for 1 week. Transformed colonies resistant to hygromycin B were resuspended in sterile distilled water and replicated to one set of replica TAP plates with or without 300 μg/L of zeocin (Duchefa Biochemie) and to the second set of replica plates, one of which was irradiated with UV-C (70 J/m²) immediately after plating (Stratalinker 2400, Stratagene). For the UV-C sensitivity screen, the plates were incubated for one day in the dark to avoid photoreactivation. Colonies sensitive to the selection conditions were maintained on TAP plates and in the subsequent 3–6 months re-screened at least twice for the sensitivity to zeocin and UV-C irradiation.

Genotoxicity tests

Sensitivity to hydroxyurea (HU) was assessed by inoculating cells on TAP plates supplemented with 50–400 mg/L HU (Sigma-Aldrich). Sensitivity to UV-C irradiation was examined by scoring cell growth on TAP media exposed to 10 J/m²–150 J/m² UV-C (Stratalinker UV Crosslinker 2400, Stratagene). Sensitivity to zeocin (50 μg/L), mitomycin C (MMC; 10 mg/L; Sigma-Aldrich) and methyl methanesulfonate (MMS; 0,005%) was tested by determining growth curves in TAP liquid media. Mutant and control strains were first grown for 3 days on TAP plates with L-arginine (100 mg/L) and hygromycin B (10 mg/L) under continuous light and then transferred to liquid TAP media supplemented with L-arginine and the tested drug. Cell density was determined in a Bürker chamber after 3 days (time point 0) and then again at the 24, 48, 72, 96 and 192 hours. Data from three independent experiments were collected and analyzed.

Nucleic acid isolation and analysis

Total genomic DNA was isolated using a modified protocol according to [25]. 50 ml of *C. reinhardtii* culture grown for 3 days (~2.10^7 cells. ml^{-1}) was pelleted by centrifugation, resuspended in 2×CTAB buffer [1.4 M NaCl, 0.1 M Tris ph 8.0, 0.02 M EDTA, 2% hexadecyltrimethylammoniumbromide (CTAB, Sigma)] and incubated for 1 h at 65°C. DNA was extracted with phenol:chloroform:isoamyl alcohol (25:24:1, v/v) and precipitated by isopropanol. RNA was removed by 30 min treatment with RNase (164 mg/L) at 37°C that was subsequently removed by phenol and chloroform:isoamyl alcohol (24:1, v/v) extractions. DNA was purified by ethanol precipitation. Genomic DNA was subjected to Southern hybridization analysis according to standard protocols [26]. A PCR fragment amplified from the pHyg3 plasmid with primers HygR_PR1_F616 5'-GAGAGCACCAACCCCG-TACTGG-3' and HygR_PR1_R1196 5'-GTGAAGTCGAC-GATCCCGGT-3' was radioactively labeled with [α ³²P] dCTP and used as a probe.

Isolation of insert-flanking regions by hairpin-PCR

2 μg of genomic DNA was digested with a restriction endonuclease according to supplier's specifications and isopropanol precipitated. 1 μg of digested DNA was ligated to a hairpin adaptor using T4 DNA ligase (Thermo Fisher Scientific) for 16 h at laboratory temperature. The hairpin adaptor [27] compatible with the DNA ends generated by restriction digest was used in a final concentration of 1.6 μM. DNA was isopropanol precipitated and 40–80 ng were used as the template in PCR. Genomic DNA flanking the site of insertion was amplified by two rounds of PCR with the primer PETRA-B (5'-CTCTAGACTGTGAGACTTG-GAGATG-3') and nested insert specific primers complementary to the *aph7"* gene (Figure 1A): NR1 (5'-CCAGTGCTCGCC-GAACAGCT-3'), NR2 (5'-TCGTTCCGCAGGCTCGCGTA-3'), NF1 (5'-GAGACTCCCGCTACAGCCTG-3') and NF2 (5'-CTGCACGACTTCGAGGTGTT CG-3'). PCR mixture consisted of 1×GoTaq Reaction Buffer (Promega), 0.2 mM dNTP, 0.5 μM PETRA-B, 0.5 μM NR1 or NF1, 40 ng–80 ng DNA, 3% DMSO, Go Taq DNA polymerase (Promega) and the reaction was performed with 35 cycles of 45 s at 94°C, 30 s at 64°C and 2 min 30 s at 72°C. 0.5 μl of PCR amplification product was used as the template for nested PCR with PETRA-B and NR2 or NF2 primers. PCR products were cloned into pGEM-T Easy Vector (Promega) or to pCR2.1-TOPO Vector (Invitrogen) and transformed to *E. coli* DH5α competent cells and sequenced using a standard procedure.

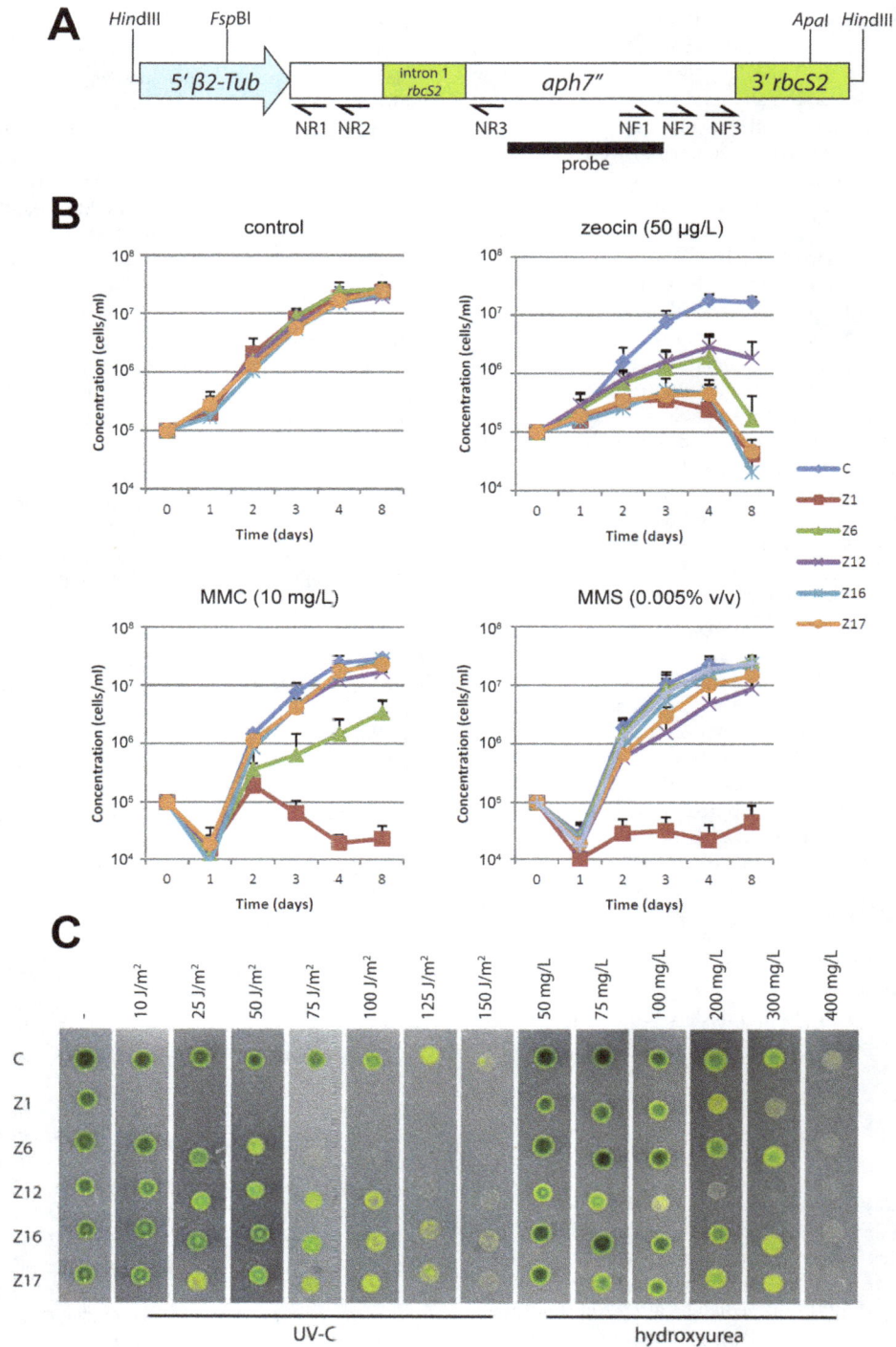

Figure 1. Sensitivity of *C. reinhardtii* mutant strains to genotoxic treatments. (A) Structure of the *aph7"* construct used for the insertional mutagenesis. Primers (arrows), restriction sites and the region used as a probe for Southern hybridization are indicated. (B) Growth curves of mutant and the parental *cw15-302 arg2* strains (denoted as C) in TAP liquid media supplemented with zeocin, MMC or MMS. (C) Growth of mutant strains on TAP plates exposed to increasing dose of UV-C or HU. Pictures were taken five days after inoculation.

Insert integration assay

C. reinhardtii strains were transformed by the glass-bead method with plasmids pUCARG7.8 and pUCBM20ΔARG [21] and transformation efficiency and HR/NHEJ ratios were calculated as previously described [28].

Results

We generated a library of *C. reinhardtii* insertional mutants by transforming the cell-wall deficient strain *cw15-302 arg2* with the 1.7 kb fragment of the pHyg3 plasmid containing the chimeric *aph7"* gene that confers resistance to hygromycin B [23] (Figure 1A). We used a glass-bead transformation technique and a low amount of DNA (500 ng) per transformation to decrease the chance of multiple insertions per genome [29]. In total, we generated 4588 transformants that were subsequently screened for growth on agar plates supplemented with zeocin (300 mg/L), an antibiotic that induces DSB, and for survival on plates exposed to UV-C irradiation (70 J/m^2). After two rounds of screening we obtained 19 clones sensitive to zeocin, 52 clones sensitive to UV-C, and 9 clones sensitive to both treatments. However, only five mutant clones retained the phenotype after 3–6 months of subculturing. All five mutant strains, which we named Z1, Z6, Z12, Z16 and Z17, were originally selected for their inability to grow on agar plates supplemented with zeocin, and this phenotype was also confirmed by cultivation in liquid media (Figure 1B). Z1, Z16 and Z17 exhibited the highest growth retardation in the presence of zeocin, while Z6 showed intermediate sensitivity and the growth of Z12 mutants was least affected (Figure 1B and Table 1). The growth of clones Z1 and Z6 was also impaired by UV-C irradiation (Figure 1C, Table 1).

We next tested sensitivity of the mutants to the genotoxic drugs MMC, methyl MMS and HU. These drugs induce DNA damage by different mechanisms: MMC causes DNA interstrand cross-linking, MMS treatment methylates DNA which is believed to lead to replication fork stalling, and HU impairs production of deoxyribonucleotides, which inhibits DNA replication and results in stalled replication forks. MMC almost fully suppressed growth of the Z1 mutant, while proliferation of Z6 was impaired to lesser extent (Figure 1B, Table 1). Other strains were unaffected. Z1 also exhibited the highest sensitivity to MMS. A slight inhibition of growth by MMS was detected in Z12 and Z17 (Figure 1B, Table 1). HU had the strongest effect on Z12, exhibiting a discernable effect already at the lowest concentration of HU. A slight inhibition was also detected in the Z1 mutant (Figure 1C, Table 1). The observation that almost all mutants were sensitive to at least two independent genotoxic treatments validated the screening approach and suggested that the mutants we obtained are indeed impaired in DDR/DSB repair. Moreover, the differential sensitivity of the 5 mutant strains to particular treatments indicated a deficiency in different aspects of DDR and DNA repair.

Insertion of exogenous DNA into a genome during transformation is governed by processes involved in DSB repair. To examine whether the efficiency of DNA integration via NHEJ or homology driven repair is altered in the isolated strains, we took advantage of the missense mutation *arg7-8* in the *ARG7* gene that encodes argininosuccinate lyase (ASL) and confers arginine auxotrophy of the *cw15-302 arg2* strain [21]. We transformed the mutant strains with plasmids carrying either the entire *ARG7* gene (pUCARG7.8), or a truncated version that lacks the 5′ region

encoding the promoter and N-terminal portion of ASL (pUCBM20ΔARG) [21,28]. Since the transformants were selected for complementation of ARG7 function by their ability to grow on plates without arginine, prototrophy of the pUCBM20ΔARG transformants was anticipated to be caused by homology driven repair that corrected the *arg7-8* mutation [21]. While transformation efficiency with the pUCARG7.8 construct was unaffected in Z1, Z6 and Z17 mutants as compared to the *cw15-302 arg2* strain, a 10-fold decrease was detected in the Z16 background suggesting deficiency in NHEJ (P = 0.063, two-tailed Student's t test; Figure 2A). Interestingly, occurrence of Arg prototrophic clones increased by one order of magnitude in Z12. Homology driven integration in *C. reinhardtii* has been reported to be 1.5–4 orders of magnitude lower than random integration and depends on the transformation method and constitution of transformed DNA [20,30,31,32]. In our conditions, the frequency of pUCBM20ΔARG transformants in the *cw15-302 arg2* strain was approximately 1000-times lower than of pUCARG7.8. Similar ratios were detected in the Z1, Z6, and Z16 strains while a slight decline in the ratio was detected in Z17, indicating a deficiency in HR (P = 0.088, two-tailed Student's t test; Figure 2B).

The high frequency of Arg prototrophs recovered from the Z12 background was unusual and warranted further verifications. We found that prototrophic colonies appeared in a transformation-independent manner and hence, represented spontaneous phenotypic reversions. These reversions are specific to the Z12 strain and were never observed in the parental or other mutant strains. The Arg prototrophic phenotype of Z12 revertant clones was stable, but they appeared to grow slower in liquid media without arginine than *ARG7* complemented *cw15-302 arg2* cells (Figure 2C). Sequencing of the *ARG7* region in several Z12 revertants excluded the possibility that the phenotypic reversions were caused by spontaneous repair of the *arg7-8* mutation (data not shown). Thus, we conclude that Z12 cells are predisposed to activate an adaptive mechanism enabling them to overcome *arg7-8* deficiency mediated arginine auxotrophy.

Next we sought to identify the loci disrupted by insertional mutagenesis. First, we determined the number of independent insertions in the genomes of the mutant strains by Southern analysis with a probe specific for the insertional cassette (Figure 1A, 3A). Restriction analysis with enzymes that digest outside of the construct (*Pst*I, *Pvu*II) revealed the presence of a single insertion in all mutants (Figure 3A). Multiple bands that appeared in the Z1 mutant after digestion with enzymes that cleave within the *aph7"* gene (*Fsp*BI, *Apa*I; Figure 1A, 3A) were suggestive of three copies of the insert in a single site. Other mutants appeared to harbour only a single copy of *aph7"* in their genome.

To localize the insertion sites, we developed a PCR-based strategy utilizing ligation of DNA hairpin adaptors to digested

Table 1. Relative sensitivity to genotoxic treatments in respect to other mutants and the control strain *cw15-302 arg2* (+++ high sensitivity, ++ intermediate sensitivity, + mild sensitivity, −no sensitivity).

Mutant	Zeocin	UV	MMC	MMS	HU
Z1	+++	+++	+++	+++	+
Z6	++	++	+	−	−
Z12	+	+	−	++	+++
Z16	+++	−	−	−	−
Z17	+++	−	−	+	−

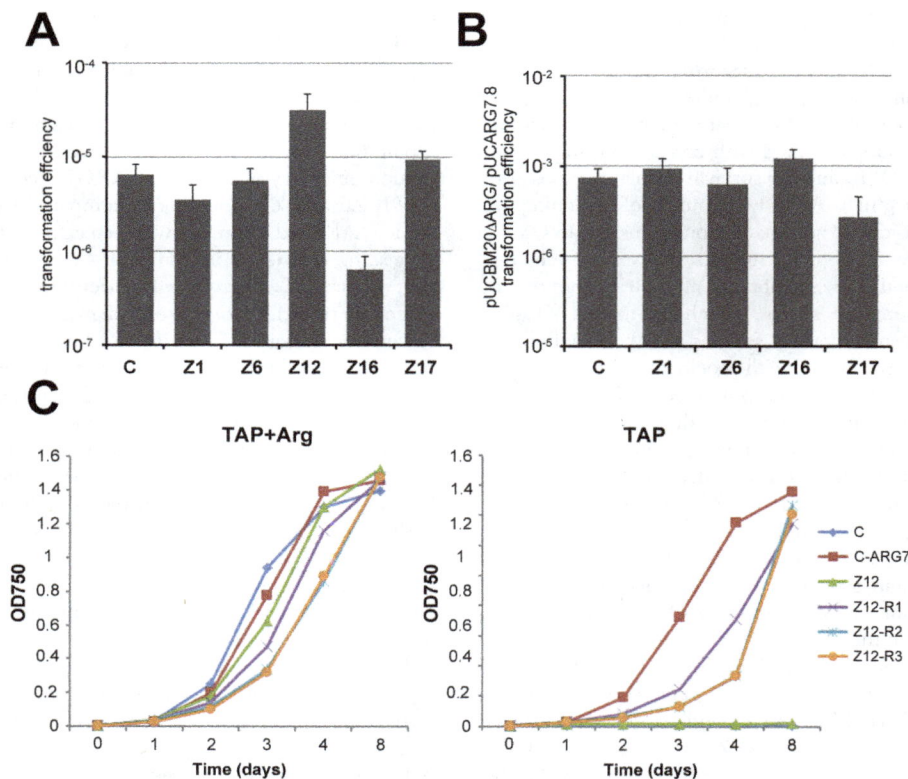

Figure 2. Transformation efficiency of *C. reinhardtii* mutants. (A) Transformation efficiency calculated as the frequency of Arg-prototrophic transformants per a total number of transformed cells normalized to 1 pmol of pUCARG7.8 construct used for transformation. (B) Efficiency of homology-driven integration estimated as a ratio between the transformation efficiencies obtained with the pUCBM20ΔARG and pUCARG7.8 constructs. (A, B) C = *cw15-302 arg2*; standard errors of three independent experiments are indicated (N = 3). (C) Growth curves of the Z12 strain and selected Z12 Arg-prototrophic revertants in TAP liquid media with or without Arg. *cw15-302 arg2* cells complemented with the *ARG7* construct were used as an Arg-prototrophic control (C-ARG7).

genomic fragments (Figure 3B). We have originally designed these adaptors to specifically amplify blunt-ended telomeres in *A. thaliana* [27]. The end of the hairpin can be designed to contain complementarity to DNA ends generated by any restriction enzyme (Figure 3B). The adaptor further harbours a short stretch of non-complementary sequence forming a bubble near the open end of the hairpin. This bubble corresponds to the sequence of a PCR primer (PETRA-B; Figure 3B) that does not amplify ligated adaptor dimers, but can only anneal to DNA strands arising from the extension of a primer specific to a gene of interest. To amplify sequences flanking the insertion site, ligation of the adaptor to digested genomic DNA was followed by two rounds of nested PCR with the PETRA-B and insert specific primers (Figures 1A and 3C). Several adaptor-ligated libraries generated by different restriction enzymes were usually tested for each mutant. The most suitable restriction enzymes for this purpose were *Pst*I, *Pau*I, *Pvu*II, *Nsb*I and *Mse*I. PCR products were sequenced and aligned to the scaffold of the *C. reinhardtii* genome. Using this approach we identified both flanking sites in Z6, Z12, Z16 and Z17 mutants. Insertions in these strains were accompanied by large deletions ranging from 11 to 30 kb that encompassed two to five genes (Figure 4). Deletions in Z6, Z16 and Z17 disrupted genes coding for proteins homologous to the known DNA repair factors DNA Pol zeta, DNA Pol theta and SAE2/COM1 nuclease, respectively. An 11 kb deletion in Z12 affected two proteins of unknown function, Cre10.g441650 and Cre10.g441700, which show no extensive sequence homology in protein databases and appear to

be specific to *Chlamydomonas* and closely related genomes, e.g. *Volvox carteri*. Only one flanking site with homology to the nuclear genome was retrieved in Z1 mutants. The other insertion border contained chloroplast DNA (Figure 4). The Z1 insertion was localized to the vicinity of genes coding for the ERCC1 endonuclease and the RAD17 DNA damage checkpoint protein. PCR analyses demonstrated that both genes are absent in the Z1 mutant, indicating that a large deletion occurred at the insertion site in this strain. We attempted to complement the zeocin sensitivity phenotypes by transforming the mutant strains with BAC clones spanning the deleted regions. However, despite extensive efforts, these experiments failed due to technical difficulties transforming *C. reinhardtii* with large DNA fragments. Therefore, we could not ascertain that the DNA damage sensitivity of mutant strains was indeed caused by disruption of genes at the insertion loci.

Discussion

The use of forward genetic screens for discovery of genes involved DNA repair and DNA damage response has been limited in plants; most plant DDR/DNA repair proteins were identified based on sequence homology with their yeast and mammalian counterparts. The major caveat of this homology driven approach is its inability to identify genes and processes that may have evolved specifically in the plant lineage to deal with genomic insults. The importance of unbiased approaches is documented by

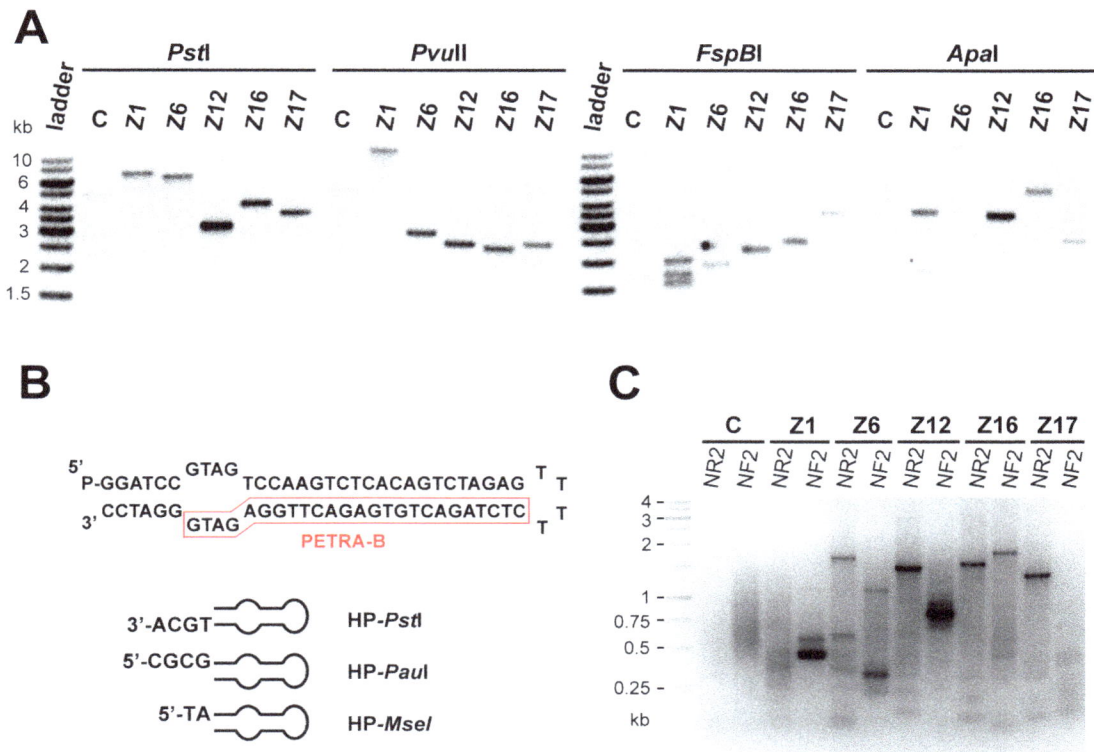

Figure 3. Molecular characterization of *aph7'* **insertions.** (A) Southern analysis of *aph7''* insertions in individual mutants. The restriction enzymes used for each experiment are indicated; positions of the restriction sites within the *aph7''* construct probe used for hybridization are depicted in Figure 1A. (B) Structure of the hairpin adaptor with a blunt end; examples of hairpins with ends compatible with selected restriction enzymes are indicated. (C) PCR products, separated in agarose gels and stained with ethidium bromide, that were generated after two rounds of nested PCR from a genomic library produced by *Pst*I digest.

a screen in *A. thaliana* that revealed novel genes linking DDR with epigenetic regulation and stress signalling pathways [13,14]. Similarly, SOG1, a transcription factor governing DDR is a plant specific protein that was discovered in a screen to identify suppressors of the radiosensitivity of a nucleotide excision repair mutant [33].

Here we exploited the unicellular green algae *C. reinhardtii* in a forward genetic screen for the discovery of DDR/DNA repair genes. By surveying a mutagenized population of *C. reinhardtii*, we have identified five transformants that exhibited stable sensitivity to the radiomimetic drug zeocin and, with exception of the mutant Z16, to at least one other genotoxic treatment. To identify the loci disrupted in these mutants, we have developed a novel PCR-mediated strategy based on ligation of hairpin adapters to genomic fragments generated by restriction endonucleases cutting within G/C-rich sequences. Inverse PCR and TAIL-PCR are the usual methods of choice for cloning sequences flanking insertion sites, but the high GC content of the *C. reinhardtii* genome renders the success of these techniques highly variable [34,35]. With the hairpin-ligation strategy we were able to readily obtain sequences flanking both insertion sites for all mutants; in fact, in some mutants we obtained these sequences several times independently using different restriction enzymes (data not shown). The forward genetic screen in *C. reinhardtii* was extremely time efficient in comparison to other plant models as the entire procedure from making the mutant library to the identification of disrupted loci took about 9 months. Validation of the DNA-damage sensitivity proved to be the most critical and time consuming step as a large proportion of isolated mutants lost the

phenotype after 3 to 6 months of subculturing. Lack of tools for efficient validation of genes identified in the screen was the major hurdle we faced in this study. Insertional mutagenesis led to deletions spanning several genes. As *C. reinhardtii* is not amenable to gene targeting, genetic complementation appears to be a method of choice for confirming association between scored phenotypes and disrupted loci. However, also this approach can be in *C. reinhardtii* troublesome. *C. reinhardtii* genes can span tens of kbs which in combination with very high CG content impedes classical cloning and PCR based approaches for making complementing constructs. For these reasons, we were unable to complement mutant phenotypes and formally confirm identity of causative genes. We anticipate that advances in genome-editing technologies utilizing TALENs or CRISPR/Cas9 will eventually solve this issue and facilitate gene discovery in *C. reinhardtii*. Nevertheless, the fact that four out of five insertion-tagged loci contained disruption of known DNA repair genes validates suitability of the experimental approach in discovery of genes involved in plant DDR.

In the case of the Z1 insertion we were able to map only one border to the nuclear genome, while the other flanking sequence consisted of a chloroplast DNA fragment. The insertion event led to deletion of genes homologous to *ERCC1* and *RAD17*. ERCC1 forms, together with XPF, a structure specific nuclease discovered for its role in nucleotide excision repair [36]. It is also involved in processing of DNA interstrand crosslink lesions and homologous recombination intermediates. ERCC1 deficiency in *A. thaliana* has been reported to cause sensitivity to MMC, MMS and UV and gamma irradiation [15]. RAD17 is a conserved replication

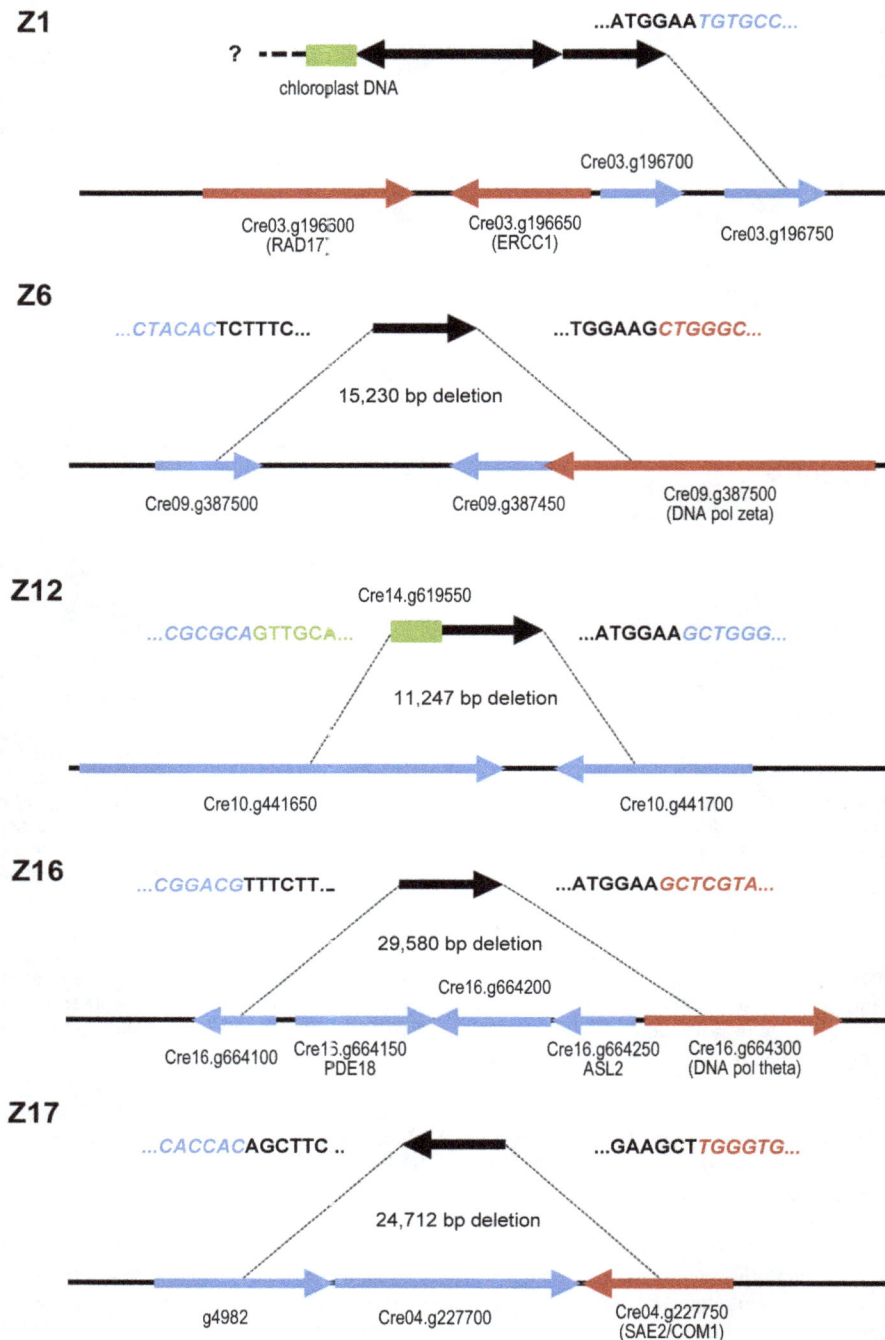

Figure 4. Structure of insertion sites. Black arrows represent the aph7″ insert, red arrows indicate homologues to known DNA repair genes, blue arrows represent genes unrelated to DNA repair, and green boxes depict DNA fragments that were co-transformed to the insertion sites from ectopic locations. Sequences at the insert borders are indicated.

checkpoint protein important for genome stability and its inactivation in *A. thaliana* renders plants sensitive to bleocin and MMC [37]. Thus, the sensitivity of *C. reinhardtii* Z1 mutants to a broad range of genotoxic treatments is likely caused by the combined absence of these DNA repair proteins.

Characterization of insertion sites in the Z6 and Z16 mutants revealed deletions disrupting genes coding for translesion synthesis DNA polymerases zeta and theta, respectively. DNA Pol zeta can bypass a range of DNA lesions including adducts induced by

cisplatin, UV and apurinic/apyrimidinic sites [38] and there is evidence suggesting its function in interstrand crosslink repair and HR [39,40]. This is consistent with the relatively mild, but broad sensitivity of the Z6 strain to genotoxic insults that include zeocin, UV and MMC; a similar spectrum of sensitivity was also observed in the *A. thaliana* DNA pol zeta mutant [41]. In contrast, the Z16 strain carrying a deletion in DNA pol theta (also known as POLQ) exhibits strong sensitivity to zeocin, but not to the other treatments. Interestingly, *A. thaliana* pol theta deficient plants display developmental defects, constitutive DNA damage response and sensitivity to MMC and MMS [42]. Thus, this polymerase may have acquired more fundamental role in DNA repair in higher plants. Mouse and human cell lines impaired in DNA pol theta are sensitive to bleomycin and ionizing radiation, implicating its function in some aspects of DSB repair [43,44]. Studies in *Drosophila melanogaster* indicated that DNA Pol theta contributes to DSB repair by facilitating microhomology-mediated DNA end-joining [45], which is in accordance with our observation of a 10-fold decrease in the transformation efficiency of the Z16 mutant.

The Z17 strain harbours a deletion that disrupts SAE2/COM1/CtIP, a protein that co-operates with the MRE11-RAD50-NBS1/Xrs2 nuclease to initiate resection of DSB during HR [46,47]. This complex also plays a central role in the repair of stalled or collapsed replication forks, and mutants described in a number of organisms are sensitive to a wide range of DNA damaging agents including MMC, ionizing radiation, hydroxyurea, bleomycin or MMS [48,49,50]. The conserved HR function of SAE2/COM1 in *C. reinhardtii* is also indicated by decreased efficiency of homology-driven integration of the pUCBM20ΔARG construct into Z17 genome. In higher plants, COM1 appears to participate in interstrand crosslink repair as *A. thaliana com1* mutant displayed sensitivity to MMC [51].

Identification of several well characterized DNA repair factors at the insertion sites in Z1, Z6, Z16 and Z17 strains demonstrated the ability of the screen to uncover genes involved in DDR and genome integrity. Nevertheless, the ultimate goal of forward genetic screens is to identify unknown genes or unanticipated links to the processes under study. Two previous studies reported on the identification of novel genes involved in DNA repair in *C. reinhardtii* by combining insertional mutagenesis with a screen for sensitivity to genotoxic treatments [52,53]. The disruption in the Z12 mutant also affects two genes encoding proteins of unknown function that appear to be specific to *C. reinhardtii* and closely related species. The Z12 mutant exhibits the strongest sensitivity to HU and MMS among all the mutants recovered in our screen, suggesting the impairment of processes associated with DNA replication. Interestingly, we observed that in addition to DNA damage sensitivity, Z12 cells are able to overcome arginine auxotrophy caused by the *arg7-8* mutation in the *ARG7* gene. Mutations in the *ARG7* have been used for decades as auxotrophic markers in *C. reinhardtii* transformation experiments for their stability and rare reversions [54,55]. However, the spontaneous phenotypic reversions to arginine prototrophy occur in Z12 at an extremely high frequency and are not accompanied by correction of the causative mutation in the *ARG7* gene. The adaptive mechanism that is activated in Z12 revertants is unknown. It is also unclear whether the propensity of Z12 cells to switch to Arg prototrophy is linked to the DNA repair deficiency phenotype and whether it associates with either of the disrupted genes. The high rate of the phenotypic reversion, along with its stability, evokes parallels with de-repression of an epigenetically suppressed mechanism. Spontaneous changes in epigenetic states can be stably inherited and occur at much higher frequencies than genetic mutations. Furthermore, malfunction of numerous DNA replication and repair factors has been linked to de-repression of epigenetically silent loci, and, *vice versa*, deficiency in chromatin regulators may lead to impaired DNA repair [13,56,57,58]. Thus, we speculate that the phenotypic switch to arginine prototrophy in Z12 cells is conditioned by a mutation in a gene that is required for both genome integrity and epigenetic inheritance. While further in depth mechanistic studies are required to decipher this phenomenon, the example of the Z12 mutant illustrates the power of forward genetics in *C. reinhardtii* in uncovering novel genes and processes involved in genome maintenance.

Acknowledgments

We thank to Sona Valuchova and Matt Watson for helpful discussions.

Author Contributions

Conceived and designed the experiments: AP MS KR. Performed the experiments: AP. Analyzed the data: AP MS KR. Contributed to the writing of the manuscript: AP KR.

References

1. Nagaria P, Robert C, Rassool FV (2013) DNA double-strand break response in stem cells: mechanisms to maintain genomic integrity. Biochim Biophys Acta 1830: 2345–2353.
2. Helle F (2012) Germ cell DNA-repair systems-possible tools in cancer research? Cancer Gene Ther 19: 299–302.
3. Riha K, McKnight TD, Griffing LR, Shippen DE (2001) Living with genome instability: plant responses to telomere dysfunction. Science 291: 1797–1800.
4. Doonan JH, Sablowski R (2010) Walls around tumours - why plants do not develop cancer. Nat Rev Cancer 10: 794–802.
5. Bray CM, West CE (2005) DNA repair mechanisms in plants: crucial sensors and effectors for the maintenance of genome integrity. New Phytol 168: 511–528.
6. Waterworth WM, Drury GE, Bray CM, West CE (2011) Repairing breaks in the plant genome: the importance of keeping it together. New Phytol 192: 805–822.
7. Mannuss A, Trapp O, Puchta H (2012) Gene regulation in response to DNA damage. Biochim Biophys Acta 1819: 154–165.
8. Schaefer DG (2001) Gene targeting in Physcomitrella patens. Curr Opin Plant Biol 4: 143–150.
9. Singh SK, Roy S, Choudhury SR, Sengupta DN (2010) DNA repair and recombination in higher plants: insights from comparative genomics of Arabidopsis and rice. BMC Genomics 11: 443.
10. Vlcek D, Sevcovicova A, Sviezena B, Galova E, Miadokova E (2008) Chlamydomonas reinhardtii: a convenient model system for the study of DNA repair in photoautotrophic eukaryotes. Curr Genet 53: 1–22.
11. De Muyt A, Pereira L, Vezon D, Chelysheva L, Gendrot G, et al. (2009) A high throughput genetic screen identifies new early meiotic recombination functions in Arabidopsis thaliana. PLoS Genet 5: e1000654.
12. Timofejeva L, Skibbe DS, Lee S, Golubovskaya I, Wang R, et al. (2013) Cytological characterization and allelism testing of anther developmental mutants identified in a screen of maize male sterile lines. G3 (Bethesda) 3: 231–249.
13. Takeda S, Tadele Z, Hofmann I, Probst AV, Angelis KJ, et al. (2004) BRU1, a novel link between responses to DNA damage and epigenetic gene silencing in Arabidopsis. Genes Dev 18: 782–793.
14. Ulm R, Revenkova E, di Sansebastiano GP, Bechtold N, Paszkowski J (2001) Mitogen-activated protein kinase phosphatase is required for genotoxic stress relief in Arabidopsis. Genes Dev 15: 699–709.
15. Hefner E, Preuss SB, Britt AB (2003) Arabidopsis mutants sensitive to gamma radiation include the homologue of the human repair gene ERCC1. J Exp Bot 54: 669–680.
16. Gutman BL, Niyogi KK (2004) Chlamydomonas and Arabidopsis. A dynamic duo. Plant Physiol 135: 607–610.
17. Merchant SS, Prochnik SE, Vallon O, Harris EH, Karpowicz SJ, et al. (2007) The Chlamydomonas genome reveals the evolution of key animal and plant functions. Science 318: 245–250.
18. Molnar A, Schwach F, Studholme DJ, Thuenemann EC, Baulcombe DC (2007) miRNAs control gene expression in the single-cell alga Chlamydomonas reinhardtii. Nature 447: 1126–1129.
19. Feng S, Cokus SJ, Zhang X, Chen PY, Bostick M, et al. (2010) Conservation and divergence of methylation patterning in plants and animals. Proc Natl Acad Sci U S A 107: 8689–8694.
20. Sodeinde OA, Kindle KL (1993) Homologous recombination in the nuclear genome of Chlamydomonas reinhardtii. Proc Natl Acad Sci U S A 90: 9199–9203.
21. Mages W, Heinrich O, Treuner G, Vlcek D, Daubnerova I, et al. (2007) Complementation of the Chlamydomonas reinhardtii arg7-8 (arg2) point mutation by recombination with a truncated nonfunctional ARG7 gene. Protist 158: 435–446.
22. Gorman DS, Levine RP (1965) Cytochrome f and plastocyanin: their sequence in the photosynthetic electron transport chain of Chlamydomonas reinhardi. Proc Natl Acad Sci U S A 54: 1665–1669.

23. Berthold P, Schmitt R, Mages W (2002) An engineered Streptomyces hygroscopicus aph7″ gene mediates dominant resistance against hygromycin B in Chlamydomonas reinhardtii. Protist 153: 401–412.

24. Kindle KL (1990) High-frequency nuclear transformation of Chlamydomonas reinhardtii. Proc Natl Acad Sci U S A 87: 1228–1232.

25. Borevitz JO, Liang D, Plouffe D, Chang HS, Zhu T, et al. (2003) Large-scale identification of single-feature polymorphisms in complex genomes. Genome Res 13: 513–523.

26. Sambrook J, Russell D (2001) Molecular Cloning: A Laboratory Manual. New Yoprk: CSHL Press.

27. Kazda A, Zellinger B, Rossler M, Derboven E, Kusenda B, et al. (2012) Chromosome end protection by blunt-ended telomeres. Genes Dev 26: 1703–1713.

28. Plecenikova A, Mages W, Andresson OS, Hrossova D, Valuchova S, et al. (2013) Studies on recombination processes in two Chlamydomonas endogenous genes, NIT1 and ARG7. Protist 164: 570–582.

29. Matsuo T, Okamoto K, Onai K, Niwa Y, Shimogawara K, et al. (2008) A systematic forward genetic analysis identified components of the Chlamydomonas circadian system. Genes Dev 22: 918–930.

30. Nelson JA, Lefebvre PA (1995) Targeted disruption of the NIT8 gene in Chlamydomonas reinhardtii. Mol Cell Biol 15: 5762–5769.

31. Zorin B, Hegemann P, Sizova I (2005) Nuclear-gene targeting by using single-stranded DNA avoids illegitimate DNA integration in Chlamydomonas reinhardtii. Eukaryot Cell 4: 1264–1272.

32. Zorin B, Lu Y, Sizova I, Hegemann P (2009) Nuclear gene targeting in Chlamydomonas as exemplified by disruption of the PHOT gene. Gene 432: 91–96.

33. Yoshiyama K, Conklin PA, Huefner ND, Britt AB (2009) Suppressor of gamma response 1 (SOG1) encodes a putative transcription factor governing multiple responses to DNA damage. Proc Natl Acad Sci U S A 106: 12843–12848.

34. Tuncay H, Findinier J, Duchene T, Cogez V, Cousin C, et al. (2013) A forward genetic approach in Chlamydomonas reinhardtii as a strategy for exploring starch catabolism. PLoS One 8: e74763.

35. Dent RM, Haglund CM, Chin BL, Kobayashi MC, Niyogi KK (2005) Functional genomics of eukaryotic photosynthesis using insertional mutagenesis of Chlamydomonas reinhardtii. Plant Physiol 137: 545–556.

36. Davies AA, Friedberg EC, Tomkinson AE, Wood RD, West SC (1995) Role of the Rad1 and Rad10 proteins in nucleotide excision repair and recombination. J Biol Chem 270: 24638–24641.

37. Heitzeberg F, Chen IP, Hartung F, Orel N, Angelis KJ, et al. (2004) The Rad17 homologue of Arabidopsis is involved in the regulation of DNA damage repair and homologous recombination. Plant J 38: 954–968.

38. Shachar S, Ziv O, Avkin S, Adar S, Wittschieben J, et al. (2009) Two-polymerase mechanisms dictate error-free and error-prone translesion DNA synthesis in mammals. EMBO J 28: 383–393.

39. Shen X, Jun S, O'Neal LE, Sonoda E, Bemark M, et al. (2006) REV3 and REV1 play major roles in recombination-independent repair of DNA interstrand cross-links mediated by monoubiquitinated proliferating cell nuclear antigen (PCNA). J Biol Chem 281: 13869–13872.

40. Wu F, Lin X, Okuda T, Howell SB (2004) DNA polymerase zeta regulates cisplatin cytotoxicity, mutagenicity, and the rate of development of cisplatin resistance. Cancer Res 64: 8029–8035.

41. Sakamoto A, Lan VT, Hase Y, Shikazono N, Matsunaga T, et al. (2003) Disruption of the AtREV3 gene causes hypersensitivity to ultraviolet B light and gamma-rays in Arabidopsis: implication of the presence of a translesion synthesis mechanism in plants. Plant Cell 15: 2042–2057.

42. Inagaki S, Suzuki T, Ohto MA, Urawa H, Horiuchi T, et al. (2006) Arabidopsis TEBICHI, with helicase and DNA polymerase domains, is required for regulated cell division and differentiation in meristems. Plant Cell 18: 879–892.

43. Goff JP, Shields DS, Seki M, Choi S, Epperly MW, et al. (2009) Lack of DNA polymerase theta (POLQ) radiosensitizes bone marrow stromal cells in vitro and increases reticulocyte micronuclei after total-body irradiation. Radiat Res 172: 165–174.

44. Higgins GS, Prevo R, Lee YF, Helleday T, Muschel RJ, et al. (2010) A small interfering RNA screen of genes involved in DNA repair identifies tumor-specific radiosensitization by POLQ knockdown. Cancer Res 70: 2984–2993.

45. Chan SH, Yu AM, McVey M (2010) Dual roles for DNA polymerase theta in alternative end-joining repair of double-strand breaks in Drosophila. PLoS Genet 6: e1001005.

46. Mimitou EP, Symington LS (2008) Sae2, Exo1 and Sgs1 collaborate in DNA double-strand break processing. Nature 455: 770–774.

47. Nicolette ML, Lee K, Guo Z, Rani M, Chow JM, et al. (2010) Mre11-Rad50-Xrs2 and Sae2 promote 5′ strand resection of DNA double-strand breaks. Nat Struct Mol Biol 17: 1478–1485.

48. Prinz S, Amon A, Klein F (1997) Isolation of COM1, a new gene required to complete meiotic double-strand break-induced recombination in Saccharomyces cerevisiae. Genetics 146: 781–795.

49. Sartori AA, Lukas C, Coates J, Mistrik M, Fu S, et al. (2007) Human CtIP promotes DNA end resection. Nature 450: 509–514.

50. Limbo O, Chahwan C, Yamada Y, de Bruin RA, Wittenberg C, et al. (2007) Ctp1 is a cell-cycle-regulated protein that functions with Mre11 complex to control double-strand break repair by homologous recombination. Mol Cell 28: 134–146.

51. Uanschou C, Siwiec T, Pedrosa-Harand A, Kerzendorfer C, Sanchez-Moran E, et al. (2007) A novel plant gene essential for meiosis is related to the human CtIP and the yeast COM1/SAE2 gene. EMBO J 26: 5061–5070.

52. Sarkar N, Lemaire S, Wu-Scharf D, Issakidis-Bourguet E, Cerutti H (2005) Functional specialization of Chlamydomonas reinhardtii cytosolic thioredoxin h1 in the response to alkylation-induced DNA damage. Eukaryot Cell 4: 262–273.

53. Cenkci B, Petersen JL, Small GD (2003) REX1, a novel gene required for DNA repair. J Biol Chem 278: 22574–22577.

54. Debuchy R, Purton S, Rochaix JD (1989) The argininosuccinate lyase gene of Chlamydomonas reinhardtii: an important tool for nuclear transformation and for correlating the genetic and molecular maps of the ARG7 locus. EMBO J 8: 2803–2809.

55. Purton S, Rochaix JD (1995) Characterisation of the ARG7 gene of Chlamydomonas reinhardtii and its application to nuclear transformation. Eur J Phycol 30: 141–148.

56. Liu J, Ren X, Yin H, Wang Y, Xia R, et al. (2010) Mutation in the catalytic subunit of DNA polymerase alpha influences transcriptional gene silencing and homologous recombination in Arabidopsis. Plant J 61: 36–45.

57. Barrero JM, Gonzalez-Bayon R, del Pozo JC, Ponce MR, Micol JL (2007) INCURVATA2 encodes the catalytic subunit of DNA Polymerase alpha and interacts with genes involved in chromatin-mediated cellular memory in Arabidopsis thaliana. Plant Cell 19: 2822–2838.

58. Kirik A, Pecinka A, Wendeler E, Reiss B (2006) The chromatin assembly factor subunit FASCIATA1 is involved in homologous recombination in plants. Plant Cell 18: 2431–2442.

7

Ecotype Diversity and Conversion in *Photobacterium profundum* Strains

Federico M. Lauro[1,2]*, Emiley A. Eloe-Fadrosh[3], Taylor K. S. Richter[3], Nicola Vitulo[4], Steven Ferriera[5], Justin H. Johnson[5], Douglas H. Bartlett[3]

1 School of Biotechnology and Biomolecular Sciences, The University of New South Wales, Sydney, New South Wales, Australia, 2 Singapore Centre on Environmental Life Sciences Engineering (SCELSE), Nanyang Technological University, Singapore, 3 Marine Biology Research Division, Scripps Institution of Oceanography, University of California San Diego, La Jolla, California, United States of America, 4 CRIBI Biotechnology Centre, University of Padua, Padova, Italy, 5 J. Craig Venter Institute, Rockville, Maryland, United States of America

Abstract

Photobacterium profundum is a cosmopolitan marine bacterium capable of growth at low temperature and high hydrostatic pressure. Multiple strains of *P. profundum* have been isolated from different depths of the ocean and display remarkable differences in their physiological responses to pressure. The genome sequence of the deep-sea piezopsychrophilic strain *Photobacterium profundum* SS9 has provided some clues regarding the genetic features required for growth in the deep sea. The sequenced genome of *Photobacterium profundum* strain 3TCK, a non-piezophilic strain isolated from a shallow-water environment, is now available and its analysis expands the identification of unique genomic features that correlate to environmental differences and define the Hutchinsonian niche of each strain. These differences range from variations in gene content to specific gene sequences under positive selection. Genome plasticity between *Photobacterium* bathytypes was investigated when strain 3TCK-specific genes involved in photorepair were introduced to SS9, demonstrating that horizontal gene transfer can provide a mechanism for rapid colonisation of new environments.

Editor: Chongle Pan, Oak Ridge National Lab, United States of America

Funding: The draft genome sequence of Photobacterium profundum 3TCK was obtained at the J. Craig Venter Institute as a part of the Moore Foundation Microbial Genome Sequencing Project (http://camera.calit2.net/microgenome/). FML is supported by a fellowship from the Australian Research Council (DE120102610). DHB is supported by grants from the National Science Foundation (EF0827051) and National Aeronautics and Space Administration (NNX11AG10G). The funders had no role in study design, data collection and analysis, decision to publish, or preparation of the manuscript.

Competing Interests: The authors have declared that no competing interests exist.

* E-mail: flauro@unsw.edu.au

Introduction

The vast majority of the earth's marine biosphere is at a relatively constant low temperature, high hydrostatic pressure and is constrained by the small amounts of refractory organic nutrients that arrive in pulses from the overlying photic zone [1]. These conditions promote and maintain a diverse microbial community as detected by culture-independent approaches [2], [3], [4]. It is still under debate the extent to which the culture-independent diversity is autochthonous [4], but generally pressure-adaptation is considered a valid criterion to discriminate against microbes recently introduced to the deep sea from shallower waters [4], [5], [6]. With few exceptions [6], [7], [8], [9] the majority of pressure-adapted isolates in culture span only a narrow phylogenetic range of gamma proteobacteria [10]. This includes the easily culturable *Photobacterium profundum* SS9 that has become a model for studying adaptations to high pressure.

With the completion of the *P. profundum* SS9 genome sequence [11], the details of physiological responses to pressure have begun to unravel: for example, microarray studies have shown that suboptimal hydrostatic pressure induces the up-regulation of chaperones and DNA repair enzymes [12], and RNA-seq analyses have revealed the differential expression of multiple ATP synthases and membrane transporters [13].

At least four separate strains of *P. profundum* (SS9, DSJ4, 3TCK and 1230sf1) have been isolated and characterized from multiple sites in the Pacific ocean [10], [14], [15], [16]. While strains SS9 and DSJ4 were isolated from deep-sea environments and are adapted to high hydrostatic pressure [14], [15], strains 1230sf1 and 3TCK were recovered from shallower waters and are inhibited by elevated hydrostatic pressure (12 and our unpublished results). These different ecotypes, which vary in their adaptation to depth in the water column, have been defined bathytypes [5]. The very existence of phylogenetically cohesive bathytypes suggests that the genetic modifications required to evolve depth-specific adaptations can be rapidly evolved [10] and bathytype conversion might occur quite frequently as a result of advective transport of microbial communities by phenomena such as up/downwelling, Ekman transport and thermohaline circulation [10], [17]. The rapid development of mutants tolerant to pressure inactivation, which has been found to extend into the gigapascal range even for non-marine bacteria [18], might aid specific taxa in rapidly adapting to the new environmental conditions while restricting others.

In this study the genome plasticity between two bathytypes of *P. profundum* is analysed in detail. The results show that no single gene is likely to restrict the environmental niche (sensu Hutchinson [19]) of each strain, but a number of genetic features specific to each strain can confer specific abilities to cope with depth-specific environmental stresses (e.g. temperature, pressure, nutrient availability). Some of these features carry signatures of horizontal gene transfer (HGT) suggesting one possible mechanism for the rapid evolution of new bathytypes.

To test the feasibility of bathytype conversion, a cluster of genes involved in the repair of ultraviolet light-induced DNA damage were transferred from the shallow to the deep bathytype. The lack of this UV protective function is predicted to restrict the colonization of shallower waters by deep bathytypes. This is the first study employing intra-specific sequence comparisons in combination with molecular genetics to address the bases of niche partitioning in piezophilic bacteria.

Materials and Methods

Strains and Growth Conditions

The bacterial strains used in this study are listed in Table 1. The strains of *P. profundum* were cultured in 75% strength 2216 Marine Medium (28 g/l; Difco Laboratories) at 15°C and 0.1 megapascal (MPa), unless otherwise specified. *E. coli* strains were grown aerobically at 37°C in Luria-Bertani (LB) medium. High-pressure growth experiments were performed by inoculating in heat-sealable plastic bulbs containing media and no gas space. The heat-sealed bulbs were placed in pressure vessels and pressurized as previously described [20], [21].

When needed, antibiotics were used in the following final concentrations: rifampicin (Rif), 100 µg/ml; kanamycin (Km), 100 µg/ml (*E. coli*) or 200 µg/ml (*P. profundum*); streptomycin (Sm), 50 µg/ml (*E. coli*) or 150 µg/ml (*P. profundum*). X-Gal (5-bromo-4-chloro-3-indolyl- [beta]-D-galactopyranoside) was added to solid medium at 40 µg/ml in N,N-dimethylformamide. The introduction of plasmids in *P. profundum* was achieved by tri-parental

conjugations using the helper *E. coli* strain pRK2073 as previously described [22].

Genome Sequencing, Assembly and Annotation

Genomic DNA was obtained from a culture of *P. profundum* strain 3TCK in mid-exponential growth. Approximately 1 liter of a liquid culture was harvested by centrifugation for 15 minutes at $5,000 \times g$ and the pellet was resuspended in 5 ml buffer A (50 mM Tris, 50 mM EDTA, pH 8.0). The suspension was incubated overnight at -20°C and thawed at room temperature with the addition of 500 µl of buffer B (250 mM Tris, pH 8.0, 10 mg/ml lysozyme). After 45 min of incubation on ice, 1 ml of buffer C (0.5% SDS, 50 mM Tris, 400 mM EDTA, pH 7.5, 1 mg/ml Proteinase K) was added and the mixture was placed in a 50°C water bath for 60 minutes. An additional 750 µl of buffer C were added followed by an additional 30 minutes of incubation at 50°C. The genomic DNA was extracted twice with 5 ml of phenol:-chloroform:isoamyl alcohol (24:24:1), and precipitated with 0.8 volumes of isopropanol. The DNA pellet was recovered by spooling on a glass rod, and rehydrated overnight at 4°C in 4 ml of buffer D (50 mM Tris, 1 mM EDTA, 200 ug/ml RNAse A, pH 8.0). Further purification was performed by extracting once with 4 ml of chloroform, then precipitating with 3.2 ml of isopropanol. The DNA pellet was recovered by centrifugation, washed once with 5 ml of 70% ethanol and stored dry at -20°C.

The dry DNA pellet was shipped to the J. Craig Venter Institute where a draft genome sequence was obtained with a conventional whole-genome sequencing approach by preparing two genomic libraries with insert sizes of 4 kb and 40 kb as described in Goldberg et al. [23] and the resulting sequences were used as input for the Celera assembler [24]. The draft genome sequence was deposited in NCBI under the BioProject accession number PRJNA13563. The reference genome sequence of *P. profundum*

Table 1. Strains and Plasmids used in this study.

Strain/plasmid	Relevant genotype or description	Reference
P. profundum		
SS9	Wild type, deep bathytype of *P. profundum*	[14], [15]
DSJ4	Wild type, deep bathytype of *P. profundum*	[15]
3TCK	Wild type, shallow bathytype of *P. profundum*	[12]
SS9R	Rif^r SS9 derivative	[22]
3TCKR	Rif^r 3TCK derivative	This study
E. coli strains		
ED8654	pRK2073 maintenance	[60]
DH5α	*recA⁻*, used for cloning	[61]
XL1-Blue	*recA⁻*, used for cloning	Stratagene
TOP10	*recA⁻*, used for cloning	Invitrogen
Plasmids		
pRK2073	Providing *tra* genes for conjugal transfer	[62]
pFL122	RSF1010 derived, broad host range cloning vector, Sm^r	[37]
pFL190	RSF1010 derived, arabinose-inducible expression vector, Sm^r	[37]
pFL303	*phr* gene cluster in pFL122, Sm^r	This study
pFL304	Δ22 deletion of *phr* gene cluster in pFL122, Sm^r	This study
pFL305	pFL304+*phr* promoter, Sm^r	This study
pFL306	pFL304+*rpoX*+*phr* promoter, Sm^r	This study
pFL307	*phr* in pFL190, Sm^r	This study

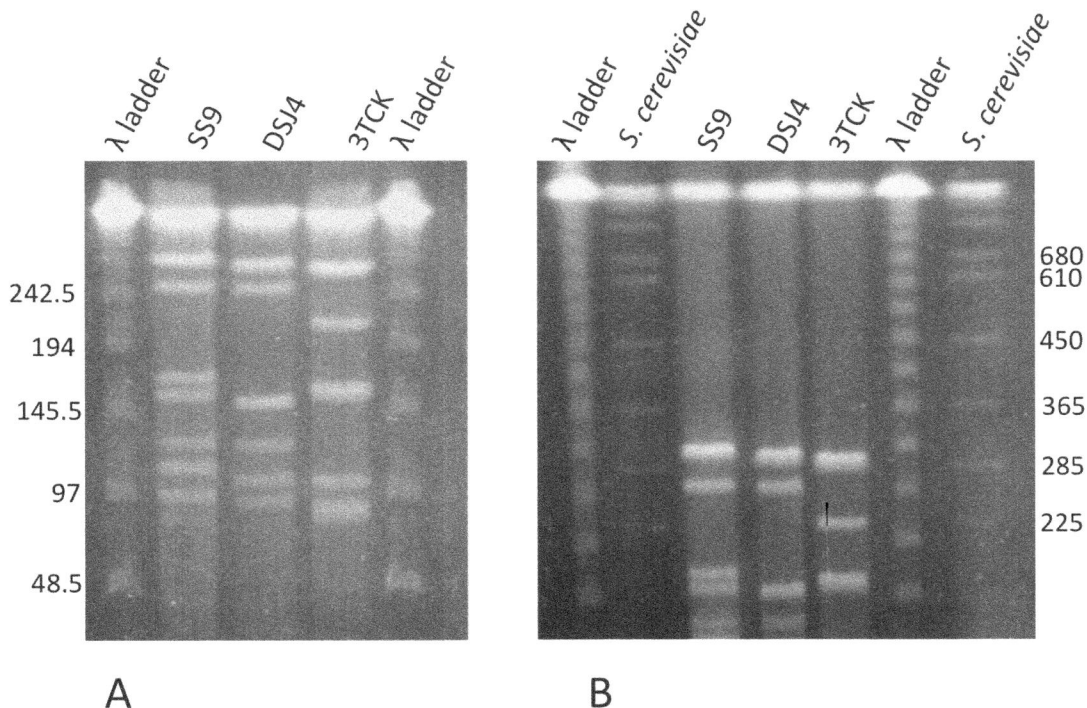

Figure 1. Pulsed field gel electrophoresis of chromosomal plugs of *P. profundum* strains SS9, DSJ4 and 3TCK. The plugs were digested overnight with *I-Ceu*I and run under the following conditions: (A) Voltage: 6 V/cm; Pulse-time: 3–18.5 s; Runtime: 18.5 h; Temperature: 14°C; Included angle: 120°; Gel: 0.95% PFGE agarose in 0.5x TBE (B) Voltage: 6 V/cm; Pulse-time: 6.7–54 s; Runtime: 28 h; Temperature: 14°C; Included angle: 120°; Gel: 0.95% PFGE agarose in 0.5x TBE.

SS9 was also retrieved from NCBI under BioProject accession number PRJNA13128. The number of ribosomal RNA operons was estimated by Pulsed Field Gel Electrophoresis of genomic DNA digested with I-*Ceu*I as previously described [25]. The draft assembly was submitted to the NCBI PGAAP (http://www.ncbi.nlm.nih.gov/genomes/static/Pipeline.html) for prediction of Open Reading Frames (ORFs) and automatic annotation. For genomic sequence comparisons the scaffolds of 3TCK were oriented and joined in alignments to the reference genome of SS9 with the 6-frame stop-codon spacer 'NNNNCACACACTTAAT-TAATTAAGTGTGTGNNNN' using the custom perl script scaffolding.pl to create a contiguous pseudomolecule. Scaffolding.pl and other custom perl scripts used in this study are available at https://github.com/flauro/3tck_comparative.

Bioinformatic Analyses

The assignment of ORFs to Clusters of Orthologous Groups (COGs) and statistical comparisons were performed as previously described [26] using the method of Rodriguez-Brito [27] with a subsample size of 4000 and 10,000 bootstraps. The average nucleotide identity (ANI) between the genomes was computed as a reciprocal two-way average with the method of Goris et al. [28] using custom perl scripts and the following parameters: fragment size 1020 bp; minimum identity 30%; minimum alignable region 714 bp.

The ratio of non-synonymous substitutions per non-synonymous site to synonymous substitutions per synonymous site (ω) was computed for every pair of orthologous genes using a custom perl pipeline. Briefly, orthologous gene-pairs were found using the reciprocal smallest distance algorithm [29], aligned with MUSCLE [30], and ω was calculated using the KaKs calculator [31]

with the YN00 method [32]. The statistical significance of orthologous pairs with $\omega > 1$ was assessed with a Fisher exact test. The time of divergence between the strains (τ) was estimated from the formula $\tau = Ks/(2*\lambda)$ where $\lambda = 8.3 \times 10^{-7}$ SNPs/site/year [33] mediated across all pairs of orthologs. Codon usage was calculated as described by Karlin [34] using a custom perl script. The relevant biases in codon usage were identified using the methods described in [35], [36]. Briefly, the codon usage of each gene is compared to the genome-wide mode of codon usage, and the significance is established using a Chi-square test. The p-value threshold was set to 0.1.

Cloning Experiments

All restriction enzymes were purchased from New England Biolabs (Beverly, MA, USA). All the PCR amplifications were performed using the Expand Long Template PCR system (Roche Applied Science, Indianapolis, IN, USA).

The genes conferring UV resistance were cloned in pFL122 as follows. A fosmid clone (GCLNU_G4) from the *P. profundum* 3TCK sequencing library containing P3TCK_10673 (*rpoX*; RNA polymerase sigma factor, ECF subfamily), P3TCK_10668 (Conserved Hypothetical Protein), P3TCK_10663 (*phr*; deoxyribodipyrimidine photolyase) was cut with *Xho*I+*Kpn*I. The 7.2 kbp band contained the genes of interest and was gel purified and ligated in pFL122 [37] cut with *Xho*I+*Kpn*I yielding pFL303. The deletion Δ22 was obtained by cutting pFL303 with *Eco*RI and re-ligating, which effectively eliminates P3TCK_10673, most of P3TCK_10668 and the region with the two divergent promoters between the two. This deletion construct was named pFL304. The promoter region was PCR amplified from pFL303 using primers PROMPHO2F (5′ – GTCGAATTCCTTTTCTTGCAGCGT-

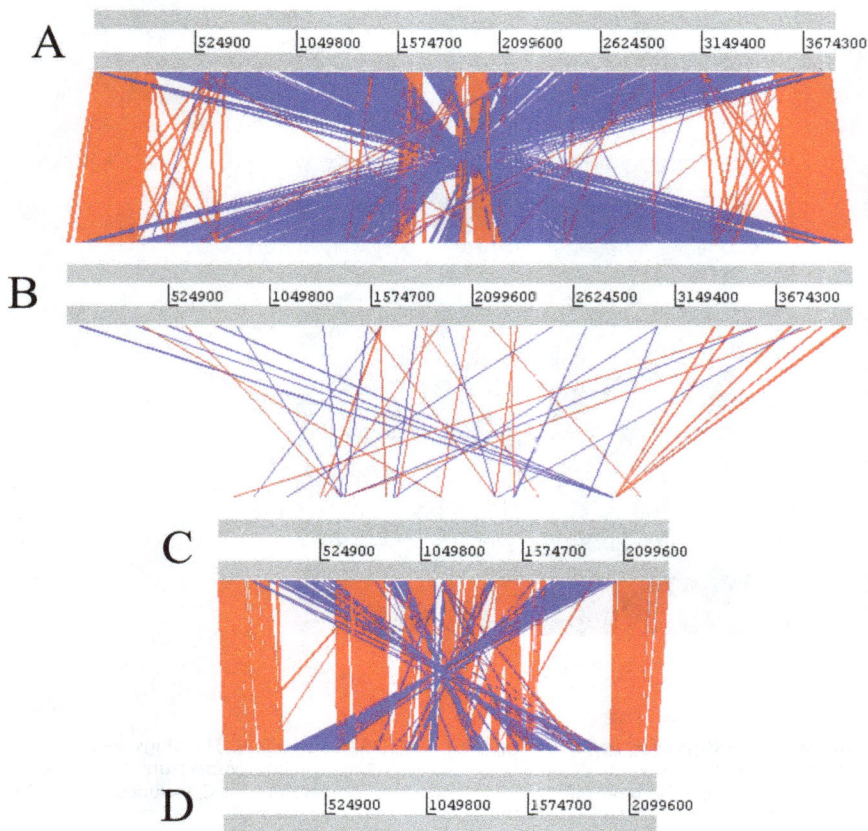

Figure 2. Chromosome organization in *P. profundum* strains. ACT nucleotide comparison [4] between the chromosomes of the two bathytypes SS9 and 3TCK. From top to bottom: (A) *P. profundum* 3TCK chromosome 1 (B) *P. profundum* SS9 chromosome 1 (C) *P. profundum* SS9 chromosome 2 (D) *P. profundum* 3TCK chromosome 2.

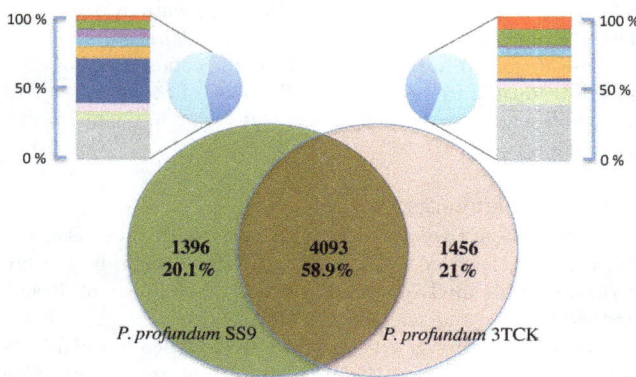

Figure 3. Gene content comparison in *P. profundum* strains. Shared gene comparison between the two bathytypes SS9 and 3TCK. The larger Venn diagram shows the number of orthologous genes shared and the number of genes unique to each bathytype. The smaller pie charts are the proportion of unique genes to each bathytype with (light blue) or without (dark blue) matches to the COG database. The matching genes were assigned to COG categories (from top to bottom): E-Amino acid transport and metabolism (red); G-Carbohydrate transport and metabolism (green); N-Motility and chemotaxis (violet); S-Function unknown (cyan); R-General function prediction only (orange); L-DNA Replication, Recombination and Repair (blue); T-Signal Transduction (pink); K-Transcription (light green); Other categories (grey).

CAGT - 3′) and PROMPHO2R (5′ – GTCGAATTCTAG-TAAGCGAATAGCAGGAC -3′).

Similarly the promoter region and the whole length P3TCK_10673 was amplified with primers SIGMAPHO2F (5′ – GTCGAATTCGTATTCAAGATGGGCACTCA – 3′) and the same reverse primer as above. Both amplicons were digested with *Eco*RI and cloned in the *Eco*RI site of pFL304 yielding pFL305 (promoter only) and pFL306 (promoter and P3TCK_10673) respectively. The directionality of the inserts was checked by PCR and confirmed by standard thermal cycle dideoxy sequencing with fluorescently labelled terminators (Applied Biosystems, Foster City, CA, USA).

For the arabinose-inducible UV resistance experiments, the *phr* gene, inclusive of its ribosome binding site, was amplified with primers expPHO2F (5′ – ATGGCCGTCTGCAA-GATCCTGTA -3′) and expPHO2R (5′ – GCTCTAGAGC-CACCCATTCATACGATGTGC – 3′), digested with *Eco*RI+ *Xba*I and cloned in the expression vector pFL190 [37] cut with the same enzymes.

In vivo Photoreactivation

The effect of ultraviolet light on the survival of *P. profundum* strains was tested as follows. Serial dilutions of late exponential cultures were plated on 75% strength 2216 Marine Agar. For each strain a triplicate dilution series was prepared: one untreated, one UV irradiated without blue light recovery, and one UV irradiated followed by a recovery period under blue light.

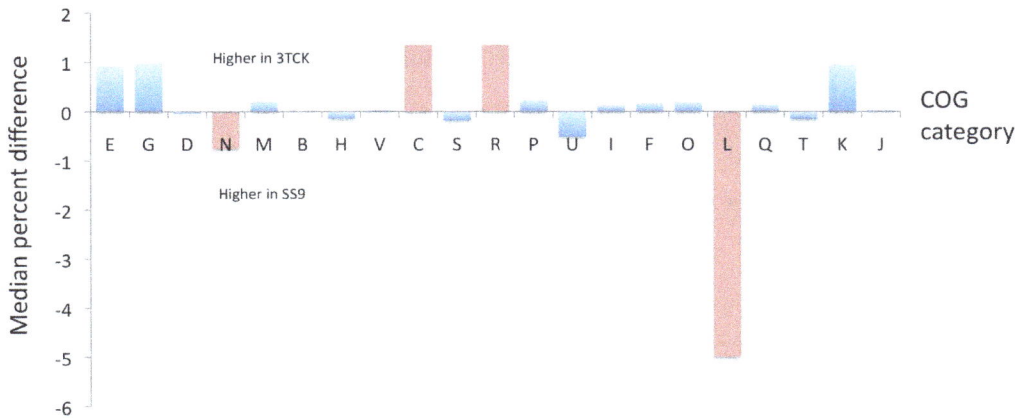

Figure 4. COG analysis comparison of the gene content in the two bathytypes. Reported are the median differences after resampling as described in material and methods. In pink the categories that were statistically over- or under-represented in one of the bathytypes at 98% significance.

The cells were irradiated uncovered using a germicidal lamp with an emission peak at 253.7 nm (Philips G25 T8), for 10 seconds at a power of 220 µW/cm2.

For the photoreactivation, the Petri dishes were then covered, to filter out the shorter wavelength radiation, and allowed to recover for 1 hour under "black" light (Philips TLD 15 W/08), emitting in the 350–400 nm range, at an irradiance of 20 µW/cm2. Irradiances were measured with a Spectroline DM-365 XA digital radiometer (Spectronics corp., Westbury, NY, USA).

The plates were wrapped in foil and grown at 15°C for 5 days after which c.f.u. were counted and the number of colonies in the irradiated samples were compared with the untreated controls to calculate the percent survival.

In all experiments, cell transfers and manipulations were performed under General Electric "gold" fluorescent light to prevent uncontrolled photorepair.

Results and Discussion

Bacteria and Archaea can be transported vertically through the water column as a result of attachment to sinking particles (see, for example [38]) and migrating metazoans or other phenomena such as advective transport [17]. Growth and survival at different

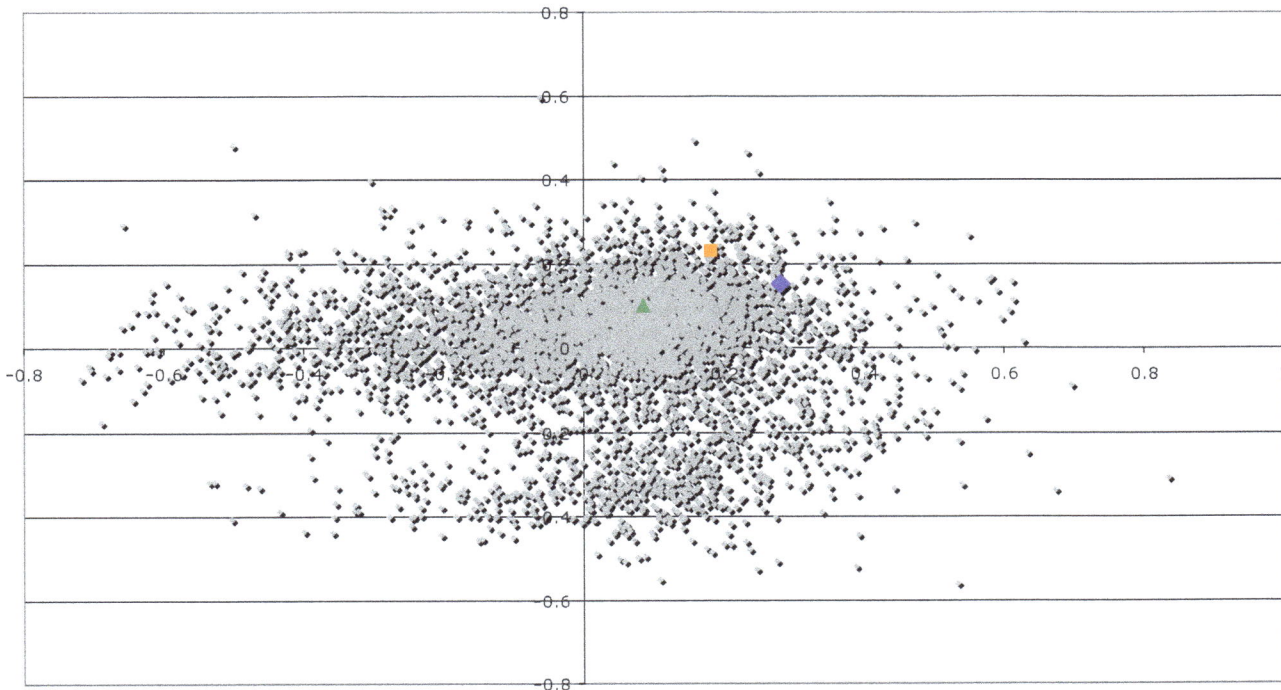

Figure 5. Correspondance analysis of codon usage for the ORFs of *P. profundum* 3TCK. The ORFs belonging to the *phr* gene cluster (in the insert) are color coded. The green triangle represents P3TCK_10668, the yellow square P3TCK_10673, and the blue diamond P3TCK_10663. The codon usage of P3TCK_10663 and P3TCK_10673 suggests their recent acquisition by horizontal gene transfer.

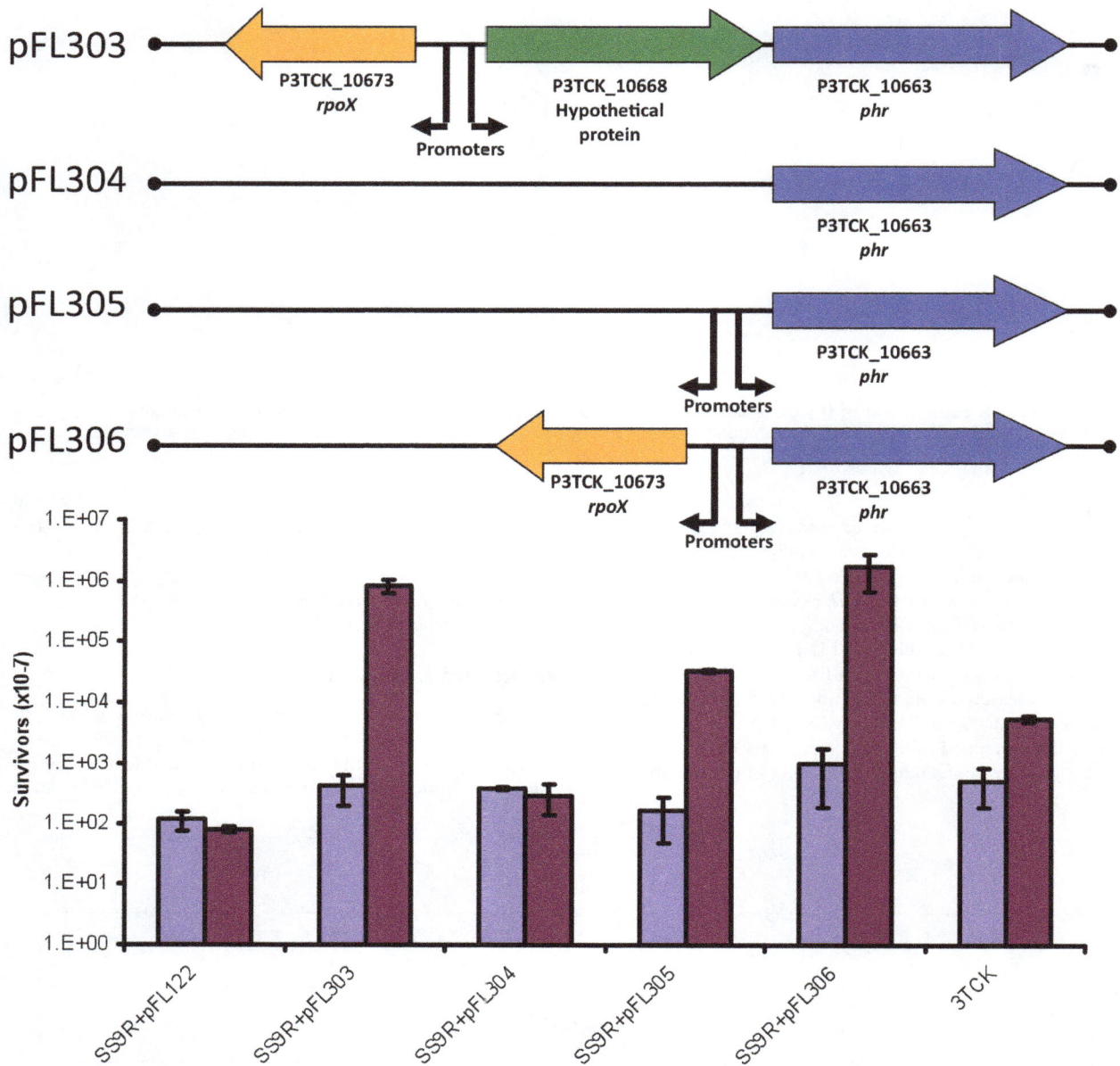

Figure 6. Introduction of the _phr_ gene cluster from the shallow bathytype 3TCK into the deep bathytype SS9 confers UV resistance. This phenotype is not observed in the deletion construct that lacks the upstream hypothetical protein and _rpoX_ gene (pFL304). The UV resistance phenotype can be partially restored by re-adding to pFL304 the promoter region of the cluster (pFL305) but is completely restored only when both the promoter and the _rpoX_ sigma factor are added (pFL306). The absence of P3TCK_10668 (Hypothetical Protein) does not affect UV sensitivity. The UV survival plots present the ratio of c.f.u. observed after UV exposure followed by blue-light photoreactivation (red) compared to the non-photoreactivated controls (blue) as described in the materials and methods.

depths requires adaptation to many depth-correlated chemo-physical parameters (e.g. light, hydrostatic pressure, organic carbon). For example, it has been shown that adapting to a higher hydrostatic pressure requires adjustments to membrane structure, DNA synthesis, translation, and protein quaternary structure [39]. Pressure also affects gene regulation at the level of transcription [12], [13] and translation [40]. The concentration of nutrients varies greatly as a function of depth and it is possible that marine bacteria use pressure as a proxy for depth in gene regulation to respond to differences in nutrient availability. The switch between different outer membrane porins as a function of

pressure, which has been observed in _P. profundum_ SS9, is likely a result of such a response. Conversely, adaptation to shallow waters would require the acquisition of novel genes to cope with unique stressors, such as UV light. Many of these features are evident from the genome comparisons of different bathytypes of _P. profundum_ presented here.

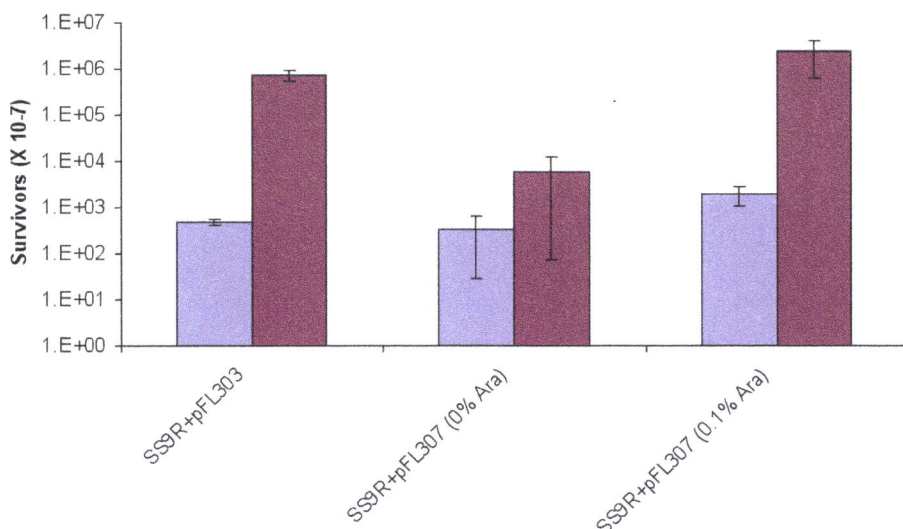

Figure 7. The UV resistant phenotype depends uniquely on the levels of expression of the *phr* gene (P3TCK_10673). Cloning of the *phr* gene under the arabinose-inducible promoter of pFL190 confers UV-resistance to the cells only when grown with 0.1% L-arabinose. The UV survival plots present the ratio of c.f.u. observed after UV exposure followed by blue-light photoreactivation (red) compared to the non-photoreactivated controls (blue) as described in the materials and methods.

General Features of the Draft Genome of *Photobacterium profundum* Strain 3TCK and Comparisons with the Genome of the Previously Sequenced Strain SS9

The draft genome of the shallow bathytype *Photobacterium profundum* 3TCK contains 11 scaffolds for a total length of 6,186,725 bp with average 41.3% GC encoding for a total of 5549 ORFs. Gene synteny plots and the existence of two different origins of replication [41] suggest that, similar to other members of the family Vibrionaceae [42], the genome is organised in two chromosomes. This size and structure is comparable to that of the previously sequenced deep bathytytpe SS9 [11], but appears to lack an 80 kb dispensable plasmid specific to SS9 [12].

The genome encodes for a complete set of tRNA synthetases and shares with SS9 the peculiarity of having the genes for the synthesis of selenocysteine and its incorporation into proteins. Like SS9 the genome also encodes for two complete F_0F_1-ATP-synthases and a type A FAS/PKS system [43] for the synthesis of polyunsaturated fatty acids such as eicosapentaenoic acid (EPA; 20: 5n−3).

P. profundum 3TCK has larger-than-average intergenic regions (~167 bp), a feature shared with most sequenced piezophiles [5], although the size of the intergenics is smaller than in the deep bathytype SS9 (~205 bp). These large intergenic regions have been shown to be transcribed and differentially expressed as a function of pressure [13], suggesting they could play an important physiological role.

P. profundum 3TCK contains at least 9 copies of the ribosomal RNA operon (rrn; Figure 1) a number which is larger than the median for microbial genomes [44]. Interestingly, SS9 still holds the record for the highest copy rrn number in a single genome with 15 copies. These operons display intragenomic variation in *P. profundum* SS9, while they are almost identical in 3TCK. The variability is concentrated in specific loops of the 16S and the 23S rRNA subunits [10], [45] and is predicted to contribute to the ribosome stability or function at high-hydrostatic pressure.

ACT comparisons [46] between the nucleotide sequences of the two strains highlights the presence of a number of insertions/

deletions, multiple inversions across the origin/terminus of both chromosomes, but only a limited number of translocations across the chromosomes (Figure 2).

The identity between the 16S gene sequences of 3TCK and SS9 is 98.73%, which suggests the strains belong to the same species, however the ANI between 3TCK and SS9 is 92.85% with a percent conserved DNA of 62.68%, which is well below the species threshold for genome-level comparisons (ANI>95%; conserved DNA>69%; [28]). The largest proportion of genes unique to each genome is located on chromosome 2 (Figure 2 and Figure 3). Within the family Vibrionaceae this chromosome has been previously implicated in gene capture for environmental adaptations during the colonization of new niches [42].

The global analysis of Ka/Ks ratios identified only four gene pairs with $\omega>1$, but none of these was statistically significant determined by a Fisher exact test (P<0.01). The median time of divergence between the strains was 126,833 years ago which is remarkably comparable to the time of establishment of the modern thermohaline circulation. This network of advective currents is the predicted cause behind the spatial separation of the 2 bathytypes.

The extreme synteny, large proportion of insertions and deletions in association with a low number of sequences with $\omega>1$ and the relatively short divergence time suggests a larger role for HGT rather than sequence substitutions in the evolution of bathytypes. Furthermore, it is a clear indication that the strains are currently undergoing adaptive radiation driven primarily by gene acquisition and loss.

COG comparisons between the two strains revealed a statistical over-representation in the shallow bathytype 3TCK of genes for energy production (COG category C,) but a significant decrease in genes for motility and chemotaxis (N) and DNA replication, recombination and repair (L) (Figure 4).

The abundance of genes for category L (DNA replication, recombination and repair) in the deep bathytype SS9 is due to the large number of transposable elements. The amplification of mobile elements seems to be a distinctive feature of all sequenced deep-sea genomes and has been observed in metagenomic surveys of different depths in the water column [3], [47]. The wide diversity of the identified transposases in the deep-sea samples that could not be accounted for by biases in community composition led DeLong et al. [47] to hypothesize that the over-representation of transposable elements relates to the slower growth and smaller effective population size of deep-sea microbial communities. Compatible with this hypothesis, the 206 COG-categorized transposable elements found in SS9 belong to 11 families, the most numerous of which (COG3436) has as many as 72 members. On the other hand, 3TCK encodes for just 3 COG-categorized transposable elements. Taken together these data support the hypothesis of intra-genomic amplification of transposable elements in the deep-sea due to habitat differences. Nevertheless, a transposon mutant with a cold-sensitive phenotype has been isolated in SS9 [48] suggesting that these mobile elements could have a functional role.

In *P. profundum* SS9 the genes for motility and flagellar assembly are arranged in two large clusters, one that is shared with 3TCK and a second one that is unique to the piezophile SS9 as a result of a large contiguous deletion in the genome of 3TCK. This deletion accounts for the under-representation of genes from COG category N (Motility and Chemotaxis) in the genome of 3TCK. This second cluster is most similar to a lateral flagella gene cluster present in some *Vibrio* strains [49] and its role in motility and chemotaxis has been previously analysed [50] comparing the piezosensitive motility of 3TCK to the piezoresistant one of SS9.

Another relevant genomic island encompasses genes PBPRA2666-PBPRA2712. This gene cluster is involved in cell-envelope biosynthesis and was initially suggested to be missing in 3TCK during a previous microarray study [12]. However, the genome sequence of 3TCK does, in fact, contain a similar and syntenic cluster (P3TCK_26512-P3TCK_26972), but with a highly divergent sequence. Interestingly, the cluster in SS9 contains at least 3 genes (PBPRA2678, PBPRA2681, PBPRA2684) that cause a cold-sensitive phenotype when inactivated [48], 8 genes that are regulated by temperature or pressure at the level of transcription (PBPRA2689, PBPRA2691, PBPRA2692, PBPRA2697, PBPRA2701, PBPRA2702, PBPRA2707, PBPRA2710) [12], [13] and 2 at the level of translation (PBPRA2686, PBPRA2687) [40].

Conversely, the over-representation in 3TCK of genes for Energy Production and Conversion (C) is caused by the expansion of COGs involved in anaerobic respiration of nitrate such as the periplasmic nitrate reductase napABCEG (COG0243, COG3043, COG3005, COG4459, COG1145), formate dehydrogenase (COG1526), cytochrome c553 (COG2863) and a two-component response regulator (COG2197) specific for nitrate reduction (P3TCK_05837). In addition chromosome 2 of 3TCK contained a 12 kb genomic island encompassing ORFs P3TCK_02186-P3TCK_02126) that encodes for the alpha (COG0804), beta (COG0832) and gamma (COG0831) subunits of urease and its accessory proteins. These genes are arranged in a single operon that also encoded for an ABC transporter. Their sequences did not display an altered GC%, GC skew or codon usage suggesting that they had been lost by SS9 rather than acquired by 3TCK. Nevertheless, their presence might be reflective of differences in the chemistry of the more eutrophic habitat of the San Diego Bay sediments (where *P. profundum* 3TCK was isolated) versus that associated with deep-sea amphipods (from which *P. profundum* SS9 was isolated) and is a further evidence of how the genome plasticity of *P. profundum* is the key to its adaptive radiation.

The Conversion of the Deep Bathytype to UV Resistance

The absence of light (apart from chemiluminescence or possibly bioluminescence) in the deepest depths of the oceans argues for a selective loss of genes associated with light tolerance. In fact the piezophile *Psychromonas* species strain CNPT3 [51] has been shown to be extremely sensitive to UV radiation [52].

The two most common types of UV-induced lesions on DNA are the generation of cyclobutane pyrimidine dimers (CPDs) and pyrimidine-pyrimidone 6–4 photoproducts (6–4PPs) [53]. This damage to DNA can be repaired by multiple pathways [53], but photoreactivation by deoxyribodipyrimidine photolyase, the product of the *phr* gene, is unique in that it requires blue light energy to split the CPDs or the 6–4PPs [54].

Because of this blue light requirement the genes for the deoxyribodipyrimidine photolyase are expected to be absent from the genomes of deep-sea microbes. In fact, the SS9 genome does not contain a *phr* gene and the metagenomic analysis of the distribution of genes in a stratified water column [47] showed significant over-representation of *phr* genes from the photic region when compared to samples from deeper waters.

In contrast to SS9, other members of the family Vibrionaceae have been shown to primarily rely on the activity of photolyases for the repair on UV-induced damage [55], [56]. The genome of *Vibrio cholerae* N16961 [57] encodes for three different members of the cryptochrome/photolyase family [56]. The first one (VCA0057) functions in repairing CPDs in dsDNA [56] the second one (VC1814) in repairing CPDs in ssDNA [55] while the function of the third one (VC1392) is still unknown. A similar array of photolyase-like ORFs can be seen in the genomes most other members of the Vibrionaceae, including the draft genome of *Photobacterium* sp. SKA34 (https://moore.jcvi.org/moore/SingleOrganism.do?speciesTag=SKA34) encoding for orthologs to all three cryptochromes/photolyases of *V. cholerae*.

The shallow bathytype 3TCK contains a *phr* gene within a three gene cluster (P3TCK_10663, P3TCK_10668, P3TCK10673) on chromosome 2 with altered codon usage (Figure 5). The *phr* gene (P3TCK_10663) and the upstream hypothetical protein (P3TCK_10668) are part of a predicted operon with a promoter upstream of P3TCK_10668 driving their expression. A different promoter in the opposite direction drives the transcription of *rpoX* (P3TCK_10673), an ECF-type sigma factor. Because of this arrangement, it was hypothesised that the *phr* gene cluster had been acquired by HGT under the selective forces provided by UV light exposure in shallow water.

The cluster was cloned on a broad host-range plasmid and introduced into SS9 (Figure 6; pFL303) resulting in approximately 1,000 fold increase in colony forming units (c.f.u.) survival after UV irradiation compared to the controls. This survival was dependant on blue-light incubation (Figure 6). A deletion encompassing *rpoX* and most of the hypothetical protein (P3TCK_10668) abolished photoreactivation (pFL304). If the promoter region is re-introduced in the right orientation into the deletion construct (pFL305), photoreactivation was partially restored, yielding approximately 100-fold more surviving c.f.u. than the untreated controls. The full restoration of the UV resistant phenotype could be obtained only by cloning, in the right orientation, both the promoter and the sigma factor (pFL306).

Based upon these results it was suggested that the gene encoding the hypothetical protein is dispensable for photoreactivation activity and the UV resistance phenotype is solely dependent on the level of expression of the *phr* gene. To test both hypotheses the *phr* gene alone was cloned in a vector (pFL190) with an arabinose-inducible promoter (pFL307). The full UV resistant phenotype was observed only after induction with 0.1% arabinose (Figure 7) indicating that high levels of expression of the *phr* gene alone are necessary and sufficient to confer UV resistance.

A noteworthy result of this experiment is that the functional activity of the *phr* gene cluster benefits from the presence of a flanking sigma factor. There is a precedent for this type of observation. Sometimes acquired genes must be obtained as clusters of functional units to overcome the barrier caused by the incapacity to transcribe the HGT gene at the appropriate level [58]. Genes providing marginal benefits, like photolyase, can also be readily lost from a population when the increased metabolic cost for replication is not balanced by selective pressure as was observed between high- and low-light-adapted *Prochlorococcus* strains [59]. These processes of gene cluster gain and loss generate and maintain the genomic diversity within bathytypes while restricting their niche.

Acknowledgments

The authors thank Verena Pfund, Alessandro Vezzi and Francesca Simonato for insightful discussions.

Author Contributions

Conceived and designed the experiments: FML DHB. Performed the experiments: FML EAE TKR NV SF JHJ. Analyzed the data: FML NV. Contributed reagents/materials/analysis tools: SF JHJ. Wrote the paper: FML DHB EAE TKR NV.

References

1. Witte U, Wenzhofer F, Sommer S, Boetius A, Heinz P, et al. (2003) In situ experimental evidence of the fate of a phytodetritus pulse at the abyssal sea floor. Nature 424: 763–766. doi:10.1038/nature01799.

2. Sogin ML, Morrison HG, Huber JA, Welch DM, Huse SM, et al. (2006) Microbial diversity in the deep sea and the underexplored "rare biosphere". Proc Natl Acad Sci USA 103: 12115–12120. doi:10.1073/pnas.0605121103.

3. Eloe EA, Fadrosh DW, Novotny M, Zeigler Allen L, Kim M, et al. (2011) Going Deeper: Metagenome of a Hadopelagic Microbial Community. PLoS ONE 6: e20388. doi:10.1371/journal.pone.0020388.

4. Brown MV, Philip GK, Bunge JA, Smith MC, Bissett A, et al. (2009) Microbial community structure in the North Pacific ocean. ISME J 3: 1374–1386. doi:10.1038/ismej.2009.86.

5. Lauro FM, Bartlett DH (2007) Prokaryotic lifestyles in deep sea habitats. Extremophiles 12: 15–25. doi:10.1007/s00792-006-0059-5.

6. Eloe EA, Malfatti F, Gutierrez J, Hardy K, Schmidt WE, et al. (2011) Isolation and characterization of a psychropiezophilic Alphaproteobacterium. Appl Environ Microbiol 77: 8145–8153. doi:10.1128/AEM.05204-11.

7. Khelaifia S, Fardeau ML, Pradel N, Aussignargues C, Garel M, et al. (2011) *Desulfovibrio piezophilus*, sp. nov., a novel piezophilic sulfate-reducing bacterium isolated from wood falls in Mediterranean Sea. Int J Syst Evol Microbiol 61: 2706–2711. doi:10.1099/ijs.0.028670-0.

8. Bale SJ, Goodman K, Rochelle PA, Marchesi JR, Fry JC, et al. (1997) *Desulfovibrio profundus* sp. nov., a novel barophilic sulphate-reducing bacterium from deep sediment layers in the Japan Sea. Int J Syst Bacteriol 47: 515–521. doi:10.1099/00207713-47-2-515.

9. Alain K, Marteinsson VT, Miroshnichenko ML, Bonch-Osmolovskaya EA, Prieur D, et al. (2002) *Marinitoga piezophila* sp. nov., a rod-shaped, thermo-piezophilic bacterium isolated under high hydrostatic pressure from a deep-sea hydrothermal vent. Int J Syst Evol Microbiol 52: 1331–1339. doi:10.1099/ijs.0.02068-0.

10. Lauro FM, Chastain RA, Blankenship LE, Yayanos AA, Bartlett DH (2007) The unique 16S rRNA genes of piezophiles reflect both phylogeny and adaptation. Appl Environ Microbiol 73: 838–845. doi:10.1128/AEM.01726-06.

11. Vezzi A, Campanaro S, D'Angelo M, Simonato F, Vitulo N, et al. (2005) Life at depth: *Photobacterium profundum* genome sequence and expression analysis. Science 307: 1459–1461. doi:10.1126/science.1103341.

12. Campanaro S, Vezzi A, Vitulo N, Lauro FM, D'Angelo M, et al. (2005) Laterally transferred elements and high pressure adaptation in *Photobacterium profundum* strains. BMC Genomics 6: 122. doi:10.1186/1471-2164-6-122.

13. Campanaro S, Pascale FD, Telatin A, Schiavon R, Bartlett DH, et al. (2012) The transcriptional landscape of the deep-sea bacterium *Photobacterium profundum* in both a toxR mutant and its parental strain. BMC Genomics 13: 567. doi:10.1186/1471-2164-13-567.

14. DeLong EF, Franks DG, Yayanos AA (1997) Evolutionary relationships of cultivated psychrophilic and barophilic deep-sea bacteria. Appl Environ Microbiol 63: 2105–2108.

15. Nogi Y, Masui N, Kato C (1998) *Photobacterium profundum* sp. nov., a new, moderately barophilic bacterial species isolated from a deep-sea sediment. Extremophiles 2: 1–7.

16. Biddle JF, House CH, Brenchley JE (2005) Enrichment and cultivation of microorganisms from sediment from the slope of the Peru Trench (ODP Site 1230). In: Jørgensen BB, D'Hondt SL, Miller DJ (Eds.) Proc ODP Sci Results 201: 1–19.

17. Wilkins D, van Sebille E, Rintoul SR, Lauro FM, Cavicchioli R (2013) Advection shapes Southern Ocean microbial assemblages independent of distance and environment effects. Nat Commun 4: 2457. doi:10.1038/ncomms3457.

18. Vanlint D, Mitchell R, Bailey E, Meersman F, McMillan PF, et al. (2011) Rapid acquisition of gigapascal-high-pressure resistance by *Escherichia coli*. mBio 2: e00130–10. doi:10.1128/mBio.00130-10.

19. Hutchinson GE (1957) Concluding remarks. Cold Spring Harbor Symposia on Quantitative Biology 22: 415–427.

20. Yayanos AA, Vanboxtel R (1982) Coupling device for quick high-pressure connections to 100 Mpa. Review of Scientific Instruments 53: 704–705.

21. Yayanos AA (2001) Deep-sea piezophilic bacteria. In: J. Paul (ed.), Marine Microbiology. Academic Press p. 615–638.

22. Chi E, Bartlett DH (1993) Use of a reporter gene to follow high-pressure signal-transduction in the deep-sea bacterium *Photobacterium* sp. strain SS9. J Bacteriol 175: 7533–7540.

23. Goldberg SMD, Johnson J, Busam D, Feldblyum T, Ferriera S, et al. (2006) A Sanger/pyrosequencing hybrid approach for the generation of high-quality draft assemblies of marine microbial genomes. Proc Natl Acad Sci USA 103: 11240–11245. doi:10.1073/pnas.0604351103.

24. Huson DH, Reinert K, Kravitz SA, Remington KA, Delcher AL, et al. (2001) Design of a compartmentalized shotgun assembler for the human genome. Bioinformatics 17: S132–S139. doi:10.1093/bioinformatics/17.suppl_1.S132.

25. Nakasone K, Masui N, Takaki Y, Sasaki R, Maeno G, et al. (2000) Characterization and comparative study of the rrn operons of alkaliphilic *Bacillus halodurans* C-125. Extremophiles 4: 209–214. doi:10.1007/PL00010713.

26. Allen MA, Lauro FM, Williams TJ, Burg D, Siddiqui KS, et al. (2009) The genome sequence of the psychrophilic archaeon, *Methanococcoides burtonii*: the role of genome evolution in cold adaptation. ISME J 3: 1012–1035. doi:10.1038/ismej.2009.45.

27. Rodriguez-Brito B, Rohwer F, Edwards RA (2006) An application of statistics to comparative metagenomics. BMC Bioinformatics. 7: 162. doi:10.1186/1471-2105-7-162.

28. Goris J, Konstantinidis KT, Klappenbach JA, Coenye T, Vandamme P, et al. (2007) DNA-DNA hybridization values and their relationship to whole-genome sequence similarities. Int J Syst Evol Microbiol 57: 81–91. doi:10.1099/ijs.0.64483-0.

29. Wall DP, Deluca T (2007) Ortholog detection using the reciprocal smallest distance algorithm. Methods Mol Biol 396: 95–110. doi:10.1007/978-1-59745-515-2_7.

30. Edgar RC (2004) MUSCLE: multiple sequence alignment with high accuracy and high throughput. Nucl. Acids Res 32: 1792–1797. doi:10.1093/nar/gkh340.

31. Zhang Z, Li J, Zhao XQ, Wang J, Wong GK, et al. (2006) KaKs Calculator: Calculating Ka and Ks through model selection and model averaging. Genomics Proteomics Bioinformatics 4: 259–263. doi:10.1016/S1672-0229(07)60007-2.

32. Yang Z, Nielsen R (2000) Estimating synonymous and nonsynonymous substitution rates under realistic evolutionary models. Mol Biol Evol 17: 32–43.

33. Mutreja A, Kim DW, Thomson NR, Connor TR, Lee JH, et al. (2011) Evidence for multiple waves of global transmission within the seventh cholera pandemic. Nature 477: 462–465. doi:10.1038/nature10392.

34. Karlin S, Mrazek J, Campbell AM (1998) Codon usages in different gene classes of the *Escherichia coli* genome. Mol Microbiol 29: 1341–1355.

35. Davis JJ, Olsen GJ (2010) Modal codon usage: assessing the typical codon usage of a genome. Mol Biol Evol 27: 800–810.

36. Davis JJ, Olsen GJ (2011) Characterizing native codon usages of a genome: an axis projection approach. Mol Biol Evol 28: 211–221.

37. Lauro FM, Eloe EA, Liverani N, Bertoloni G, Bartlett DH (2005) Conjugal vectors for cloning, expression, and insertional mutagenesis in gram-negative bacteria. Biotechniques 38: 708–712.

38. Tamburini C, Garcin J, Gregori G, Leblanc K, Rimmelin P, et al. (2006) Pressure effects on surface Mediterranean prokaryotes and biogenic silica dissolution during a diatom sinking experiment. Aquat Microb Ecol 43: 267–276. doi:10.3354/ame043267.

39. Bartlett DH (2002) Pressure effects on in vivo microbial processes. Biochim Biophys Acta 1595: 367–381.

40. Le Bihan T, Rayner J, Roy MM, Spagnolo L (2013) *Photobacterium profundum* under pressure: a MS-based label-free quantitative proteomics study. PLoS ONE 8: e60897. doi:10.1371/journal.pone.0060897.

41. Egan ES, Waldor MK (2003) Distinct replication requirements for the two *Vibrio cholerae* chromosomes. Cell 114: 521–530. doi:10.1016/S0092-8674(03)00611-1.

42. Okada K, Iida T, Kita-Tsukamoto K, Honda T (2005) Vibrios commonly possess two chromosomes. J Bacteriol 187: 752–757. doi:10.1128/JB.187.2.752-757.2005.

43. Shulse CN, Allen EE (2011) Widespread occurrence of secondary lipid biosynthesis potential in microbial lineages. PLoS ONE 6: e20146. doi:10.1371/journal.pone.0020146.

44. Větrovský T, Baldrian P (2013) The variability of the 16S rRNA gene in bacterial genomes and its consequences for bacterial community analyses. PLoS ONE 8: e57923. doi:10.1371/journal.pone.0057923.

45. Pei A, Nossa CW, Chokshi P, Blaser MJ, Yang L, et al. (2009) Diversity of 23S rRNA Genes within Individual Prokaryotic Genomes. PLoS ONE 4: e5437. doi:10.1371/journal.pone.0005437.

46. Carver TJ, Rutherford KM, Berriman M, Rajandream MA, Barrell BG, et al. (2005) ACT: the Artemis Comparison Tool. Bioinformatics 21 3422–3423. doi:10.1093/bioinformatics/bti553.

47. DeLong EF, Preston CM, Mincer T, Rich V, Hallam SJ, et al. (2006) Community genomics among stratified microbial assemblages in the ocean's interior. Science 311: 496–503. doi:10.1126/science.1120250.

48. Lauro FM, Tran K, Vezzi A, Vitulo N, Valle G, et al. (2008) Large-scale transposon mutagenesis of Photobacterium profundum SS9 reveals new genetic loci important for growth at low temperature and high pressure. J Bacteriol 190: 1699–1709. doi:10.1128/JB.01176-07.

49. McCarter LL (2004) Dual flagellar systems enable motility under different circumstances. J. Mol. Microbiol. Biotechnol. 7: 18–29. doi:10.1159/000077866.

50. Eloe EA, Lauro FM, Vogel RF, Bartlett DH (2008) The deep-sea bacterium Photobacterium profundum SS9 utilizes separate flagellar systems for swimming and swarming under high-pressure conditions. Appl Environ Microbiol 74: 6298–6305. doi:10.1128/AEM.01316-08.

51. Lauro FM, Stratton TK, Chastain RA, Ferriera S, Johnson J, et al. (2013) Complete Genome Sequence of the Deep-Sea Bacterium Psychromonas Strain CNPT3. Genome Announc 1: e00304–13. doi:10.1128/genomeA.00304-13.

52. Lutz L, Yayanos AA (1986) UV Repair in Deep-Sea Bacteria. Federation Proceedings 45: 1784–1784.

53. Yasui A, McCready SJ (1998) Alternative repair pathways for UV-induced DNA damage. Bioessays 20: 291–297.

54. Todo T (1999) Functional diversity of the DNA photolyase/blue light receptor family. Mutat Res 434: 89–97.

55. Selby CP, Sancar A (2006) A cryptochrome/photolyase class of enzymes with single-stranded DNA-specific photolyase activity. Proc Natl Acad Sci USA 103: 17696–17700. doi:10.1073/pnas.0607993103.

56. Worthington EN, Kavakli IH, Berrocal-Tito G, Bondo BE, Sancar A (2003) Purification and characterization of three members of the photolyase/cryptochrome family blue-light photoreceptors from Vibrio cholerae. J Biol Chem 278: 39143–39154. doi:10.1074/jbc.M305792200.

57. Heidelberg JF, Eisen JA, Nelson WC, Clayton RA, Gwinn ML, et al. (2000) DNA sequence of both chromosomes of the cholera pathogen Vibrio cholerae. Nature 406: 477–483. doi:10.1038/35020000.

58. Kurland CG, Canback B, Berg OG (2003) Horizontal gene transfer: a critical view. Proc Natl Acad Sci USA 100: 9658–9662. doi:10.1073/pnas.1632870100.

59. Rocap G, Larimer FW, Lamerdin J, Malfatti S, Chain P, et al. (2003) Genome divergence in two Prochlorococcus ecotypes reflects oceanic niche differentiation. Nature 424: 1042–1047. doi:10.1038/nature01947.

60. Murray NE, Brammar WJ, Murray K (1977) Lambdoid phages that simplify the recovery of in vitro recombinants. Mol Gen Genet 150: 53–61.

61. Hanahan D (1983) Studies on transformation of Escherichia coli with plasmids. J Mol Biol 166: 557–580.

62. Better M, Helinski DR (1983) Isolation and characterization of the recA gene of Rhizobium meliloti. J Bacteriol 155: 311–316.

8

Comparative DNA Damage and Repair in Echinoderm Coelomocytes Exposed to Genotoxicants

Ameena H. El-Bibany, Andrea G. Bodnar, Helena C. Reinardy*

Molecular Discovery Laboratory, Bermuda Institute of Ocean Sciences, St. George's, Bermuda

Abstract

The capacity to withstand and repair DNA damage differs among species and plays a role in determining an organism's resistance to genotoxicity, life history, and susceptibility to disease. Environmental stressors that affect organisms at the genetic level are of particular concern in ecotoxicology due to the potential for chronic effects and trans-generational impacts on populations. Echinoderms are valuable organisms to study the relationship between DNA repair and resistance to genotoxic stress due to their history and use as ecotoxicological models, little evidence of senescence, and few reported cases of neoplasia. Coelomocytes (immune cells) have been proposed to serve as sensitive bioindicators of environmental stress and are often used to assess genotoxicity; however, little is known about how coelomocytes from different echinoderm species respond to genotoxic stress. In this study, DNA damage was assessed (by Fast Micromethod) in coelomocytes of four echinoderm species (sea urchins *Lytechinus variegatus*, *Echinometra lucunter lucunter*, and *Tripneustes ventricosus*, and a sea cucumber *Isostichopus badionotus*) after acute exposure to H_2O_2 (0–100 mM) and UV-C (0–9999 J/m^2), and DNA repair was analyzed over a 24-hour period of recovery. Results show that coelomocytes from all four echinoderm species have the capacity to repair both UV-C and H_2O_2-induced DNA damage; however, there were differences in repair capacity between species. At 24 hours following exposure to the highest concentration of H_2O_2 (100 mM) and highest dose of UV-C (9999 J/m^2) cell viability remained high (>94.6±1.2%) but DNA repair ranged from 18.2±9.2% to 70.8±16.0% for H_2O_2 and 8.4±3.2% to 79.8±9.0% for UV-C exposure. Species-specific differences in genotoxic susceptibility and capacity for DNA repair are important to consider when evaluating ecogenotoxicological model organisms and assessing overall impacts of genotoxicants in the environment.

Editor: Sam Dupont, University of Gothenburg, Sweden

Funding: The authors would like to thank a Bermuda Charitable Trust for the financial support to conduct this study. AHEB was supported by the National Science Foundation's Research Experiences for Undergraduates (REU, Grant No. 1266880). The funders had no role in study design, data collection and analysis, decision to publish, or preparation of the manuscript.

Competing Interests: The authors have declared that no competing interests exist.

* Email: helena.reinardy@bios.edu

Introduction

There has been much interest to integrate assessment of genetic effects into environmental studies to broaden the understanding of ecotoxicological impacts on organisms and populations [1–4]. Maintenance of DNA integrity is essential for proper cellular and organismal function, and the capacity to withstand genotoxic challenge is important to avoid long-term genetic instability and population vulnerability [5]. Unrepaired DNA damage can lead to mutations, cellular senescence, apoptosis, progression of cancer [6], and the process of aging [7]. Of particular concern in ecotoxicology is the potential for chronic effects and trans-generational impacts on populations by transfer of damaged DNA to offspring [8]. To minimize the harmful consequences of DNA damage, organisms are equipped with a variety of cellular defense and DNA repair mechanisms.

DNA is constantly damaged by both endogenous and exogenous sources, and genotoxicity can be considered as an imbalance between DNA damage and DNA repair mechanisms. Two major model genotoxicants are ultraviolet (UV) radiation and hydrogen peroxide (H_2O_2), which each induce different forms of DNA lesions. UV-C (<280 nm) is absorbed by the ozone in the earth's atmosphere and UV-B is the main component of UV radiation of environmental concern [9]; however, both UV-B and UV-C induce formation of cyclobutane-pyrimidine dimers (CPDs) and 6-4 photoproducts (6-4PPs) [10], in addition to DNA strand breaks [11]. UV-C induces high levels of DNA damage [12] and is commonly used as a model genotoxicant to investigate biological effects of UV irradiation [13,14]. H_2O_2 is produced as a byproduct of metabolic processes and cellular defense mechanisms [15], and is an important reactive oxygen species (ROS) involved in exogenously-induced oxidative DNA damage [16]. Antioxidant activity can restrict oxidative DNA lesions to several hundred per day, but excess ROS or a deficiency in antioxidants can lead to increased base oxidation and DNA strand breaks [17]. UV- and H_2O_2-induced DNA damage are primarily repaired by nucleotide excision repair (NER) and base excision repair (BER), respectively [18,14]. Investigation of DNA damage and repair after exposure to these two genotoxicants can inform on susceptibility to both oxidative damage and UV-induced DNA lesions, in addition to the capacity for both BER and NER.

Marine invertebrates have been extensively studied as bioindicators of environmental stress [19], and the sea urchin embryo test has served as a sensitive indicator of pollutant genotoxicology, embryo-toxicity, and teratogenicity [20–23]. Activation of DNA damage checkpoints, DNA repair, and apoptosis in sea urchin embryos have been demonstrated in response to genotoxicants such as methyl methanesulfonate, bleomycin, and exposure to ultraviolet radiation [24–26]. Despite the fact that sea urchin embryos are frequently used in toxicity testing, little is known of the effects of genotoxicants on the cells of adult sea urchins. Information about the cellular response of adult sea urchins to environmental stress is valuable for ecotoxicological studies and would increase understanding of the life history traits of these animals. Life history studies show that different species of sea urchins exhibit a very large range of reported lifespans (from approximately 3 to more than 100 years) [27–30], there is little evidence of senescence [31], and few reported cases of neoplasia [32,33]. Investigating DNA damage and DNA repair in cells of different sea urchin species would provide valuable information on selection of appropriate bioindicator species, allow assessments of environmental stress on different species, and shed light on mechanisms underlying life history traits of these animals.

The open circulatory system of echinoderms is comprised of coelomic fluid containing different cells types, collectively termed 'coelomocytes'. Coelomocytes fall into one of three categories: phagocytes, spherule cells (red and colorless), and vibratile cells, with further sub-categories within each cell type [34]. Coelomocytes play an integral role in immune cell functions such as fighting microbial infections and wound healing [34]. Damage to coelomocytes can compromise these essential functions, directly affecting the health of organisms and stability of populations. Coelomocytes (or circulating cells) from a variety of terrestrial and aquatic organisms (e.g. earthworms, bivalves, fish) have been useful bioindicators of environmental stress and are frequently used to assess genotoxicity [35–40]. Changes in the number and/or composition of coelomocytes have been reported in sea urchins from contaminated environments and those exposed to elevated $pCO2$ or increased temperature, suggesting that sea urchin coelomocytes may also serve as sensitive indicators of environmental stress [9,41–44]. However, another study showed that DNA from coelomocytes of the sea urchin *Lytechinus variegatus* is relatively resistant to genotoxicants [45]. Understanding susceptibility to DNA damage and DNA repair capacity of coelomocytes from different echinoderm species would be useful in assessing the value of coelomocytes as bioindicator cells and understanding the overall impacts of genotoxicants on these organisms. Persistent genotoxic damage is dependent on the balance between repair and replacement of damaged cells. Studies on echinoderms indicate a low level of cell turnover in the coelomocyte population ($<1.5\%$ BrdU incorporation in 3 hours [46] or 16 hours [47] in star fish) and low levels of apoptosis following acute exposures to UV-B [9], UV-C, hydrogen peroxide, methylmethane sulfonate and benzo [a]pyrene [45]; however, the DNA repair capacity of coelomocytes from different echinoderm species has not been investigated.

The objectives of this study are to assess the capacity to which cells from different echinoderm species are able to repair different types of DNA damage after exposure to two model genotoxicants, UV-C and H_2O_2. The specific aims are to comparatively evaluate the DNA damage and DNA repair capabilities in coelomocytes of four echinoderm species (sea urchins *L. variegatus*, *Echinometra lucunter lucunter*, *Tripneustes ventricosus*, and sea cucumber *Isostichopus badionotus*). We hypothesize that echinoderm coelomocytes will be able to repair some level of DNA damage, and the extent of genotoxicity sensitivity and DNA repair capacity will differ among species.

Materials and Methods

Animal collection and maintenance

All animals were collected and maintained in strict accordance with the Collecting and Experimental Ethics Policy (CEEP) of the Bermuda Institute of Ocean Sciences. All experiments complied with the ethical policy of the CEEP committee and did not require specific approval. All experiments were carried out on coelomocytes extracted from animals with minimal impact, except for a single small *E. l. lucunter* which was sacrificed in order to collect sufficient coelomic fluid for the experiment, and all efforts were made to minimize suffering. Except as mentioned above, all animals showed no adverse behavioral effects of the coelomocyte sampling procedure, all animals survived the procedure, and all animals were returned to their collection location.

Collection of animals complied with the collection policy of CEEP, no species were endangered, and no animals were collected from protected locations. Collection numbers of *L. variegatus*, *I. badionotus*, and *T. ventricosus* were within the CEEP collection limits and no specific collection permission was required. Collection of *E. l. lucunter* was carried out under a Department of Environmental Protection special permit (permit no, 131002, Bermuda Government), approved by the Director of Environmental Protection. All species were collected from the shallow sublittoral zone (less than 2 m depth at low tide), September–October, 2013, in Bermuda. *L. variegatus* and *I. badionotus* were collected from Harrington Sound (32°19.4'N, 64°43.6'W), *T. ventricosus* were collected from Fort St. Catherine beach (32°23.3'N, 64°40.3'W), and *E. l. lucunter* were collected from Castle Harbor (32°21.2'N, 64°39.8'W) and Gravelly Bay (32°19.1'N, 64°42.8'W). Animal husbandry and maintenance complied with CEEP policy. Sea urchins were maintained in flow-through aquaria with ambient temperature and light, and were left to acclimate for a minimum of 1 week after collection. *I. badionotus* were maintained in an outdoor flow-through aquarium with a layer of sediment on the bottom, and were left to acclimate for 1 week. Sea urchins were fed weekly with fresh sea grass, and sediment was replenished fortnightly in the *I. badionotus* aquarium.

Coelomocyte collection and treatment

Unless otherwise specified, all chemicals were sourced from Sigma-Aldrich (Sigma-Aldrich Co., St. Louis, MO, USA). Sea urchin test diameter was measured with calipers, and 2–6 ml coelomic fluid was extracted by syringe with an 18-guage needle inserted through the peristomial membrane surrounding the Aristotle's lantern. Sea cucumber size was estimated by weight, width, and length measurements, and 6–10 ml coelomic fluid was extracted by syringe with a 21-guage needle inserted laterally in the mid-body region. The experiments were designed to include a single coelomocyte collection per animal, division of the coelomic fluid for UV-C or H_2O_2 treatment, and proceeding concurrently with exposure and recovery period of both sets of treatment samples. Cell concentration, cell viability, and differential cell counts (red and other coelomocytes) were calculated after 1:1 dilution with trypan blue [0.8% trypan blue in calcium-magnesium-free seawater (CMFSW: 460 mM NaCl, 10 mM KCL, 7 mM Na_2SO_4, 2.4 mM $NaHCO_3$, pH 7.4) containing 30 mM EDTA] using a haemocytometer (Neubauer Bright Line haemocytometer). The volume for 50,000 cells per assay reaction (in triplicate or quadruplicate) was estimated and aliquoted into microcentrifuge tubes for each exposure. From the species selected for this study, cell aggregation was not a considerable factor in the experimental set-up. Coelomocytes from *L. variegatus*, *T. ventricosus*, and *I. badionotus* did not exhibit a strong agglutination reaction and could easily be dissociated to single cell suspensions by gently pipetting or vortexing. *E. l. lucunter* coelomocytes did exhibit some aggregation but clumps of cells were avoided when sample aliquots were taken. Differential cell counts and cell viability were estimated on all control and highest-exposed (9999 J/m^2 and 100 mM for UV-C and H$_2$O$_2$, respectively) samples after 24 hours recovery.

For the UV-C (254 nm) treatment, coelomocyte samples (25–132 μl volume) were irradiated (0, 250, 1000, 3000, or 9999 J/m^2) in 0.5 ml open microcentrifuge tubes in a Stratalinker UV Crosslinker 1800 (Stratagene, La Jolla, CA, USA). The recovery period was timed to begin immediately after dose delivery, and samples were left to recover for 0, 1, 3, 6, and 24 hours in the dark at room temperature. At each recovery timepoint, samples were placed on ice to halt DNA repair and processed for the Fast Micromethod assay.

For the H$_2$O$_2$ treatment, coelomocyte samples were exposed in 1.5 ml microcentrifuge tubes. H$_2$O$_2$ stock dilutions were prepared in CMFSW and added to coelomocyte samples to give the following final concentrations: 0, 0.1, 1, 10, or 100 mM H$_2$O$_2$. Samples were left in the dark for 10 min followed by 5 min centrifugation (8000 g) at room temperature. H$_2$O$_2$ exposure was halted by removal of supernatant after centrifugation, and cells were re-suspended in cell-free coelomic fluid (CFCF, prepared by collection of supernatant after centrifugation, 13000 g for 5 min, of coelomic fluid to remove cells) and the recovery period was started. At each recovery timepoint, samples were placed on ice to halt DNA repair, and processed for the Fast Micromethod assay.

Fast Micromethod for estimation of DNA damage

The method for fluorescent detection of alkaline DNA unwinding was carried out as described by Schröder et al. [48], with minor modifications. In brief, samples were assayed after respective periods of recovery and coelomocyte sample volume was adjusted with CFCF to make up to 50,000 cells per reaction. Samples were assayed in triplicate or quadruplicate by loading 20 μl sample to each replicate well on a black-walled 96-well microplate (USA Scientific, Inc., Ocala, FL, USA), and 20 μl of suitable blank (CMFSW or CFCF) were added to control wells. In some instances for *L. variegatus*, fewer cells were used per reaction when the cell concentration in coelomic fluid was low. Cells were lysed by adding 20 μl of lysing solution (9.0 M urea, 0.1% SDS, 0.2 M EDTA) containing 1:49 PicoGreen (Life Technologies, Grand Island, NY, USA, P7581), and left in the dark on ice for 40 min. DNA unwinding solution (20 mM EDTA, 1 M NaOH) was added (200 μl) to initiate alkaline unwinding (pH 12.4±0.02), fluorescence was detected (kinetic mode, excitation 480 nm, emission 520 nm, SpectraMax M2 Microplate Reader, Molecular Devices, CA, USA), and relative fluorescent units (RFU) was recorded every 5 min for a 30-min period. DNA unwinding was carried out at room temperature.

DNA damage was calculated according to the strand scission factor (SSF) equation [48]: SSF = log (% dsDNA$_{sample}$/% dsDNA$_{control}$)×(−1), where dsDNA$_{sample}$ are the treated samples and dsDNA$_{control}$ are the unexposed samples, and percentages are calculated from RFU after 20-min unwinding compared with initial (0 min unwinding) RFU, after subtracting respective blank RFU (CMFSW or CFCF). Due to high background fluorescence in CFCF from *I. badionotus*, RFU for that species were blanked with CMFSW RFU, but other species' RFU were blanked with individual CFCF RFU.

Analyses

Both treatments (UV-C and H$_2$O$_2$) were conducted concurrently on a single coelomocyte sample per animal, and different animals (*T. ventricosus* n = 5, *L. variegatus* n = 12, *E. l. lucunter* n = 8, and *I. badionotus* n = 8) were considered biological replicates in all analyses. DNA damage estimation by Fast Micromethod included technical replicates (n = 3–4) of each sample to give an overall SSF per sample for each animal, and all biological replicates were combined for analyses of coelomocyte parameters, initial dose/concentration response, and DNA damage profiles over the 24-hour period of recovery.

Statistical analyses were performed in Statgraphics Centurion XVI.I (StatPoint Technologies, Inc., VA, USA). Intraspecific effects of size on DNA damage (SSF) during the 24-hour recovery period was tested by general linear model (GLM) with test diameter (average length for *I. badionotus*), dose/concentration, and recovery time as quantitative variables. To investigate intraspecific effects of concentration/dose and time, all individuals within a species were combined and DNA damage (SSF) was tested by GLM with concentration/dose and time as quantitative independent variables; dose/concentration differences from controls after 24 hours recovery were tested by one-way ANOVA or Kruskal-Wallis (for normally distributed or non-normally distributed data, respectively), with *post-hoc* Fisher's least significant difference (LSD) test at the 95% confidence level. Differences in DNA repair between species were tested by GLM, with species as a categorical factor, and concentration/dose and time as quantitative independent variables; species differences were established by *post-hoc* multiple range tests. Additionally, DNA repair was estimated as the percentage of DNA damage after 24 hours recovery compared with initial (0-hours recovery) DNA damage for each individual and for each exposure level, following the equation: % DNA repair = 100−((T$_{24}$ SSF/T$_0$ SSF)×100), where T$_{24}$ SSF is SSF after 24 hours recovery, and T$_0$ SSF is the initial (0-hours recovery) SSF; negative DNA repair values indicated no DNA repair and were set to zero. DNA repair (%) data was arcsine transformed to test for intraspecific differences in repair capacity (ANOVA, *post-hoc* multiple range tests). DNA repair capacity was categorized as follows: low (<25% DNA repair), moderate (25–50% DNA repair), high (50–75% DNA repair), or very high (>75% DNA repair).

Results

No anti-coagulant was used for collection of coelomic fluid and there was minimal or no cell aggregation in coelomocytes from *L. variegatus*, *T. ventricosus*, and *I. badionotus*. A proportion of coelomocytes from *E. l. lucunter* aggregated within the first few minutes after collection, clumps were disaggregated by gently pipetting before analysis, and persistent clumps were avoided. Coelomocytes isolated from the different species were evaluated for cell concentration, proportion of white to red cells, and cell viability. *E. l. lucunter* and *I. badionotus* had significantly higher total coelomocyte concentrations compared with the other species (Kruskal-Wallis and multiple range test, p<0.05), and no red coelomocytes were observed in any sample from *I. badionotus* (Table 1). There was no significant cell death in any of the coelomocyte samples over the course of the study, with cell viability >94% 24 hours after exposure to UV-C or H$_2$O$_2$ (Table 1). A slight reduction in overall coelomocyte size was

Table 1. Number of individuals, size ranges, coelomocyte characterization, and cell viability after 24-hours recovery from highest levels exposures to UV-C and H_2O_2 of all echinoderms tested.

Species	n	Test diameter range (mm)	Pre-treatment		Viability 24 hours recovery after treatment (% of total cells)	
			Coelomocyte concentration (cells/μl)	Red coelomocytes (% of total cells)	UV-C (9999 J/m²)	H_2O_2 (100 mM)
T. ventricosus	5	101–115	1855±280 A	7.6±2.4 C	99.8±0.2	94.6±1.2
L. variegatus	12	47–85	1957±220 A	9.1±1.6 C	98.5±0.4	98.4±0.6
E. l. lucunter	8	27–71	4565±745 B	8.0±3.1 B	99.4±0.3	99.7±0.1
I. badionotus	8	87–258*	4386±839 B	0 D	99.2±0.3	98.1±0.5

*Length (mm, average of several measurements during sampling) was measured for I. badionotus. Different letters (A/B/C/D) denote significant interspecific difference in means (multiple range tests, p<0.05).
Data are mean ± s.e.m.

observed after 24 hours recovery from the highest levels of UV-C and H_2O_2.

Coelomocytes from all species showed an increase in DNA damage with increasing concentration or dose of genotoxicant (Figure 1). Patterns of dose responses indicated higher sensitivity in *T. ventricosus* and lower sensitivity in *E. l. lucunter* coelomocytes exposed to H_2O_2. *I. badionotus* had a lower magnitude of DNA damage after both genotoxicant treatments compared with the sea urchin species, and there was considerable inter-individual variation. The different sea urchin species responses to UV-C exposure were similar, with a slight indication of higher DNA damage at the highest doses in *T. ventricosus*.

Individuals of each species varied in size but there was no significant size effect over the 24-hour period of recovery after exposure to either H_2O_2 or UV-C in *T. ventricosus*, *L. variegatus*, *E. l. lucunter* (H_2O_2 only) or *I. badionutus* (GLM, p>0.05). There was a significant effect of size of DNA damage in *E. l. lucunter* after exposure to UV-C; however, the sample size was small and only 3 large individuals were collected therefore the biological significance is unknown and all individuals were grouped for further analyses.

Each species had a different response in reduction in DNA damage over a 24-hour period of recovery after exposure to UV-C, however *L. variegatus* and *E. l. lucunter* were not different from each other after exposure to H_2O_2 (Figure 2, GLM p<0.05, *post-hoc* multiple range test). The temporal pattern of DNA damage over time was consistent among species, with clear DNA repair for most treatment levels for both exposures only evident after 6–24 hours recovery, and *I. badionotus* had greater inter-individual variation compared with the sea urchin species (Figure 2). None of the sea urchin species showed very high repair of DNA damage in the highest two exposures (10 and 100 mM H_2O_2, and 3000 and 9999 J/m² UV-C) after 24 hours recovery, however *I. badionotus* showed high (>55%) or very high (>75%) repair of DNA damage at all exposure levels after 24 hours recovery (Figure 2, *post-hoc* Fisher's LSD, p<0.05, Table 2). *T. ventricosus* had highest DNA repair 24 hours after exposure to 0.1 mM H_2O_2 (59%) and 250 J/m² UV-C (20%), compared with controls, but *L. variegatus* and *E. l. lucunter* had high (>65%) DNA repair up to 10 mM H_2O_2, and *E. l. lucunter* had moderate (38%) DNA repair at 3000 J/m² UV-C.

There was a trend in overall DNA repair capacity (% DNA repair) between species: *T. ventricosus*<*L. variegatus*<*E. l. lucunter*<*I. badionotus* (Table 2). *E. l. lucunter* and *I. badionotus* had moderate (42%) and high (71%) repair of DNA damage, respectively, 24 hours following exposure to the highest concentration of H_2O_2 (100 mM), and high (53%) and very high (80%) repair of DNA damage, respectively, 24 hours following exposure to the highest dose of UV-C (9999 J/m²); these values contrast with low (<25%) repair in *L. variegatus* and *T. ventricosus* for the highest levels of both UV-C and H_2O_2. *I. badionotus* had high or very high DNA repair at all levels of exposure, and *E. l. lucunter* had high or very high levels of DNA repair after exposure to concentrations of H_2O_2 up to 10 mM. *T. ventricosus* had moderate or low DNA repair at all levels of exposure, except 0.1 mM H_2O_2 (59%), and both *T. ventricosus* and *L. variegatus* had reduced DNA repair at high concentrations or doses. There was an indication among all species for higher DNA repair capacity for H_2O_2-induced DNA damage, compared with UV-C-induced DNA damage.

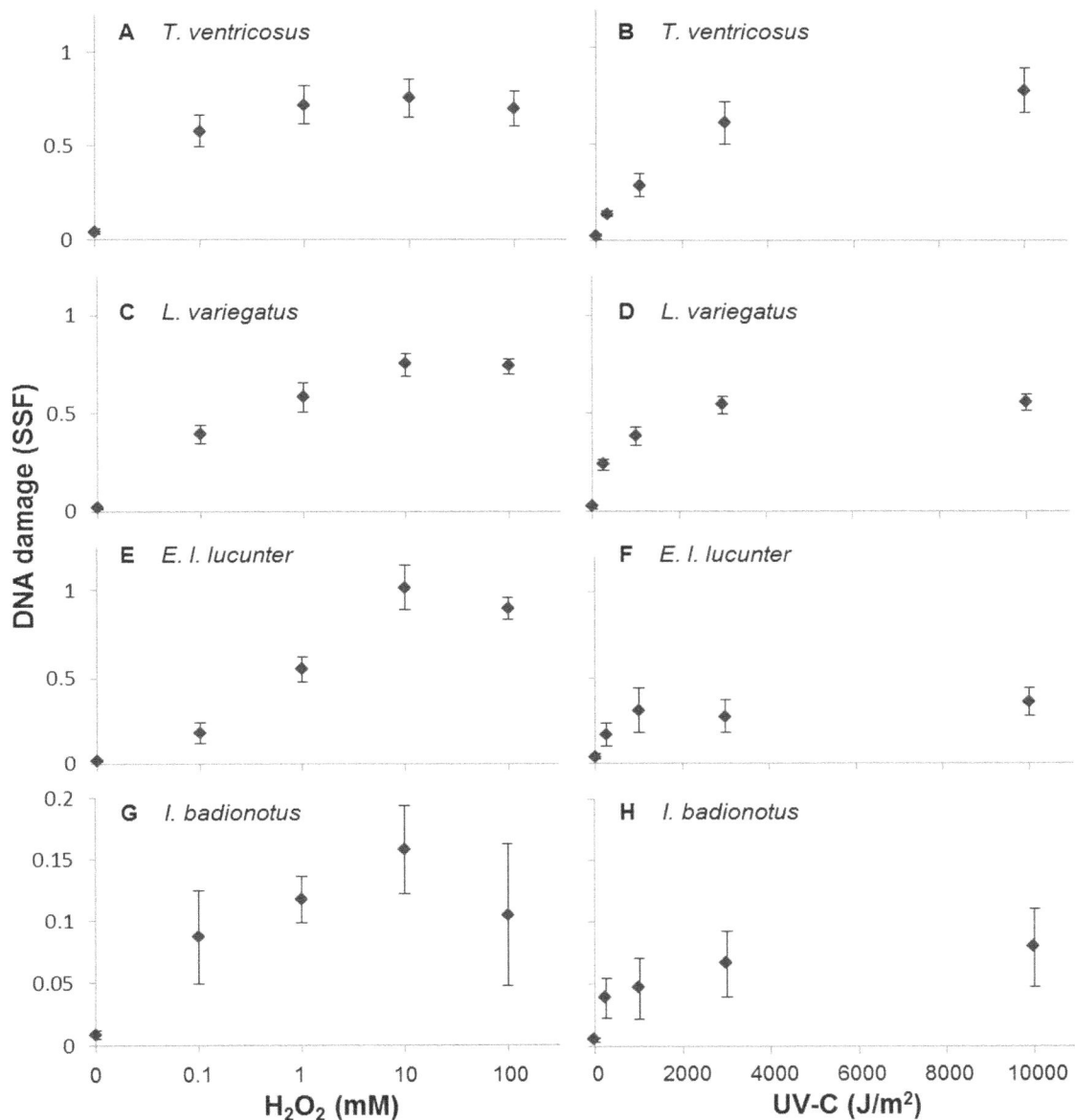

Figure 1. Dose/concentration response in echinoderm coelomocytes. Increase in DNA damage (strand scission factor, SSF, Fast Micromethod) with increasing concentration of H_2O_2 (**A**, **C**, **E**, and **G**) or dose of UV-C (**B**, **D**, **F**, and **H**) after acute exposure of coelomocytes from *T. ventricosus* (**A** and **B**, n = 5), *L. variegatus* (**C** and **D**, n = 11–12), *E. l. lucunter* (**E** and **F**, n = 6–7), and *I. badionotus* (**G** and **H**, n = 8). Data are means ± s.e.m.

Discussion

The objective of this study was to comparatively evaluate DNA damage and DNA repair capabilities of coelomocytes from four echinoderm species (*L. variegatus*, *E. l. lucunter*, *I. badionotus*, and *T. ventricosus*). Investigating DNA damage and DNA repair in cells of these different species can provide information on the value of coelomocytes as bioindicator cells and increase understanding of the overall impacts of genotoxicants on these organisms. Coelomocytes were chosen to evaluate the response to DNA damaging agents because they are well characterized cells involved in immunity and wound healing that have been proposed to be sensitive indicator cells for environmental stress [9,35–38,43–44], yet little is known of their response to genotoxicants. The

coelomocyte populations differed between species with *E. l. lucunter* and *I. badionotus* having higher cell concentrations than *L. variegatus* and *T. ventricosus*. There were no differences in the percentage of red spherule cells in the coelomocytes of the three sea urchin species; however, no red spherule cells were identified in the coelomic fluid of *I. badionotus*. This is consistent with a study on the sea cucumber *Apostichopus japonicus* which identified six cell types, none of which were red spherule cells [49]. Because little is known about the DNA repair capacity of various coelomocyte types, it is unknown whether differences in composition of coelomic fluid among species play a role in the ability for coelomocytes to repair damaged DNA. In addition, differences in coelomocyte composition between individuals may be a potential source of inter-individual variations observed in both treatment

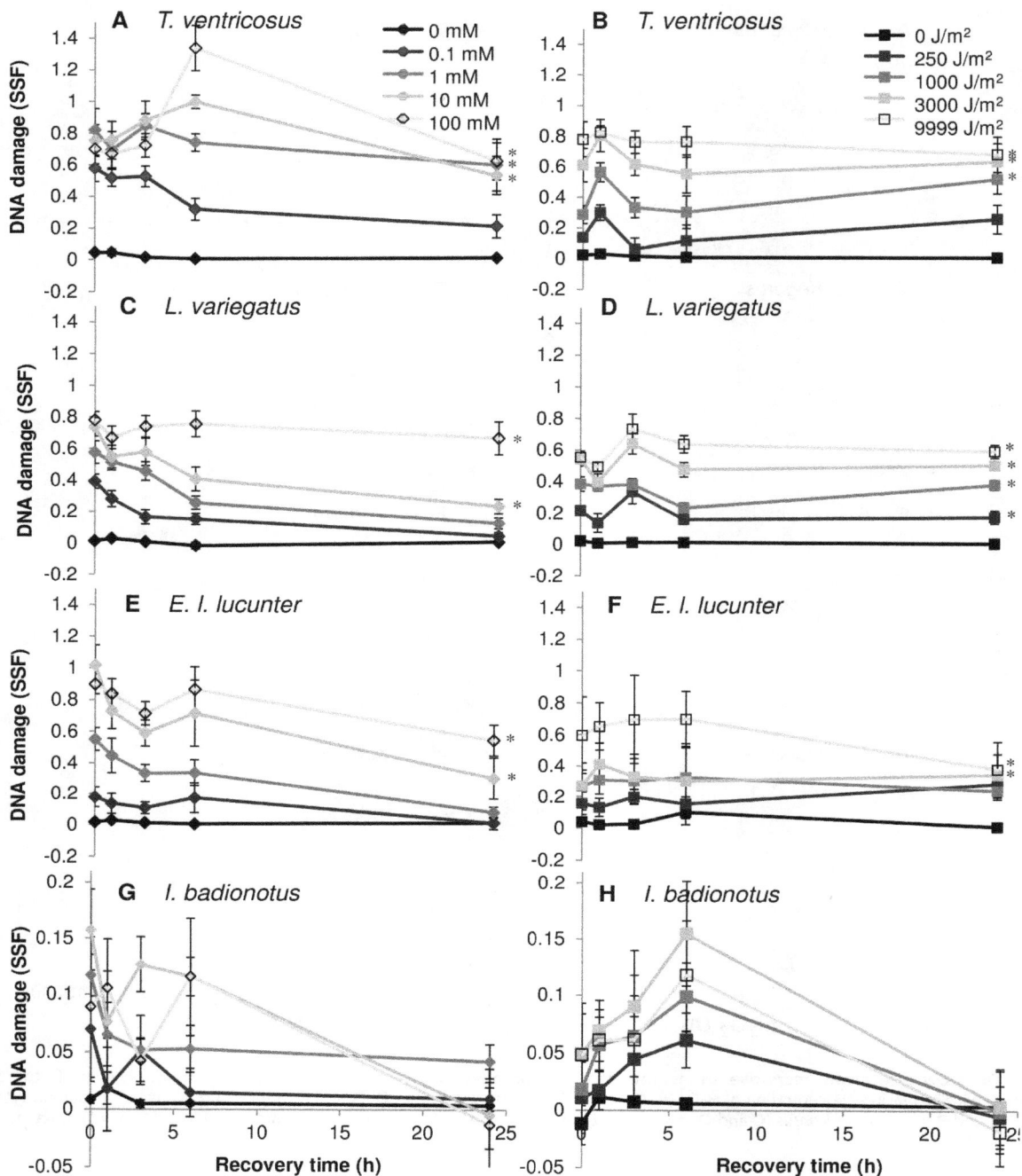

Figure 2. DNA repair in echinoderm coelomocytes. DNA repair [reduction in DNA damage (SSF)] over a 24-hour period of recovery after acute exposure to H_2O_2 (**A**, **C**, **E**, and **G**) or UV-C (**B**, **D**, **F**, and **H**) in coelomocytes from *T. ventricosus* (**A** and **B**, n = 5), *L. variegatus* (**C** and **D**, n = 12), *E. l. lucunter* (**E** and **F**, n = 8), and *I. badionotus* (**G** and **H**, n = 8). Data are means ± s.e.m. *Significantly higher than controls, indicating incomplete repair (within 24-hour timepoint, Fisher's LSD, p<0.05).

groups, in particular for *I. badionotus*. Apoptosis has been reported in sea urchin embryos exposed to UV radiation [26], but high coelomocyte viability 24 hours post-exposure over the course of this study suggests that apoptosis was not a factor contributing to the levels of DNA damage. This is consistent with the report of low levels of apoptosis in coelomocytes of the sea urchin *P. lividus* exposed to up to 2000 J/m² of UV-B (312 nm) [9]. Despite little cell death, 24 hours after exposure to the highest

levels of H_2O_2 and UV-C, coelomocytes were observed to be smaller in size. A study on cultured mouse myotubes found that 24 hours of chronic exposure to H_2O_2 significantly reduced myotube diameter *in vitro* [50]; however, it is unknown whether this decrease in cell size may have an impact on DNA repair activity in the nucleus.

In this study, DNA damage was detected by the Fast Micromethod, as recommended for high-throughput genotoxic

Table 2. Percent DNA repair (DNA damage at 24 hours recovery compared with initial DNA damage[†]) in echinoderm coelomocytes after 24 hours recovery from acute exposure to H_2O_2 or UV-C.

		T. ventricosus (n = 5)	L. variegatus (n = 12)	E. l. lucunter (n = 7–8)	I. badionotus (n = 8)
H_2O_2 (mM)	0.1	58.6±12.2	81.8±5.4	59.6±17.6	57.8±17.0
	1	33.4±19.8	73.2±6.8	79.8±5.2	59.0±13.0
	10	32.6±8.2	65.0±7.2	71.2±7.8	83.4±12.2
	100	18.2±9.2	24.8±6.6*	41.8±8.4	70.8±16.0
UV-C (J/m²)	250	20.0±20.0	35.2±10.4	23.6±11.8	54.8±17.6
	1000	11.0±11.0	13.2±4.4	27.0±15.8	67.0±16.2
	3000	15.6±6.6	13.6±3.8	38.0±18.0	61.2±17.0
	9999	16.2±15.8	8.4±3.2	53.2±15.8	79.8±9.0

[†]% DNA repair = $100 - ((T_{24}\ SSF/T_0\ SSF) \times 100)$, where T_{24} SSF is strand scission factor (SSF) after 24 hours recovery and T_0 SSF is the initial (0-hour) recovery SSF; negative % DNA repair values indicated no DNA repair and were set to zero.
*Significant reduction in DNA repair within a species (arcsine transformed, ANOVA, $p < 0.05$, *post-hoc* multiple range test).
Data are means ± s.e.m. from individually calculated % DNA repair.

analyses [13] and comparable with the comet assay for DNA strand break detection and sensitivity [48]. There was a clear concentration- and dose-dependent increase in DNA damage for all echinoderm species tested. DNA damage levels in coelomocytes from *I. badionotus* appeared to be much lower than those for the sea urchin species; however, CMFSW blanks (not CFCF blanks) were subtracted from *I. badionotus* samples due to high relative fluorescent units in the CFCF from this species, which may underestimate the amount of DNA damage. Further investigation is needed to determine whether differences in the overall magnitude of SSF values of *I. badionotus* reflect high genotoxicity resistance in this species, and interspecific comparisons of overall levels of DNA damage with this species are carried out with caution. Based on the response over a similar concentration range of H_2O_2, the sensitivities of echinoderm coelomocytes are similar to that reported for zebrafish larvae exposed to H_2O_2 *in vivo*, where DNA damage (as estimated by comet assay) reached a plateau in the response curve between 100–200 mM H_2O_2 [51]. Other marine invertebrates such as shrimp (embryo and larvae exposures) and mussels (*in vitro* haemocyte exposures) have high levels of reported DNA damage at concentrations of H_2O_2 below 1 mM [37,52–53]. These interspecific differences highlight the need for consideration of suitable genotoxic bioindicator species. Genotoxic exposure of HeLa cells, mouse lymphoma cells, and peripheral blood mononuclear cells resulted in SSF values in a similar range to the levels of initial damage induced in coelomocytes of sea urchins [48]. However, comparable treatments of HeLa cells exposed to 1000 J/m² UV-C resulted in a SSF of 1.196 [48], considerably higher than the SSFs values of 0.28, 0.38, and 0.26 from coelomocytes of *T. ventricosus, L. variegatus* and *E. l. lucunter*, respectively, exposed to the same dose. This is consistent with the observation that LD_{50} values for sea urchin coelomocytes (*L. variegatus*) exposed to H_2O_2 and UV-C are much higher than those of mammalian cells [45,54–56] and suggests that echinoderm coelomocytes are generally more resistant to genotoxicity than mammalian cells.

Comparisons of SSF values and DNA repair capacity revealed clear differences between the four species after exposure to UV-C and H_2O_2. It is thought that shallow coastal marine species may be readily exposed to genotoxicants and therefore evolutionarily well-adapted to repair DNA damage [57]. It is clear from the present results that coelomocytes from all species were able to repair some level of DNA damage from both genotoxic treatments, resulting in

reduction in DNA damage levels within 24 hours. The time profile and temporal delay in reduction of SSF within the first 6 hours of recovery could be indicative of direct DNA repair activity as both NER and BER pathways involve removal of a nucleotide or base which temporarily produces a single-strand break in the DNA [11,14]. The lower levels of DNA damage and pattern of a peak in DNA damage 1–6 hours after acute exposure to UV-C might indicate the relative lack of direct DNA strand breaks induced initially by UV-C exposure, and NER-induced strand breaks during repair [11,58–59]. Clear indication of DNA repair in coelomocytes indicates that these cells are active in the DNA damage response system of echinoderms and supports the need for further studies of the biology of these cells.

Variability after 24 hours recovery between the four species in the present study highlights important differences in DNA repair capacity even among species that share similar habitats and presumably similar exposure to genotoxicants. Sediment-dwelling species including sea cucumbers are thought to be more susceptible to genotoxicant exposure due to direct contact with the sediment [57]; however, the results of *I. badionotus* coelomocytes indicated the species was the most effective of the selected species in DNA repair, with very high repair of DNA damage after 24 hours of recovery. Phylogenetic relationships among the echinoderms reveal *T. ventricosus* and *L. variegatus* to belong to the family Toxopneustidea, whereas, *E. l. lucunter* belongs to the family Echinometridea; both families belong to the class Echinoidea [60]. All four echinoderm species belong to the same subphylum Echinozoa. Because *T. ventricosus* and *L. variegatus* are more closely related to each other than to *E. l. lucunter*, and even less so to *I. badionotus*, it is striking that there are differences in DNA repair capacity between the two, suggesting factors more significant than ancestry are involved in determining repair capacity. One determining factor may be the lifespan of the species, and the four echinoderm species included in this study vary in their natural lifespan. Life history data indicate that *T. ventricosus* and *L. variegatus* are relatively short-lived species (<4 years) [27,30] while *E. l. lucunter* is a longer-lived species with an estimated maximum lifespan of approximately 50 years [61]. There are very few studies of life history traits of sea cucumbers and no specific information is available for *I. badionotus* growth, survival, and longevity. However, growth data of other sea cucumber species suggest that sea cucumbers are slow-growing and long-lived. It is estimated that *Cucumaria frondosa* may take more than 25 years to reach a harvestable size [62] and modeled growth of *Holothuria nobilis* suggests that it may live for several decades [63]. DNA repair capacity (% DNA repair) after H_2O_2 exposure was greater in *E. l. lucunter* and *L. variegatus* than in the shorter-lived *T. ventricosus*. Additionally, percentage repair of UV-C-induced DNA damage indicated greater repair in the longer-lived *E. l. lucunter* group than in both other shorter-lived sea urchin species. A link between longevity and resistance to genotoxic stress has also been shown in bivalves with varying natural lifespans [64–65], and a greater repair capacity in longer-lived sea urchin species supports the idea that longer-lived species invest greater energy in cellular maintenance and repair [66–67]. Lack of lifespan information for *I. badionotus* restricts comparison between the species with regards to lifespan, but their highly efficient DNA repair capacity supports the speculation that they may be relatively long-lived in concordance with other sea cucumber species [62–63].

In conclusion, coelomocytes from different echinoderm species showed distinct differences in their sensitivity to DNA-damaging agents and their ability to repair damaged DNA over a 24-hour recovery period, therefore the choice of a single 'sensitive' species for ecotoxicological studies must be made with caution and consideration of differences within and between species It is clear that coelomocytes from all species tested show some capacity for DNA repair, indicating involvement of these cells in the DNA damage response system of echinoderms; these results warrant further investigation into the biology of the DNA damage response and immune cell system in echinoderms. There was a trend for longer-lived echinoderms to have a greater DNA repair capacity

compared with shorter-lived species, and it would be interesting to investigate this further with more species over a great range of natural life spans. Complete DNA repair after 24 hours recovery from exposure to both H_2O_2 and UV-C was evident for *I. badionotus*, while *T. ventricosus* (with the shortest estimated lifespan) had the lowest overall capacity for DNA repair. Interspecific variability in echinoderms, however, must be taken into account when considering suitable model organisms for ecotoxicological investigations, and life history characteristics such as longevity may be important determinants for species vulnerability to environmental genotoxicity.

Supporting Information

File S1   Supplementary data.

Acknowledgments

The authors would like to thank Thomas Ebert at Oregon State University for assistance with life history information of various echinoderm species and many thanks to the editor and reviewers for helpful suggestions to improve this manuscript.

Author Contributions

Conceived and designed the experiments: HCR AGB. Performed the experiments: AHEB. Analyzed the data: HCR AHEB. Contributed reagents/materials/analysis tools: AGB HCR. Wrote the paper: AHEB HCR AGG.

References

1. Belfiore NM, Anderson SL (2001) Effects of contaminants on genetic patterns in aquatic organisms: a review. Mutat Res 489: 97–122.

2. Kleinjans JCS, van Schooten FJ (2002) Ecogenotoxicology: the evolving field. Environ Tox Pharm 11: 173–179.

3. Klerks PL, Xie L, Levinton JS (2011) Quantitative genetics approaches to study evolutionary processes in ecotoxicology; a perspective from research on the evolution of resistance. Ecotoxicology 20: 513–523.

4. Ribeiro R, Lopes I (2013) Contaminant driven genetic erosion and associated hypotheses on alleles loss, reduced population growth rate and increases susceptibility to future stressors: an essay. Ecotoxicology 22:889–899.

5. Wurgler FE, Kramers PGN (1992) Environmental effects of genotoxins (ecogenotoxicology). Mutagenesis 7:321–327.

6. Enoch T, Norbury C (1995) Cellular responses to DNA damage: cell-cycle checkpoints, apoptosis and the roles of p53 and ATM. Trends Biochem Sci 20:426–430.

7. Cooke MS, Evans MD, Dizdaroglu M, Lunec J (2003) Oxidative DNA damage: mechanisms, mutation, and disease. FASEB J 17:1195–1214.

8. Dubrova YE (2003) Radiation-induced transgenerational instability. Oncogene 22:7087–7093.

9. Matranga V, Pinsino A, Celi M, Di Bella G, Natoli A (2006) Impacts of UV-B radiation on short-term cultures of sea urchin coelomocytes. J Mar Biol 149:25–34.

10. Sinha RP, Hader DP (2002) UV-induced DNA damage and repair: a review. Photochem Photobiol Sci 1:225–236.

11. Rastogi RP, Richa, Kumar A, Tyagi MB, Sinha RP (2010) Molecular mechanisms of ultraviolet radiation-induced DNA damage and repair. J Nucleic Acids 2010.

12. Misovic M, Milenkovic D, Martinovic T, Ciric D, Bumbasirevic V et al. (2013) Short-term exposure to UV-A, UV-B, and UV-C irradiation induces alteration in cytoskeleton and autophagy in human keratinocytes. Ultrastruct Pathol 37:241–248.

13. Bihari N, Batel R, Jaksic Z, Muller WEG, Waldmann P, et al. (2002) Comparison between the comet assay and fast micromethod for measuring DNA damage in HeLa cells. Croat Chem Acta 75:793–804.

14. Ramos-Espinosa P, Rojas E, Valverde M (2012) Differential DNA damage response to UV and hydrogen peroxide depending of differentiation stage in a neuroblastoma model. Neurotoxicology 33:1086–1095.

15. Henle ES, Linn S (1997) Formation, prevention, and repair of DNA damage by iron/hydrogen peroxide. J Biol Chem 272:19095–19098.

16. Valavanidis A, Vlahogianni T, Dassenakis M, Scoullos M (2006) Molecular biomarkers of oxidative stress in aquatic organisms in relation to toxic environmental pollutants. Ecotoxicol Environ Saf 64:178–189.

17. Azqueta A, Shaposhnikov S, Collins AR (2009) DNA oxidation: investigating its key role in environmental mutagenesis with the comet assay. Mutat Res 674:101–108.

18. Friedberg EC (2003) DNA damage and repair. Nat 421:436–440.

19. Jha AN (2004) Genotoxicological studies in aquatic organisms: an overview. Mut Res 552:1–17.

20. Anderson SL, Wild GC (1994) Linking genotoxic responses and reproductive success in ecotoxicology. Environ Health Perspect 102:9–12.

21. Bay S, Burgess R, Nacci D (1993) Status and applications of echinoid (Phylum Echinodermata) toxicity test methods. In: Landis WG, Hughes JS, Lewis MA, editors. Environmental Toxicology and Risk Assessment, pp. 281–302.

22. Hose JE (1985) Potential uses of sea-urchin embryos for identifying toxic chemicals - description of a bioassay incoporating cytologic, cytogenetic and embryologic endpoints. J Appl Toxicol 5:245–254.

23. Saco-Alvarez L, Duran I, Ignacio Lorenzo J, Beiras R (2010) Methodological basis for the optimization of a marine sea-urchin embryo test (SET) for the ecological assessment of coastal water quality. Ecotoxicol Environ Saf 73:491–499.

24. Lamare MD, Barker MF, Lesser MP, Marshall C (2006) DNA photorepair in echinoid embryos: effects of temperature on repair rate in Antarctic and non-Antarctic species. J Exp Biol 209:5017–5028.

25. Le Bouffant R, Cormier P, Cueff A, Belle R, Mulner-Lorillon O (2007) Sea urchin embryo as a model for analysis of the signaling pathways linking DNA damage checkpoint, DNA repair and apoptosis. Cell Mol Life Sci 64:1723–1734.

26. Lesser MP, Kruse VA, Barry TM (2003) Exposure to ultraviolet radiation causes apoptosis in developing sea urchin embryos. J Exp Biol 206:4097–4103.

27. Beddingfield SD, McClintock JB (2000) Demographic characteristics of *Lytechinus variegatus* (Echinoidea: Echinodermata) from three habitats in a North Florida Bay, Gulf of Mexico. Mar Ecol 21:17–40.

28. Ebert TA, Southon JR (2003) Red sea urchins (*Strongylocentrotus franciscanus*) can live over 100 years: confirmation with A-bomb (14)carbon. Fishery Bulletin 101:915–922.

29. Moore HB, Jutare T, Bauer JC, Jones JA (1963) The biology of *Lytechinus variegatus*. Bull Mar Sci 23–53.

30. Pena MH, Oxenford HA, Parker C, Johnson A (2010) Biology and fishery management of the white sea urchin, *Tripneustes ventricosus*, in the eastern Caribbean. Rome: Food and Agriculture Organization of the United Nations. FAO Fisheries and Aquaculture Circular No 1056. p 43.

31. Ebert TA (2008) Longevity and lack of senescence in the red sea urchin *Strongylocentrotus franciscanus*. Exp Gerontol 43:734–738.

32. Jangoux M (1987) Diseases of Echinodermata. 4. Structural abnormalities and general considerations on biotic diseases. Dis Aquat Organ 3:221–229.

33. Robert J (2010) Comparative study of tumorigenesis and tumor immunity in invertebrates and nonmammalian vertebrates. Dev Comp Immunol 34:915–925.

34. Smith LC, Ghosh J, Buckley KM, Clow KA, Dheilly NM, et al. (2010) Echinoderm Immunity. In: Söderhäll K, editor. Invertebrate Immunology. Springer Science+Business Media, LLC, Landes Bioscience pp. 260–301.

35. Bolognesi C, Hayashi M (2011) Micronucleus assay in aquatic animals. Mutagenesis 26:205–213.

36. Canty MN, Hutchinson TH, Brown RJ, Jones MB, Jha AN (2009) Linking genotoxic responses with cytotoxic and behavioural or physiological consequences: differential sensitivity of echinoderms (Asterias rubens) and marine molluscs (Mytilus edulis). Aquat Toxicol 94:68–76.

37. Dallas IJ, Bean TP, Turner A, Lyons BP, Jha AN (2013) Oxidative DNA damage may not mediate Ni-induced genotoxicity in marine mussels: assessment of genotoxic biomarkers and transcriptional responses of key stress genes. Mutat Res Genet Toxicol Environ Mutagen 754:22–31.

38. Kolarevic S, Knezevic-Vukcevic J, Paunovic M, Kracun M, Vasiljevic B, et al. (2013) Monitoring DNA damage in haemocytes of freshwater mussel Sinanodonta woodiana sampled from the Velika Morava River in Serbia with the comet assay. Chemosphere 93:243–251.

39. Muangphra P, Gooneratne R (2011) Comparative genotoxicity of cadmium and lead in earthworm coelomocytes. Applied and Environmental Soil Science 2011:1–7.

40. Reinecke SA, Reinecke AJ (2004) The comet assay as biomarker of heavy metal genotoxicity in earthworms. Arch Environ Contam Toxicol 46:208–215.

41. Branco PC, Borges JCS, Santos MF, Jensch BE, da Silva JRMC (2013) The impact of rising sea temperature on innate immune parameters in the tropical subtidal sea urchin Lytechinus variegatus and the intertidal sea urchin Echinometra lucunter. Mar Environ Res 92:95–101.

42. Dupont S, Thorndyke M (2012) Relationship between CO2-driven changes in extracellular acid-base balance and cellular immune response in two polar echinoderm species. J Exp Mar Bio Ecol 424:32–37.

43. Pinsino A, Della Torre C, Sammarini V, Bonaventura R, Amato E, et al. (2008) Sea urchin coelomocytes as a novel cellular biosensor of environmental stress: a field study in the Tremiti Island Marine Protected Area, Southern Adriatic Sea, Italy. Cell Biol Toxicol 24:541–552.

44. Matranga V, Toia G, Bonaventura R, Muller WE (2000) Cellular and biochemical responses to environmental and experimentally induced stress in sea urchin coelomocytes. Cell Stress Chaperones 5:113–120.

45. Loram J, Raudonis R, Chapman J, Lortie M, Bodnar A (2012) Sea urchin coelomocytes are resistant to a variety of DNA damaging agents. Aquat Toxicol 124–125:133–138.

46. Holm K, Dupont S, Sköld H, Stenius A, Thorndyke M, et al (2008) Induced cell proliferation in putative haematopoietic tissues of the sea star, Asterias rubens (L.). J Exp Biol 211:2551–2558.

47. Hernroth B, Farahani F, Brunborg G, Dupont S, Dejmek A, et al (2010) Possibility of mixed progenitor cells in sea star arm regeneration. J Exp Zool B Mol Dev Evol 341B:457–468.

48. Schröder HC, Batel R, Schwertner H, Boreiko O, Müller WEG (2006) Fast micromethod DNA single-strand-break assay. In: Henderson DS, editor. Methods in Molecular Biology: DNA Repair Protocols: Mammalian Systems, 2nd ed. Humana Press Inc, Totowa, NJ, pp 287–305.

49. Xing K, Yang HS, Chen MY (2008) Morphological and ultrastructural characterization of the coelomocytes in Apostichopus japonicus. Aquat Biol 2:85–92.

50. McClung JM, Judge AR, Talbert EE, Powers SK (2009) Calpain-1 is required for hydrogen peroxide-induced myotube atrophy. Am J Physiol Cell Physiol 296:C363–371.

51. Reinardy HC, Dharamshi J, Jha AN, Henry TB (2013) Changes in expression profiles of genes associated with DNA repair following induction of DNA damage in larval zebrafish Danio rerio. Mutagenesis 28:601–608.

52. Cheung VV, Depledge MH, Jha AN (2006) An evaluation of the relative sensitivity of two marine bivalve mollusc species using the Comet assay. Mar Environ Res 62:S301–S305.

53. Hook SE, Lee RF (2004) Genotoxicant induced DNA damage and repair in early and late developmental stages of the grass shrimp Paleomonetes pugio embryo as measured by the comet assay. Aquat Toxicol 66:1–14.

54. Long AC, Colitz CMH, Bomser JA (2004) Apoptotic and necrotic mechanisms of stress-induced human lens epithelial cell death. Exp Biol Med 229:1072–1080.

55. Murakami S, Salmon A, Miller RA (2003) Multiplex stress resistance in cells from long-lived dwarf mice. FASEB J 17:1565–1566.

56. Salmon AB, Akha AAS, Buffenstein R, Miller RA (2008) Fibroblasts from naked mole-rats are resistant to multiple forms of cell injury, but sensitive to peroxide, ultraviolet light, and endoplasmic reticulum stress. J Gerontol A Biol Sci Med Sci 63:232–241.

57. Depledge MH (1998) The ecotoxicological significance of genotoxicity in marine invertebrates. Mut Res 399:109–122.

58. Collins AR (2014) Measuring oxidative damage to DNA and its repair with the comet assay. Biochim Biophys Acta 1840:794–800.

59. Azqueta A, Langie SAS, Slyskova J, Collins AR (2013) Measurement of DNA base and nucleotide excision repair activities in mammalian cells and tissues using the comet assay – a methodological overview. DNA Repair 12:1007–1010.

60. WoRMS (2013) Echinozoa, World Register of Marine Species. Available: http://www.marinespecies.org/aphia.php?p=taxdetails&id=148744.

61. Ebert TA, Russell MP, Gamba G, Bodnar A (2008) Growth, survival, and longevity estimates for the rock-boring sea urchin Echinometra lucunter lucunter (Echinodermata, Echinoidea) in Bermuda. Bull Mar Sci 82:381–403.

62. So JJ, Hamel JF, Mercier A (2010) Habitat utilisation, growth and predation of Cucumaria frondosa: implications for an emerging sea cucumber fishery. Fish Manag Ecol 17:473–484.

63. Uthicke S, Welch D, Benzie JAH (2004) Slow growth and lack of recovery in overfished holothurians on the Great Barrier Reef: evidence from DNA fingerprints and repeated large-scale surveys. Conservation Biol 18:1395–1404.

64. Ungvari Z, Ridgeway I, Philipp EER, Campbell CM, McQuary P, et al. (2011) Extreme longevity is associated with increased resistance to oxidative stress in Arctica islandica, the longest-living non-colonial animal. J Gerontol A Biol Sci Med Sci 66:741–750.

65. Ungvari Z, Sosnowska D, Mason JB, Gruber H, Lee SW, et al. (2013) Resistance to genotoxic stress in Arctica islandica, the longest living noncolonial animal: is extreme longevity associated with a multistress resistance phenotype? J Gerontol A Biol Sci Med Sci 68:521–529.

66. Kirkwood TBL (2005) Understanding the odd science of aging. Cell 120:437–447.

67. Bodnar AG (2009) Marine invertebrates as models for aging research. Exp Gerontol 44:477–484.

9

The Progeroid Phenotype of Ku80 Deficiency Is Dominant over DNA-PK$_{CS}$ Deficiency

Erwin Reiling[1,2], Martijn E. T. Dollé[1], Sameh A. Youssef[3], Moonsook Lee[4], Bhawani Nagarajah[1], Marianne Roodbergen[1], Piet de With[1], Alain de Bruin[3], Jan H. Hoeijmakers[2], Jan Vijg[4], Harry van Steeg[1,5], Paul Hasty[6]*

1 National Institute for Public Health and the Environment, Bilthoven, The Netherlands, 2 Department of Cell Biology and Genetics, Center for Biomedical Genetics, Erasmus MC, Rotterdam, The Netherlands, 3 Faculty of Veterinary Medicine, Department of Pathobiology, Utrecht University, Utrecht, The Netherlands, 4 Department of Genetics, Albert Einstein College of Medicine, Bronx, New York, United States of America, 5 Department of Toxicogenetics, Leiden University Medical Center, Leiden, The Netherlands, 6 Department of Molecular Medicine and Institute of Biotechnology, Barshop Institute for Longevity and Aging Studies, Cancer Therapy and Research Center, University of Texas Health Science Center at San Antonio, San Antonio, Texas, United States of America

Abstract

Ku80 and DNA-PK$_{CS}$ are both involved in the repair of double strand DNA breaks via the nonhomologous end joining (NHEJ) pathway. While $ku80^{-/-}$ mice exhibit a severely reduced lifespan and size, this phenotype is less pronounced in $dna\text{-}pk_{cs}^{-/-}$ mice. However, these observations are based on independent studies with varying genetic backgrounds. Here, we generated $ku80^{-/-}$, $dna\text{-}pk_{cs}^{-/-}$ and double knock out mice in a C57Bl6/J*FVB F1 hybrid background and compared their lifespan, end of life pathology and mutation frequency in liver and spleen using a lacZ reporter. Our data confirm that inactivation of Ku80 and DNA-PK$_{CS}$ causes reduced lifespan and bodyweights, which is most severe in $ku80^{-/-}$ mice. All mutant mice exhibited a strong increase in lymphoma incidence as well as other aging-related pathology (skin epidermal and adnexal atrophy, trabacular bone reduction, kidney tubular anisokaryosis, and cortical and medullar atrophy) and severe lymphoid depletion. LacZ mutation frequency analysis did not show strong differences in mutation frequencies between knock out and wild type mice. The $ku80^{-/-}$ mice had the most severe phenotype and the Ku80-mutation was dominant over the DNA-PK$_{CS}$-mutation. Presumably, the more severe degenerative effect of Ku80 inactivation on lifespan compared to DNA-PK$_{CS}$ inactivation is caused by additional functions of Ku80 or activity of free Ku70 since both Ku80 and DNA-PK$_{CS}$ are essential for NHEJ.

Editor: Robert W. Sobol, University of Pittsburgh, United States of America

Funding: This work was supported by a National Institutes of Health grant (P01 AG17242). The funders had no role in study design, data collection and analysis, decision to publish, or preparation of the manuscript.

Competing Interests: The authors have declared that no competing interests exist.

* E-mail: hastye@uthscsa.edu

Introduction

DNA damage is known to be involved in tumorigenesis and aging. There are multiple DNA repair pathways that specialize in repairing a specific DNA lesion. One such pathway is nonhomologous end joining (NHEJ), which plays an import role in the repair of double strand breaks (DSBs). Ku80 and DNA-PK$_{CS}$ are both components of the NHEJ pathway. Ku80 forms a heterodimer with Ku70, known as Ku, which binds to DNA ends at a DSB [1]. DNA-dependent protein kinase (DNA-PK) is a holo enzyme formed by a complex between Ku and DNA-depended protein kinase catalytic subunit (DNA-PK$_{CS}$). DNA-PK$_{CS}$, together with Artemis, Xrcc4-DNA ligase and Xrcc4-like factor, process DNA overhangs and ligation [2–4].

Cells with deletion of any of these NHEJ components show severe combined immunodeficiency (SCID) due to defects in assembling variable (diverse) joining (V(D)J) segments of antigens, genetic instability and hypersensitivity to DSB inducing agents [5–9]. However, there is some phenotypic variation. For instance, deletion of Ku70 and Ku80 results in reduced size, severely decreased lifespan, neuronal apoptosis and accelerated aging.

Strikingly, in previous experiments using mixed backgrounds, these mice seem to be protected against tumor development, although they show a low level of thymic lymphomas [10,11]. In contrast, such a phenotype is less pronounced in DNA-PK$_{CS}$ knock out mice. Compared to the $ku70^{-/-}$ and $ku80^{-/-}$ mice, the early aging phenotype for the $dna\text{-}pk_{cs}^{-/-}$ mice is less severe and best observed with shortened telomeres [12–14]. Yet, these studies were performed in different genetic backgrounds in different labs. Therefore, phenotypic differences could be the result of inconsistent genetic backgrounds and environments. On the other hand, some phenotype differences can be explained by additional functions of the deleted NHEJ components or that a non-deleted NHEJ component is deleterious in absence of the deleted protein. Such deleterious effects have been observed previously [15,16]. For instance, DNA ligase IV deficiency is lethal in presence of Ku80 but lethality is rescued by deletion of Ku80 [16]. Furthermore, the prenatal lethal phenotype of Xrcc4 deficient mice is partly rescued by p53 knock out [15], that negates neuronal apoptosis.

Here, we generated cohorts of $ku80^{-/-}$, $dna\text{-}pk_{cs}^{-/-}$, double knockout and wild type mice in an identical genetic background and compared lifespan, development of body weight, end-of-life pathology and accumulation of genetic insults (LacZ mutant frequencies in liver and spleen). We find that $ku80^{-/-}$ mice exhibit a more sever phenotype than $dna\text{-}pk_{cs}^{-/-}$ mice while the double mutant mice are as severe as the $ku80^{-/-}$ mice. These observations suggest that Ku80 has additional functions that do not require DNA-PK$_{CS}$ or that free Ku70 has a deleterious affect in the absence of Ku80. Both possibilities are in line with our previous reports that show Ku70 and Ku80 have functions independent of the Ku heterodimer [17].

Results

Previous reports suggest that $ku80^{-/-}$ mutant mice exhibit a more severe phenotype than $dna\text{-}pk_{CS}^{-/-}$ mice. This was not predicted since both Ku80 and DNA-PK$_{CS}$ are essential for NHEJ. Thus, the phenotypic differences could be due to divergent function or divergent genetic background and environment. Alternatively, one of the components could be toxic in the absence of the deleted protein. To better understand the reason for these different phenotypes we generated $ku80^{-/-}$ and $dna\text{-}pk_{cs}^{-/-}$ and double mutant cohorts in the same genetic background using double heterozygous breeders ($ku80^{+/-}$ $dna\text{-}pk_{cs}^{+/-}$) as shown in figure 1. The breeding pairs were composed of C57Bl6/J-pUR288 males and FVB females so the cohorts were F1 brothers and sisters raised in the same cages. Thus, all cohorts are controlled for genetic background and environment so any phenotypic difference between the $ku80^{-/-}$ and $dna\text{-}pk_{cs}^{-/-}$ cohorts must be due to divergent function. The double mutant mice will also help determine if one component assumes a toxic activity after the other is deleted.

During the longevity study, all mice were weighed every two weeks. As can be seen in figure 2, all mutant mice were smaller compared to wild type mice. $ku80^{-/-}$ and double knock out mice weighted between 10 and 20 grams, while $dna\text{-}pk_{cs}^{-/-}$ mice weighted between 15 and 35 grams and were therefore considerably heavier than $ku80^{-/-}$ and double knock out mice. Due to differences in body size, organ weights differed significantly

between the different genotypes (Figure S1). However, after correction for bodyweight, most organ weights were comparable. The exception is the testis, which was significantly heavier in $dna\text{-}pk_{cs}^{-/-}$ mice compared to the other knock out mice. Marginal significant differences were observed for males in relative organ weight for kidney, liver, spleen and heart. The wild type cohort was terminated after the final mutant mouse died since their life span appeared much longer; therefore, no end of life organ weight data were collected for this cohort. Thus, deletion of either Ku80 or DNA-PK$_{CS}$ reduced body weight with Ku80-deletion being more severe and dominant to DNA-PK$_{CS}$ -deletion.

All mutant mice displayed a severely reduced lifespan compared to wild type mice (Fig. 3). Previously, we showed that $dna\text{-}pk_{cs}^{-/-}$ mice were loner-lived than $ku80^{-/-}$ mice for both males and females ($p = 0.01$ and $p = 8.7*10^{-6}$ respectively) [17]. Here we also show that the double knock out mice had the same short life span as $ku80^{-/-}$ mice (Fig. 3). Thus, deletion of either Ku80 or DNA-PK$_{CS}$ reduced life span with Ku80-deletion being more severe and dominant to DNA-PK$_{CS}$ -deletion.

We screened for aging pathology in a selected set of animals (n = 4–5). Results are shown in table 1. Evidence of aging was found in the kidney. Mild increased renal tubular anisokaryosis was observed in all knock out models. Moderate to severe renal tubulonephrosis was present in all mutant cohorts compared to wild type animals. Analysis of skin revealed moderate to severe epidermal and adnexal atrophy. Dermis and epidermis thickness was reduced in all mutant mice. Finally, all mutant mice showed mild to moderate reduction in trabecular bone thickness, which is indicative of osteopenia. There were no differences between genotypes in anisokaryosis or lipofuscin accumulation in liver, which was previously observed in aged wild type mice [18]. Thus, deletion of either Ku80 or DNA-PK$_{CS}$ caused an early onset of many but not all aging characteristics previously reported for wild type mice.

Mutant cohorts showed a high tumor incidence at the time of death compared to age-matched controls (table 2). Most tumors were observed in the thymus followed by liver tumors. Pathological examination showed that tumors were almost exclusively

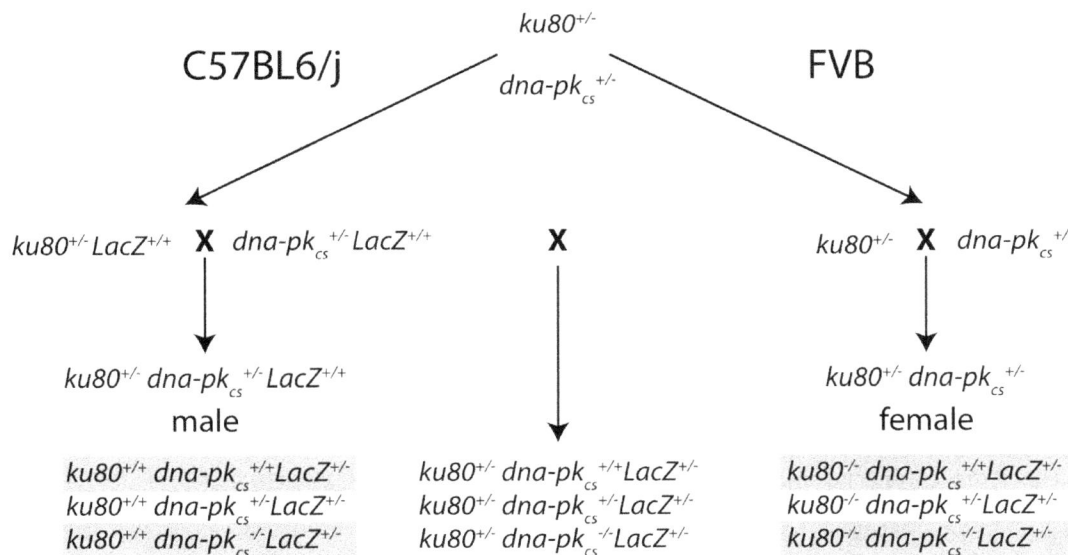

Figure 1. Breeding strategy to generate the knockout cohorts. $Ku80^{+/-}$ and $dna\text{-}pk_{cs}^{+/-}$ animals are backcrossed for multiple generations to C57Bl6/J-pUR288 (left of dashed line) and FVB (right of dashed line) background. Double heterozygous knock out C57BL6/J male mice are crossed with double heterozygous knock out FVB female mice. Using this strategy all four desired cohorts are generated as F1 hybrids (grey boxes).

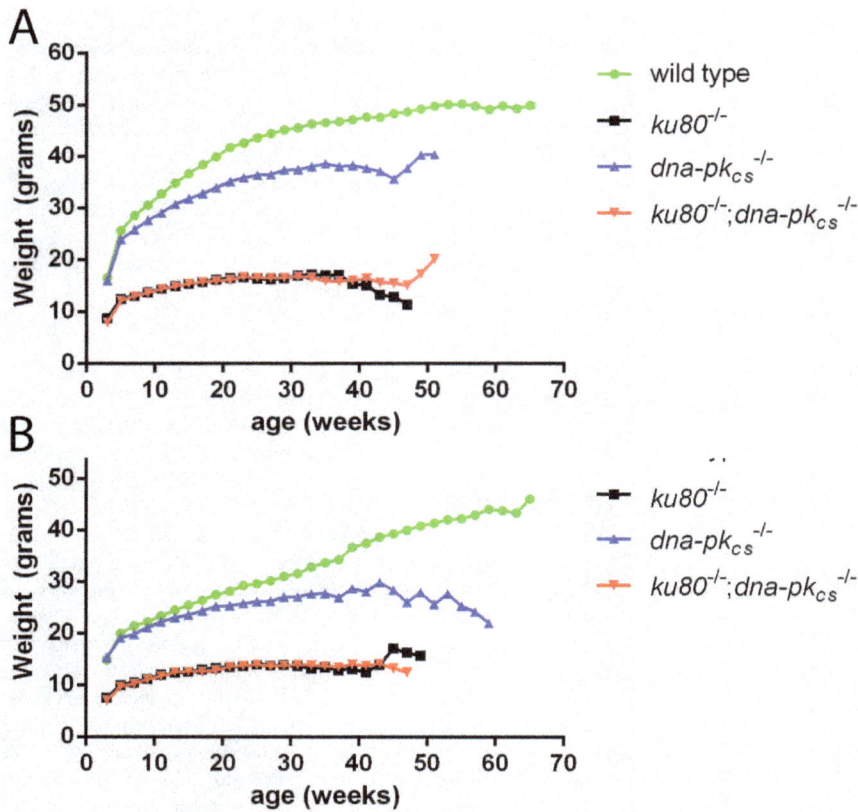

Figure 2. Mean body weights. (A) Males. (B) Females.

lymphomas. CD3 staining showed lymphomas to be of t-cell origin (three mice analyzed for each genotype). There were also rare hepatic adenomas (n = 6), a hepatic carcinoma (n = 1) and a bronchiolo-alveolar carcinoma (n = 1), which might be caused by the C57Bl6/J background according to the Mouse Tumor Database (http://tumor.informatics.jax.org/mtbwi/index.do). Lymphomas were also present in other organs than thymus with a high frequency in liver, kidney, spleen and lymph nodes. Lymphoid tissues of mutant animals, not affected by lymphomas, showed severe lymphoid depletion (table 1) as expected due to the failure to complete V(D)J recombination. This was observed in spleen and mesenterial lymph nodes and was severe in most mutant animals, with exception of female $dna\text{-}pk_{cs}^{-/-}$ mice, which showed a mild to moderate lymphoid depletion in mesenterial lymph nodes. Animals affected by lymphoid depletion also showed an increased myeloid/erythroid ratio. Although tumor incidence was slightly higher in $ku80^{-/-}$ mice compared to $dna\text{-}pk_{cs}^{-/-}$ and the highest in double knock out mice, this did not reach statistical significance. No synergistic effects of DNA-PK$_{CS}$ deletion in addition to Ku80 deletion were found nor did deletion of one protein ameliorate the phenotype for deletion of the other protein.

Since Ku80 and DNA-PK$_{CS}$ are DNA damage repair proteins, it would be expected that deletion of these proteins affect mutant frequency. In order to test the LacZ mutant frequencies in liver and spleen, 5 $ku80^{-/-}$, $dna\text{-}pk_{cs}^{-/-}$ and double knock males were killed at 38.7, 41.3 and 33.7 weeks, which corresponds approximately with their median lifespan (figure 3). Because the wild type cohort was not maintained beyond median age, we selected the 5 oldest wild type males available at the same observation time (mean age is 37.3 weeks). No significant differences in mutant

frequency were observed between genotypes in both liver as spleen (figure 4). We did observe a trend towards increased mutant frequency in double knock out compared to wild type mice in spleen, but this did not reach statistical significance ($p = 0.07$).

Discussion

Previously, it was reported that Ku80 inactivation resulted in reduced cancer and accelerated aging [12,14,19] while a similar but less pronounced phenotype was observed with DNA-PK$_{CS}$ inactivation [12–14]. This phenotypic variance is unexpected since both Ku80 and DNA-PK$_{CS}$ are essential for NHEJ. Therefore, these different phenotypes could be due to differences in genetic background and/or environment. However, our experiment presented here does not support either possibility since we controlled for both genetic background and environment but still observed a disparity in phenotypes. Alternatively, $ku80^{-/-}$ mice could exhibit a more severe phenotype than $dna\text{-}pk_{CS}^{-/-}$ mice, if DNA-PK$_{CS}$ is toxic in the absence of Ku80. There is president for this possibility since Ku80-deletion rescued embryonic lethality for *DNA ligase IV*-null mice showing that Ku80 was toxic in the absence of DNA ligase IV. However, deleting DNA-PK$_{CS}$ did not ameliorate the Ku80-mutant phenotype. Instead the double-mutant mice exhibited the same severe phenotype as the Ku80-mutant mice. Therefore, our experiment suggests that Ku80 has extra-NHEJ function or that free Ku70 has a toxic activity in the absence of Ku80.

Our previously published data supports both possibilities. First we found that Ku80 has extra-NHEJ activity in mice by comparing p53-mutant mice deleted for Ku80 or Ku70 or both

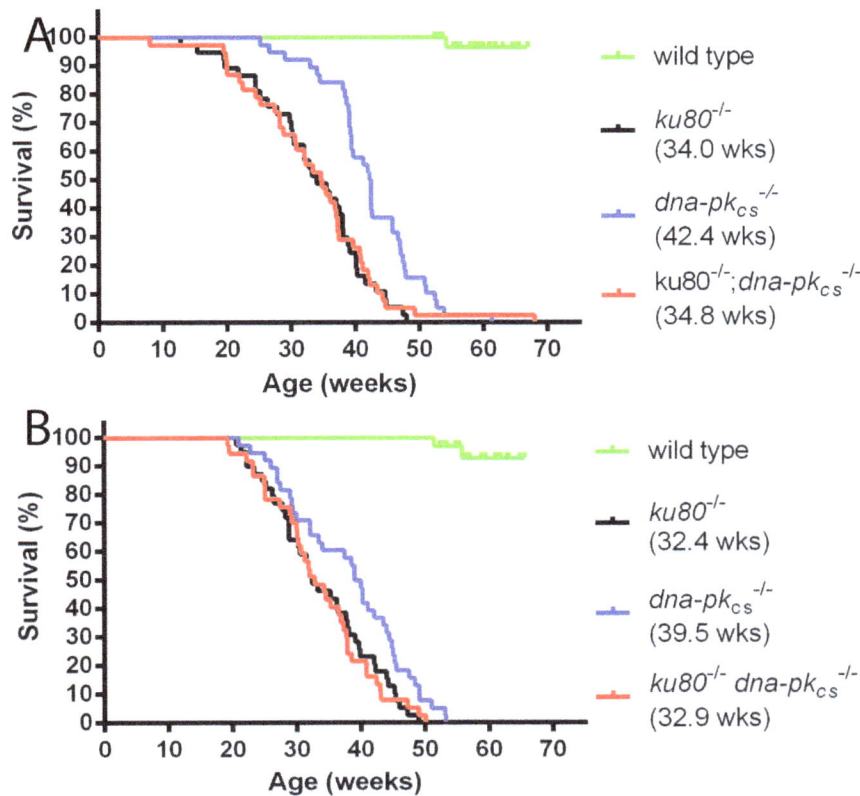

Figure 3. Survival curves. Median survivals are shown in the figure legends in parentheses. dna-$pk_{cs}^{-/-}$ mice live significantly longer compared to $ku80^{-/-}$ mice in both male ($p = 0.01$) and females ($p = 8.7*10^{-6}$) mice. Survival of double knock out mice is identical to $ku80^{-/-}$. (A) Males. (B) Females.

[20]. For this experiment, we found that $ku70^{-/-}$ $p53^{-/-}$ mice lived longer than $ku80^{-/-}$ $p53^{-/-}$ mice. This life span extension required Ku80 since the triple-mutant mice had the same life span as the $ku80^{-/-}$ $p53^{-/-}$ mice. The $ku70^{-/-}$ $p53^{-/-}$ mice lived longer because they had a lower incidence of pro-B cell lymphoma that was restored with deletion of Ku80. Therefore, these experiments suggest that Ku80 has extra-NHEJ activity. In addition, we showed that free Ku70 and free Ku80 bind to apurinic/apyrimidinic (AP) sites and that free Ku70 inhibits AP endonuclease 1 [17]. Another group also showed that Ku and DNA-PK$_{CS}$ inhibited AP site cleavage by APE1 [21]. Thus, Ku80 and Ku70 have activity that is separable from the Ku heterodimer and NHEJ that could exacerbate the Ku80-mutant phenotype and account for the phenotypic disparity between deleting Ku80 and DNA-PK$_{CS}$.

Ku80 as well as DNA-PK$_{CS}$ inactivation accelerated aging. These characteristics included skin atrophy, femur osteopenia, renal tubular anisokaryosis and cortical and medullar atrophy. Accelerated aging was observed previously in Ku80 and DNA-PK$_{CS}$ negative mice in different genetic backgrounds and environments suggesting that early aging is not sensitive to genetic and environmental changes [10,22] [19].

Deletion of either Ku80 or DNA-PK$_{CS}$ also increased cancer risk. Pathological examination revealed that tumors were predominantly lymphomas. Most lymphomas were observed in the thymus. Lymphomas in non-lymphoid tissues (e.g. liver and lung) are presumably metastases.

By contrast to this report, our prior reports show that $ku80^{-/-}$ and $ku70^{-/-}$ mice exhibited low levels of cancer [10,11].

Furthermore, deleting Ku80 in a familial adenomatous polyposis mouse model (APC^{MIN}) reduced intestinal tumors and increased life span [23]. We proposed that Ku80-deletion reduced cancer levels due to constitutive activation of the p53 DNA damage response and possibly other responses [24]. In support, deleting p53 greatly enhanced pro-B cell lymphomas and medulloblastoma in $ku80^{-/-}$ and $ku70^{-/-}$ mice [20,25,26]. Thus, difference in cancer incidences between our prior reports and this report suggest genetic background influence cancer incidence in the $ku80^{-/-}$ mice. This is possible since the former reports described mice derived from a 129*C57Bl6/J hybrid background while the current report describes mice derived from a C57Bl6/J*FVB F1 hybrid background.

Furthermore, we observed an increased myeloid/erythroid ratio that suggests true myeloid hyperplasia. However, complete blood counts are needed to confirm this. By contrast, we observed severe lymphoid depletion in all knock out cohorts. Lymphoid depletion is a known phenotype in SCID mice, which have a defective V(D)J recombination [27,28]. Since $ku80^{-/-}$ and dna-$pk_{cs}^{-/-}$ are SCID [9,22], it is not surprising that lymphoid depletion was found in the knock out cohorts. Therefore, this element of the phenotype is not age-related. The increased myeloid/erythroid ratio could be the result of a compensatory response to this lymphoid depletion, triggering myeloid hyperplasia and lymphoma development.

Since Ku80 and DNA-PK$_{CS}$ mutant mice have a defective DSB repair system, one expects accumulation of DNA damage. However, LacZ mutant frequency analysis could not confirm the previously observed increased mutant frequency in spleen [29]. Although our data do show a weak trend into the same direction,

Table 1. Histopathology non-neoplastic lesions.

Lesion	ku80$^{-/-}$		dna-pk$_{cs}$$^{-/-}$		k-/-;d-/-		Wild type	
Median pathology score (min – max)	**Male (41w)**	**Female (42w)**	**Male (45w)**	**Female (45w)**	**Male (42w)**	**Female (40w)**	**Male (37w)**	**Female (40w)**
Liver anisokaryosis	2 (1-3)	1.5 (1-3)	3 (1-3)	2* (2-3)	2 (2-3)	2.5 (1-3)	1 (1-2)	1 (1-2)
Liver lipofuscin	1 (1-1)	1 (1-1)	1 (1-1)	1 (1-2)	1 (1-1)	1 (1-1)	1 (1-1)	1 (1-1)
Renal tubular anisokaryosis	2 (1-3)	2 (1-3)*	2 (1-3)	2 (2-3)	3* (2-3)	2 (1-3)	1 (1-2)	1 (1-1)
Renal tubulonephrosis	3*** (1-3)	3*** (2-4)	3*** (1-3)	4*** (2-4)	2*** (1-3)	3*** (1-3)	0 (0-0)	0 (0-0)
Epidermal atrophy	2** (1-3)	2* (1-3)	3** (2-4)	3* (2-3)	3** (1-4)	3* (2-3)	0 (0-0)	0 (0-0)
Adnexal atrophy	1** (1-2)	1* (0-1)	2.5** (1-4)	3** (1-3)	2* (0-4)	2** (1-3)	0 (0-0)	0 (0-0)
Lymphoid depletion spleen	4* (3-4)	4* (3-5)	3* (3-3)	3* (3-3)	4* (3-5)	3* (2-4)	0 (0-0)	0 (0-0)
Lymphoid depletion mes ln	4** (4-5)	3.5** (2-5)	2* (1-2)	1* (1-2)	4** (4-4)	4** (4-4)	0 (0-0)	0 (0-0)
Tissue thickness in μm (St. dev.)								
Dermis and epidermis	253.8* (56.9)	151.3* (11 7)	222.0** (38.3)	172.8 (58.7)	252.0* (99.0)	150.0 (06.3)	374.3 (50.6)	247.3 (43.4)
Trabecular bone	130.8* (36.7)	131.0* (38.2)	124.5* (18.8)	139.8* (25.5)	138.4 (43.1)	118.8** (10.6)	184.0 (19.4)	185.3 (22.4)

4-5 animals examined per group. All selected mutant mice are at end of life, wild type mice are aged matched to mutant mice.
Mean age in weeks is shown (w).
mes ln: mesenterial lymph node.
k-/-;d-/-: ku80$^{-/-}$; dna-pk$_{cs}$$^{-/-}$.
St. Dev.: Standard deviation.
p-values based on comparison with wild type animals.
* p<0.05.
** p<0.01.
*** p<0.001.

Table 2. Histopathology neoplastic lesions.

Organ	Tumor	ku80 −/−		dna-pk$_{cs}$ −/−		k-/-;d-/-	
		Male (n=25)	Female (n=15)	Male (n=27)	Female (n=10)	Male (n=24)	Female (n=21)
Liver	lymphoma	9	5	4	7	5	2
	adenoma	1	1			3	1
	carcinoma					1	
Lung	lymphoma	1					
	carcinoma		1				
Pancreas	lymphoma					1	
Kidney	lymphoma	5	2	3	6	4	5
Thymus	lymphoma*	23	14	21	9	19	15
ovary/uterus	lymphoma		1		2		
Spleen	lymphoma	4		2	3	3	
mes. lm. nd.	lymphoma	3	2	5	5	3	
ax. lm. nd.	lymphoma	3	3	5	5	5	
Skin	lymphoma				1		
Testis	lymphoma	1					
bone marrow**	lymphoma	2	1	2	2	1	1
pituitary gland	lymphoma				1		
Tumor incidence (%)		67	39	55	24	75	51

Mentioned n-numbers are tumor bearing animals available for microscopical examination.

k-/-;d-/-: ku80-/-; dna-pk$_{cs}$ −/−.

mes. lm. nd.: mesenterial lymph node.

ax. lm. nd.: axillary lymph node.

*~5 thymus tumor preparations were microscopically examined per group and found to be all lymphomas. All other macroscopic neoplastic lesions observed in thymus are assumed to be lymphomas as well.

** analyzed in femur.

Tumor incidence based on all animals from longevity cohort, including those not microscopically examined.

A

MF liver

B

MF spleen

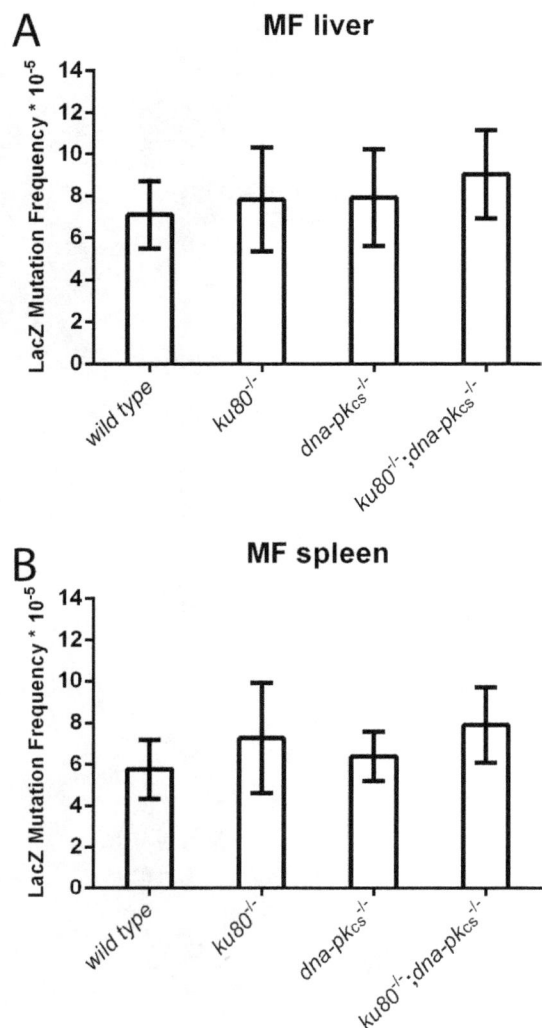

Figure 4. LacZ mutant frequency. Error bars represent standard deviation. (A) Mutant frequency in liver. (B) Mutant frequency in spleen.

variation is too large and effect sizes too small to achieve statistical significance. Previously we had to separate the small mutations from the big mutations before a difference could be seen since the former was reduced while the latter was increased.

In conclusion, we have shown that deletion of the NHEJ components Ku80 and DNA-PK$_{CS}$ in a C57Bl6/J*FVB F1 hybrid background resulted in a accelerated aging phenotype and a strongly increased incidence of lymphomas and that simultaneous deletion does not further enhance these characteristics. Yet, Ku80-deletion was more severe than DNA-PK$_{CS}$ –deletion. This observation is consistent with the possibility that Ku80 has an extra-NHEJ function or that free Ku70 is toxic in the absence of Ku80.

Materials and Methods

Ethics Statement

All animal work was approved by the ethics committee of the National Institutes for Public Health and the Environment (RIVM), Antonie van Leeuwenhoeklaan, Bilthoven, The Netherlands, IACUC protocol #:99047x.

Mice breeding

Ku80 null mice are not viable in a pure C57BL6/J background [30]. Therefore, we generated F1 hybrid animals with a C57Bl6/J*FVB background. This resulted in viable offspring and prevented C57BL6/J related ulcerative dermatitis [31]. $ku80^{-/-}$ mice [9] and $dna\text{-}pk_{cs}^{-/-}$ mice [32] were imported in our SPF facility, rederived and back crossed to C57Bl6/JIco (Charles River, France) and FVB/NHanHsd (Harlan, Germany) using a speed congenics approach [33]. Knock out mice in C57Bl6/JIco background were maintained by backcrossing with C57Bl6/J-pUR288 (LacZ locus integrated at chromosome 13 [34]) generating heterozygous knock out animals which are homozygous for pUR288. Knock out mice on FVB background were maintained by backcrossing with pure FVB wild type animals. Cohorts with single and double knock out animals were generated using double heterozygous knock out breeders ($ku80^{+/-}$; $dna\text{-}pk_{cs}^{+/-}$). All male breeders were on C57Bl6/J-pUR288 and all female breeders on FVB background. Using this breeding strategy, all genotypes could be generated from the same breeding colony and all experimental mice were C57Bl6/J*FVB F1 hybrids carrying one LacZ allele. The breeding scheme is depicted in figure 1.

This study was carried out in strict accordance with institutional guidelines and regulations. All animal work was approved by the ethics committee of the National Institutes for Public Health and the Environment (RIVM), Antonie van Leeuwenhoeklaan, Bilthoven, The Netherlands, IACUC protocol #:99047x. These were survival studies; therefore, mice were monitored every day without intervention. Moribund mice were sacrificed with ketamine/xylazine anesthesia followed by cervical dislocation and all efforts were made to minimize suffering and discomfort. Criteria for moribund were >15% weight loss within 2 weeks, not responsive to touch, prominent appearance of ribs, spine and hips, hunch body position, matted fur, or a visible tumor.

Study setup

Each cohort consisted of 45 males and 45 females. All animals were maintained under specific pathogen free conditions and were fed ad libitum using CRM pelleted maintenance diet (Special Diet Services, UK). A 12 hr/12 hr dark/light cycle was maintained and temperature was 20°C. Animals were maintained until death or moribund. Moribund animals were euthanized by exsanguinations and major organs were isolated and partly stored in formaldehyde and partly snap frozen in liquid nitrogen. Results from $ku80^{-/-}$ and $dna\text{-}pk_{cs}^{-/-}$ longevity cohorts were also used in a study comparing Ku70 and Ku80 function [17]. In addition, 5 male animals of all genotypes were euthanized at approximately median lifespan for LacZ mutant frequency analysis.

Pathology

Pathology lesions in mutant animals were compared with those in age matched wild type littermates. Representative sections from the liver, kidney, thymus, spleen, axillary and mesenteric lymph nodes, femur, vertebra, uterus, ovaries, testes, and accessory male genital glands were processed (n = 4–5 per genotype), stained with Hematoxylin and Eosin, and microscopically examined for the presence of histopathologic lesions. All tumors macroscopically identified during necropsy were also prepared for histopathologic examination Severity score of all recorded lesions was semi-quantitatively assessed. Scores were given as absent (0), subtle (1), mild (2), moderate (3), severe (4), and massive (5) for each criteria. Digital images from the femur cortical bone at mid-shaft area, and skin were taken for morphometric analysis using Labsense image analysis software (Olympus). The thickness of femur cortical bone thickness, and skin thickness (dermis and epidermis with exclusion of subcutaneous fat) were measured using arbitrary line option.

CD3 staining

After deparaffinization, rehydration and citrate buffer pretreatment, sections were incubated with rabbit polyclonal antibody to the human CD3 molecule (DAKO Corporation- Code-Nr. A 0452), diluted 1:250 in 10% normal goat serum, for 30 minutes at room temperature. The remainder of the procedure was accomplished according to the manufacturer's instructions. Normal goat serum was substituted for the primary antibody as a negative control. Mouse lymphoid tissue was used as a positive control.

LacZ mutant frequency analysis

LacZ mutant frequency was assessed in liver and spleen using the pUR288(lacZ) reporter. This assay has been described previously [34]. In short, total DNA was extracted from liver and spleen using a phenol/chloroform/iso-amyl alcohol extraction (25:24:1) and digested with the endonuclease HindIII in the presence of magnetic beads, coated with lacI-lacZ fusion protein. After digestion, magnetic beads were washed and DNA fragments eluted using isopropyl-L-thio-B-D-galactopyranoside. DNA fragments were circularized with T4 DNA ligase and transformed into *E. Coli* (ΔlacZ, galE-) bacteria. Of the transformed bacteria, one thousandth was plated on x-gal plates and the remainder on selective p-gal plates. The mutant frequency was defined as the amount of colonies on the selective plate divided by the amount of colonies on the X-gal plate multiplied by 1000 (dilution factor).

Statistical analysis

Survival curves were analyzed using a log rank test (Graphpad Prism 5.04, GraphPad Software Inc, La Jolla, CA, USA). Pathological scores were compared using a two-tailed Mann-Whitney U test and mutant frequencies were compared using a T-Test (IBM SPSS Statistics 19, SPSS Inc - IBM, New York, USA).

Supporting Information

Figure S1   a. Relative organ weights in males. b. Relative organ weights in females. c. Absolute organ weights in males. d. Absolute organ weights in females.

Acknowledgments

We thank Dr. G.E.Taccioli for providing DNA-PK$_{CS}$ knock out mice and S. Imholz and A. Verhoef for technical support.

Author Contributions

Conceived and designed the experiments: ER METD JHH JV HVS PH. Performed the experiments: ER SAY ML BN MR PDW ADB. Analyzed the data: ER METD JHH JV HVS PH. Wrote the paper: ER PH.

References

1. Liang F, Romanienko PJ, Weaver DT, Jeggo PA, Jasin M (1996) Chromosomal double-strand break repair in Ku80-deficient cells. Proc Natl Acad Sci U S A 93: 8929–8933.
2. Ahnesorg P, Smith P, Jackson SP (2006) XLF Interacts with the XRCC4-DNA Ligase IV Complex to Promote DNA Nonhomologous End-Joining. Cell 124: 301–313.
3. Grawunder U, Wilm M, Wu X, Kulesza P, Wilson TE, et al. (1997) Activity of DNA ligase IV stimulated by complex formation with XRCC4 protein in mammalian cells. Nature 388: 492–495.
4. Ma Y, Pannicke U, Schwarz K, Lieber MR (2002) Hairpin opening and overhang processing by an Artemis/DNA-dependent protein kinase complex in nonhomologous end joining and V(D)J recombination. Cell 108: 781–794.
5. Biedermann KA, Sun JR, Giaccia AJ, Tosto LM, Brown JM (1991) scid mutation in mice confers hypersensitivity to ionizing radiation and a deficiency in DNA double-strand break repair. Proc Natl Acad Sci U S A 88: 1394–1397.
6. Fulop GM, Phillips RA (1990) The scid mutation in mice causes a general defect in DNA repair. Nature 347: 479–482.
7. Gu Y, Jin S, Gao Y, Weaver DT, Alt FW (1997) Ku70-deficient embryonic stem cells have increased ionizing radiosensitivity, defective DNA end-binding activity, and inability to support V(D)J recombination. Proc Natl Acad Sci U S A 94: 8076–8081.
8. Taccioli GE, Gottlieb TM, Blunt T, Priestley A, Demengeot J, et al. (1994) Ku80: product of the XRCC5 gene and its role in DNA repair and V(D)J recombination. Science 265: 1442–1445.
9. Zhu C, Bogue MA, Lim DS, Hasty P, Roth DB (1996) Ku86-deficient mice exhibit severe combined immunodeficiency and defective processing of V(D)J recombination intermediates. Cell 86: 379–389.
10. Li H, Vogel H, Holcomb VB, Gu Y, Hasty P (2007) Deletion of Ku70, Ku80, or both causes early aging without substantially increased cancer. Mol Cell Biol 27: 8205–8214.
11. Vogel H, Lim DS, Karsenty G, Finegold M, Hasty P (1999) Deletion of Ku86 causes early onset of senescence in mice. Proc Natl Acad Sci U S A 96: 10770–10775.
12. Espejel S, Martin M, Klatt P, Martin-Caballero J, Flores JM, et al. (2004) Shorter telomeres, accelerated ageing and increased lymphoma in DNA-PKcs-deficient mice. EMBO Rep 5: 503–509.
13. Gu Y, Sekiguchi J, Gao Y, Dikkes P, Frank K, et al. (2000) Defective embryonic neurogenesis in Ku-deficient but not DNA-dependent protein kinase catalytic subunit-deficient mice. Proc Natl Acad Sci U S A 97: 2668–2673.
14. Gao Y, Chaudhuri J, Zhu C, Davidson L, Weaver DT, et al. (1998) A targeted DNA-PKcs-null mutation reveals DNA-PK-independent functions for KU in V(D)J recombination. Immunity 9: 367–376.
15. Frank KM, Sharpless NE, Gao Y, Sekiguchi JM, Ferguson DO, et al. (2000) DNA ligase IV deficiency in mice leads to defective neurogenesis and embryonic lethality via the p53 pathway. Mol Cell 5: 993–1002.
16. Karanjawala ZE, Adachi N, Irvine RA, Oh EK, Shibata D, et al. (2002) The embryonic lethality in DNA ligase IV-deficient mice is rescued by deletion of Ku: implications for unifying the heterogeneous phenotypes of NHEJ mutants. DNA Repair (Amst) 1: 1017–1026.
17. Choi YJ, Li H, Son MY, Wang XH, Fornsaglio JL, et al. (2014) Deletion of Individual Ku Subunits in Mice Causes an NHEJ-Independent Phenotype Potentially by Altering Apurinic/Apyrimidinic Site Repair. PLoS ONE 9: e86358.
18. Dolle ME, Kuiper RV, Roodbergen M, Robinson J, de Vlugt S, et al. (2011) Broad segmental progeroid changes in short-lived Ercc1(-/Delta7) mice. Pathobiol Aging Age Relat Dis 1.
19. Espejel S, Klatt P, Menissier-de Murcia J, Martin-Caballero J, Flores JM, et al. (2004) Impact of telomerase ablation on organismal viability, aging, and tumorigenesis in mice lacking the DNA repair proteins PARP-1, Ku86, or DNA-PKcs. J Cell Biol 167: 627–638.
20. Li H, Choi YJ, Hanes MA, Marple T, Vogel H, et al. (2009) Deleting Ku70 is milder than deleting Ku80 in p53-mutant mice and cells. Oncogene 28: 1875–1878.
21. Ilina ES, Lavrik OI, Khodyreva SN (2008) Ku antigen interacts with abasic sites. Biochim Biophys Acta 1784: 1777–1785.
22. Holcomb VB, Vogel H, Hasty P (2007) Deletion of Ku80 causes early aging independent of chronic inflammation and Rag-1-induced DSBs. Mech Ageing Dev 128: 601–608.
23. Holcomb VB, Rodier F, Choi Y, Busuttil RA, Vogel H, et al. (2008) Ku80 deletion suppresses spontaneous tumors and induces a p53-mediated DNA damage response. Cancer Res 68: 9497–9502.
24. Hasty P (2008) Is NHEJ a tumor suppressor or an aging suppressor? Cell Cycle 7: 1139–1145.
25. Lim DS, Vogel H, Willerford DM, Sands AT, Platt KA, et al. (2000) Analysis of ku80-mutant mice and cells with deficient levels of p53. Mol Cell Biol 20: 3772–3780.
26. Holcomb VB, Vogel H, Marple T, Kornegay RW, Hasty P (2006) Ku80 and p53 suppress medulloblastoma that arise independent of Rag-1-induced DSBs. Oncogene 25: 7159–7165.
27. Shinkai Y, Rathbun G, Lam KP, Oltz EM, Stewart V, et al. (1992) RAG-2-deficient mice lack mature lymphocytes owing to inability to initiate V(D)J rearrangement. Cell 68: 855–867.
28. Mombaerts P, Iacomini J, Johnson RS, Herrup K, Tonegawa S, et al. (1992) RAG-1-deficient mice have no mature B and T lymphocytes. Cell 68: 869–877.
29. Busuttil RA, Munoz DP, Garcia AM, Rodier F, Kim WH, et al. (2008) Effect of Ku80 deficiency on mutation frequencies and spectra at a LacZ reporter locus in mouse tissues and cells. PLoS ONE 3: e3458.
30. Reliene R, Goad ME, Schiestl RH (2006) Developmental cell death in the liver and newborn lethality of Ku86 deficient mice suppressed by antioxidant N-acetyl-cysteine. DNA Repair (Amst).

31. Sundberg JP, Rozell B, Everts H (2011) Association between hair-induced oronasal inflammation and ulcerative dermatitis in C57BL/6 mice. Comp Med 61: 204–205; author reply 205.

32. Taccioli GE, Amatucci AG, Beamish HJ, Gell D, Xiang XH, et al. (1998) Targeted disruption of the catalytic subunit of the DNA-PK gene in mice confers severe combined immunodeficiency and radiosensitivity. Immunity 9: 355–366.

33. Wakeland E, Morel L, Achey K, Yui M, Longmate J (1997) Speed congenics: a classic technique in the fast lane (relatively speaking). Immunol Today 18: 472–477.

34. Dolle ME, Martus HJ, Gossen JA, Boerrigter ME, Vijg J (1996) Evaluation of a plasmid-based transgenic mouse model for detecting in vivo mutations. Mutagenesis 11: 111–118.

Dot1-Dependent Histone H3K79 Methylation Promotes the Formation of Meiotic Double-Strand Breaks in the Absence of Histone H3K4 Methylation in Budding Yeast

Mohammad Bani Ismail, Miki Shinohara, Akira Shinohara*

Institute for Protein Research, Graduate School of Science, Osaka University, Suita, Osaka, Japan

Abstract

Epigenetic marks such as histone modifications play roles in various chromosome dynamics in mitosis and meiosis. Methylation of histones H3 at positions K4 and K79 is involved in the initiation of recombination and the recombination checkpoint, respectively, during meiosis in the budding yeast. Set1 promotes H3K4 methylation while Dot1 promotes H3K79 methylation. In this study, we carried out detailed analyses of meiosis in mutants of the *SET1* and *DOT1* genes as well as methylation-defective mutants of histone H3. We confirmed the role of Set1-dependent H3K4 methylation in the formation of double-strand breaks (DSBs) in meiosis for the initiation of meiotic recombination, and we showed the involvement of Dot1 (H3K79 methylation) in DSB formation in the absence of Set1-dependent H3K4 methylation. In addition, we showed that the histone H3K4 methylation-defective mutants are defective in SC elongation, although they seem to have moderate reduction of DSBs. This suggests that high levels of DSBs mediated by histone H3K4 methylation promote SC elongation.

Editor: Dean S. Dawson, Oklahoma Medical Research Foundation, United States of America

Funding: This work was supported by a Grant-in-Aid from the Ministry of Education, Science, Sport and Culture to A.S. and M.S., as well as grants from the Asahi-Glass Science Foundation, the Uehara Science Foundation, the Mochida Medical Science Foundation and the Takeda Science Foundation to A.S. M.S. was supported by the Japanese Society for the Promotion of Science through the Next Generation World-Leading Researchers program (NEXT). The funders had no role in study design, data collection and analysis, decision to publish, or preparation of the manuscript.

Competing Interests: The authors have declared that no competing interests exist.

* E-mail: ashino@protein.osaka-u.ac.jp

Introduction

Germ cells undergo meiosis to generate haploid gametes. Meiosis involves two consecutive chromosome segregations following one round of DNA replication. During meiosis I, homologous chromosomes segregate to opposite poles, and during meiosis II, as in mitosis, sister chromatids are separated [1]. Physical linkages between the homologous chromosomes ensure the proper segregation of the chromosomes during meiosis I. This physical linkage is cytologically visualized as the chiasma. The formation of chiasmata requires exchanges between parental homologous chromosomes, products of homologous recombination during meiosis [2].

Meiotic recombination occurs at distinct regions of the genome, called recombination hotspots [3,4]. The hotspots are distributed non-randomly along chromosomes. The recombination is initiated by the formation of double-strand breaks (DSBs) at the hotspot by a meiosis-specific topoisomerase II-like protein, Spo11, and its associated partner proteins [3,4]. Meiotic DSB formation in yeast often occurs in intergenic regions, which are depleted in nucleosomes [5,6]. Meiotic recombination hotspots are marked with histone post-translational modifications such as histone H3K4 methylation in budding yeast and mammals, and histone H3K9 acetylation in fission yeast [7–9]. Histone H3K4 methylation at the hotspot is catalyzed by Set1 and Prdm9 methyltransferases in budding yeast and in mammals, respectively [8,10]. Deletion of Set1 in the yeast reduces DSB formation and changes its

distribution, and Prdm9 knockout in mouse changes the distribution of DSBs across the genome [11,12]. Indeed, substitution of histone H3K4 modulates DSB formation, as seen in the *set1* mutant [12,13]. Moreover, Spp1, a component of the Set1 complex (COMPASS), recognizes H3K4 methylation through its PHD finger and binds to a Spo11 partner, Mer2, by tethering the hotspot located in chromatin loops to the chromosome axis-associated DSB machinery [12,13]. Importantly, the *set1* mutant of yeast and *prdm9* mutant mice still show significant residual DSB formation, and therefore show meiotic recombination [7,11]. The yeast *set1* mutant affects DSB distribution with creation of new recombination hotspots [7], suggesting the presence of an alternative pathway for DSB formation. How the formation of these residual DSBs is promoted in the absence of H3K4 methylation remains unsolved.

DSBs are processed to generate 3′-OH over-hanged single-stranded DNAs (ssDNA). This ssDNA is engaged in the interaction with intact duplex DNA on a homologous chromosome. Once homology is matched, the ssDNA invades the duplex DNA to form a recombination intermediate; this is called single-stranded invasion (SEI) [14]. The homology search and strand exchange is dependent on two RecA homologs, Rad51 and Dmc1, particularly Dmc1 [15–17]. The SEI is then converted into an intermediate with two Holliday junctions, called a double-Holliday junction (dHJ) [18]. The dHJ is then preferentially resolved into a crossover product for a chiasma. Non-crossovers are formed

A

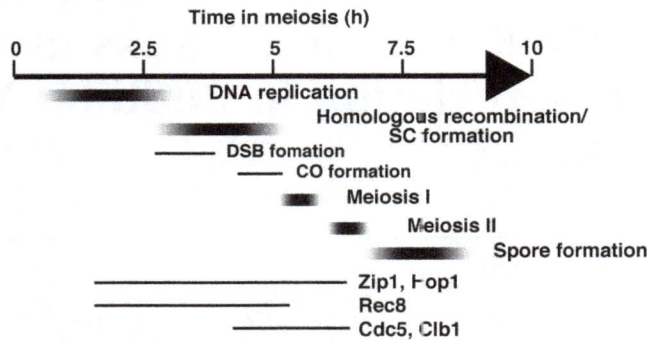

B

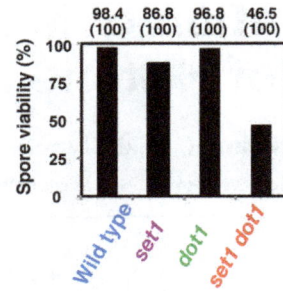

C

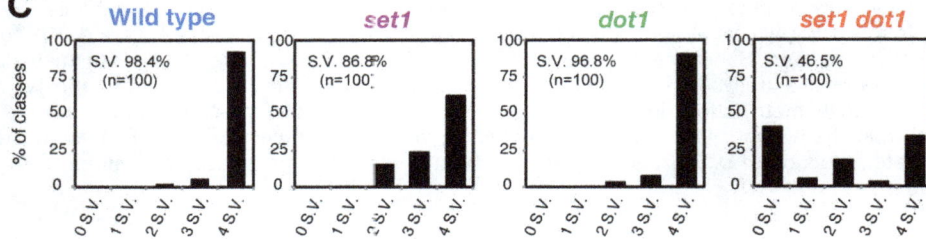

D

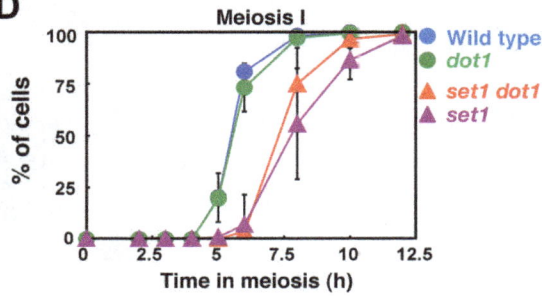

E

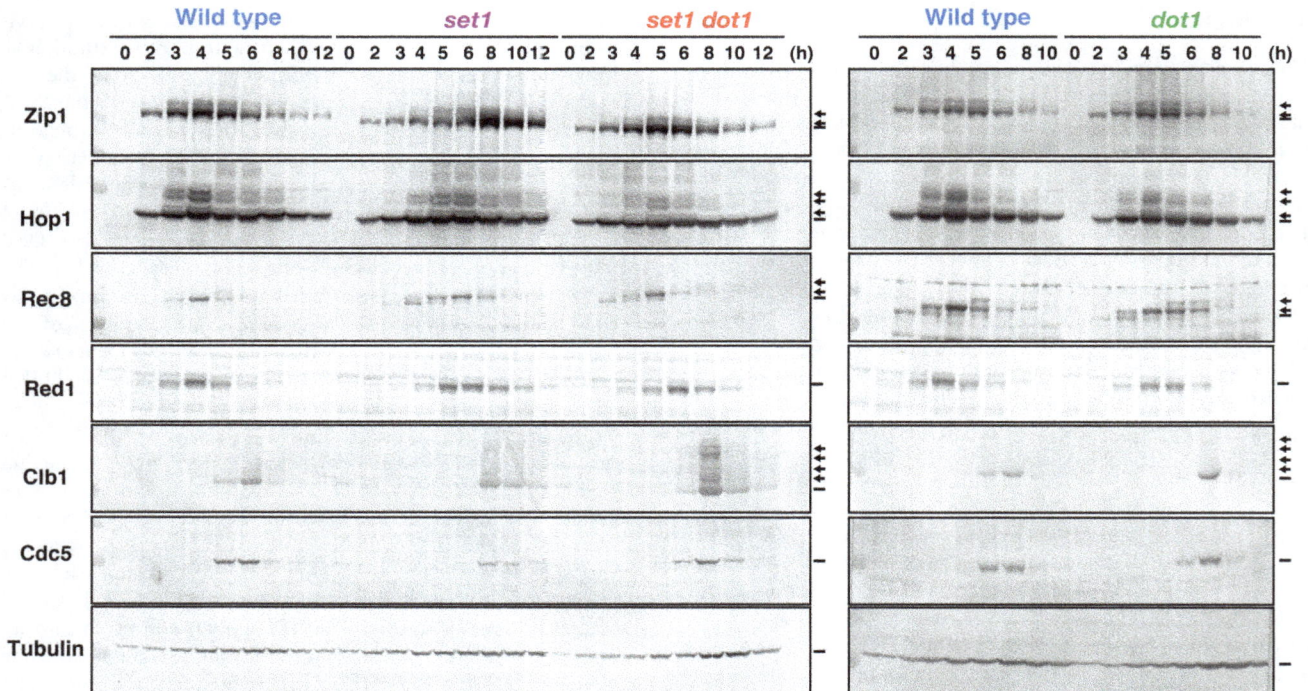

Figure 1. Dot1 plays a meiotic role in the absence of Set1. (A) Schematic representation of events during meiosis. (B, C) Spore viability of various strains was measured by dissecting spores. Spores were incubated at 30°C for 3 days. Each bar indicates percentage of spore viability and actual number of total dissected tetrads (parenthesis). Distribution of viable spores per tetrad is shown in (C). Wild type, NKY1303/1543; *set1* mutant, MBY015/016; *dot1* mutant, MBY005/006; *set1 dot1* double mutant, MBY037/039. (D) Meiotic cell division I was analyzed by DAPI staining of wild type (blue circles; NKY1303/1543), *dot1* (green circles; MBY005/006), *set1* (purple triangles; MBY005/006 and *set1 dot1* (red triangle; MBY037/039) mutant cells. At least 150 cells were counted by DAPI staining for each time point. Plotted values are the mean values with standard deviation (S.D.) from four independent time courses. (E) Expression of various meiotic proteins was verified by western blotting. At each time point, cells were fixed with TCA and cell lysates were subject to the analysis. Representative images are shown. Phosphorylated species of Zip1, Hop1, Rec8, and Clb1 are shown by arrows. Wild type, NKY1303/1543; *set1*, MBY015/016; *dot1*, MBY005/006; *set1 dot1* double mutant, MBY037/039.

through an early-branched pathway prior to SEI and dHJ formation [14,19].

During this recombination, a meiotic cell undergoes drastic changes in chromosome structures [20]. One prominent meiosis-specific chromosome structure is the synaptonemal complex (SC), which has a zipper-like morphology. Two sister chromatids are tightly connected to form a chromosome axis. In the SC, 2 chromosome axes from homologous chromosomes pair with each other through transverse filaments between the axes. Chromosome axis structures in SCs are referred to as axial/lateral elements. The formation of SCs is tightly coupled with ongoing recombination in the budding yeast [21].

It has been proposed that meiotic recombination and possibly SC formation are subject to surveillance. One of the surveillance mechanisms is a coupling mechanism of the meiotic events with cell cycle progression, which is often referred as to the pachytene checkpoint or the recombination checkpoint [22]. This surveillance mechanism has been studied extensively using mutants defective in meiotic recombination and/or SC formation; e.g., the *dmc1* mutant for meiotic recombination and *zip1* mutant for SC formation [15,23]. These mutants show delay or arrest in entry into meiosis I. When recombination is defective, meiotic cells cannot exit the middle of the pachytene phase. This is due to an inability of the mutant cell to express the Ndt80 transcriptional activator [24], which promotes the expression of so-called "middle sporulation" genes such as Cdc5 polo-like kinase and Clb1 cyclin for exit from the pachytene phase [25]. Increased Cdc5 as well as increased Cdk1 activities are key to exiting the mid-pachytene phase for SC disassembly and resolution of dHJs [26,27]. Genetic screens have identified several mutations that suppress meiotic cell progression delay/arrest by the *dmc1* or *zip1* mutations. Mutations of the *DOT1(PCH1)* and *PCH2* genes have been found to alleviate arrest in the *zip1* mutant [28,29]. The *PCH2* gene encoding a meiosis-specific AAA⁺ ATPase is also involved in chromosome morphogenesis and recombination [28,30,31]. The *DOT1* gene encodes a histone H3K79 methyltransferase which is required for gene silencing and control of some DNA damage repair pathways in mitosis [32–36]. Interestingly, both Set1-dependent H3K4 methylation and Dot1-dependent H3K79 methylation are promoted by the Rad6/Bre1-dependent ubiquitination of the histone H2BK123 [37,38]. In meiosis, H2BK123 ubiquitination is also important for DSB formation and for timely entry into meiosis I [39].

In this study, we analyzed the role of Set1 and Dot1 histone H3 methyltransferases in DSB formation and SC formation during meiosis. Consistent with previous studies, the *set1* mutant reduces DSBs on the genome as revealed by immunostaining studies for Rad51 foci. Surprisingly, *set1* deletion or H3K4 methylation-defective mutants still retain two-thirds the levels of Rad51 foci, and thus presumably DSBs, compared to those in the wild type. This suggests the presence of additional determinants in hotspots for DSB formation. Indeed, we find that Dot1-dependent H3K79 methylation is critical for the efficient formation of DSBs in the absence of Set1. Therefore, there might be multiple histone modifications controlling the formation of meiotic DSBs. These studies reinforce the importance of histone posttranslational modifications for chromosome dynamics during meiosis.

Materials and Methods

Strains and plasmids

All strains described here are derivatives of the *S. cerevisiae* SK1 diploid strain NKY1551 (*MATα/MATa*, *HO::LYS2/"*, *lys2/"*, *ura3/"*, *leu2::hisG/"*, *his4X-LEU2-URA3/his4B-LEU2*, *arg4-nsp/arg4-bgl*). The genotypes of each strain used in this study are described in Table S1. The *hht1-K4R hht2-K4R* mutant was constructed by PCR-based mutagenesis. Briefly, wild-type *HHT1* and *HHT2* genes were cloned onto pBluescript II KS+ (Stratagene). PCR-based site-directed mutagenesis using mutant primers was carried out and the presence of the mutation was confirmed by DNA sequencing. The *hht1-K4R hht2-K4R* mutant genes were cloned into YIp*lac*211 and pRS406, respectively. After digestion with *Kpn*I, the DNA was integrated by transformation. The *URA3* gene was popped-out by counter-selection for the *ura⁻* phenotype on a 5-FOA plate. Mutant sequences were verified by DNA sequencing using genomic DNAs for candidates. The *hht1-K79R hht2-K79R* strain was a generous gift from Dr. Takehiko Usui. The primers for strain construction are shown in Table S2.

Cytological analysis and antibodies

Immunostaining was conducted as described [40]. Stained samples were observed using an epifluorescence microscope (BL51; Olympus, Tokyo, Japan) with a 100× objective (NA1.4). Images were captured by CCD camera (Cool Snap; Photometrics) at room temperature, and then processed using iVision (Sillicon, California) software. Pseudo-coloring was performed using Photoshop (Adobe) software. At each timepoint, about 100 spreads were analyzed for counting foci. Primary antibodies directed against Rad51 (guinea pig, 1:500 dilution), Dmc1 (rabbit, 1:500 dilution), Zip1 (rabbit, 1:1000 dilution), Red1 (chicken, 1:400 dilution), and Rec8 (rabbit, 1:1000 dilution) were used. Secondary antibodies (Alexa-fluor-488 and -594, Molecular Probes, Carlsbad, CA) directed against primary antibodies from the different species were used at a 1:2000 dilution. Open-reading frames of Hop1 were PCR-amplified and inserted into a pET21a plasmid (Novagen) in which the C-terminus was tagged with hexahistidine (His6). Fusion proteins with His6 were affinity-purified on nickel/cobalt columns, which was performed by the manufacturers, and used for immunization of guinea pig (MBL Co. Ltd, Nagoya, Japan). The resulting antibody preparation was used at a 1:1000 dilution for western blotting and at a 1:500 dilution for immunostaining. A monoclonal antibody directed against the α-subunit of rat tubulin was also used (AbD Serotec, Oxford, UK). Meiotic time course analysis for cytology was carried out 3 times and a representative result is shown.

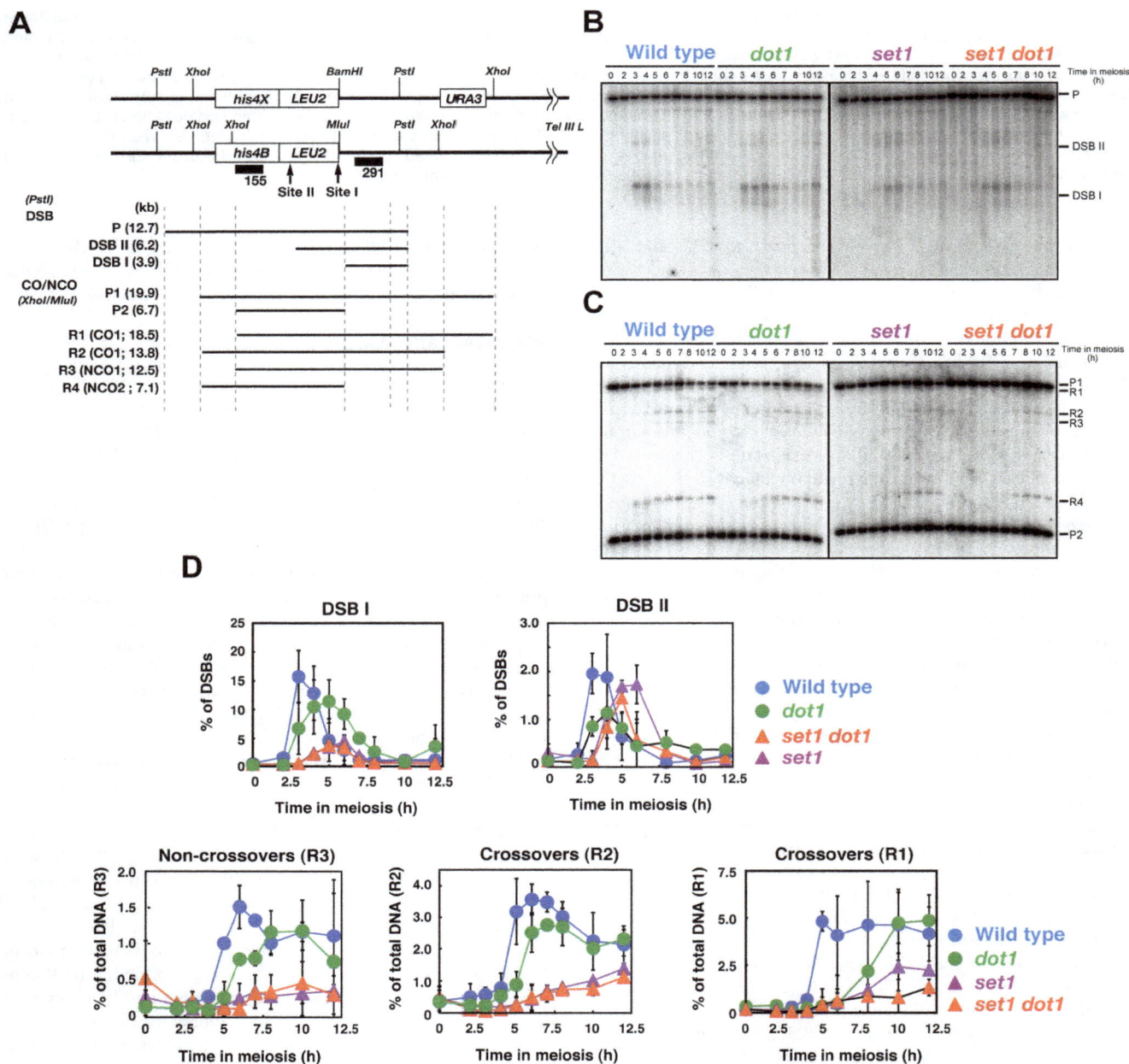

Figure 2. Set1 is necessary for meiotic recombination. (A) Schematic representation of the *HIS4-LEU2* locus. Sizes of fragments for DSB and recombinant analysis are shown with lines below. (B) DSB formation and repair at the *HIS4-LEU2* locus in different strains were verified by Southern blotting. The experiments were independently performed several times and representative blots are shown. Genomic DNAs were digested with *Pst*I. (C) Formation of crossovers and no-crossovers was also analyzed. The experiments were independently performed several times and representative blots are shown. Genomic DNAs were digested with *Mlu*I and *Xho*I. (D) The bands of DSB I (top left) and DSB II (top right), R1 (crossovers; CO; bottom right), R2 (CO; bottom middle) and R3 (non-crossovers; NCO; bottom left) and were quantified. The symbols represent the wild type (blue circles; NKY1303/1543), *dot1* mutant (green circles; MBY005/006), *set1* mutant (purple triangles; MBY015/016) and *set1 dot1* mutant (red triangle; MBY037/039). Plotted values are the mean values with standard deviation (S.D.) from three independent time courses.

Southern and western blotting

For western blotting, cell precipitates were washed twice with 20% (w/v) trichloroacetic acid (TCA) and then disrupted using a bead beater (Yasui Kikai Co. Ltd., Osaka, Japan). Precipitated proteins were recovered by centrifugation and then suspended in sodium dodecyl sulfate polyacrylamide gel electrophoresis (SDS-PAGE) sample buffer. After adjusting the pH to 8.8, samples were

incubated at 95°C for 2 min. Antibodies against Cdc5 (sc-33625, SantaCruz), Clb1 (sc-50440, SantaCruz), Hop1, Zip1, Rec8, Red1, and the α-subunit of rat tubulin (Serotec, UK) were used. Antibodies against histone H3K4-me3 (ab8580) and H3K79-me3 (ab2621) were from Abcam (Cambridge, UK).

For Pulse Field Gel Electrophoresis (PFGE), DNAs were prepared in agarose plugs as described [41], and run under the

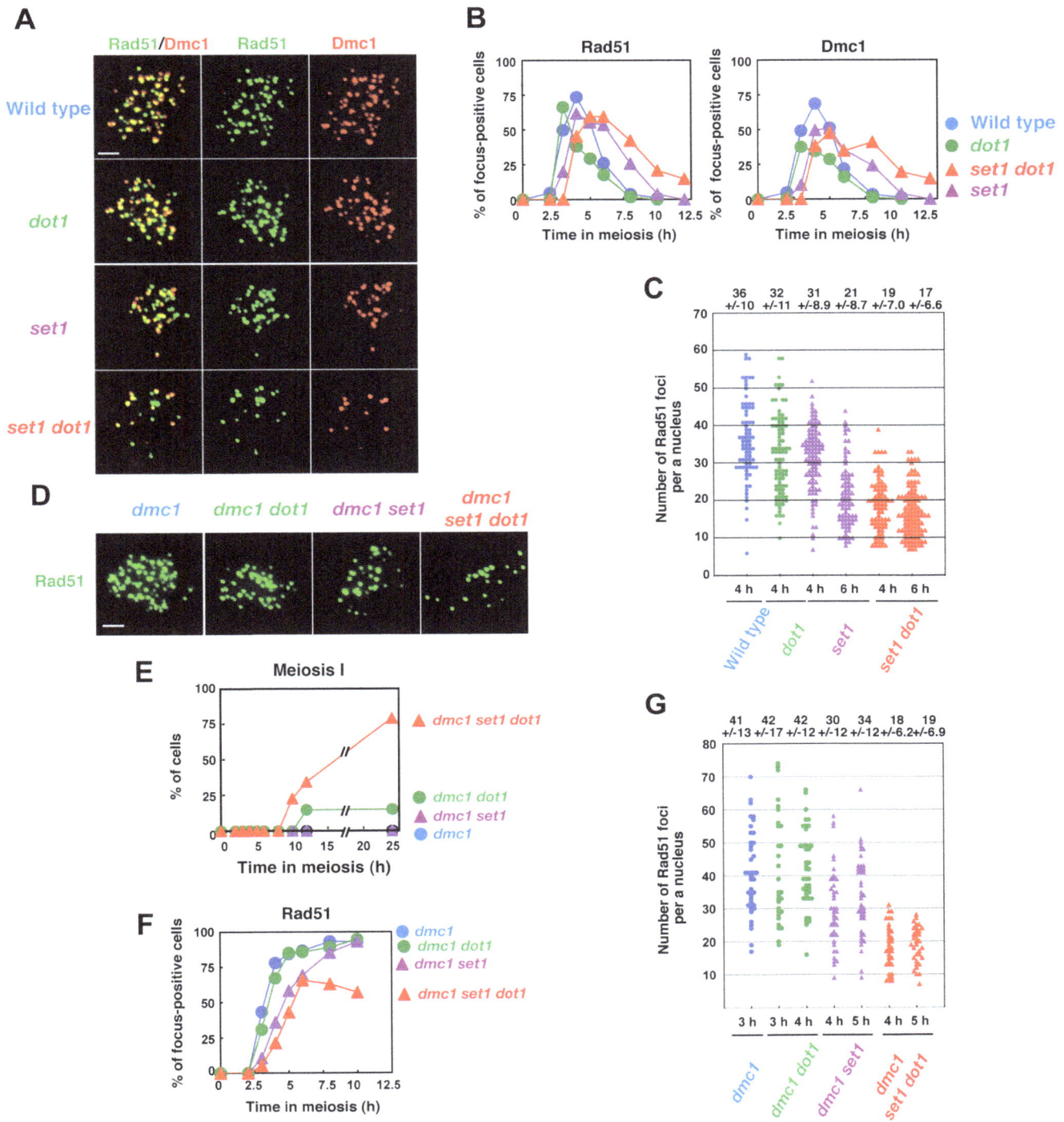

Figure 3. Dot1 promotes Rad51-focus formation in the absence of Set1. (A) Immunostaining analysis of Rad51 (green) and Dmc1 (red) for wild type (NKY1303/1543), *dot1* (MBY005/006), *set1* (MBY015/016) and *set1 dot1* (MBY037/039) mutant strains was carried out. The bar indicates 2 μm. (B) Kinetics of Rad51 (left) or Dmc1 (right)-focus positive cells in various strains. A focus-positive cell was defined as a cell with more than 5 foci. More than 100 nuclei were counted at each time point. The symbols represent the wild type (blue circles; NKY1303/1543), *dot1* mutant (green circles; MBY005/006), *set1* mutant (purple triangles; MBY015/016), and *set1 dot1* mutant (red triangle; MBY037/039). (C) A number of foci of Rad51 were counted in different strains. The symbols represent the wild type (blue circles; NKY1303/1543), *dot1* mutant (green circles; MBY005/006), *set1* mutant (purple triangles; MBY015/016), and *set1 dot1* mutant (red triangle; MBY037/039). The average number of foci per positive nucleus with S.D. is shown on top. (D) Immunostaining analysis of Rad51 (green) for the *dmc1* mutant (MBY009/010), *dmc1 dot1* mutant (MBY003/004), *dmc1 set1* mutant MBY021/022), and *dmc1 set1 dot1* mutant (MBY282/285) was carried out. The bar indicates 2 μm. (E) Meiotic cell division I was analyzed by DAPI staining of the *dmc1* mutant (blue circles; MBY009/010), *dmc1 dot1* mutant (green circles; MBY003/004), *dmc1 set1* mutant (purple triangles; MBY021/022), and *dmc1 set1 dot1* mutant (red triangle; MBY282/285) cells. At least 150 cells were counted by DAPI staining for each time point. (F) Kinetics of Rad51-focus positive cells in the *dmc1* mutant (blue circles; MBY009/010), *dmc1 dot1* mutant (green circles; MBY003/004), *dmc1 set1* mutant (purple triangles; MBY021/022), and *dmc1 set1 dot1* mutant (red triangle; MBY282/285) strains. A focus-positive cell was defined as a cell with more than 5

foci. More than 100 nuclei were counted at each time point. (G) The number of Rad51 foci was counted in different strains as described above. The average numbers of foci per a Rad51-foci positive nucleus with S.D. is shown on top.

condition (120°, 14C°, 46 h at 6 V/cm) by CHEF DR-III (BioRad). Switching time was 25 to 125 seconds.

Southern blotting was performed as described previously [42,43]. For the *HIS4-LEU2* locus, genomic DNA was digested using *Mlu*I and *Xho*I (for CO and NCO) or *Pst*I (for meiotic DSB). For the *YCR048W* locus, the DNA was digested with *Bgl*II. Probes for Southern blotting were "Probe 155" for CO/NCO, and "Probe 291" for DSB detection at the *HIS4-LEU2* locus [43]. For DSBs at the *YCR047C/048W* locus, a probe for the *YCR052W* locus (215426-216686) was used. For DSBs along chromosome III and VII, *CHA1* and *CUP2* were used as a probe, respectively. Image Gauge software (Fujifilm Co. Ltd., Tokyo, Japan) was used to quantify bands.

Results

Set1 and Dot1 play differential roles during meiosis

Previous studies established the role of Set1-mediated histone H3K4 methylation in DSB formation and the role of Dot1-mediated histone H3K79 methylation in signaling for defective SC formation [10,29]. To understand the role of these methyltransferases in events during meiosis, we characterized the meiotic phenotypes of the *set1* and *dot1* single mutants, and the *set1 dot1* double mutant in the SK1 background, which confers synchronous meiosis (Figure 1A). As shown previously [29], the *dot1* single mutant exhibits wild-type spore viability. On the other hand, the *set1* single mutant shows a slight reduction to 86.8%, compared to 98.4% in the wild type (Figure 1B). This is different from a published result in which spore viability in the *set1* deletion mutant is not different from that in wild type [12]. The *set1 dot1* double mutant shows a synergistic decrease in viability to 46.5% compared to either single mutant, indicating that Set1 and Dot1 work independently in meiosis. Importantly, the distribution of viable spores per tetrad indicated that the double mutant is more biased towards 4-, 2-, and 0-viable spores rather than 3- and 1-viable spores (Figure 1C), suggesting non-disjunction of homologous chromosomes at meiosis I, which is caused by a defect in meiotic prophase-I. However, among 122 2-spore-viable tetrads, we found that only 1 spore was non-mater, which is indicative for non-disjunction of chromosome III, indicating that non-disjunction of chromosome III is not elevated in the mutant (see Discussion).

4′,6-Diamidino-2-phenylindole (DAPI) staining reveals that the *dot1* mutant shows wild-type like entry into meiosis I (Figure 1D). As reported [10,12], the *set1* single mutant delays the entry of meiosis I by 2 h compared to wild type, which is mainly caused by delay in the meiotic S-phase [10]. The *set1 dot1* double mutant cells exhibit similar delay to the *set1* single mutant, although the double mutant is more heterogeneous in synchronous progression of the meiotic division than the *set1* single mutant.

We also studied the expression of various proteins in the meiotic prophase, including the SC components Zip1, Hop1, Red1, and Rec8, as well as the pachytene marker proteins Clb1 cyclin and Cdc5 polo-like kinase (Figure 1E). Consistent with the DAPI analysis described above, western blot analysis showed that, in wild type, the appearance of Clb1 and Cdc5 is consistent with decrease of Rec8 level, which is roughly consistent with the entry into MI. The *dot1* mutant shows similar expression pattern of Hop1, Red1, Cdc5 and Clb1 to wild type with slight delayed disappearance of Rec8. The *set1* mutant shows normal appearance of Hop1 and

Zip1, but a ~1-h delay in the appearance of phosphorylated Hop1 and phosphorylated Zip1, a ~3-h delay in the appearance of Cdc5 and Clb1, a ~3-h delay in the disappearance of Rec8 and more than 3-h delay in the disappearance of Zip1, compared to the wild type. Like the *set1* mutant, the *set1 dot1* double mutant shows normal appearance of Hop1 and Zip1, but a ~1-h delay in the appearance of phosphorylated Hop1 and phosphorylated Zip1. Importantly, the double mutant exhibits ~1-h delay in appearance of Cdc5 and Clb1 compared to wild type, but about 2 h earlier appearance than the *set1* mutant. Consistent with this, disappearance of Zip1, Rec8, and phospho-Hop1 in the double mutant is earlier than the *set1* single mutant. These could be explained by the role of Dot1 in coupling of recombination with exist of pachytene in the absence of Set1 (see below). This is consistent with the role of Dot1 in the pachytene checkpoint when the recombination is perturbed [28,29].

In order to know the role of Set1 and Dot1 in meiotic recombination, we studied DSB repair and recombinant formation at a recombination hotspot, the *HIS4-LEU2* locus (Figure 2A) [44]. In the wild type, DSB starts at 2 h, peaks at 3 h, and then gradually disappears (Figure 2B and 2D). The *dot1* mutant exhibits slight delay in the formation of DSBs and delay in the DSB repair relative to the wild type. As reported [10], the *set1* mutant shows a delay in DSB appearance by ~2 h and a peak at 5 h with reduced steady-state levels of DSBs at site I to 18% of the levels seen in the wild type (at 3 h vs. 5 h in the *set1*). This confirms the role of Set1 in efficient DSB formation [7,10]. The *set1 dot1* double mutant exhibits similar kinetics to those seen in the *set1* single mutant. The double mutant shows a similar level of steady-state DSBs as seen in the *set1* single mutant, suggesting that Dot1 does not play a role in DSB formation at the *HIS4-LEU2* locus in the absence of Set1.

Using restriction site polymorphisms present on 2 parental DNAs, formation of both CO (R1 and R2) and NCO (R3) was assessed at the *HIS4-LEU2* locus [43] (Figure 2A). The *dot1* mutant exhibits delayed formation of both COs and NCOs by 1–2 h relative to wild type, but the CO and NCO levels in the mutant are almost similar to those in the wild type (Figure 2C and 2D). The *set1* single mutant shows a delay in the formation of recombinants by 3 h and decreases COs (R2) to ~35% and NCOs (R3) to ~25% of the levels in the wild type (at 7 h), supporting a role for Set1 in efficient meiotic recombinant formation. The levels of the 2 recombinants in the *set1 dot1* double mutant are almost indistinguishable from those in the *set1* single mutant (Figure 2D).

Dot1 plays a role in DSB formation in the absence of Set1

To address the role of Set1 and Dot1 in the formation and repair of DSBs across the genome, we carried out immunostaining analysis for Rad51, a RecA homolog [40], involved in both mitotic and meiotic recombination, and the meiosis-specific RecA homolog, Dmc1 [15]. The collaboration of Rad51 and Dmc1 is key to interhomolog recombination [16,45,46]. As shown previously [47], Rad51 shows punctate staining on meiotic chromosomes (Figure 3A). Rad51 foci correspond with sites of ongoing recombination [40,48]. Counting of nuclei positive for Rad51 foci (more than 5 foci) shows the kinetics of DSB repair (Figure 3B). The *dot1* mutant shows the similar kinetics of Rad51-focus appearance as seen in the wild type. However, the disappearance of Rad51 foci occurs earlier in the *dot1* mutant than the wild type. The appearance of Rad51 foci in the *set1* mutant is delayed by

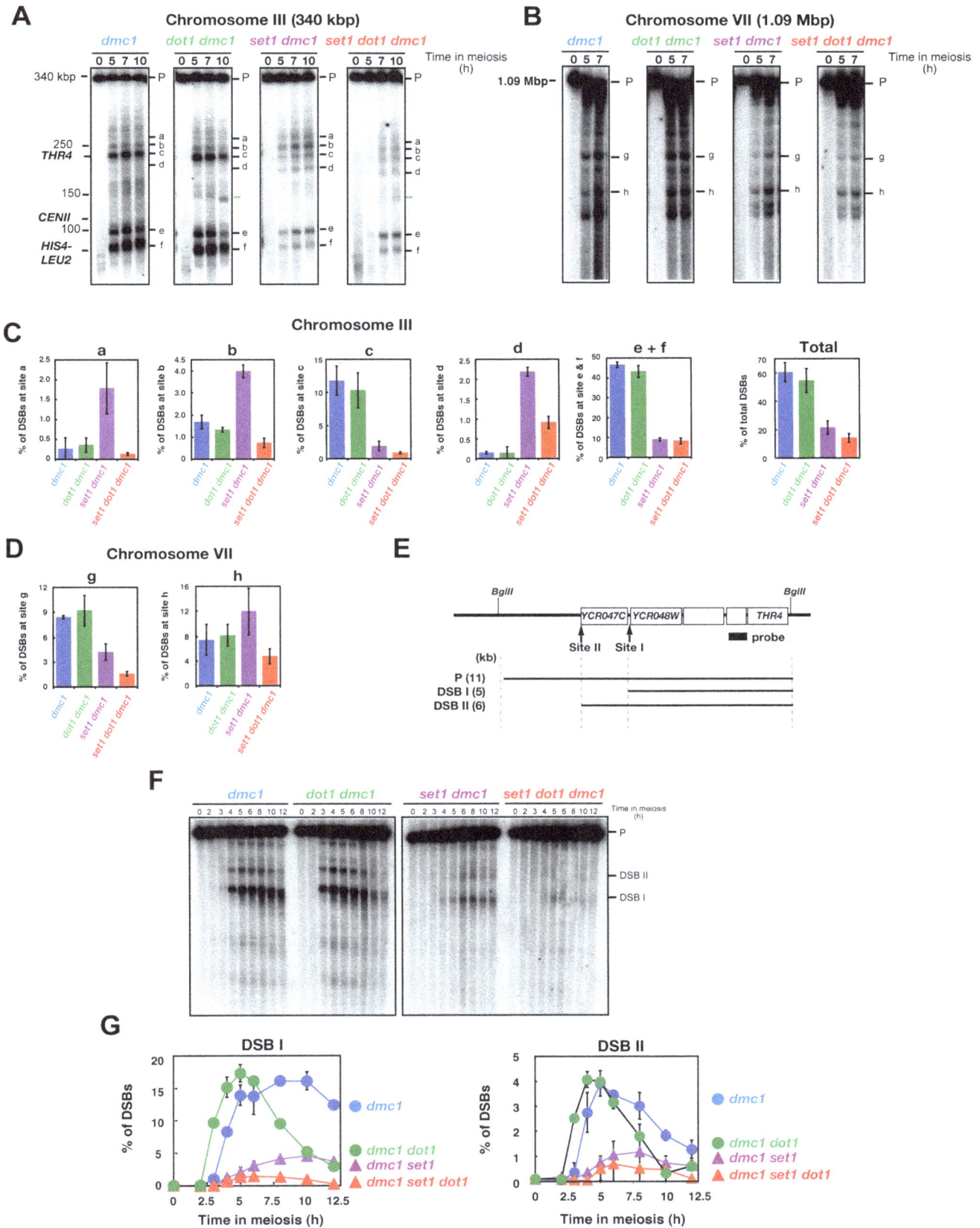

A Chromosome III (340 kbp)

B Chromosome VII (1.09 Mbp)

C Chromosome III

D Chromosome VII

E

F

G

Figure 4. Dot1 promotes the formation of DSBs in the absence of Set1. (A) Distribution of DSBs along Chromosome III was analyzed by PFGE followed by indirect labeling of one chromosome end using the *CHA1*. Samples from meiotic time courses of the *dmc1* mutant (MSY2630/2632), *dmc1 dot1* mutant (MBY003/004), *dmc1 set1* mutant (MBY021/022), and *dmc1 set1 dot1* mutant (MBY282/285) were analyzed. Band positions used for the quantification in (C) are shown on right. On left, approximate size of chromosomes and the position of two recombination hotspots, *HIS4-LEU2* and *THR4* are indicated. Green bars on the right side are a possible "Dot1"-dependent DSB bands. (B) Distribution of DSBs along Chromosome VII was analyzed by PFGE followed by indirect labeling of one chromosome end using the *CUP2*. Band positions used for the quantification in (D) are shown on right. (C) Quantification of DSB frequencies at defined positions on chromosome III were carried out for the *dmc1* mutant (Blue bars, MSY2630/ 2632), *dmc1 dot1* mutant (Green Bars, MBY003/004), *dmc1 set1* mutant (purple bars, MBY021/022), and *dmc1 set1 dot1* mutant (red bars, MBY282/ 285). Total amounts of DSBs along the chromosome are also shown in right. Plotted values are the mean values with standard deviation (S.D.) at 7 h from three independent time courses. (D) Quantification of DSB frequencies at defined positions on chromosome VII were carried out as shown in (C). Plotted values are the mean values standard deviation (S.D.) at 7 h from two independent time courses. (E) Schematic representation of the *YCR047C/ CR048W* locus. Sizes of fragments for DSB are shown with lines below. (F) DSB formation at the *YCR047C/CR048W* locus in different strains was verified by Southern blotting. Genomic DNAs were digested with *Bgl*II. (G) The bands of DSBs I (left) and II (right) at the *YCR047C/CR048W* locus were quantified. The experiments were independently performed three times and representative blots are shown. The symbols represent the *dmc1* (blue circles; MSY2630/2632), *dmc1 dot1* mutant (green circles; MBY003/004), *dmc1 set1* mutant (purple triangles; MBY021/022) and *dmc1 set1 dot1* mutant (red triangle; MBY282/285). Plotted values are the mean values with standard deviation (S.D.) from three independent time courses.

about 2 h relative to the wild type, consistent with the delay of the onset of the pre-meiotic S phase in the mutant. On the other hand, disappearance of Rad51 foci in the mutant shows a ~3-h delay compared to the wild type. If the S-phase delay is accounted for [10], the *set1* mutant delays Rad51-focus disassembly by about 1 h, suggesting a role for Set1 in DSB repair in meiosis. The *set1 dot1* double mutant shows delayed appearance of Rad51 similar to the *set1* single mutant. However, the disappearance of Rad51 foci in the double mutant is 1 h later than that in the single mutant. This suggests a role for Dot1 in meiotic DSB repair in the absence of Set1.

Numbers of Rad51 foci per focus-positive cell are indicative of the steady-state numbers of DSBs in a cell (Figure 3C). The average numbers of foci in the wild type and the *dot1* mutant at 4 h are 36 ± 10 (n = 78) and 32 ± 12 (n = 89), respectively (P value = 0.01; Mann-Whitney's *U*-test). The *set1* mutant shows a slightly reduced number of foci (31 ± 8.9; n = 95; 86% of wild-type level) at 4 h and more reduced number (21 ± 8.8; n = 79; 58% of wild-type level at 4 h) at 6 h, consistent with the reduction of DSBs in this mutant (wild type; P value = 0.0041, 2.2×10^{-15}, respectively, Mann-Whitney's *U*-test). Reduced focus number at 6 h compared to that at 4 h may be due to disassembly of Rad51 from chromosomes by DSB repair in the mutant. Moreover, the *set1 dot1* double mutant shows a reduced Rad51-focus number (17 ± 6.6; n = 116) at 6 h, which is much lower than that in the *set1* single mutant (at 6 h; P value = 5.3×10^{-5}, as well as at 4 h (19 ± 7.0; n = 78; versus at 4 h in the *set1*, P value = 2.2×10^{-16}), suggesting a significant role for Dot1 in DSB formation in the absence of Set1. Mutations in the *SET1* and/or *DOT1* genes show similar effects in the kinetic analysis of Dmc1 as those seen with Rad51 foci (Figure 3B).

Given the critical role of Set1-dependent histone H3K4 tri-methylation in DSB formation across genome [7,12,13], relatively high numbers of Rad51 foci, thus DSBs, in the *set1* single mutant are a bit surprising. Consistent with significant DSB formation in the *set1* mutant, we observed that steady levels of Hop1 phosphorylation, which depends on DSBs through the activation of Mec1/Tel1 kinases [49], in the *set1* mutant were comparable to those in wild type (shifted bands [arrows] in Figure 1E). To confirm the results, we also counted the number of Rad51 foci in the background of the *dmc1* mutant (Figure 3D), which is defective in the repair of DSBs and, as a result, accumulates the foci [15]. As reported [29], the *dot1* mutation weakly suppresses *dmc1*-induced cell cycle arrest (Figure 3E). Interestingly, the combination of *set1* and *dot1* mutations alleviates *dmc1* arrest to a greater extent than does the *dot1* mutation alone. As expected, all 4 strains with the *dmc1* mutation accumulate Rad51-focus positive cells (Figure 3D and 3F). As with Rad51-focus counting, we analyzed early time

points up to 5 h when Rad51 is in the assembly stage. In the *dmc1* and *dot1 dmc1* mutants, the average numbers of Rad51 foci at 3 h are 41 ± 13 (n = 41) and 42 ± 17 (n = 30), respectively (Figure 3G). The *set1 dmc1* mutant shows a reduced number of foci (34 ± 12; n = 42) at 5 h (P value = 1.5×10^{-4}, versus at 3h in *dmc1*, Mann-Whitney's *U*-test), confirming the role of Set1 in DSB formation, although the effect of the *set1* mutant is 30% reduction compared to wild type. Again, the *set1 dot1 dmc1* triple mutant shows a decreased number of foci, to 19 ± 6.9 (n = 52) at 5 h, which is much lower than that in the *set1 dmc1* mutant (P value = 5.6×10^{-6}, versus *dmc1 set1* at 5 h, Mann-Whitney's *U*-test). This supports a role for Dot1 in DSB formation without Set1.

Dot1 plays a role in DSB formation at some regions of chromosomes

The above cytological analysis of Rad51 foci across the genome indicates a role of Dot1 for DSB formation in the absence of Set1. On the other hand, physical analysis at the artificial recombination hotspot, *HIS4-LEU2*, did not support this idea (Figure 2D). To know the role of Dot1 in genome-wide DSB formation, we studied the distribution of DSBs on single chromosomes by using pulse-field gel electrophoresis (PFGE) [41]. We analyzed DSB distribution on chromosome III as a representative of small chromosomes of yeast and VII as a long chromosome in the *dmc1* mutant background (Figure 4A and 4B). For the mapping, we used the *dmc1* rather than the *rad50S*, in which Tel1/ATM kinase is activated to down-regulate DSB formation, particularly on long chromosomes [50]. The DSB mapping showed region-specific enrichment of DSBs on chromosome III and VII in the *dmc1* mutant. We also chose several regions hot for DSB formation and quantified the amounts of DSBs at the regions (Figure 4C and 4D). The *dot1* mutant (with the *dmc1*) shows similar patterns of DSB distributions on both chromosomes III and VII with similar DSB formation efficiencies to the control wild type. The *set1* mutant greatly reduces DSB formation along chromosomes with variation in its effect. This is consistent with previous observations by ChIP(Chromatin Immunoprecipitation)-chip [7,12]. The *set1* mutation not only reduces DSB formation but also increases DSB formation at several loci [7,12]. Increased levels of DSBs in the *set1 dmc1* mutant are seen at regions, a, b and d on chromosome III compared to the *dmc1* mutant (Figure 4C). Although the *dot1* mutation did not affect DSB formation at regions c and e + f of chromosome III in the absence of the *SET1*, the mutation reduced DSB formation at regions a, b, and d on chromosome III, and g and h on chromosome VII. The positive role of Dot1 in DSB formation is clearly seen in regions in which DSB formation is increased in the absence of Set1. Interestingly, the *dot1* (and *set1 dot1*) mutant shows novel DSB hotspots at a late

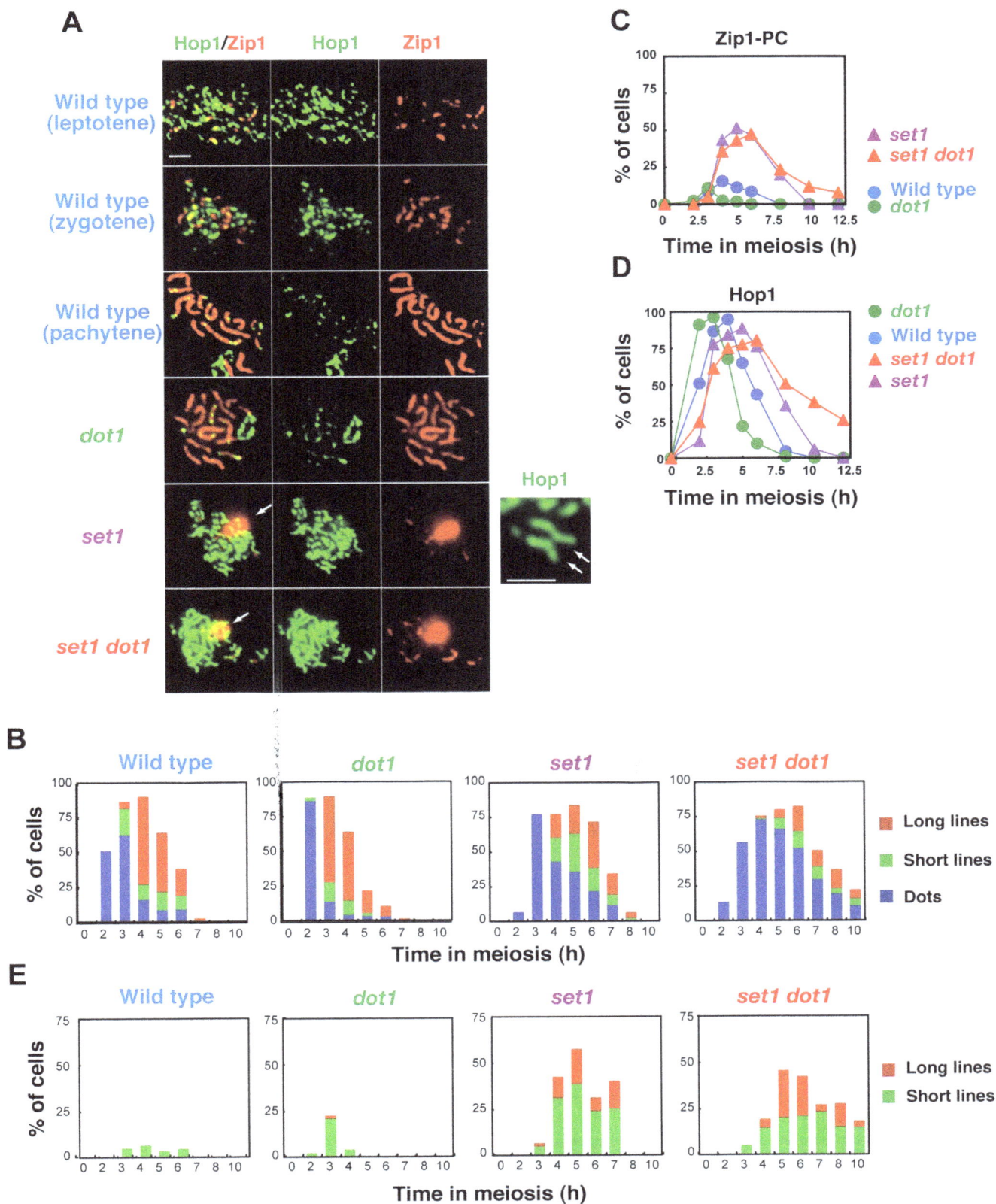

Figure 5. Set1 promotes the formation of synaptonemal complex. (A) Immunostaining analysis of chromosome proteins, Zip1 (red) and Hop1 (green), was carried out for wild type and different mutant strains. Representative images are shown for each strain. Representative images for parallel Hop1 lines in the *set1* mutants are shown in right. White arrows indicate polycomplexes of Zip1. Wild type, NKY1303/1543; *set1* mutant, MBY015/016; *dot1* mutant, MBY005/006; *set1 dot1* double mutant, MBY037/039. The bar indicates 2 μm. (B) Zip1 staining in wild type and mutant

strains was classified as follows: dot (dots I, blue), partial linear (short lines, green), full SC (long lines, red). More than 100 nuclei were counted at each time point. Wild type, NKY1303/1543; *set1* mutant, MBY015/016; *dot1* mutant, MBY005/006; *set1 dot1* double mutant, MBY037/039. (C) Kinetics of spreads with Zip1-PCs were analyzed. Wild type, blue circles; *set1* mutant, green circles; *dot1* mutant, purple triangles; *set1 dot1* double mutant, red triangles. (D) Kinetics of spreads positive for Hop1 were verified in different strains. Wild type, blue circles; *set1* mutant, green circles; *dot1* mutant, purple triangles; *set1 dot1* double mutant, red triangles. (E) Hop1-staining in different strains was classified: short lines (green) and long lines (red). Positive cells for each class were counted. More than 100 nuclei were counted at each time point.

time point (green bars around ~150 kb region of chromosome III, right green bar in Figure 4A), supporting a possible role of Dot1 in DSB formation even in normal meiosis.

We also analyzed the role of Dot1 on DSB formation at a single locus. We focused on a hotspot, *YCR047C/YCR048W* locus on chromosome III (Figure 4E). The *dmc1* mutant accumulates DSBs at the *YCR047C/YCR048W* locus (Figure 4F and 4G). On the other hand, DSBs in the *dot1 dmc1* double mutant accumulate as in the *dmc1* mutant but gradually reduce during further incubation. This might be due to either more resection of DSB ends or DNA repair [15]. The *set1 dmc1* double mutant shows decreased DSB levels (for DSB I) to 23% of wild type (at 6 h). Importantly, the *set1 dot1 dmc1* triple mutant reduced the DSB levels to 48% for DSB I) compared to the *set1 dmc1* double mutant ($P = 0.0089$, Student's *t*-test). These results support the idea that Dot1 is involved in DSB formation in the absence of Set1.

Set1 and Dot1 play a role in the formation of synaptonemal complex

Previously, the role of 2 histone H3 methyltransferases in the formation of meiotic chromosome structures had not been described well. We first examined the formation of the synaptonemal complex (SC) by immunolocalization of Zip1, which is a component in the central region of the SC [23]. Zip1 staining was classified into 3 classes: dots, partial lines and full lines, which may roughly correspond with the leptotene, zygotene, and pachytene stages, respectively (Figure 5A). Immunostaining reveals unique contributions of Set1 and Dot1 to SC formation. The *dot1* single mutant shows near wild-type kinetics for Zip1 assembly and disassembly, except that dotty staining of Zip1 appears earlier and Zip1 assembly disappears a bit earlier in the *dot1* cells relative to wild type (Figure 5B).

The *set1* single mutant shows clear defects in SC assembly (Figure 5A and 5B). The appearance of Zip1 dotty staining is delayed by ~1 h, probably due to a delay in the S-phase.

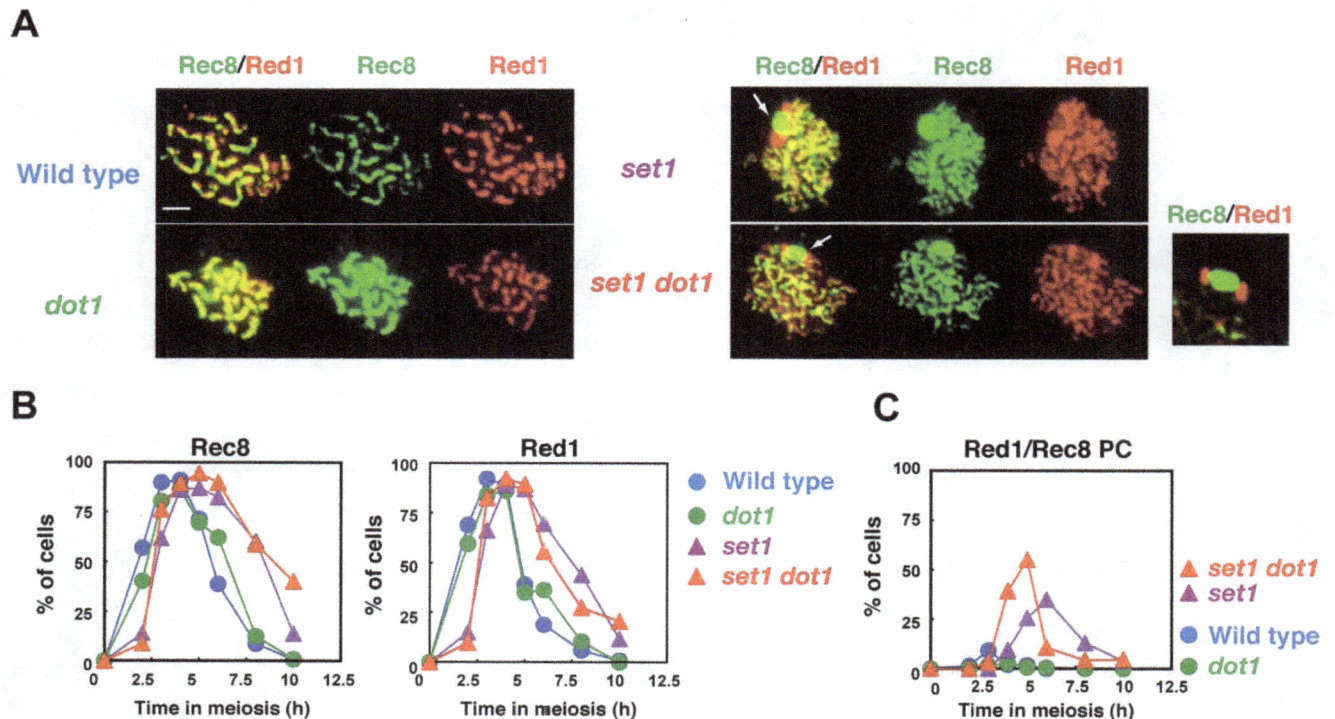

Figure 6. Set1 promotes normal assembly of chromosome axes. (A) Immunostaining analysis of chromosome proteins, Red1 (red) and Rec8 (green), were carried out for wild type and different mutant strains. Representative images for pachytene (wild type *dot1*) and pseudo-pachytene (*set1* and *set1 dot1*) stages are shown for each strain; wild type 5 h; the *dot1*, 5h; the *set1*, 6 h; the *set1 dot1*, 6 h. White arrows in the *set1* or *set1 dot1* mutants shows Rec8/Red1 aggregates. An image for Rec8/Red1 aggregates in the *set1 dot1* mutant is enlarged and shown in right. Wild type, NKY1303/1543; *set1* mutant, MBY015/016; *dot1* mutant, MBY005/006; *set1 dot1* double mutant, MBY037/039. The bar indicates 2 μm. (B) Kinetics of spreads positive for Red1 (right) and Rec8 (left) were verified in different strains. The symbols indicate the wild type (blue circles; NKY1303/1543), *dot1* mutant (green circles; MBY005/006), *set1* mutant (purple triangles; MBY015/016), and *set1 dot1* mutant (red triangle; MBY037/039). (C) Kinetics of spreads with Red1/Rec8-PCs were analyzed. Wild type, blue circles; *set1* mutant, green circles; *dot1* mutant, purple triangles; *set1 dot1* double mutant, red triangles.

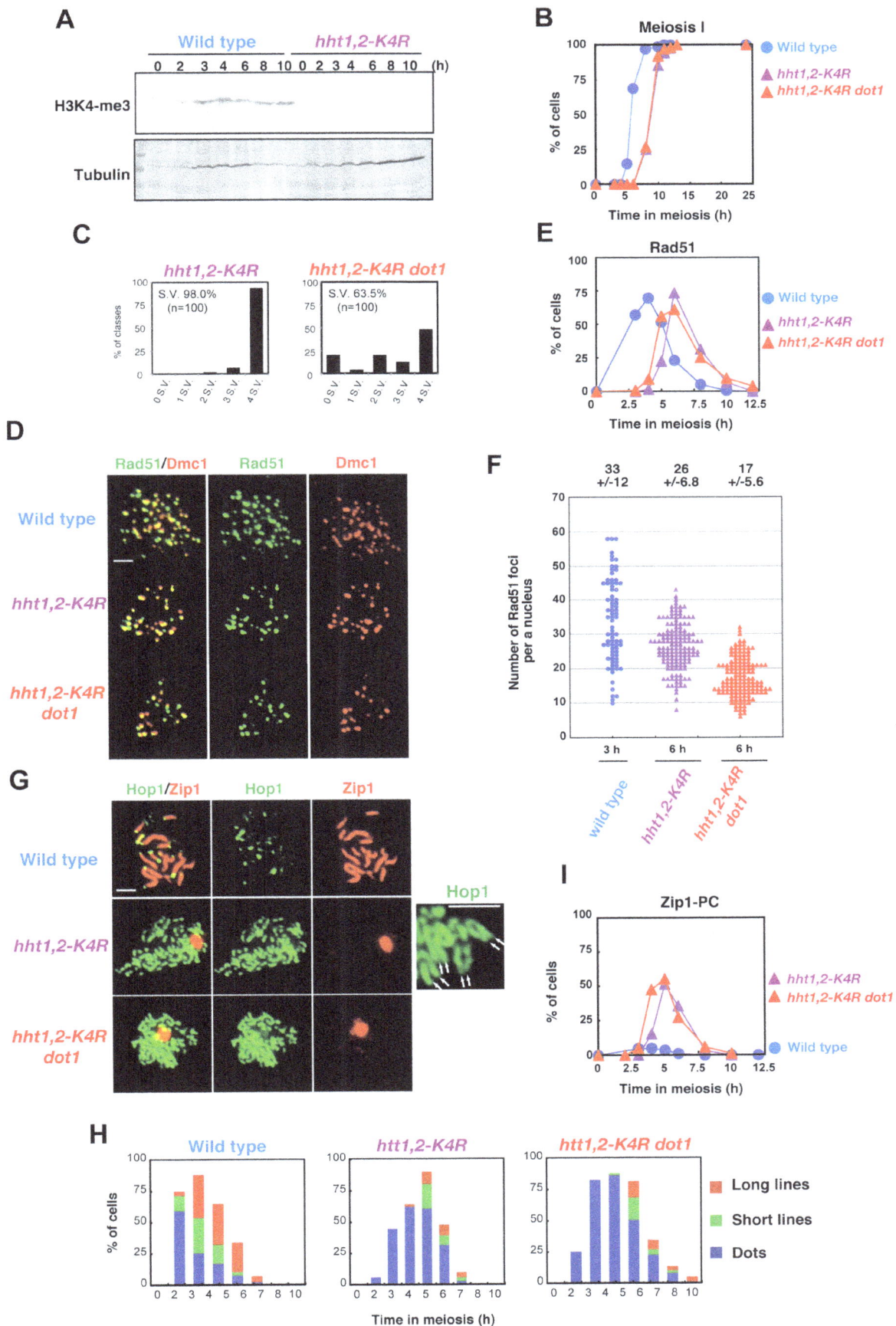

Figure 7. Histone H3K4 is critical for DSB and SC formation. (A) Expression of histone H3K4 trimethylation during meiosis. Western blotting analysis for wild type (NKY1303/1543), hht1,2-K4R (MBY211/218) was carried out using anti-histone H3K4-me3. (B) Meiotic cell division I was analyzed by DAPI staining of wild-type (blue circles; NKY1303/1543), hht1-K4R (purple triangles; MBY211/218), and hht1,2-K4R dot1 (red triangles; MBY233/237) strains. At least 150 cells were counted by DAPI staining for each time point. (C) Distribution of viable spores per tetrad in wild-type and hht1,2-K4R dot1 (MBY233/237) strains. For each strain, 100 tetrads were dissected. (D) Immunostaining analysis of Rad51 (green) and Dmc1 (red) for wild type (NKY1303/1543), hht1,2-K4R (MBY211/218), and hht1,2-K4R dot1 (MBY233/237) strains was carried out. The bar indicates 2 μm. (E) Kinetics of Rad51 focus-positive cells in various strains. A focus-positive cell was defined as a cell with more than 5 foci. More than 100 nuclei were counted at each time point. The symbols indicate the wild type (blue circles; NKY1303/1543), hht1,2-K4R (purple triangles; MBY211/218), and hht1,2-K4R dot1 (red triangles; MBY233/237) strains. (F) Rad51 focus numbers per nucleus were counted in different strains. The symbols indicate the wild type (blue circles; NKY1303/1543), hht1,2-K4R (purple triangles; MBY211/218), and hht1,2-K4R dot1 (red triangle; MBY233/237) strains. The average number of foci is shown per positive nucleus. (G) Representative images for staining of Zip1(red) and Hop1(green) in wild-type and mutant strains are shown. Hop1 parallel lines the hht1,2-K4R mutant are shown in a pair of arrows on the right. The bar indicates 2 μm. (H) Zip1-staining was classified into 3 classes: dot (dots, blue), partial linear (short lines, green), and full SC (long lines, red). More than 100 nuclei were counted at each time point Kinetics of spreads (Zip1 polycomplexes) were analyzed. Wild type, NKY1303/1543; hht1,2-K4R, MBY211/218; hht1,2-K4R dot1, MBY233/237. (I) Kinetics of Zip1-PC in different strains. The number of spreads containing Zip1-PC was counted in each strain. The symbols indicate the wild type (blue circles; NKY1303/1543), hht1,2-K4R (purple triangles; MBY211/218), and hht1.2-K4R dot1 (red triangle; MBY233/237) strains.

Moreover, the mutant shows reduced frequencies of full-length SCs. Furthermore, Zip1 disassembly occurs ~1 h later than in wild type, even after compensating for the delay in assembly. Consistent with the defect in Zip1 assembly in the *set1* mutant, the mutant accumulates an aggregate of Zip1, referred to as a polycomplex (PC; Figure 5A and 5C). This confirms previous observation that the *set1* mutant is defective in SC formation [51]. The *set1 dot1* double mutant exhibits more defects in Zip1 elongation (reduced pachytene cells) and a greater delay in Zip1-disassembly than does the *set1* single mutant, suggesting a role for Dot1 in SC formation in the absence of Set1. This SC-defect in the double mutant may be caused by the repair defect and/or reduced DSB formation in the mutant (see above).

To analyze SC defects seen in the *set1* mutant in more detail, we also examined the localization of Hop1 (Figure 5A), which is a component of the chromosome axis and is required for SC formation as well as meiotic recombination [52]. In the wild type, Hop1 appearance occurs as early as 2 h after the induction of meiosis, and Hop1 disappearance takes place around the pachytene stage; e.g., 5 h (Figure 5D). In wild-type cells, Hop1 shows punctate staining in early meiotic prophase I and reduced staining during late prophase (Figure 5A). The *dot1* single mutant shows very similar Hop1 staining patterns to those seen in the wild type, although, as seen for Zip1, Hop1-loading occurs earlier in the mutant than in wild type. Importantly, the *set1* single mutant shows a 1-h delay in the assembly of Hop1 foci relative to the wild type, and a 3-h delay in the disassembly. Moreover, in addition to dotty staining of Hop1, *set1* cells show elongated lines of Hop1, which is rarely seen in the wild type (Figure 5A and 5E). In some nuclei, 2 lines of Hop1 are aligned side-by-side (Figure 5A, shown by a pair of arrows), suggesting that the pairing of homologous chromosomes takes place normally, but full synapsis is impaired in the *set1* mutant. The *dot1 set1* double mutant exhibits very similar patterns with the exception of greater proportions of long Hop1 lines and delayed disappearance of Hop1 from chromosomes relative to the *set1* mutant, consistent with a role for Dot1 in SC formation without Set1.

Consistent with a previous observation [30], double staining of Zip1 and Hop1 clearly shows that Hop1-enriched regions lack strong Zip1 signals in all strains including wild type (Figure 5A), confirming the previous idea that Hop1 in yeast is disassembled along with Zip1 elongation as seen in other eukaryotic organisms, such as Hormad1 in mammals [53] and Asy1 in plants [54]. The accumulated localization of Hop1 along the chromosomes in the *set1* mutant is possibly consistent with the fact that Set1 is required for Zip1 elongation.

We also analyzed the localization of another axis protein, Red1 [55], as well as that of the meiosis-specific kleisin, Rec8, which is a component of the cohesion complex [56](Figure 6). Red1 works together with Hop1 as well as with Mek1/Mre4 in both meiotic recombination and chromosome morphogenesis [57,58]. Red1 initially appears as focal staining like Hop1, but later, unlike Hop1, it forms discontinuous lines as the SC elongates (Figure 6A and 6B) [59]. In the *set1* and *set1 dot1* mutants, both of which shows delay in the assembly and disassembly of Red1 (Figure 6B), there are little thick Red1 lines, consistent with defective SC elongation in the mutants. Rec8 localization is similar to that of Red1 in wild type and *dot1* strains. On the other hand, the *set1* and *set1 dot1* mutants rarely form thick lines of Rec8 as seen in the wild-type and *dot1* mutant cells, consistent with the lack of full SCs in the mutants (Figure 5A). Notably, we observed aggregates of Red1 and Rec8 in the *set1* and *set1 dot1* mutants (Figure 6A). At 6 h, about 35% of *set1* mutant cells contain Red1 and Rec8 aggregates (Figure 6C). This number is increased to 55% at 5 h in the *set1 dot1* double mutant. In the PC-like structure, Red1 shows bipolar staining on a large Rec8-block (Figure 6A). It is important to point out that these Red1 and Rec8 aggregates are not formed in other SC-deficient mutants such as the *dmc1* mutant. These results show that Set1 is important for SC elongation and that Dot1 plays a role in SC formation only in the absence of Set1. Set1 may be important for the organization of the chromosome axis containing Red1 and Rec8 for synapsis. On the other hand, recent ChIP-chip study shows that the *set1* mutant shows wild-type distribution of Rec8 along chromosomes [12]. Therefore, more higher order structure of chromosome axes might be compromised in the mutant.

The histone H3K4 mutant is defective in SC formation

Set1 is a H3K4 methyltransferase [38]. To confirm the role of Set1 in meiotic chromosome metabolism through this histone modification, we constructed a *hht1-K4R hht2-K4R* double mutant (hereafter, *hht1,2-K4R*) at the native chromosomal loci. This strain construct is different from a previous strain, in which both *HHT1* and *HHT2* were deleted, but an *ARS-CEN* plasmid with the *hht1-K4R hht2-K4R* mutations were present [13]. The absence of H3K4 tri-methylation was confirmed by western blotting (Figure 7A). The *hht1,2-K4R* double mutant shows wild-type spore viability (Figure 7C). This is different from slight reduction of spore viability of the *set1* mutant. The *hht1,2-K4R* cells show a greater delay (~3 h) in the entry into meiosis I than does the *set1* single mutant with ~2 h delay (Figure 7B). The *hht1,2-K4R* almost recapitulates the meiotic phenotype of the *set1* single mutant. The *hht1,2-K4R* mutant shows reduced DSBs and is defective in SC assembly. Rad51/Dmc1 staining (Figure 7D and 7E) shows that steady state number of Rad51 foci in the *hht1,2-K4R* mutant is, on average, 26±6.8 (n = 143) at 6 h (Figure 7F; statistically significant from numbers at 3 h in wild type, P value = 1.4×10^{-5}, Mann-

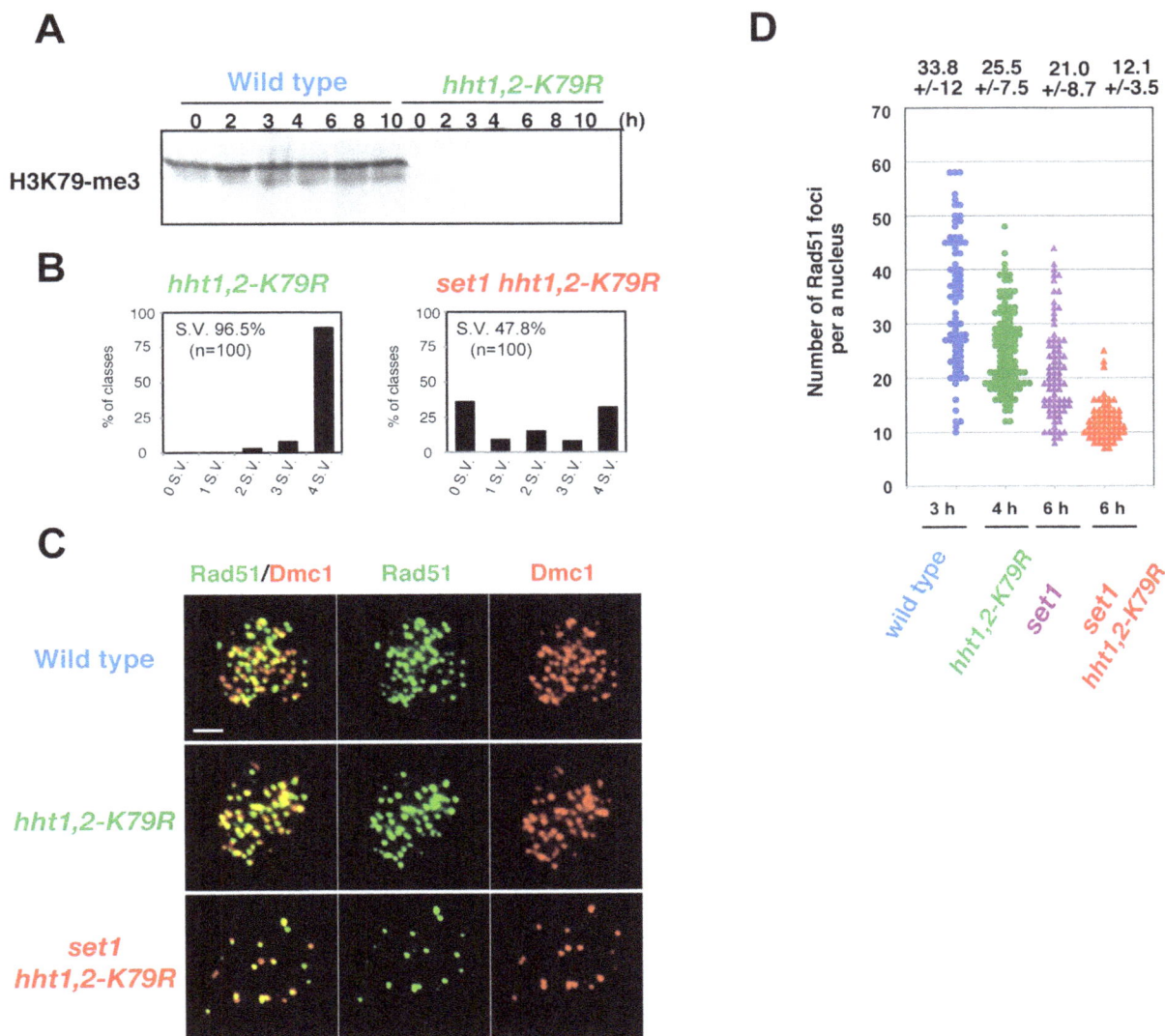

Figure 8. Histone H3K79 is critical for DSB without Set1. (A) Expression of histone H3K79-methylation during meiosis. Western blotting analysis for the wild-type (NKY1303/1543) and *hht1,2-K79R* (MBY151/152) strains was carried out using anti-H3K79-methylation. (B) Distribution of viable spores per tetrad in the *hht1,2-K79R* (MBY151/152) and *set1 hht1,2-K79R* (MBY219/221) strains. For each strain, 100 tetrads were dissected. (C) Immunostaining analysis of Rad51 (green) and Dmc1 (red) in wild-type cells at 3 h (NKY1303/1543) and *set1 hht1,2-K79R* (MBY219/221) cells at 6 h. The bar indicates 2 μm. (D) Rad51 focus numbers per nucleus were counted in different strains. Wild type (blue circles; NKY1303/1543), *hht1,2-K79R* (green circles; MBY219/221), the *set1* (purple triangles; MBY015/016) and *hht1,2-K79R set1* (red triangle; MBY219/221). Both the *set1* and *set1 hht1,2-K79R* mutants show delayed appearance of Rad51 foci on chromosomes. Thus, focus numbers of Rad51 at 6 h was measured. The number for the *set1* is the same as that in Figure 3C. The average number of foci with SD is shown per positive nucleus.

Whitney's *U*-test), indicating the role of H3K4 in DSB formation. Moreover, the *hht1,2-K4R* mutant with the *dot1* mutation, with a greater reduction in spore viability (63.5%) compared to the wild type, shows a greater reduction in Rad51-focus number with 17 ± 5.6 (n = 140) at 6 h (versus wild type; P value $\approx$ 0, Mann-Whitney's *U*-test). Rad51 focus number in the *hht1,2-K4R dot1* mutant is 65% of the number in the *hht1,2-K4R* mutant. These support the notion that the Dot1 plays a role in DSB formation in the absence of H3K4 methylation.

The *hht1,2-K4R* double mutant is also defective in Zip1 elongation, and therefore in SC formation like the *set1* mutant (Figure 7G and 7I). The *hht1,2-K4R dot1* mutant shows more delay in SC disassembly compared to *hht1,2-K4R* (Figure 7G and 7H). We also found that the *hht1,2-K4R* double mutant often shows 2

parallel Hop1 lines like the *set1* mutant (Figure 7G). These strongly suggest a role for Set1 in SC formation as well as DSB formation through the methylation of histone H3K4.

Histone H3K79 is critical for DSB formation in the absence of *SET1*

In order to know the involvement of histone H3K79-methylation in DSB formation, we also used a strain with histone H3K79R mutations at the native chromosomal loci (*hht1-K79R hht2-K79R*, hereafter *hht1,2-K79R*; Takehiko Usui and A.S., unpublished). The absence of H3K79 methylation was confirmed by western blot analysis using an anti-histone H3K79 methylation antibody (Figure 8A). The *hht1,2-K79R* mutant shows wild-type spore viability (Figure 8B). Importantly, when *hht1,2-K79R* was combined with the *set1* deletion, the triple mutant shows 47.8% spore viability (Figure 8B), similar to the *set1 dot1* double mutant (see Figure 1). We found that the *set1 hht1,2-K79R* mutant shows a decreased number of Rad51 foci (Figure 8C and 8D; 12.1±3.5; n = 62), which is more reduced in the *set1* mutant. This supports the idea that Dot1-dependent histone H3K79 methylation promotes meiotic DSB formation in the absence of Set1-dependent histone H3K4 methylation. The Rad51 focus number in the *set1 hht1,2-K79R* mutant is smaller than that in the *set1 dot1* mutant (Figure 8C and 8D; versus 6 h, P value = 1.5×10^{-5}, Mann-Whitney's *U*-test). This may be due to culture-to culture difference.

Discussion

Previous studies have shown that 2 histone-modifications, H3K4 methylation and H2BK123 ubiquitination, play a critical role in the formation of meiotic DSBs [7,12,13,39]. The effect of H2BK123 ubiquitination seems to be indirect since this mark promotes H3K4 methylation *in trans* [38]. In this study, we have demonstrated the role of Dot1 H3K79 methyltransferase in DSB formation in the absence of Set1.

Cytological analysis of Rad51 foci, which mark sites of ongoing recombination [47], showed that, even in the absence of Set1-dependent H3K4-methylation, meiotic cells form significant numbers of DSBs, about 2/3–3/4 of the levels seen in wild type, consistent with high spore viability of the mutants defective in H3K4 methylation. In contrast, previous studies using whole genome mapping showed a large reduction of DSBs at some regions in the absence of either *SET1* or H3K4 methylation [7]. In addition, the lack of Set1-dependent H3K4 methylation changes the distribution of DSBs along chromosomes [7,12]. However, such studies used ChIP-chip for mapping using *dmc1* or *rad50S* (*sae2*) mutants which block recombination; therefore, it is very difficult to quantify/estimate how much DSBs are dependent on the specific histone marks. Our results of Southern blotting for individual loci (Figure 2D and 4G) confirm the previous results. It is reported that the *set1* mutant showed increased DSBs at a specific chromosomal locus [7,12], suggesting the presence of a backup system for DSB formation. We used counting of Rad51 foci to get a rough estimate of DSB numbers in a nucleus. Our ongoing research showed that the number of Rad51 foci is roughly proportional to the number of DSBs (M. S., unpublished). We found that the *set1* mutant showed a mild reduction of Rad51 focus numbers along the genome. Indeed, DSB mapping on individual chromosomes in the mutant support the idea (Figure 4). Thus, we believe that the contribution of Set1-dependent H3K4 methylation to DSB formation is weaker than expected, at least in the budding yeast. These data suggest the presence of other critical determinants for DSB formation. Indeed, we found that elimination of Dot1-dependent H3K79 methylation reduces DSB levels to about half of that seen in the *set1* deletion mutant. This indicates the involvement of the histone post-translational modification in DSB formation. In the fission yeast, H3K9 acetylation is known to promote DSB formation, while H3K4 methylation is not involved in DSB formation [9]. Moreover, in mice, meiosis-specific Prdm9-dependent H3K4 methylation shapes hotspot activity for recombination. Interestingly, even the absence of Prdm9 methyltransferase changes the distribution of the hotspot by creating new spots [11]. In Prdm9 KO mice, Prdm9-INdependent H3K4 methylation might be responsible for this activity [11]. These studies confirm that multiple histone post-translational modifications determine the site of initiation of meiotic recombination. We want to stress that, even in the absence of both Set1 and Dot1, mutant cells form 40–50% of the wild-type levels of Rad51 foci,

likely DSBs, suggesting the presence of other determinant(s) for hotspot activity. Recently, it is shown that, in a plant, *Arabidopsis thaliana*, a histone H2A variant, H2A.Z, plays a role in recombination hotspot activity during meiosis [60].

Our results suggest that H3K79 methylation plays a role in DSB formation. H3K79 methylation is recognized by the Tudor domain of Rad9 in yeast [61]. Since the *rad9* mutant is proficient in meiosis [62], it is unlikely that Rad9 plays a role in DSB formation. However, we need to analyze a *rad9* mutant with the *set1* deletion to know the exact role of this protein in DSB formation, since the effect of the *dot1* is only seen in the absence of the *SET1*. Alternatively, the other protein involved in DSB formation may recognize this mark. Recent reports suggest a role for Dot1-dependent H3K79-methylation in the recombination checkpoint during meiosis [63]. In the recombination checkpoint, Dot1-dependent H3K79 methylation promotes the efficient binding of the Hop1 protein in the *zip1* mutant. This could be interpreted simply as that H3K79 methylation is bound to Hop1. However, this is unlikely, since we showed that the *dot1* mutant is proficient in Hop1 binding at least in the wild-type background. In wild type meiotic cells, there is another pathway to recruit Hop1 in a Dot1-independent manner. Recently, Dot1 has been shown to play a role in the Tel1/ATM pathway in meiotic recombination [31,63], which somehow controls DSB formation [49,50,64]. If this is true in wild-type meiosis, the role of Dot1 in DSB formation described here is indirect; e.g. signaling. Indeed, recent genome-wide mapping showed that H3K79 methylation is less in promoter regions than coding regions [65]. This suggests that the Dot1-dependent H3K79 methylation play a negative role rather than a positive role in the DSB formation. We suggest that meiotic chromosomes adapt different alternatives to create the recombination hotspot, possibly using different histone marks. This kind of multiple alternatives or flexibility may contribute to the rapid evolution of the recombination hotspots.

The effect of the *dot1* mutation on DSB formation is clearly seen in the absence of H3K4 methylation. In this line, it is interesting to see subtle effects of the *dot1* mutation on DSB formation in the presence of H3K4 methylation. This includes altered kinetics of DSB repair (Figure 2D), slight but significant reduction of Rad51 focus number (Figure 3C) and the appearance of late DSBs formation (Figure 4A) in the *dot1* mutant. We need further careful evaluation on the role of the Dot1-dependent H3K79 mutation on DSB formation.

Although our studies described here suggest a direct link of Dot1 with DSB formation in the absence of Set1, we cannot exclude the possibility that the effect of *dot1* mutation is indirect through the transcription [37]. We also need more careful evaluation on the role of any histone posttranslational modifications in meiotic recombination such as DSB formation.

We also revealed a role for Set1-dependent H3K4 methylation in chromosome morphogenesis in meiosis, such as SC formation. Both the *set1* and *hht1,2-K4R* mutants produce viable spores and retain high levels of DSBs relative to wild type, as judged from the number of Rad51 foci. However, the 2 mutants are almost defective in SC elongation. The SC elongation defect in the *set1* and *hht1,2-K4R* mutants reflects persistent loading of Hop1, which often forms a linear line. The defect in SC elongation in the mutants may be caused by abnormal assembly of chromosome axes. This idea is supported by the accumulation of several axis proteins such as Rec8 and Red1 as abnormal aggregates. This abnormal assembly of the axis proteins has not been seen in other mutants defective in synapsis; e.g., *zip1* or *dmc1* mutants [59]. Thus, the recombination defect cannot account for the accumulation of this aggregate. Set1-dependent H3K4 methylation may promote the assembly of Red1 or Rec8 in the context of meiotic chromosomes.

Alternatively, reduced DSBs might be directly linked with a defect in SC elongation. In this scenario, excess DSBs in wild-type cells are necessary for normal levels of chromosome synapsis rather than recombination. This idea is somehow consistent with previous proposal of 2 types of DSBs; one for synapsis and the other for recombination [66,67]. Indeed, a moderate reduction in DSBs does not affect the frequency of COs due to CO homeostasis [68].

Supporting Information

Table S1   Strain list.

References

1. Petronczki M, Siomos MF, Nasmyth K (2003) Un menage a quatre: the molecular biology of chromosome segregation in meiosis. Cell 112: 423–440.
2. Kleckner N (2006) Chiasma formation: chromatin/axis interplay and the role(s) of the synaptonemal complex. Chromosoma 115: 175–194.
3. Borde V, de Massy B (2013) Programmed induction of DNA double strand breaks during meiosis: setting up communication between DNA and the chromosome structure. Curr Opin Genet Dev 23: 147–155.
4. Keeney S (2001) Mechanism and control of meiotic recombination initiation. Curr Top Dev Biol 52: 1–53.
5. Lichten M, de Massy B (2011) The impressionistic landscape of meiotic recombination. Cell 147: 267–270.
6. Yamada T, Ohta K (2013) Initiation of meiotic recombination in chromatin structure. J Biochem.
7. Borde V, Robine N, Lin W, Bonfils S, Geli V, et al. (2009) Histone H3 lysine 4 trimethylation marks meiotic recombination initiation sites. EMBO J 28: 99–111.
8. Buard J, Barthes P, Grey C, de Massy B (2009) Distinct histone modifications define initiation and repair of meiotic recombination in the mouse. EMBO J 28: 2616–2624.
9. Yamada S, Ohta K, Yamada T (2013) Acetylated Histone H3K9 is associated with meiotic recombination hotspots, and plays a role in recombination redundantly with other factors including the H3K4 methylase Set1 in fission yeast. Nucleic Acids Res 41: 3504–3517.
10. Sollier J, Lin W, Soustelle C, Suhre K, Nicolas A, et al. (2004) Set1 is required for meiotic S-phase onset, double-strand break formation and middle gene expression. EMBO J 23: 1957–1967.
11. Brick K, Smagulova F, Khil P, Camerini-Otero RD, Petukhova GV (2012) Genetic recombination is directed away from functional genomic elements in mice. Nature 485: 642–645.
12. Sommermeyer V, Beneut C, Chaplais E, Serrentino ME, Borde V (2013) Spp1, a member of the Set1 Complex, promotes meiotic DSB formation in promoters by tethering histone H3K4 methylation sites to chromosome axes. Mol Cell 49: 43–54.
13. Acquaviva L, Szekvolgyi L, Dichtl B, Dichtl BS, de La Roche Saint Andre C, et al. (2013) The COMPASS subunit Spp1 links histone methylation to initiation of meiotic recombination. Science 339: 215–218.
14. Hunter N, Kleckner N (2001) The single-end invasion: an asymmetric intermediate at the double-strand break to double-holliday junction transition of meiotic recombination. Cell 106: 59–70.
15. Bishop DK, Park D, Xu L, Kleckner N (1992) DMC1: a meiosis-specific yeast homolog of E. coli recA required for recombination, synaptonemal complex formation, and cell cycle progression. Cell 69: 439–456.
16. Cloud V, Chan YL, Grubb J, Budke B, Bishop DK (2012) Rad51 is an accessory factor for Dmc1-mediated joint molecule formation during meiosis. Science 337: 1222–1225.
17. Shinohara A, Ogawa H, Ogawa T (1992) Rad51 protein involved in repair and recombination in S. cerevisiae is a RecA-like protein. Cell 69: 457–470.
18. Schwacha A, Kleckner N (1994) Identification of joint molecules that form frequently between homologs but rarely between sister chromatids during yeast meiosis. Cell 76: 51–63.
19. Allers T, Lichten M (2001) Differential timing and control of noncrossover and crossover recombination during meiosis. Cell 106: 47–57.
20. Zickler D, Kleckner N (1999) Meiotic chromosomes: integrating structure and function. Annu Rev Genet 33: 603–754.
21. Borner GV, Kleckner N, Hunter N (2004) Crossover/noncrossover differentiation, synaptonemal complex formation, and regulatory surveillance at the leptotene/zygotene transition of meiosis. Cell 117: 29–45.
22. Hochwagen A, Amon A (2006) Checking your breaks: surveillance mechanisms of meiotic recombination. Curr Biol 16: R217–228.
23. Sym M, Engebrecht JA, Roeder GS (1993) ZIP1 is a synaptonemal complex protein required for meiotic chromosome synapsis. Cell 72: 365–378.
24. Xu L, Ajimura M, Padmore R, Klein C, Kleckner N (1995) NDT80, a meiosis-specific gene required for exit from pachytene in Saccharomyces cerevisiae. Mol Cell Biol 15: 6572–6581.
25. Chu S, Herskowitz I (1998) Gametogenesis in yeast is regulated by a transcriptional cascade dependent on Ndt80. Mol Cell 1: 685–696.
26. Clyne RK, Katis VL, Jessop L, Benjamin KR, Herskowitz I, et al. (2003) Polo-like kinase Cdc5 promotes chiasmata formation and cosegregation of sister centromeres at meiosis I. Nat Cell Biol5: 480–485.
27. Sourirajan A, Lichten M (2008) Polo-like kinase Cdc5 drives exit from pachytene during budding yeast meiosis. Genes Dev 22: 2627–2632.
28. San-Segundo PA, Roeder GS (1999) Pch2 links chromatin silencing to meiotic checkpoint control. Cell 97: 313–324.
29. San-Segundo PA, Roeder GS (2000) Role for the silencing protein Dot1 in meiotic checkpoint control. Mol Biol Cell 11: 3601–3615.
30. Borner GV, Barot A, Kleckner N (2008) Yeast Pch2 promotes domainal axis organization, timely recombination progression, and arrest of defective recombinosomes during meiosis. Proc Natl Acad Sci U S A 105: 3327–3332.
31. Ho HC, Burgess SM (2011) Pch2 acts through Xrs2 and Tel1/ATM to modulate interhomolog bias and checkpoint function during meiosis. PLoS Genet 7: e1002351.
32. Conde F, Ontoso D, Acosta I, Gallego-Sanchez A, Bueno A, et al. (2010) Regulation of tolerance to DNA alkylating damage by Dot1 and Rad53 in Saccharomyces cerevisiae. DNA Repair (Amst) 9: 1038–1049.
33. Conde F, Refolio E, Cordon-Preciado V, Cortes-Ledesma F, Aragon L, et al. (2009) The Dot1 histone methyltransferase and the Rad9 checkpoint adaptor contribute to cohesin-dependent double-strand break repair by sister chromatid recombination in Saccharomyces cerevisiae. Genetics 182: 437–446.
34. Conde F, San-Segundo PA (2008) Role of Dot1 in the response to alkylating DNA damage in Saccharomyces cerevisiae: regulation of DNA damage tolerance by the error-prone polymerases Polzeta/Rev1. Genetics 179: 1197–1210.
35. Levesque N, Leung GP, Fok AK, Schmidt TI, Kobor MS (2010) Loss of H3 K79 trimethylation leads to suppression of Rtt107-dependent DNA damage sensitivity through the translesion synthesis pathway. J Biol Chem 285: 35113–35122.
36. Tatum D, Li S (2011) Evidence that the histone methyltransferase Dot1 mediates global genomic repair by methylating histone H3 on lysine 79. J Biol Chem 286: 17530–17535.
37. Nguyen AT, Zhang Y (2011) The diverse functions of Dot1 and H3K79 methylation. Genes Dev 25: 1345–1358.
38. Shilatifard A (2012) The COMPASS family of histone H3K4 methylases: mechanisms of regulation in development and disease pathogenesis. Annu Rev Biochem 81: 65–95.
39. Yamashita K, Shinohara M, Shinohara A (2004) Rad6-Bre1-mediated histone H2B ubiquitylation modulates the formation of double-strand breaks during meiosis. Proc Natl Acad Sci U S A 101: 11380–11385.
40. Shinohara M, Gasior SL, Bishop DK, Shinohara A (2000) Tid1/Rdh54 promotes colocalization of Rad51 and Dmc1 during meiotic recombination. Proc Natl Acad Sci U S A 97: 10814–10819.
41. Farmer S, Leung WK, Tsubouchi H (2011) Characterization of meiotic recombination initiation sites using pulsed-field gel electrophoresis. Methods Mol Biol 745: 33–45.
42. Shinohara M, Sakai K, Ogawa T, Shinohara A (2003) The mitotic DNA damage checkpoint proteins Rad17 and Rad24 are required for repair of double-strand breaks during meiosis in yeast. Genetics 164: 855–865.
43. Storlazzi A, Xu L, Cao L, Kleckner N (1995) Crossover and noncrossover recombination during meiosis: timing and pathway relationships. Proc Natl Acad Sci U S A 92: 8512–8516.
44. Cao L, Alani E, Kleckner N (1990) A pathway for generation and processing of double-strand breaks during meiotic recombination in S. cerevisiae. Cell 61: 1089–1101.

Table S2   Primer list.

Acknowledgments

We thank Drs. Doug Bishop and Neil Hunter for discussion. We are also indebted to members of the Shinohara lab, particularly Dr. Saori Mori for technical help and Dr. Takehiko Usui for the hht1,2-K79R strain.

Author Contributions

Conceived and designed the experiments: MBI MS AS. Performed the experiments: MBI MS. Analyzed the data: MBI MS AS. Contributed reagents/materials/analysis tools: MBI MS AS. Wrote the paper: MBI MS AS.

45. Schwacha A, Kleckner N (1997) Interhomolog bias during meiotic recombination: meiotic functions promote a highly differentiated interhomolog-only pathway. Cell 90: 1123–1135.

46. Shinohara A, Gasior S, Ogawa T, Kleckner N, Bishop DK (1997) *Saccharomyces cerevisiae recA* homologues *RAD51* and *DMC1* have both distinct and overlapping roles in meiotic recombination. Genes Cells 2: 615–629.

47. Bishop DK (1994) RecA homologs Dmc1 and Rad51 interact to form multiple nuclear complexes prior to meiotic chromosome synapsis. Cell 79: 1081–1092.

48. Miyazaki T, Bressan DA, Shinohara M, Haber JE, Shinohara A (2004) In vivo assembly and disassembly of Rad51 and Rad52 complexes during double-strand break repair. Embo J 23: 939–949.

49. Carballo JA, Panizza S, Serrentino ME, Johnson AL, Geymonat M, et al. (2013) Budding Yeast ATM/ATR Control Meiotic Double-Strand Break (DSB) Levels by Down-Regulating Rec114, an Essential Component of the DSB-machinery. PLoS Genet 9: e1003545.

50. Argunhan B, Farmer S, Leung WK, Terentyev Y, Humphryes N, et al. (2013) Direct and indirect control of the initiation of meiotic recombination by DNA damage checkpoint mechanisms in budding yeast. PLoS One 8: e65875.

51. Trelles-Sticken E, Bonfils S, Sollier J, Geli V, Scherthan H, et al. (2005) Set1- and Clb5-deficiencies disclose the differential regulation of centromere and telomere dynamics in *Saccharomyces cerevisiae* meiosis. J Cell Sci 118: 4985–4994.

52. Hollingsworth NM, Goetsch L, Byers B (1990) The *HOP1* gene encodes a meiosis-specific component of yeast chromosomes. Cell 61: 73–84.

53. Wojtasz L, Daniel K, Roig I, Bolcun-Filas E, Xu H, et al. (2009) Mouse HORMAD1 and HORMAD2, two conserved meiotic chromosomal proteins, are depleted from synapsed chromosome axes with the help of TRIP13 AAA-ATPase. PLoS Genet 5: e1000702.

54. Armstrong SJ, Caryl AP, Jones GH, Franklin FC (2002) Asy1, a protein required for meiotic chromosome synapsis, localizes to axis-associated chromatin in Arabidopsis and Brassica. J Cell Sci 115: 3645–3655.

55. Rockmill B, Roeder GS (1988) *RED1*: a yeast gene required for the segregation of chromosomes during the reductional division of meiosis. Proc Natl Acad Sci U S A 85: 6057–6061.

56. Klein F, Mahr P, Galova M, Buonomo SB, Michaelis C, et al. (1999) A central role for cohesins in sister chromatid cohesion, formation of axial elements, and recombination during yeast meiosis. Cell 98: 91–103.

57. Hollingsworth NM, Ponte L (1997) Genetic interactions between *HOP1*, *RED1* and *MEK1* suggest that *MEK1* regulates assembly of axial element components during meiosis in the yeast *Saccharomyces cerevisiae*. Genetics 147: 33–42.

58. Leem SH, Ogawa H (1992) The *MRE4* gene encodes a novel protein kinase homologue required for meiotic recombination in *Saccharomyces cerevisiae*. Nucleic Acids Res 20: 449–457.

59. Smith AV, Roeder GS (1997) The yeast Red1 protein localizes to the cores of meiotic chromosomes. J Cell Biol 136: 957–967.

60. Choi K, Zhao X, Kelly KA, Venn O, Higgins JD, et al. (2013) *Arabidopsis* meiotic crossover hot spots overlap with H2A.Z nucleosomes at gene promoters. Nat Genet 45: 1327–1336.

61. Grenon M, Costelloe T, Jimeno S, O'Shaughnessy A, Fitzgerald J, et al. (2007) Docking onto chromatin via the *Saccharomyces cerevisiae* Rad9 Tudor domain. Yeast 24: 105–119.

62. Lydall D, Nikolsky Y, Bishop DK, Weinert T (1996) A meiotic recombination checkpoint controlled by mitotic checkpoint genes. Nature 383: 840–843.

63. Ontoso D, Acosta I, van Leeuwen F, Freire R, San-Segundo PA (2013) Dot1-dependent histone H3K79 methylation promotes activation of the Mek1 meiotic checkpoint effector kinase by regulating the Hop1 adaptor. PLoS Genet 9: e1003262.

64. Gray S, Allison RM, Garcia V, Goldman AS, Neale MJ (2013) Positive regulation of meiotic DNA double-strand break formation by activation of the DNA damage checkpoint kinase Mec1(ATR). Open Biol 3: 130019.

65. Zhang L, Ma H, Pugh BF (2011) Stable and dynamic nucleosome states during a meiotic developmental process. Genome Res 21: 875–884.

66. Stahl FW, Foss HM, Young LS, Borts RH, Abdullah MF, et al. (2004) Does crossover interference count in *Saccharomyces cerevisiae*? Genetics 168: 35–48.

67. Zalevsky J, MacQueen AJ, Duffy JB, Kemphues KJ, Villeneuve AM (1999) Crossing over during *Caenorhabditis elegans* meiosis requires a conserved MutS-based pathway that is partially dispensable in budding yeast. Genetics 153: 1271–1283.

68. Martini E, Diaz RL, Hunter N, Keeney S (2006) Crossover homeostasis in yeast meiosis. Cell 126: 285–295.

11

Analysis of the Role of Homology Arms in Gene-Targeting Vectors in Human Cells

Ayako Ishii[1,⚑], **Aya Kurosawa**[1,⚑], **Shinta Saito**[1], **Noritaka Adachi**[1,2*]

1 Graduate School of Nanobioscience, Yokohama City University, Yokohama, Japan, 2 Advanced Medical Research Center, Yokohama City University, Yokohama, Japan

Abstract

Random integration of targeting vectors into the genome is the primary obstacle in human somatic cell gene targeting. Non-homologous end-joining (NHEJ), a major pathway for repairing DNA double-strand breaks, is thought to be responsible for most random integration events; however, absence of DNA ligase IV (LIG4), the critical NHEJ ligase, does not significantly reduce random integration frequency of targeting vector in human cells, indicating robust integration events occurring via a LIG4-independent mechanism. To gain insights into the mechanism and robustness of LIG4-independent random integration, we employed various types of targeting vectors to examine their integration frequencies in LIG4-proficient and deficient human cell lines. We find that the integration frequency of targeting vector correlates well with the length of homology arms and with the amount of repetitive DNA sequences, especially SINEs, present in the arms. This correlation was prominent in LIG4-deficient cells, but was also seen in LIG4-proficient cells, thus providing evidence that LIG4-independent random integration occurs frequently even when NHEJ is functionally normal. Our results collectively suggest that random integration frequency of conventional targeting vectors is substantially influenced by homology arms, which typically harbor repetitive DNA sequences that serve to facilitate LIG4-independent random integration in human cells, regardless of the presence or absence of functional NHEJ.

Editor: Kefei Yu, Michigan State University, United States of America

Funding: This work was supported by grants from Yokohama City University (Strategic Research Promotion G2201/G2301/G2401/S2501) and by Grants-in-Aid from the Ministry of Education, Culture, Sports, Science, and Technology (MEXT) of Japan. The funders had no role in study design, data collection and analysis, decision to publish, or preparation of the manuscript.

Competing Interests: The authors have declared that no competing interests exist.

* Email: nadachi@yokohama-cu.ac.jp

⚑ These authors contributed equally to this work.

Introduction

Gene targeting via homologous recombination provides a powerful means for studying gene function by a reverse genetic approach. In gene-targeting experiments, cells are transfected with targeting vector, which is typically designed and constructed so as to contain a selection marker (drug-resistance) gene flanked with two genomic DNA fragments, called 5'- and 3'-homology arms (or simply 5' and 3' arms, or left and right arms) [1]. After transfection, these two arms should be homologously recombined with the target genome sequence in the cell to achieve successful genetic modification of the chromosomal locus [2]. In human somatic cells, the frequency of such targeted integration is at least two to three orders of magnitude lower than that of random integration [3] (depicted in Figure 1). It therefore seems reasonable to expect that reducing random integration events would enhance gene targeting by increasing the ratio of targeted to random integration.

Non-homologous end-joining (NHEJ), which repairs DNA double-strand breaks (DSBs) in a Ku-dependent manner [4], is responsible for nearly all random integration events in lower eukaryotes, such as *Neurospora crassa*, and thus NHEJ deficiency dramatically enhances gene targeting [5,6]. Unfortunately, however, this is not the case for human somatic cells, as apparently NHEJ is not the sole mechanism of random integration [7,8] (Figure 1). Although earlier studies have suggested a substantial role of NHEJ in random integration [9,10], we have previously observed robust random integration events in cells lacking DNA ligase IV (LIG4, a critical NHEJ factor [11]). This finding strongly suggests the involvement of other DSB repair pathways, most presumably alternative end-joining, in causing random integration [7]. The molecular mechanism of alternative end-joining in DSB repair is not yet fully understood; however, recent work has established that alternative end-joining is mechanistically distinct from NHEJ (i.e., Ku/LIG4-dependent NHEJ) [12,13] and requires DNA ligase I or III, not LIG4, in repair of DSBs [14-16]. Additionally, it is now accepted that initiation of alternative end-joining requires end resection of the broken DNA to produce single-stranded DNA that is used for strand annealing, a mechanism similar to initiation of homologous recombination involving single-strand annealing (SSA) [17,18]. By virtue of this mechanism, alternative end-joining is believed to favor micro-homologies (≥ 4 nt) for the joining of DNA ends, unlike NHEJ that typically joins DNA ends with short or no homology (0–4 nt) [4,19,20]. However, there exists an alternative end-joining mechanism with no apparent microhomologies [21,22].

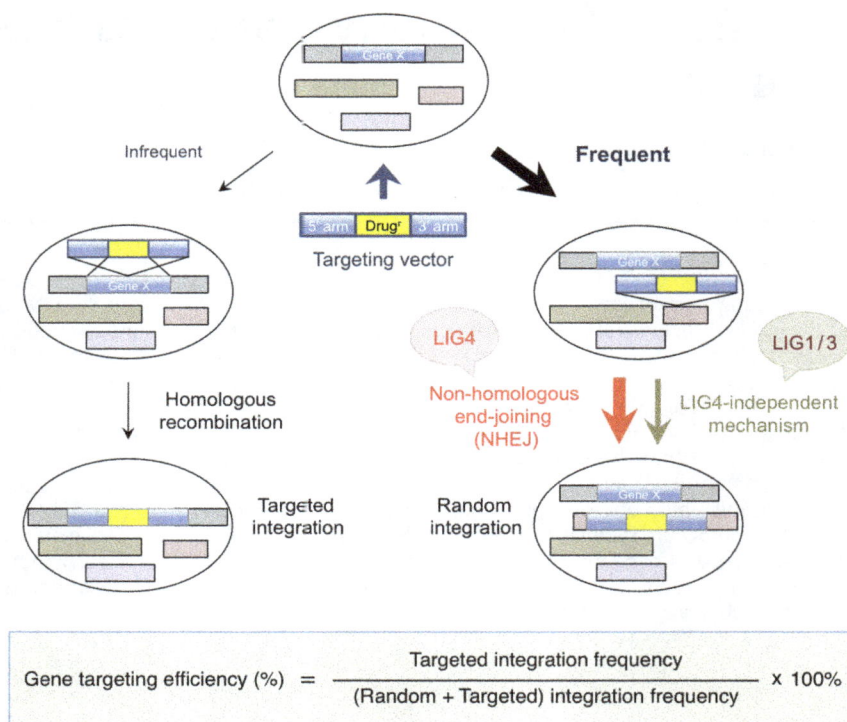

Figure 1. Gene targeting is inefficient in human somatic cells. When targeting vector is transfected into human cells, random integration by non-homologous recombination occurs at least 2 to 3 orders of magnitude more frequently than homologous recombination-mediated targeted integration. The LIG4-dependent NHEJ pathway has been thought to be responsible for random integration, but recent evidence indicates a contribution from LIG4-independent mechanisms that rely on LIG1/3 (DNA ligase I or IIIα). The gene-targeting efficiency is calculated by dividing the number of targeted clones with that of drug-resistant clones analyzed (see Materials and Methods for details).

Despite its biological and medical importance, the precise mechanism of random integration in human somatic cells remains largely unclear, and at least two distinct mechanisms exist — LIG4-dependent NHEJ and LIG4-independent alternative end-joining. Recently, Suzuki et al. [23] have reported that chromosomal integration of plasmid DNA (a non-targeting vector) transfected into mouse embryonic stem (ES) cells is mostly a complex reaction with frequent terminal deletions of the plasmid and the genome. Interestingly, sequence analysis of those random integrants revealed a frequent use of microhomologies at the plasmid-genome junctions [23]. This finding may suggest the occurrence of random integration events via alternative end-joining, given the aforementioned microhomology preference of this mechanism in repair of DSBs. However, the view that NHEJ avoids using microhomologies may not be entirely correct [19]. In addition, mouse ES cells express low levels of DNA-PKcs and thus may not be fully competent for NHEJ [24]. Hence it is yet uncertain whether alternative end-joining is involved in random integration events occurring in human cells with normal NHEJ capacity.

The human genome, unlike the genomes of lower eukaryotes, is large in size (3×10^9 bp) and contains a huge amount of repetitive DNA sequences; among these, short interspersed nucleotide elements (SINE) such as Alu and long interspersed nucleotide elements (LINE) occur in $\sim 1-2 \times 10^6$ copies per genome [25,26]. Thus, SINE and LINE sequences occupy 36% of the human genome. The intact human SINEs and LINEs are ~ 300 bp and 6 kb, respectively, but some SINEs and most LINEs are fragments and can be as short as 100 bp or less. As mentioned above, conventional gene-targeting vectors possess two homology arms

whose DNA sequence is identical to the target genome sequence. In designing and constructing a targeting vector, it is generally unavoidable to incorporate a repetitive DNA sequence(s) into homology arms (especially when one intends to make long arms) because of the high abundance of repetitive DNA fragments in the genome. Thus, most if not all targeting vectors contain a repetitive DNA fragment(s) in their homology arms.

We have previously shown that although LIG4-deficient human cells exhibit reduced integration frequencies when transfected with non-targeting vectors having no homology to the host genome, such a decrease was not observed when targeting vectors were employed [7]. This suggests that homology arms present in the vector somehow facilitated random integration in a LIG4-independent fashion. In this study, we generated various types of human HPRT targeting vectors to analyze the relationship between the frequency of LIG4-dependent and LIG4-independent random integration and the length (or presence) of homology arms. For this purpose, we employed the human pre-B leukemia cell line Nalm-6 (NHEJ-competent) and its LIG4-null (NHEJ-deficient) cells. Additionally, using these cell lines and targeting vectors for more than ten different human genes, we performed a detailed analysis on random integration frequency and homology arms or repetitive DNA sequences. Our data collectively suggest that integration frequency of targeting vector correlates well with the lengths of homology arms and repetitive DNA sequences, which likely facilitate alternative end-joining-mediated random integration even in the presence of functional NHEJ.

Results and Discussion

Long homology arms stimulate LIG4-independent random integration of the targeting vector

We first examined the integration frequency of four *HPRT* targeting vectors pHPRT8.9-Puro(+), pHPRT8.9-Puro(−), pHPRT2.2-Puro(+) and pHPRT2.2-Puro(−), and pPGK-Puro (a non-targeting vector with no homology arms) in Nalm-6 wild-type and *LIG4*-null cells [27]. As shown in Figure 2A and B, these targeting vectors were designed to disrupt exon 3 of the *HPRT* gene. The pHPRT8.9-Puro vectors have a 3.8-kb 5' arm and a 5.1-kb 3' arm (the two arms are adjacent to one another in the genome), whereas the pHPRT2.2-Puro vectors have shorter homology arms (1.1 kb each; located 0.7 kb apart in the genome) (Figure 2A–C and Figure S1 in File S1). In these vectors, 5' and 3' arms flank a drug-resistance gene cassette (*Puro^r*) present in forward (+) or reverse (−) orientation to the gene. As shown in Figure 2D, integration frequency of pPGK-Puro was significantly lower in *LIG4*-null cells than in wild-type cells (P<0.00001; n = 11), thus confirming the contribution of NHEJ to random integration. Very similar results were obtained with vectors containing other drug-resistance genes (data not shown; [7]). In contrast, integration frequency of pHPRT8.9-Puro(+) and pHPRT8.9-Puro(−) was not decreased, but rather slightly increased in *LIG4*-null cells (Figure 2D). Interestingly, vectors with shorter homology arms, pHPRT2.2-Puro(+) and pHPRT2.2-Puro(−), displayed decreased integration frequency in *LIG4*-null cells (P<0.05 for pHPRT2.2-Puro(+) and P<0.01 for pHPRT2.2-Puro(-); n = 8), a result similar to that of non-targeting vectors. These results suggest that the presence of long homology arms serves to facilitate LIG4-independent random integration of the targeting vector.

As the pHPRT2.2-Puro vectors harbor relatively short arms, we next examined whether pHPRT2.2-Puro(−) functioned as a genuine targeting vector. We thus picked puromycin-resistant colonies derived from pHPRT2.2-Puro(−)-transfected cells and confirmed that correctly targeted clones are actually obtained (1 out of 127 clones in wild-type cells and 3 out of 45 clones in *LIG4*-null cells) (Figure 3A). The increased targeting efficiency associated with *LIG4* deficiency is consistent with previous studies using human cells [7,28]. Of note, although pHPRT2.2-Puro(−) exhibited low targeting efficiencies compared to pHPRT8.9-Puro(-), gene-targeting enhancement associated with the *LIG4* deficiency was more pronounced for pHPRT2.2-Puro(−) (~10-fold increase) than for pHPRT8.9-Puro(−) (~2–3-fold increase)(-Figure 3A). Consistent with the aforementioned data, random integration frequency of pHPRT2.2-Puro(−) was reduced in *LIG4*-null cells, while that of pHPRT8.9-Puro(−) was not (Figure 3B).

LIG4-independent random integration is decreased when a homology arm with repetitive DNA sequences is deleted from the targeting vector

To further investigate the relationship between the presence/length of homology arms and the frequency of LIG4-independent random integration, we then generated "imperfect" targeting vectors lacking either a 5' or 3' arm by using the four types of *HPRT* targeting vectors described above. Each vector was transfected into Nalm-6 wild-type and *LIG4*-null cells to calculate the integration frequency and the ratio between the two cell lines. As shown in Figure 4A, the absence of either arm significantly decreased the integration frequency of pHPRT8.9-Puro(+) and pHPRT8.9-Puro(−) in *LIG4*-null cells relative to wild-type cells. In contrast, deleting either arm of pHPRT2.2-Puro(+) and

pHPRT2.2-Puro(−) had a marginal effect on the ratio of integration frequency in *LIG4*-null to wild-type cells (Figure 4B). These results suggest that integration frequency in *LIG4*-null cells is roughly proportional to the total length of homology arms present in the vector, further supporting the notion that homology arms facilitate LIG4-independent targeting-vector integration.

It is interesting to note that our results also suggest that the direction or position of the drug-resistance gene relative to the homology arm may affect LIG4-independent random integration. In the experiments using pHPRT8.9-Puro-derived imperfect vectors, the absence of an arm located downstream of *Puro^r* (i.e., the 5' arm for (−) vectors and the 3' arm for (+) vectors) had a greater impact on LIG4-independent random integration (i.e., the ratio of integration frequency in *LIG4*-null to wild-type cells). Likewise, in the experiments using pHPRT2.2-Puro-derived imperfect vectors, the impact of 5'-arm deletion on LIG4-independent random integration was slightly more prominent when the 5' arm was originally located downstream of *Puro^r*. This position effect, however, was not observed in 3' arm-deleted pHPRT2.2-Puro vectors. It should be mentioned that pHPRT2.2-Puro 3' arm harbors essentially no repetitive sequences, whereas the 5' arm contains a large number of SINE/LINE sequences (78.4% occupancy)(see Figure 2C; in pHPRT8.9-Puro vectors, the two arms similarly contain repetitive sequences). Together, these findings imply that repetitive DNA sequences present in the homology arms facilitate LIG4-independent targeting-vector integration, especially when these repetitive sequences are present downstream of the drug-resistance gene cassette, which has a promoter. Although the precise mechanism of this possible position effect is currently unclear, we speculate that transcription of the marker gene (i.e., transient expression occurring before or during vector DNA integration) may lead to a partially denatured state of the downstream region and then a denatured repetitive sequence(s) present in the arm could serve to facilitate LIG4-independent integration of the vector into the genome.

Integration frequency of targeting vector correlates with the lengths of homology arms and repetitive DNA sequences regardless of NHEJ status

The above results using various types of *HPRT* vectors suggest a correlation between the length of arms and the absolute frequency of random integration. Importantly, this correlation was observed in wild-type cells, though less prominent than in *LIG4*-null cells; for instance, the integration frequency of pHPRT8.9-Puro(−) was ~2-fold higher than that of pHPRT2.2-Puro(−)(Figure 3B), suggesting that LIG4-independent random integration does occur in cells with normal NHEJ function. To test this directly, we knocked down the expression of DNA ligase I or IIIα in wild-type and *LIG4*-null cells. Because the LIG4-independent integration mechanism does not rely on LIG4, either or both of DNA ligase I and IIIα should be involved in this mechanism. As shown in Figure S2 in File S1, transfection of *LIG1* siRNA or *LIG3* siRNA had an effect on reducing random integration frequency in wild-type cells, even though the siRNA-mediated knockdown was somewhat incomplete. These results suggest that DNA ligase I and IIIα are both involved in LIG4-independent random integration events, consistent with recent findings on DNA ligase usage in DSB repair [14–16].

We next performed a comprehensive analysis using various gene-targeting vectors to examine whether the correlation between arm length and integration frequency is actually observed in wild-type cells. For this purpose, we used twelve different gene-targeting vectors (Figure S3 in File S1). In these vectors, the lengths of 3'

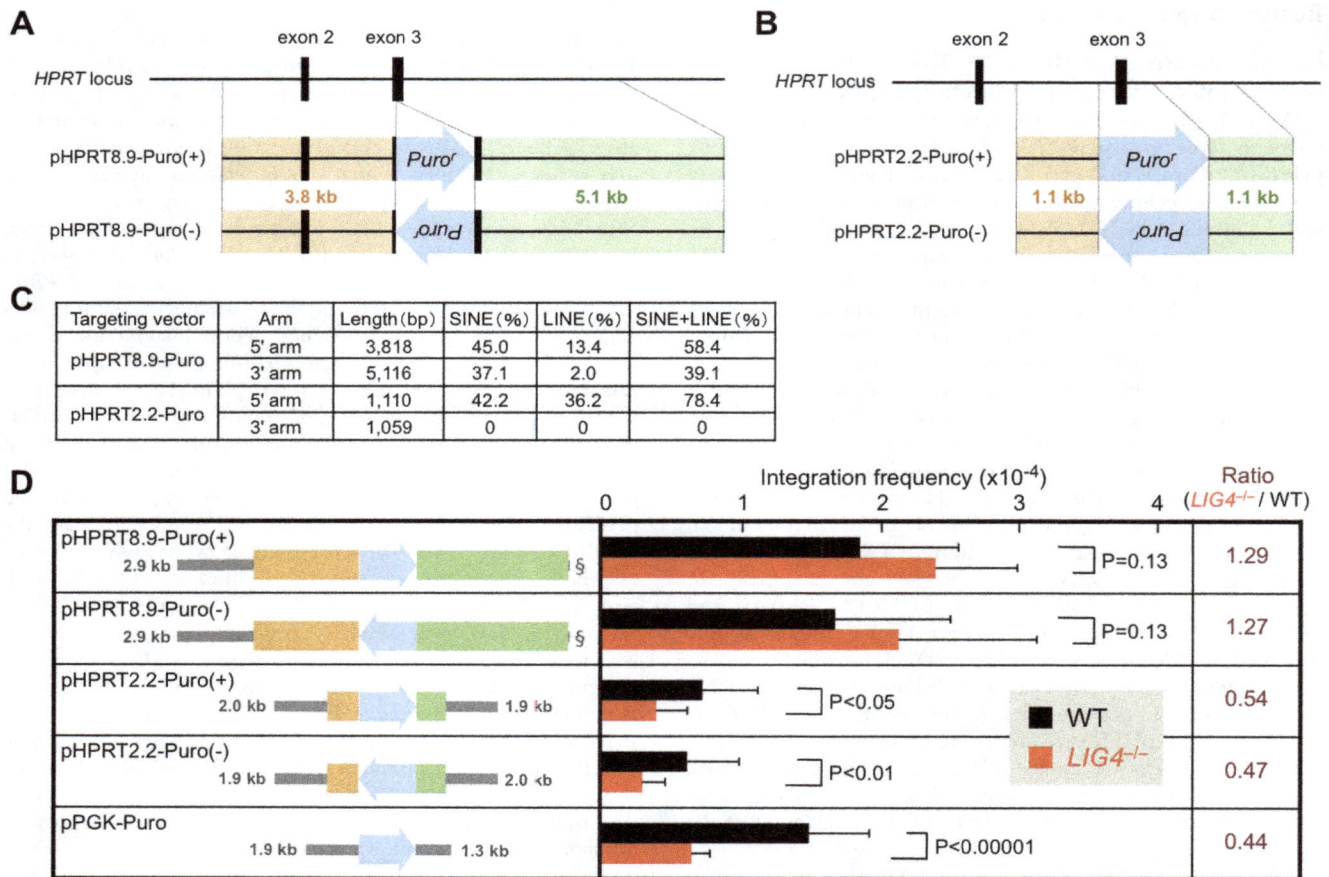

Figure 2. *HPRT* targeting vectors with long, but not short, homology arms stimulate NHEJ-independent random integration. (A) Schematic representation of *HPRT* targeting vectors pHPRT8.9-Puro(+) and pHPRT8.9-Puro(−). (B) Schematic representation of *HPRT* targeting vectors pHPRT2.2-Puro(+) and pHPRT2.2-Puro(−). (C) Structural features of the *HPRT* targeting vectors. (D) Integration frequency of *HPRT* targeting vectors and pPGK-Puro (a non-targeting vector) in human Nalm-6 wild-type and *LIG4*-null cells. The ratio of integration frequency in *LIG4*-null to wild-type cells is indicated in the right column. At least six independent experiments were performed for each vector. Note that pPGK-Puro harbors little or no homology to the human genome. Grey lines indicate the lengths of plasmid backbones, and § denotes a 14-bp sequence.

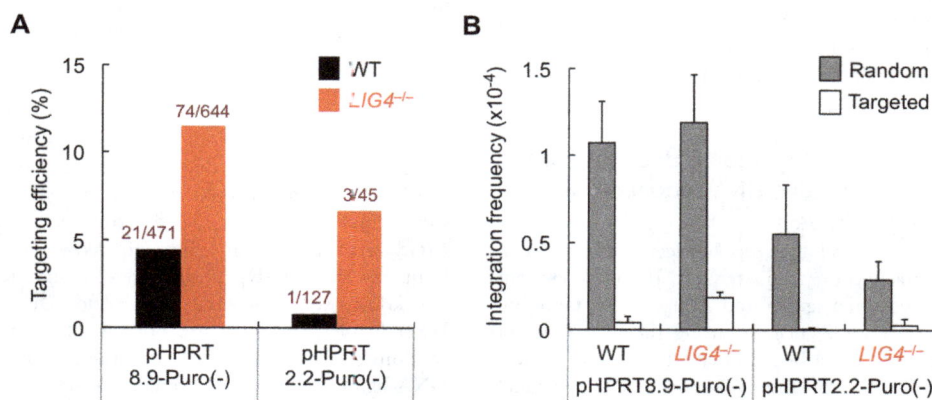

Figure 3. The short-arm vector pHPRT2.2-Puro(−) functions as a genuine targeting vector. (A) Gene-targeting efficiency of pHPRT8.9-Puro(−) and pHPRT2.2-Puro(−) in wild-type and *LIG4*-null cells. (B) Random and targeted integration frequencies of pHPRT8.9-Puro(−) and pHPRT2.2-Puro(−) in wild-type and *LIG4*-null cells. At least three independent experiments were performed for each vector.

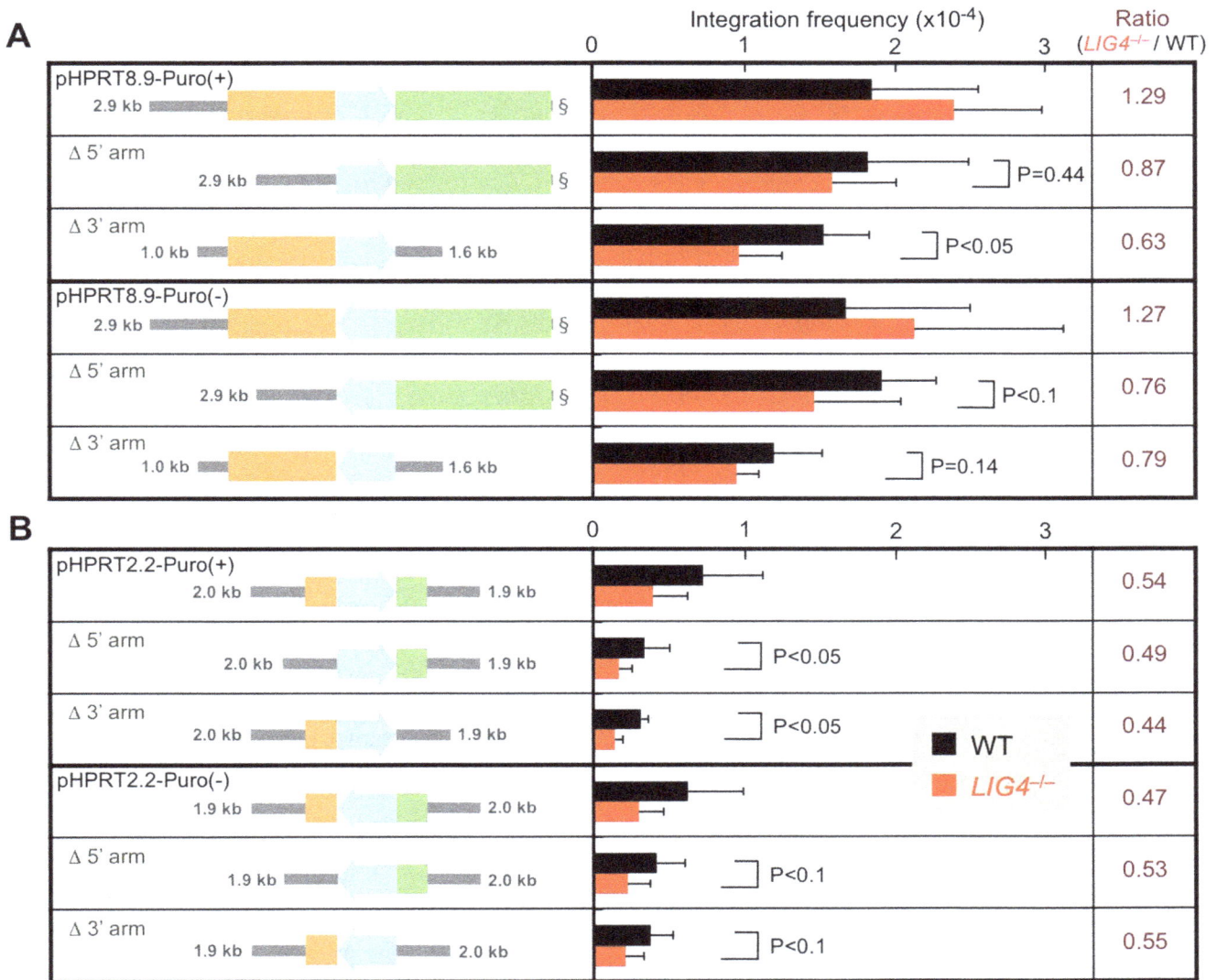

Figure 4. NHEJ-independent random integration is significantly decreased when a homology arm is deleted from the *HPRT* targeting vector. (A) Integration frequency of pHPRT8.9-Puro vectors and their derivatives in Nalm-6 wild-type and *LIG4*-null cells. The ratio of integration frequency in *LIG4*-null to wild-type cells is indicated in the right column. At least five independent experiments were performed for each vector. The data for the arm-proficient vectors are the same as that in Figure 2D. (B) Integration frequency of pHPRT2.2-Puro vectors and their derivatives in Nalm-6 wild-type and *LIG4*-null cells. The ratio of integration frequency in *LIG4*-null to wild-type cells is indicated in the right column. At least three independent experiments were performed for each vector. Symbols are as in Figure 2D, and the data for the arm-proficient vectors are the same as that in Figure 2D.

arms vary, while 5' arms of most of the vectors are roughly similar in size ($\sim$2.5–3.2 kb) but contain different lengths of repetitive DNA sequences (i.e., SINE and LINE sequences), and drug-resistance gene cassettes are placed in the same (forward) direction. From the observations described above, we predicted that the absolute integration frequency would be proportional to the length of the 3' arm (or both arms) and also to the total length of repetitive DNA sequences present in the vector. As shown in Figure 5, this was indeed the case. We found that the length of arms, especially that of 3' arm, had a positive, albeit weak, correlation with the integration frequency in wild-type cells (total arm length, $R^2 = 0.38$; 3'-arm length, $R^2 = 0.42$)(Figure 5A–C). We also found a weak correlation between the integration frequency and the length of SINE/LINE sequences present in the arms (total SINE/LINE length, $R^2 = 0.26$; 3'-arm SINE/LINE length, $R^2 = 0.25$)(Figure 5D–F). Since this correlation was not fully statistically significant and seemed a bit lower than we expected, we analyzed the characteristics of those targeting vectors whose integration frequency was deviating largely from the approximation. This analysis led us to notice that vectors containing an extremely short 3' arm (*ARTEMIS* and *APTX*) conferred low integration frequencies, whereas a vector with a relatively long ($\sim$0.6 kb) LINE fragment in the distal end of the 5' arm (*RAG1*) gave a high integration frequency. When these "exceptional" vectors were excluded from the data set to redraw a fitted curve, a stronger correlation ($R^2 = 0.47$) was observed between the total SINE/LINE length and the integration frequency (Figure S5 in File S1). Even though this R^2 value is still not high enough to show a statistical significance, our results collectively suggest that the length of repetitive sequences within

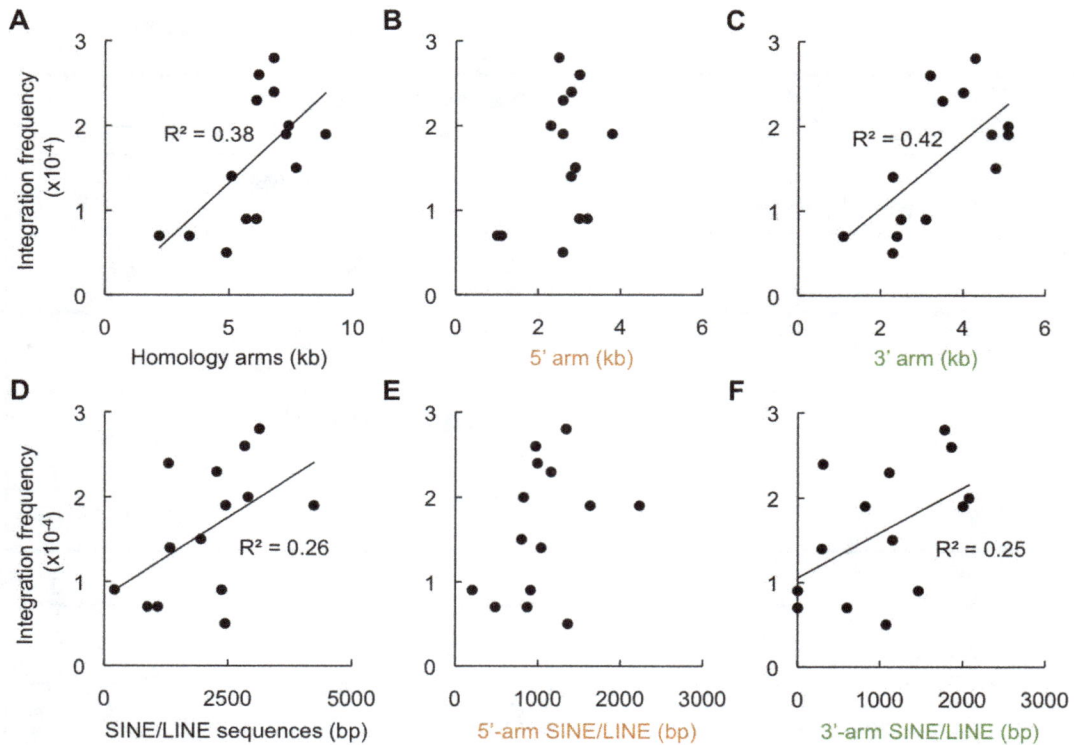

Figure 5. Integration frequency of targeting vector correlates with the lengths of homology arms and repetitive DNA sequences.
Integration frequencies of pHPRT8.9-Puro(+), pHPRT2.2-Puro(+), and twelve other (non-*HPRT*) gene-targeting vectors are shown as a function of the total length of homology arms (A), 5'-arm length (B), 3'-arm length (C), the total length of SINE/LINE sequences present in the arms (D), 5'-arm SINE/LINE length (E), and 3'-arm SINE/LINE length (F). The R^2 values shown in the graphs were calculated from the fourteen points. See also Figures S4 and S5 in File S1 for details.

arms as well as arm length may be a major determinant of the integration frequency of targeting vectors.

We then examined which interspersed element, SINE or LINE, contributed to facilitating random integration. As shown in Figure S4 in File S1, the length of SINE sequences had roughly the same impact on integration frequency as the length of SINE/LINE sequences. In contrast, the length of LINE sequences did not appear to have a clear correlation with the integration frequency. Previous work has shown, however, that human LINEs contribute to chromosomal integration of exogenous DNA [29]. Indeed, we observed that the *RAG1* targeting vector with a 579-bp LINE fragment in its short 5' arm, showed a higher integration frequency than expected as described above. In the human genome, the copy number of SINEs is higher than LINEs [30], and thus SINE-containing sequences may gain easier access to the genome sequence. Moreover, SINEs tend to distribute in gene-rich GC-rich regions, while LINEs in gene-poor AT-rich regions [30]. In this regard, recent work has reported that integration of exogenous DNA into mouse ES cell chromosomes shows preference into genes [23]. Alternatively or additionally as LINEs present in the targeting vectors are short fragments (<500 bp)(Figure S1 in File S1), these LINE fragments may lack the ability to facilitate random integration.

Finally, we set out to directly compare the frequencies of LIG4-dependent and independent random integration in wild-type and *LIG4*-null cells by using seven different gene-targeting vectors. For this purpose, we subtracted the integration frequency in *LIG4*-null cells from that in wild-type cells to estimate the frequency of LIG4-dependent integration. The subtracted values are only an

approximation, but should reflect the frequency of LIG4-dependent integration, although likely underestimated, given that the LIG4-independent mechanism may not be fully active in NHEJ-competent cells. As shown in Figure 6A and B, the integration frequency in *LIG4*-null cells was directly proportional to the arm length and the amount of repetitive sequence ($R^2 = 0.82$ and $R^2 = 0.88$, respectively), indicating a statistically significant correlation between LIG4-independent random integration frequency and the length of homology arms or repetitive sequences of the targeting vector. Importantly, a similar, but less pronounced, correlation was also found in wild-type cells ($R^2 = 0.36$ and $R^2 = 0.45$, respectively), consistent with the aforementioned data shown in Figure 5. In sharp contrast, the subtracted values had no obvious correlation with the lengths of homology arms or repetitive DNA sequences, as shown in Figure 6C and D. These data provide further evidence that LIG4-independent integration does occur when NHEJ is functionally normal, and that LIG4-independent random integration, but not LIG4-dependent random integration, is substantially affected by homology arms and repetitive sequences of the targeting vector.

As the mechanism of random integration remains largely unclear, how homology arms with repetitive sequences facilitate targeting-vector random integration is even more enigmatic. Almost undoubtedly, random integration of non-targeting vectors in *LIG4*-null cells is mediated by alternative end-joining, as the involvement of NHEJ or SSA is highly unlikely (and this is also true for joining at the arm-deleted side of aforementioned *HPRT* vectors). When a homology arm(s) are present, an increase in integration frequency was observed in our experiments. This

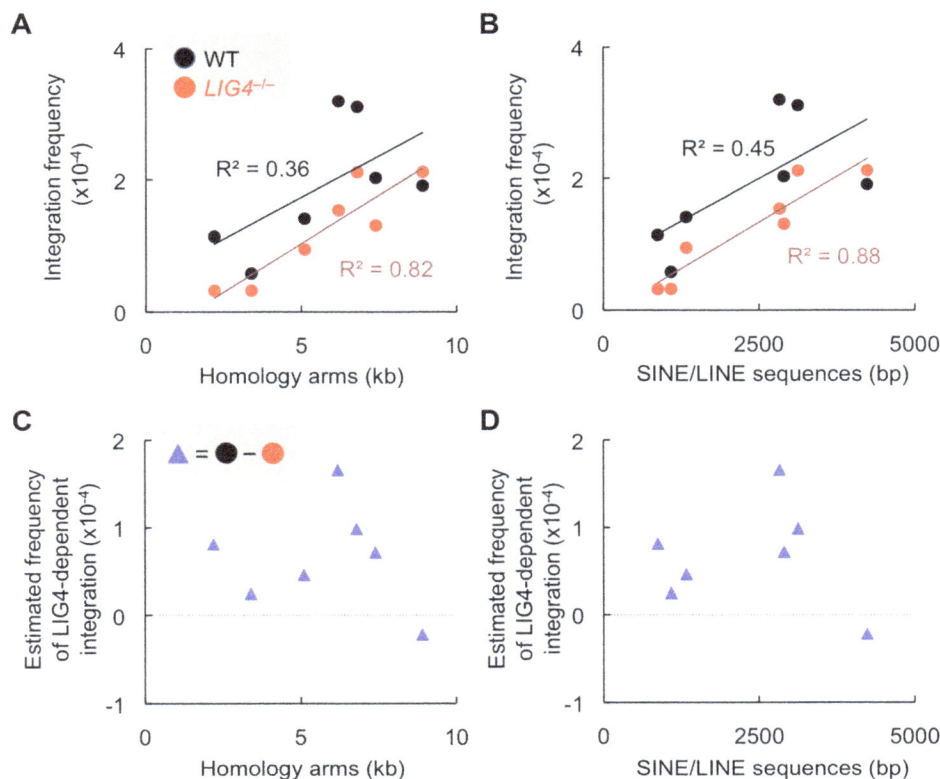

Figure 6. Integration frequency of targeting vector correlates well with the lengths of homology arms and repetitive DNA sequences, particularly in the absence of LIG4. (A, B) Integration frequencies of seven gene-targeting vectors in wild-type and *LIG4*-null cells are shown as a function of the total length of homology arms (A) and the total length of SINE/LINE sequences present in the arms (B). (C, D) Estimated frequencies of LIG4-dependent integration are shown as a function of the total length of homology arms (C) and the total length of SINE/LINE sequences present in the arms (D). The values were calculated by subtracting the integration frequency in *LIG4*-null cells from that in wild-type cells. See text for more details.

increase most likely depends on alternative end-joining, given that the presence of SINE/LINE fragments increases the amount of sequences with microhomology to the genome. The possibility of SSA involvement cannot be fully excluded, however, as most SINE/LINE fragments have enough length of homology to carry out homologous recombination, even though perfect homology may not be available due to the heterogeneity of these repetitive sequences [26]. Indeed, recent work by Escribano-Díaz *et al.* showed that knockdown of 53BP1/RIF1 (which act to stimulate NHEJ) led to a very similar increase in alternative end-joining and SSA at I-*Sce*I-induced DSBs [31]. Importantly, for SSA or microhomology-mediated alternative end-joining to bring about random integration, not only must the end of the vector be extensively deleted, but also the target genome should have a pre-existing DSB at or near a SINE/LINE fragment. This is likely enough, given the observed frequent terminal deletions of transfected plasmid [23] and the high abundance of SINE/LINE fragments in the genome. However, a more likely possibility for LIG4-independent integration may be a synthesis-dependent microhomology-mediated end-joining (SD-MMEJ) mechanism, as proposed by Yu and McVey [22]. This model well explains frequent terminal modifications observed at plasmid-genome junctions in random integrants from mouse ES cells [23] as well as little or no apparent junctional microhomologies in *LIG4*-null cells ([21]; our unpublished observations).

Concluding remarks

In this paper, we have shown, to our knowledge for the first time, that targeting vectors with long homology arms tend to confer high random-integration frequencies in human cells, most likely by virtue of the presence of repetitive DNA sequences. Perhaps conflicting with this finding, it is generally accepted that lengthening homology arms is beneficial for increasing gene-targeting efficiency [3,32]. Indeed, the targeting efficiency of pHPRT8.9-Puro(−) was greater than that of pHPRT2.2-Puro(−) in our experiments (Figure 3A). It should be noted, however, that pHPRT8.9-Puro(-) conferred higher random and targeted integration frequencies than did pHPRT2.2-Puro(−) (Figure 3B). Additionally, we examined and compared the targeting efficiency in Nalm-6 cells using more than ten different gene-targeting vectors, and did not observe a clear correlation between the arm length and the targeting efficiency (Figure S6 in File S1), although this could simply be due to the unavoidable lack of uniformity of the targeting constructs used in those experiments. It is therefore suggested that the length of homology arms affects both random and targeted integration, and hence the presence of long homology arms may allow for enhanced targeting efficiency but simultaneously increase the absolute frequency of random integration. We have also shown in this study that LIG4-independent end-joining mechanisms contribute to random integration events, even when NHEJ is functionally normal, and that repetitive DNA sequences, especially SINEs, present in targeting-vector arms serve to facilitate LIG4-independent random integration. These results imply that constructing a targeting vector that contains no repetitive DNA sequence should be a promising way to minimize random integrants. This may particularly be the case when an artificial nuclease-based gene targeting system, such as TALEN or CRISPR, is being employed, in which a targeting vector does not require long homology arms. Even more importantly, the identification of a factor(s) specifically involved in LIG4-independent integration mechanisms will be of great value for enhancing gene targeting, assuming that simultaneous suppression of LIG4-dependent and independent mechanisms would have a dramatic effect on reducing random integrants and if this strategy does not disturb viability or genome integrity of human somatic cells.

Materials and Methods

Vectors

The long-arm *HPRT* targeting vectors pHPRT8.9-Puro(+) and pHPRT8.9-Puro(−) were constructed as previously described [7]. Other gene-targeting vectors, including the short-arm *HPRT* targeting vectors pHPRT2.2-Puro(+) and pHPRT2.2-Puro(−), were constructed using the MultiSite Gateway system (Life Technologies, Carlsbad, CA) to assemble two homology arms and a drug-resistance gene cassette, as previously described [1,27]. Genomic fragments for homology arms were PCR amplified using Nalm-6 genomic DNA as template with primers listed in Figures S7 and S8 in File S1. Imperfect vectors lacking either arm of pHPRT8.9-Puro(+) and pHPRT8.9-Puro(−) were constructed as shown in Figure S9 in File S1. Imperfect vectors lacking either arm of pHPRT2.2-Puro(+) and pHPRT2.2-Puro(−) were constructed by digestion with SacI (for 5'-arm deletion) or SalI (for 3'-arm deletion), followed by self-ligation of the linearized plasmid DNA. All the plasmid vectors were purified with Qiagen Plasmid Maxi Kits (Qiagen K.K., Tokyo) and linearized with an appropriate restriction enzyme prior to transfection [1].

Cell Culture

The human pre-B leukemia cell line Nalm-6 [21] and its derivative (the *LIG4*−/− cell line; [27]) were maintained in a 5% CO_2 incubator at 37°C in ES medium (Nissui Seiyaku Co., Tokyo, Japan) supplemented with 10% calf serum (Hyclone, Logan, UT) and 50 µM 2-mercaptoethanol. The $LIG4^{-/-}$ cells were generated as described [27].

Integration Assays and Gene-targeting Experiments

Transfection of vector DNA or siRNA was performed as described previously [1,7,33]. For integration assays, 4×10^6 cells were electroporated with 4 µg of linearized plasmid DNA, cultured for 22 hr, and replated at a density of $0.5–1 \times 10^6$ cells per 90-mm dish into agarose medium containing 0.5 µg/ml puromycin (Wako Pure Chemical, Osaka, Japan). Meanwhile, small aliquots of the transfected cells were replated into drug-free agarose medium to determine the plating efficiency. After cultivation for 2–3 weeks, the resulting colonies were counted, and the total integration frequency was calculated by dividing the number of drug-resistant colonies with that of surviving cells. For gene-targeting experiments, each targeting vector was linearized and transfected into wild-type or $LIG4^{-/-}$ cells. After a 2–3 week incubation, genomic DNA was prepared from drug-resistant colonies and subjected to PCR and Southern blot analysis as described [7]. The gene-targeting efficiency was calculated by dividing the number of targeted clones with that of drug-resistant clones analyzed (Figure 1). The targeted integration frequency was calculated by multiplying the total integration frequency by the targeting efficiency. The random integration frequency was calculated by subtracting the targeted integration frequency from the total integration frequency.

Supporting Information

File S1   Figures S1–S9. Figure S1. Schematic representation of repetitive DNA sequences present in the *HPRT* vectors used in this study. The location and length (bp) of each SINE/LINE fragment is based on the UCSC Genome Browser Database: Update 2006 (Nucleic Acids Res. 34:D590–D598, 2006). Figure S2. Impact of siRNA-mediated knockdown of DNA ligase I or IIIα on integration frequency. (A) The nucleotide sequence of *LIG1* and *LIG3* siRNA. These siRNAs were designed as reported previously (Nucleic Acids Res. 36: 3297–3310, 2008). (B) Western blot analysis for DNA ligase I and IIIα in siRNA-transfected Nalm-6 wild-type and *LIG4*-null cells. M, mock-transfected. (C, D) Integration frequency in wild-type and *LIG4*-null cells treated with *LIG1* siRNA (C) or *LIG3* siRNA (D). A non-targeting vector (pβactin-His; Nucleic Acids Res. 36: 6333–6342, 2008) was used for transfection. The integration frequency in mock-transfected wild-type cells was taken as 1, and the relative integration frequency was calculated. Figure S3. Structural features of gene-targeting vectors used for the analysis of integration frequency. (A) Fundamental structure of targeting vectors. In all the vectors, 5' and 3' arms flank the drug-resistance gene cassette (*Puro^r*), which is placed in the forward direction. (B) Structural features of the fourteen gene-targeting vectors used. Shown are the lengths of 5' and 3' arms and SINE/LINE sequences within each arm and the integration frequency. The length of SINE/LINE is based on the UCSC Genome Browser Database: Update 2006 (Nucleic Acids Res. 34:D590–D598, 2006). Figure S4. Integration frequency of targeting vector as a function of the length of repetitive DNA sequences. Integration frequencies of pHPRT8.9-Puro(+), pHPRT2.2-Puro(+), and twelve other gene-targeting vectors are shown as a function of the total length of SINE sequence (A), 5'-arm SINE length (B), 3'-arm SINE length (C), the total length of LINE sequence (D), 5'-arm LINE length (E), and 3'-arm LINE length (F). See also Figure 5. Figure S5. Correlation between the integration frequency and repetitive DNA sequences. (A) Integration frequencies of pHPRT8.9-Puro(+), pHPRT2.2-Puro(+), and twelve other gene-targeting vectors as a function of the total length of SINE/LINE sequences. Note that this graph is the same as Figure 5D. (B) Same as (A), except that the three vectors have been omitted from the data set. Note that the redrawn fitted curve reveals a stronger correlation between the total SINE/LINE length and the integration frequency. See text and Figure 5 for details. Figure S6. Gene-targeting efficiency is little affected by the length of homology arms. Targeting efficiencies are shown as a function of the total length of homology arms of the targeting vector. The twelve non-*HPRT* gene-targeting vectors were used for the analysis (see Figure S3B in File S1). Figure S7. PCR primers used to amplify the homology arms of pHPRT2.2-Puro vectors. The restriction sites used to construct the arm-deleted vectors are shown in red (SacI) or blue (SalI). Figure S8. PCR primers used to amplify the homology arms of non-*HPRT* targeting vectors. Red denotes *attB* sequences. Figure S9. Schematic representation of construction of imperfect pHPRT8.9-Puro vectors lacking the 3' arm (A) or 5' arm (B).

Acknowledgments

We thank Susumu Iiizumi, Eriko Toyoda, Sairei So, Koichi Uegaki, Yuji Nomura, Hideki Koyama, Shiho Makino, Mikako Mori, Kana Ito, and Haruna Kamekawa for helpful discussions and technical assistance.

Author Contributions

Conceived and designed the experiments: AI AK NA. Performed the experiments: AI AK SS. Analyzed the data: AI AK NA. Wrote the paper: AI AK NA.

References

1. Adachi N, Kurosawa A, Koyama H (2008) Highly proficient gene targeting by homologous recombination in the human pre-B cell line Nalm-6. Methods Mol Biol 435: 17–29.
2. Kan Y, Ruis B, Lin S, Hendrickson EA (2014) The mechanism of gene targeting in human somatic cells. PLoS Genet 10: e1004251.
3. Vasquez KM, Marburger K, Intody Z, Wilson JH (2001) Manipulating the mammalian genome by homologous recombination. Proc Natl Acad Sci U S A 98: 8403–8410.
4. Lieber MR (2008) The mechanism of human nonhomologous DNA end joining. J Biol Chem 283: 1–5.
5. Ishibashi K, Suzuki K, Ando Y, Takakura C, Inoue H (2006) Nonhomologous chromosomal integration of foreign DNA is completely dependent on MUS-53 (human Lig4 homolog) in Neurospora. Proc Natl Acad Sci U S A 103: 14871–14876.
6. Ninomiya Y, Suzuki K, Ishii C, Inoue H (2004) Highly efficient gene replacements in Neurospora strains deficient for nonhomologous end-joining. Proc Natl Acad Sci U S A 101: 12248–12253.
7. Iiizumi S, Kurosawa A, So S, Ishii Y, Chikaraishi Y, et al. (2008) Impact of non-homologous end-joining deficiency on random and targeted DNA integration: implications for gene targeting. Nucleic Acids Res 36: 6333–6342.
8. Fattah FJ, Lichter NF, Fattah KR, Oh S, Hendrickson EA (2008) Ku70, an essential gene, modulates the frequency of rAAV-mediated gene targeting in human somatic cells. Proc Natl Acad Sci U S A 105: 8703–8708.
9. Sado K, Ayusawa D, Enomoto A, Suganuma T, Oshimura M, et al. (2001) Identification of a mutated DNA ligase IV gene in the X-ray-hypersensitive mutant SX10 of mouse FM3A cells. J Biol Chem 276: 9742–9748.
10. Jeggo PA, Smith-Ravin J (1989) Decreased stable transfection frequencies of six X-ray-sensitive CHO strains, all members of the xrs complementation group. Mutat Res 218: 75–86.
11. Adachi N, Ishino T, Ishii Y, Takeda S, Koyama H (2001) DNA ligase IV-deficient cells are more resistant to ionizing radiation in the absence of Ku70: Implications for DNA double-strand break repair. Proc Natl Acad Sci U S A 98: 12109–12113.
12. Decottignies A (2013) Alternative end-joining mechanisms: a historical perspective. Front Genet 4: 48.
13. Chiruvella KK, Liang Z, Wilson TE (2013) Repair of double-strand breaks by end joining. Cold Spring Harb Perspect Biol 5: a012757.
14. Paul K, Wang M, Mladenov E, Bencsik-Theilen A, Bednar T, et al. (2013) DNA ligases I and III cooperate in alternative non-homologous end-joining in vertebrates. PLoS One 8: e59505.
15. Wang H, Rosidi B, Perrault R, Wang M, Zhang L, et al. (2005) DNA ligase III as a candidate component of backup pathways of nonhomologous end joining. Cancer Res 65: 4020–4030.
16. Oh S, Harvey A, Zimbric J, Wang Y, Nguyen T, et al. (2014) DNA ligase III and DNA ligase IV carry out genetically distinct forms of end joining in human somatic cells. DNA Repair (Amst), in press.
17. Truong LN, Li Y, Shi LZ, Hwang PY, He J, et al. (2013) Microhomology-mediated End Joining and Homologous Recombination share the initial end resection step to repair DNA double-strand breaks in mammalian cells. Proc Natl Acad Sci U S A 110: 7720–7725.
18. Adachi N, Saito S, Kurosawa A (2013) Repair of accidental DNA double-strand breaks in the human genome and its relevance to vector DNA integration. Gene Technology 3: 1000e1107.
19. Pannunzio NR, Li S, Watanabe G, Lieber MR (2014) Non-homologous end joining often uses microhomology: Implications for alternative end joining. DNA Repair (Amst) 17: 74–80.
20. Cortizas EM, Zahn A, Hajjar ME, Patenaude AM, Di Noia JM, et al. (2013) Alternative end-joining and classical nonhomologous end-joining pathways repair different types of double-strand breaks during class-switch recombination. J Immunol 191: 5751–5763.
21. So S, Adachi N, Lieber MR, Koyama H (2004) Genetic interactions between BLM and DNA ligase IV in human cells. J Biol Chem 279: 55433–55442.
22. Yu AM, McVey M (2010) Synthesis-dependent microhomology-mediated end joining accounts for multiple types of repair junctions. Nucleic Acids Res 38: 5706–5717.
23. Suzuki K, Ohbayashi F, Nikaido I, Okuda A, Takaki H, et al. (2010) Integration of exogenous DNA into mouse embryonic stem cell chromosomes shows preference into genes and frequent modification at junctions. Chromosome Res 18: 191–201.
24. Bañuelos CA, Banath JP, MacPhail SH, Zhao J, Eaves CA, et al. (2008) Mouse but not human embryonic stem cells are deficient in rejoining of ionizing radiation-induced DNA double-strand breaks. DNA Repair (Amst) 7: 1471–1483.
25. Treangen TJ, Salzberg SL (2012) Repetitive DNA and next-generation sequencing: computational challenges and solutions. Nat Rev Genet 13: 36–46.
26. Callinan PA, Batzer MA (2006) Retrotransposable elements and human disease. Genome Dyn 1: 104–115.
27. Iiizumi S, Nomura Y, So S, Uegaki K, Aoki K, et al. (2006) Simple one-week method to construct gene-targeting vectors: application to production of human knockout cell lines. Biotechniques 41: 311–316.
28. Oh S, Wang Y, Zimbric J, Hendrickson EA (2013) Human LIGIV is synthetically lethal with the loss of Rad54B-dependent recombination and is required for certain chromosome fusion events induced by telomere dysfunction. Nucleic Acids Res 41: 1734–1749.
29. Watson JE, Slorach EM, Maule J, Lawson D, Porteous DJ, et al. (1995) Human repeat-mediated integration of selectable markers into somatic cell hybrids. Genome Res 5: 444–452.
30. Lander ES, Linton LM, Birren B, Nusbaum C, Zody MC, et al. (2001) Initial sequencing and analysis of the human genome. Nature 409: 860–921.
31. Escribano-Diaz C, Orthwein A, Fradet-Turcotte A, Xing M, Young JT, et al. (2013) A cell cycle-dependent regulatory circuit composed of 53BP1-RIF1 and BRCA1-CtIP controls DNA repair pathway choice. Mol Cell 49: 872–883.
32. Mae S, Shono A, Shiota F, Yasuno T, Kajiwara M, et al. (2013) Monitoring and robust induction of nephrogenic intermediate mesoderm from human pluripotent stem cells. Nat Commun 4: 1367.
33. So S, Nomura Y, Adachi N, Kobayashi Y, Hori T, et al. (2006) Enhanced gene targeting efficiency by siRNA that silences the expression of the Bloom syndrome gene in human cells. Genes Cells 11: 363–371.

12

Low-Dose Formaldehyde Delays DNA Damage Recognition and DNA Excision Repair in Human Cells

Andreas Luch[1], Flurina C. Clement Frey[2], Regula Meier[2], Jia Fei[2], Hanspeter Naegeli[2*]

1 Department of Product Safety, German Federal Institute for Risk Assessment (BfR), Berlin, Germany, 2 Institute of Pharmacology and Toxicology, University of Zürich-Vetsuisse, Zürich, Switzerland

Abstract

Objective: Formaldehyde is still widely employed as a universal crosslinking agent, preservative and disinfectant, despite its proven carcinogenicity in occupationally exposed workers. Therefore, it is of paramount importance to understand the possible impact of low-dose formaldehyde exposures in the general population. Due to the concomitant occurrence of multiple indoor and outdoor toxicants, we tested how formaldehyde, at micromolar concentrations, interferes with general DNA damage recognition and excision processes that remove some of the most frequently inflicted DNA lesions.

Methodology/Principal Findings: The overall mobility of the DNA damage sensors UV-DDB (ultraviolet-damaged DNA-binding) and XPC (xeroderma pigmentosum group C) was analyzed by assessing real-time protein dynamics in the nucleus of cultured human cells exposed to non-cytotoxic ($<100~\mu M$) formaldehyde concentrations. The DNA lesion-specific recruitment of these damage sensors was tested by monitoring their accumulation at local irradiation spots. DNA repair activity was determined in host-cell reactivation assays and, more directly, by measuring the excision of DNA lesions from chromosomes. Taken together, these assays demonstrated that formaldehyde obstructs the rapid nuclear trafficking of DNA damage sensors and, consequently, slows down their relocation to DNA damage sites thus delaying the excision repair of target lesions. A concentration-dependent effect relationship established a threshold concentration of as low as 25 micromolar for the inhibition of DNA excision repair.

Conclusions/Significance: A main implication of the retarded repair activity is that low-dose formaldehyde may exert an adjuvant role in carcinogenesis by impeding the excision of multiple mutagenic base lesions. In view of this generally disruptive effect on DNA repair, we propose that formaldehyde exposures in the general population should be further decreased to help reducing cancer risks.

Editor: Robert W. Sobol, University of Pittsburgh, United States of America

Funding: This research was supported by the German Federal Institute for Risk Assessment (BfR) (project FK-1329-038); http://www.bfr.bund.de/en/home.html. The funders had no role in study design, data collection and analysis, decision to publish, or preparation of the manuscript.

Competing Interests: The authors have declared that no competing interests exist.

* E-mail: naegelih@vetpharm.uzh.ch

Introduction

In aging populations, the incidence of chemoresistant malignancies continues to rise. For example, the ultraviolet (UV) radiation of sunlight is a primary risk factor for skin cancer [1]. As a major interface separating the body from the environment, skin cells provide an effective barrier against physical and chemical insults. Besides the photoprotectant melanin, a dedicated nucleotide excision repair (NER) activity mitigates the carcinogenicity of sunlight by excising UV-induced photoproducts from DNA before they are converted to genetic mutations [2,3]. Another defense line, known as base excision repair (BER), removes concurrent oxidative base lesions [3–5]. However, the skin and other tissues frequently come in contact with chemicals that react with cellular components like chromatin in the close vicinity of DNA, thus potentially interfering with DNA-repairing enzymes. Formaldehyde is a compound of particular concern because of its ubiquitous distribution, widespread human exposure and verified human carcinogenicity. This reactive aldehyde appears in automotive emissions or tobacco smoke and is added to many industrial and medicinal products. Formaldehyde-containing resins are used in the manufacture of plywood, paper, textile fibers, plastics, paints, lubricants and dyes. Formaldehyde is also employed in furniture, upholstery, carpeting, drapery and other household products [6–9]. Cosmetics are another important source as they are often preserved with formaldehyde donors [10].

At high doses, formaldehyde generates DNA-protein crosslinks (DPXs) [11,12] and induces nasal carcinomas in rodents [13]. In view of the documented risk of nasopharyngeal, sinonasal, lymphatic and hematopoietic cancer in occupationally exposed workers [14–16], it has been categorized as a human carcinogen [17,18]. This current classification does not consider formaldehyde as a possible risk factor for cutaneous cancer, although very substantial levels of this compound (65% of the dose) are found in the skin after topical application to experimental animals [19]. This pronounced ability of formaldehyde to penetrate the skin raises the concern that cutaneous cancer may represent another adverse endpoint. In an animal model of carcinogenesis, formaldehyde exhibits no tumorigenicity by dermal exposure on its own but nevertheless displays a deleterious effect by dramatically reducing the latency time of carcinogen-initiated skin tumors [20]. Such an adjuvant role during the carcinogenesis process, detected in animal experiments, is also supported by the observation that morticians, who used formaldehyde as an embalming fluid, display an elevated mortality due to skin cancer [21].

Previous studies suggested an interference of formaldehyde with the proper processing of various forms of DNA damage [22,23]. In the case of global-genome NER activity, the detection of UV lesions depends on a specific accessory complex, known as UV-DDB (for UV-damaged DNA-binding), which consists of a damage sensor (DDB2) and a regulatory subunit (DDB1) [24]. To explore the basis of the observed adjuvant effect in skin carcinogenesis, we tested how formaldehyde, at non-cytotoxic concentrations, influences the activity of this critical DNA damage recognition complex. We thereby identified a novel mechanism by which NER activity is inhibited in formaldehyde-exposed cells.

Materials and Methods

Expression constructs

The XPC complementary DNA [25] was cloned into pGFP-N3 (Clontech) using the restriction enzymes XmaI and KpnI. The DDB2 complementary DNA was transferred from pGFP-DDB2-C1 (Dr. S. Linn, University of California, Berkeley, California) to pmRFP1-C3 using its BamHI sites. The pGFP-OGG1-C1 vector was obtained from Dr. P. Radicella, Institut de Radiologie Cellulaire et Moleculaire, Fontenay-aux-Roses, France, and the pGFP-APE1-C1 vector from Dr. H. Lans, Erasmus University Rotterdam, The Netherlands.

Cell culture

All culture media and supplements were from Invitrogen. Simian virus 40-transformed fibroblasts (GM00637) were obtained from the Coriell Institute for Medical Research (Camden, NJ). They were grown in a humidified incubator at 37°C and 5% CO_2 using Dulbecco's modified Eagle's medium (DMEM) supplemented with 10% heat-inactivated fetal calf serum, 100 units/ml penicillin G and 100 µg/ml streptomycin.

Transfections

Fibroblasts (~500,000) were seeded into 6-well plates. After 24 h, at a confluence of 80–85%, the cells were transfected with 1 µg expression vector using 4 µl FuGENE HD reagent (Roche). Following a 4-h incubation, the transfection mixture was replaced by complete medium and the cells were incubated for another 18 h at 37°C.

Exposures

Solutions of 37% aqueous formaldehyde (Fluka) or acetaldehyde (Sigma) were each serially diluted in complete medium. All formaldehyde and acetaldehyde stocks were made fresh and maintained on ice to prevent evaporation. Cisplatin (cis-diamine-platinum-II, Sigma) was dissolved in dimethyl sulfoxide. After transfections, the cells were incubated with fresh complete medium containing again the indicated concentrations of formaldehyde or acetaldehyde. For irradiation, the medium was temporarily removed and the cells were rinsed with phosphate-buffered saline (PBS) before exposure to UV-C light from a germicidal lamp (257 nm wavelength). This wavelength results in the formation of cyclobutane pyrimidine dimers and (6–4) pyrimidine-pyrimidone photoproducts, which constitute the most prevalent forms of DNA damage induced by solar light [26].

Protein dynamics

High-resolution fluorescence recovery after photobleaching (FRAP) analyses were carried out under a controlled environment at 37°C and with a CO_2 supply of 5% using a Leica TCS SP5 confocal microscope equipped with an Ar^+ laser (488 nm) and a 63× oil immersion lens (numerical aperture of 1.4) as illustrated in **Figure 1A**. A region of interest of ~4 µm^2 was photo-bleached for 2.3 s at 80% laser intensity. Fluorescence recovery within the region of interest was monitored 200 times using 115-ms intervals followed by 30 frames at 250 ms and 20 frames at 500 ms. Simultaneously, a reference area of the same size was monitored throughout all time points to correct for overall bleaching. Finally, the data were normalized to the pre-bleach intensity.

Chromatin-binding assay

Salt extraction and micrococcal nuclease (MNase) digestion was used to analyze the binding of UV-DDB to chromatin [25]. After electrophoretic separation, the samples were transferred to a polyvinylidene (PVDF) membrane (BioRad) blocked by incubation for 2 h at room temperature with Tris-buffered saline containing 0.05% (v/v) Tween-20 and 5% (w/v) nonfat dry milk. Antibodies against the following proteins were used in Western blots: DDB2 (dilution 1:50, ab51017, Abcam), GAPDH (1:4′000, No. 4300, Ambion), H3 (1:10′000, No. 07-690, Millipore). HRP-conjugated secondary antibodies were diluted 10,000-fold. Reactions were developed with SuperSignal West Pico (Pierce), recorded with a FUJI LAS-3000 imaging system and quantified using the Quantity One software (BioRad).

Induction of UV foci

Human fibroblasts were grown on glass cover slips (20 mm diameter) and transfected with DDB2-EGFP or XPC-EGFP constructs. After 18-h incubations, the cell culture medium was removed and the cells were rinsed with PBS. UV foci were induced by irradiation through the 5-µm pores of polycarbonate filters (Millipore) using a UV-C source (257 nm, 150 J/m^2). After irradiation, the filters were removed and the cells incubated for 15 min at 37°C in complete DMEM. Immunocytochemistry was carried out as described [25].

Colony-forming assay

Fibroblasts were treated with formaldehyde as described above, UV-irradiated at increasing doses, seeded in different dilutions and incubated in cell culture medium for 7 days at 37°C to allow for colony formation. Colonies were stained with 0.5% (w/v) crystal violet in 80% ethanol and counted.

DNA repair assays

The pGL3 and phRL-TK vectors were from Promega. The pGL3 vector was UV-irradiated (257 nm, 1000 J/m^2) on ice in 10 mM Tris-HCl (pH 8) and 1 mM EDTA. Alternatively, pGL3 was incubated with 5.4 µM cisplatin for 24 h at 37°C. The modified plasmids were recovered by ethanol precipitation. To generate 8-oxo-2′-deoxyguanosine (8-oxo-dG) lesions, pGL3 DNA was mixed with 2 µM methylene blue and irradiated for 60 min on ice with visible light using a tungsten lamp (75 W) at a distance of 20 cm. Subsequently, the dye was extracted with 1% (w/v) sodium dodecyl sulfate and Tris-EDTA-saturated 1-butanol [27]. Human fibroblasts, grown to a confluence of 80% in 6-well plates, were transfected with 0.45 µg pGL3 DNA and 0.05 µg phRL-TK. After a 4-h incubation, the transfection reagent was replaced by complete medium. The cells were lysed after a further 18-h period using 0.5 ml Passive Lysis Buffer (Promega) according to the

A

Prebleach   Bleach   Post-bleach recovery ⟶

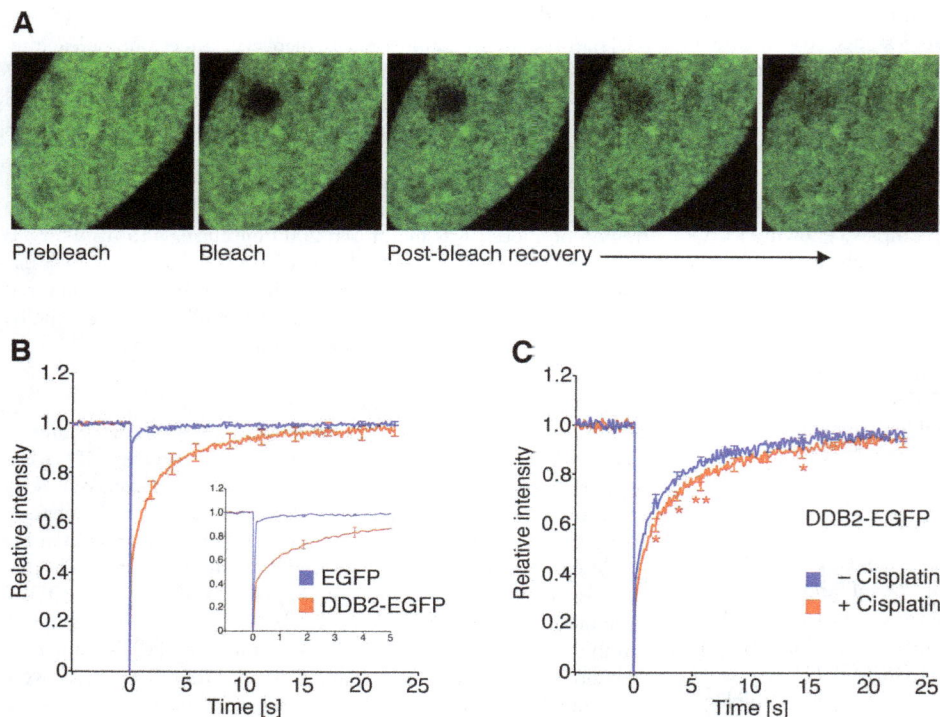

Figure 1. Analysis of repair protein dynamics in living cells. (**A**) A region of interest of ∼4 μm² was photo-bleached and the fluorescence recovery within this area was monitored over a time frame of 23 seconds. Simultaneously, a reference area of the same size was monitored to correct for overall bleaching and the resulting data were normalized to the pre-bleach intensity. (**B**) Quantitative FRAP recordings determined in fibroblasts transfected with expression vectors coding for the DDB2-EGFP fusion or the EGFP moiety alone (N = 30; error bars, S.E.M.). (**C**) Recognition of cisplatin-DNA adducts by UV-DDB revealed by FRAP analyses. Human fibroblasts transfected with the DDB2-EGFP construct were pre-incubated with 5 μM cisplatin. The resulting FRAP curves were compared with those of untreated controls. Asterisks, statistically significant differences between cisplatin-treated cells and untreated controls (N = 50; *p<0.05; **p<0.01).

manufacturer's instructions. *Photinus* and *Renilla* luciferase activity was determined in a Dynex microtiter plate luminometer using the Dual-Luciferase Assay System (Promega). For the assessment of DNA repair activities, mean values were calculated from the ratios between *Photinus* and *Renilla* luciferase activity [28]. Antibodies against UV lesions (MBL International) were used in a slot-blot assay following the manufacturer's instructions.

Statistical procedures

All results were analyzed with Prism 5 (GraphPac Software) using the Student's *t*-test for comparisons. A value of p<0.05 was considered statistically significant. The number of independently repeated experiments (N) is indicated in each figure legend.

Results

Analysis of UV-DDB dynamics in living cells

To test whether formaldehyde disturbs the cellular trafficking of the critical UV-DDB sensor, we transfected human skin fibroblasts with a construct that drives the expression of DDB2 conjugated to enhanced green-fluorescent protein (EGFP). The DDB2-EGFP fusion, located in the nucleus, was subjected to protein dynamics studies by FRAP, which is a powerful real-time method to monitor the *in situ* mobility of DNA repair subunits in living cells [29–31]. A nuclear area (∼4 μm²) of human skin fibroblasts expressing low levels of DDB2-EGFP is bleached with a 488-nm wavelength laser that does not produce DNA damage. Subsequently, the replacement of bleached DDB2 molecules with non-bleached counter-

parts, diffusing from surrounding nuclear regions, leads to a progressive recovery of fluorescence in the target area (**Figure 1A**). Besides the diffusion properties of each protein, the rate by which this fluorescence intensity returns to pre-bleach levels depends on possible protein associations with DNA, chromatin fibers or nuclear scaffolds. As the EGFP tag itself undergoes minimal such interactions [32], the fluorescence signal associated with the DDB2-EGFP fusion recovers much slower than that of EGFP alone (**Figure 1B**).

Although UV-DDB is thought to be required only for the repair of UV lesions [24,25,30], previous biochemical assays showed that it also binds to other forms of DNA damage, including base adducts induced by the antitumor agent cisplatin [33]. To validate this FRAP assay as a probe of UV-DDB interactions with chemically induced DNA damage, fibroblasts transfected with DDB2-EGFP were pre-incubated with 5 μM cisplatin representing a non-cytotoxic drug concentration [34]. The bleached area ultimately regained the initial fluorescence intensity within ∼25 s, reflecting rapid movements of EGFP-tagged UV-DDB complexes within the nuclei. However, cisplatin exposure led to a transient delay in fluorescence recovery (**Figure 1C**), indicating that the nuclear trafficking of UV-DDB is slowed down by interactions with cisplatin-DNA adducts. From these findings, we concluded that FRAP provides a suitable method to monitor the interaction of UV-DDB with chemically damaged chromatin.

Reduced UV-DDB trafficking by low-dose formaldehyde

Next, human fibroblasts expressing DDB2-EGFP were exposed for 18 h to a formaldehyde concentration (75 μM) that was selected to remain below the cytotoxic range but clearly above the reported physiological blood content, ranging between 13 and 20 μM [35]. The extended-time treatment was carried out to reflect the long-term exposure to formaldehyde in the population. Compared to the untreated control cells, the fluorescence recovery of DDB2-EGFP was retarded in formaldehyde-treated fibroblasts (**Figure 2A**), demonstrating a restrained movement of UV-DDB. Conversely, the fast fluorescence recovery observed with the EGFP moiety remained unchanged after formaldehyde treatment (**Figure 2B**), implying that interactions between DDB2 itself and formaldehyde-damaged chromatin are responsible for the reduced mobility. The same experiment was repeated with acetaldehyde at concentrations (≥3.6 mM) that have been shown to induce at least as many DPXs as 125 μM formaldehyde [36]. However, even at the highest dose, acetaldehyde did not interfere with the nuclear trafficking of UV-DDB (**Figure 2C**), pointing to a specific DDB2 response to formaldehyde damage. The distinct chemistry of the covalent linkage, the varying DNA structure at the site of the linkage or the particular pattern of covalently linked proteins [37] might explain this different effect of the two crosslinkers.

Because UV-DDB is implicated in DNA repair of UV lesions, we also monitored its mobility in cells transfected with DDB2-EGFP and then exposed to UV radiation, or in cells subjected to a dual treatment with both formaldehyde and UV light. The joint exposure did not further slow down the DDB2 movements compared to UV radiation alone (**Figure 2D**), and this lack of additive interaction suggests that the binding of UV-DDB to formaldehyde lesions and UV photoproducts is mutually exclusive. This view is supported below by the observation that low-dose formaldehyde hinders UV-DDB from recognizing the UV lesions.

Non-covalent binding to formaldehyde-damaged chromatin

The FRAP experiments revealed that UV-DDB remains attached to formaldehyde-damaged chromatin, such that its nuclear movement is inhibited. To confirm this finding, chromatin associations were tested biochemically as outlined in **Figure 3A**. First, free UV-DDB not bound to chromatin was removed by salt (0.3 M NaCl) extraction. Second, the resulting nucleoprotein complexes were solubilized by MNase, which liberates chromatin by cleavage into short nucleosomal repeats [38]. The fractions of free proteins (released by 0.3 M salt) and chromatin-bound proteins (released by MNase digestion) were analyzed using

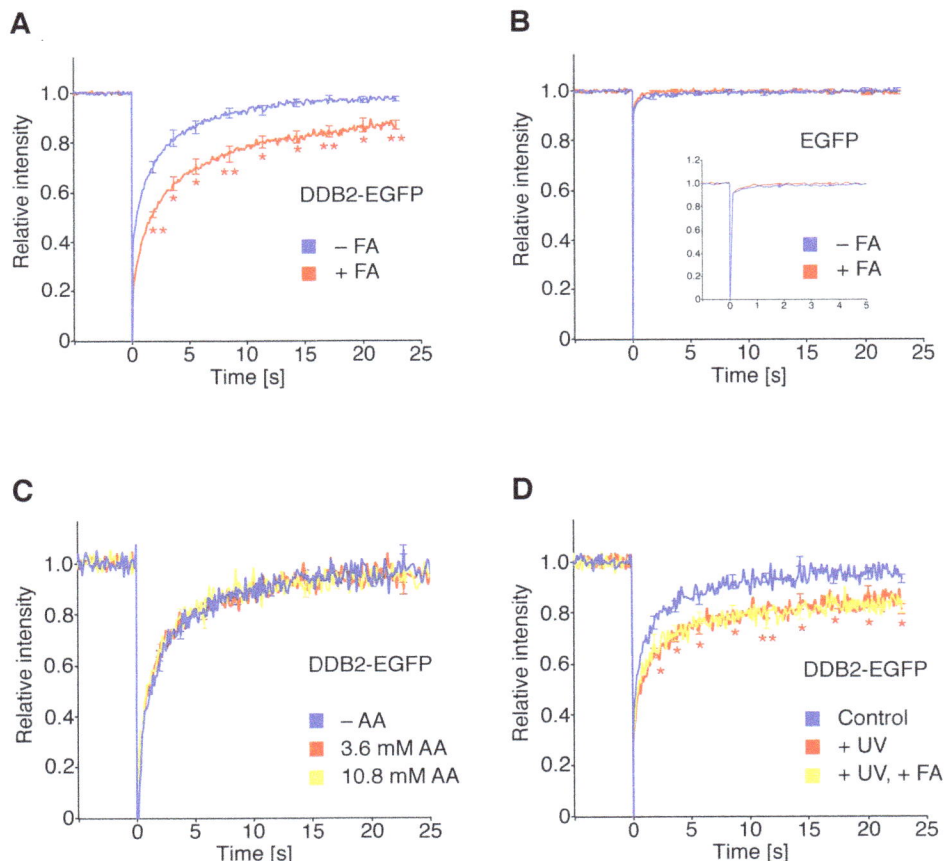

Figure 2. Delayed nuclear trafficking of UV-DDB. (A) FRAP analysis in human fibroblasts. Cells were transfected with DDB2-EGFP, incubated for 18 h with 75 μM formaldehyde (FA) and analyzed by FRAP (N = 50; error bars, S.E.M.). The fluorescence recovery curves were compared to those of untreated controls (*p<0.05; **p<0.01). **(B)** FRAP studies (N = 15; error bars, S.E.M.) demonstrating that EGFP movements are not affected by formaldehyde (18 h, 75 μM). **(C)** FRAP analysis with analogous acetaldehyde (AA) treatments (N = 50). **(D)** Combined formaldehyde and UV treatment. Transfected human fibroblasts were exposed to 75 μM formaldehyde for 18 h, UV-irradiated (30 J/m²) and subjected to FRAP analysis (N = 30). The asterisks indicate significant differences between UV-damaged fibroblasts and untreated controls (*p<0.05; **p<0.01).

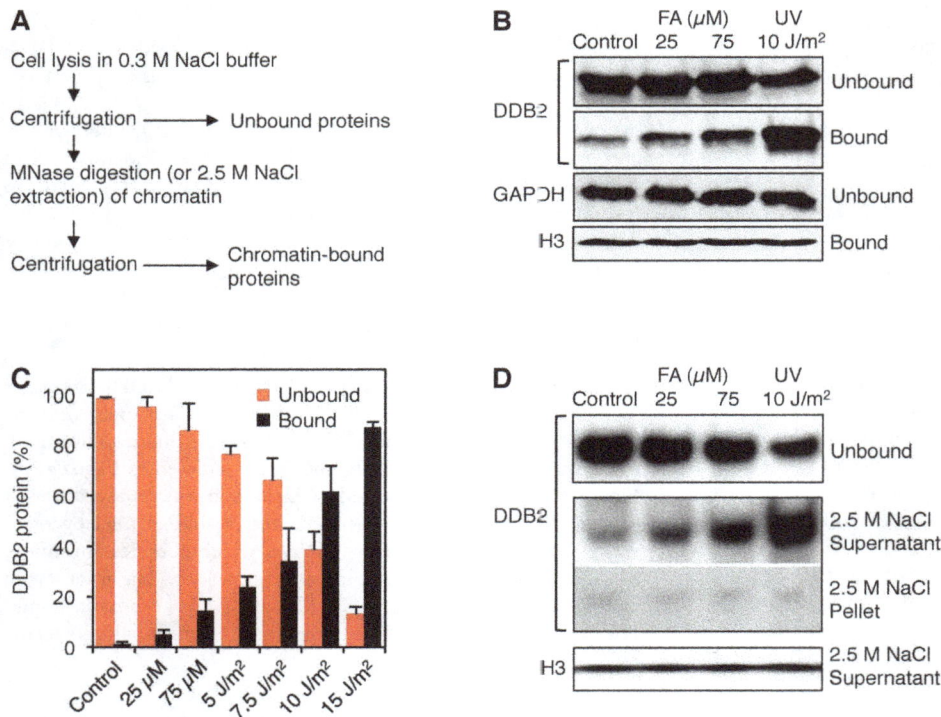

Figure 3. Association of UV-DDB with damaged chromatin. (A) Flow diagram illustrating how chromatin was dissected to monitor the binding of UV-DDB. Unbound proteins were released by salt (0.3 M NaCl) extraction and the remaining chromatin was solubilized by MNase digestion. **(B)** Western blot visualization of the chromatin partitioning of UV-DDB using antibodies against DDB2. GAPDH (glyceraldehyde 3-phosphate dehydrogenase), marker of unbound proteins; histone H3, marker of chromatin. Human fibroblasts were exposed for 18 h to formaldehyde or UV-irradiated at the indicated doses. **(C)** Quantification of three independent binding assays demonstrating the differential interaction of DDB2 with formaldehyde- and UV-damaged chromatin (error bars, S.D.). **(D)** Release of chromatin-bound DDB2 and histone H3 by high-salt extraction. After incubation with 0.3 M NaCl buffer, the chromatin was dissolved with 2.5 M NaCl, thus liberating non-covalently bound chromatin proteins.

antibodies against endogenous DDB2 and different markers to compare their distribution following each treatment.

Upon exposure of human fibroblasts to increasing doses of UV light, a growing proportion of the cellular DDB2 pool translocated to chromatin and essentially all of this chromatin-bound DDB2 is released by MNase digestion (**Figure 3B and 3C**). In contrast, 18-h formaldehyde exposures induced a comparably weak binding of DDB2 to damaged chromatin. Even a 75-μM formaldehyde treatment, which in the FRAP experiments (**Figure 2**) reduced the protein mobility as much as a UV dose of 30 J/m², resulted in considerably less chromatin association of DDB2. Importantly, the residual DDB2 that remained in chromatin after the 0.3 M salt treatment was completely released, in the absence of MNase digestion, by high salt (2.5 M NaCl) extraction (**Figure 3D**). These biochemical analyses demonstrate, therefore, that the association of UV-DDB, of which DDB2 constitutes the DNA-binding subunit, with formaldehyde-damaged chromatin occurs through transient non-covalent interactions. It is important to note that standard protocols that employ this reactant as a biochemical crosslinking agent, for example in chromatin immune-precipitation studies, use nearly 5,000-fold higher concentrations.

Inhibition of XPC trafficking by formaldehyde

We next tested the movement of XPC, a major interaction partner of UV-DDB [24,25,39]. Unlike other NER factors, XPC displays a constitutive binding to native DNA that retards its nuclear dynamics, thus leading to an incomplete fluorescence

recovery even in undamaged cells [31]. Again, FRAP experiments were performed on nuclear areas of ~4 μm² in human skin fibroblasts expressing low levels of XPC-EGFP. The validity of this approach for monitoring the binding of XPC to chemically damaged DNA is demonstrated by a 5-μM cisplatin treatment, whereby the mobility of XPC-EGFP is reduced (**Figure 4A**) reflecting its ability to recognize cisplatin-damaged DNA. Unlike the DDB2 response, however, the dynamics of the XPC complex is not influenced by 18-h formaldehyde (75 μM) treatments either alone (**Figure 4B**) or in combination with UV light (**Figure 4C**). A reduction of XPC mobility is nevertheless detected in cells that, in addition to endogenous DDB2, express ectopic DDB2 conjugated to red-fluorescent protein (DDB2-RFP; **Figure 4D**). This finding indicates that the XPC complex is indirectly immobilized by formaldehyde damage through its association with UV-DDB. Thus, in view of its low constitutive nuclear mobility, it is necessary to raise the level of the DDB2 interaction partner to visualize a sequestration of XPC protein in low-dose formaldehyde-damaged chromatin.

Impaired recognition of UV lesions

The reduced nuclear trafficking of UV-DDB and XPC indicates that formaldehyde may hinder recognition of their common targets. To test this hypothesis, we examined the UV damage recognition function of UV-DDB and XPC complexes in living fibroblasts. First, cells expressing low levels of the DDB2-EGFP construct were UV-irradiated through the pores of polycarbonate

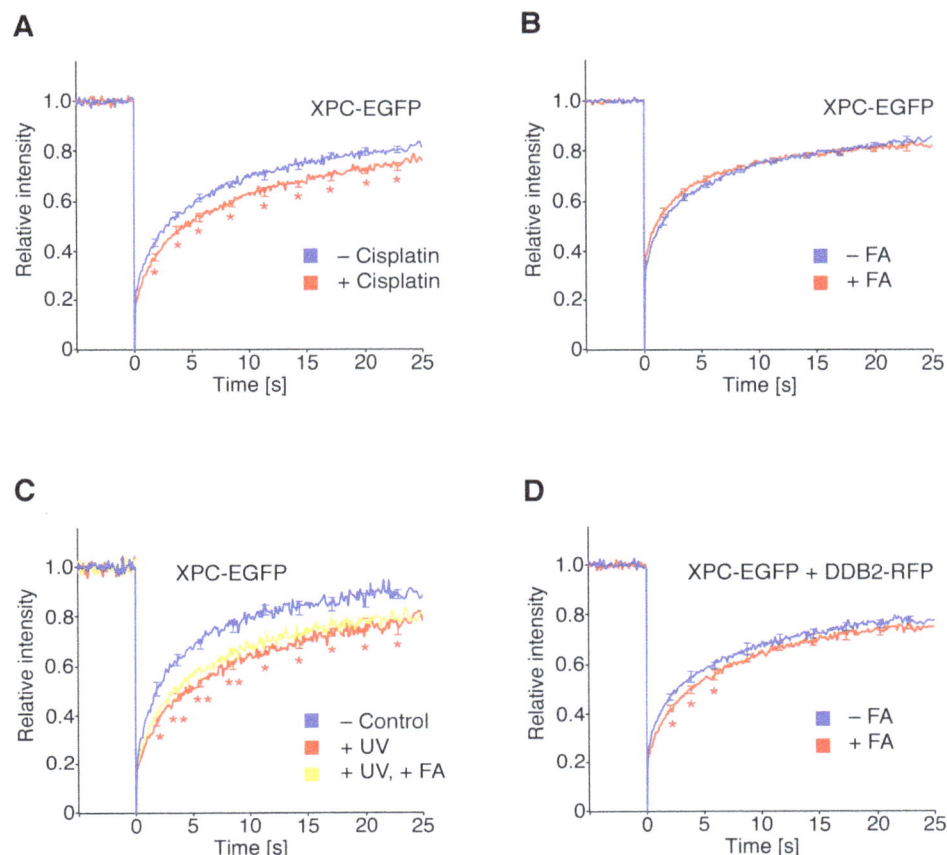

Figure 4. Indirect formaldehyde-induced reduction of XPC mobility. (**A**) Protein dynamics studies of XPC in the nuclei of human fibroblasts. Cells were transfected with the XPC-EGFP construct, incubated with 5 μM cisplatin and subjected to FRAP analysis (N = 30; error bars, S.E.M.). The resulting fluorescence recovery curves were compared to those of untreated controls (*p<0.05). (**B**) FRAP studies (N = 50) demonstrating that the fluorescence recovery curves of XPC-EGFP are not affected by an 18-h formaldehyde treatment (75 μM). (**C**) FRAP analysis of transfected fibroblasts demonstrating that the 18-h formaldehyde treatment (75 μM) does not further reduce the delayed XPC-EGFP trafficking in UV-irradiated cells (30 J/m²; N = 50). (**D**) Combined formaldehyde treatment and DDB2 overexpression. The transfected fibroblasts were exposed to 75 μM formaldehyde for 18 h and subjected to FRAP analysis. The presence of DDB2-RFP resulted in a slightly delayed fluorescence recovery curve of XPC-EGFP upon formaldehyde exposure (N = 30). Asterisks, significant differences between formaldehyde-treated and untreated fibroblasts (*p<0.05).

filters, thus focalizing DNA damage to narrow nuclear spots. After a 15-min recovery, DNA damage recognition was demonstrated by recording the co-localization of DDB2, detected by fluorescence measurements, and UV lesions detected by immunocytochemistry.

Figure 5A shows that the UV-dependent nuclear redistribution of DDB2-EGFP induces bright green spots accompanied by a reduced overall nuclear fluorescence. This pattern reflects an efficient accumulation of UV-DDB at damaged sites with concomitant depletion of the protein from undamaged nuclear regions containing no lesions. In 75-μM formaldehyde-treated cells, UV-DDB still relocated to UV-irradiated sites, but less efficiently, resulting in weaker spots of green fluorescence signals over the surrounding nuclear background (**Figure 5B**). A similarly reduced relocation to UV lesion spots was observed in cells transfected with XPC-EGFP and exposed to formaldehyde (**Figure 5C**). The quantitative comparison of three experiments confirmed that formaldehyde impairs the UV-dependent redistribution of both UV-DDB and XPC, thus inhibiting their translocation from undamaged nuclear areas to UV-irradiated sites. In each case, the ratio of fluorescence intensity at UV lesion

spots against the surrounding background was reduced by low-dose formaldehyde (**Figure 5D**).

Differential effects on BER enzymes

Another permanent trigger of mutagenesis are oxidative base lesions such as 8-oxo-dG [3–5]. Therefore, we tested whether formaldehyde affects the nuclear dynamics of 8-oxo-dG-DNA glycosylase 1 (OGG1), which initiates BER by recognizing and removing 8-oxo-dG from DNA leaving apurinic sites. The FRAP experiments of **Figure 6A** revealed that, in fibroblasts transfected with OGG1-EGFP, the fluorescence recovery is delayed upon incubation with 75 μM formaldehyde, indicating that the constitutive movement of this DNA glycosylase is disturbed by low-dose formaldehyde. Then, we probed the nuclear dynamics of an immediately downstream enzyme in the BER pathway, i.e., apurinic/apyrimidinic endonuclease 1 (APE1). This follow-up BER subunit displayed much faster movements in FRAP assays that were refractory to the 75-μM formaldehyde treatment (**Figure 6B**).

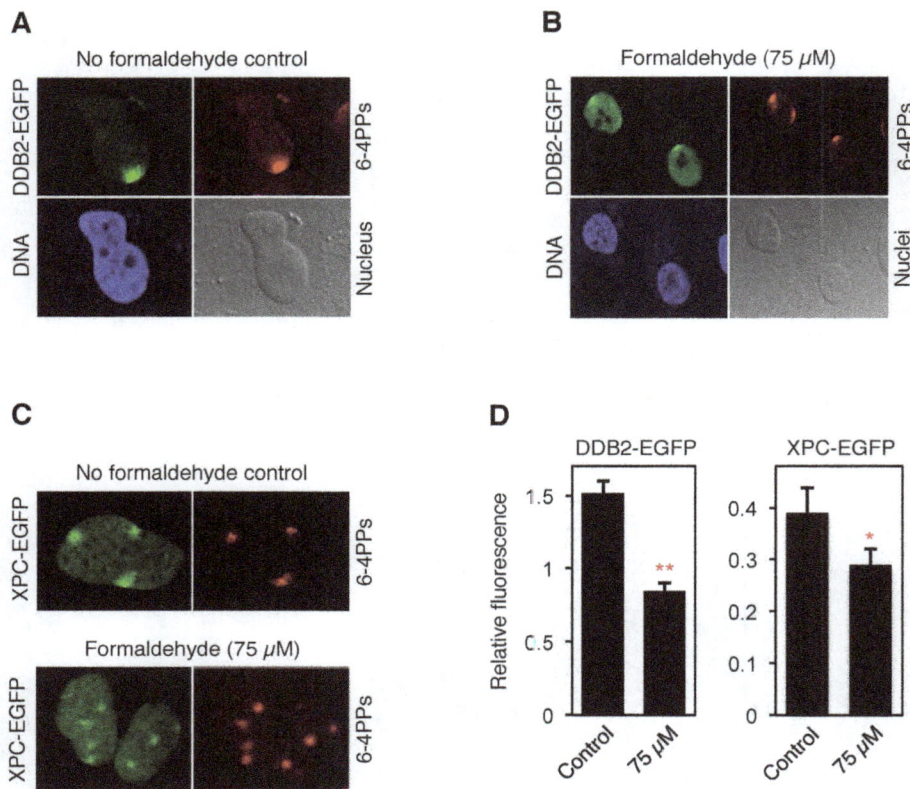

Figure 5. Accumulation of damage recognition factors on UV lesions. (A) Representative image illustrating the redistribution of DDB2 to UV lesion sites visualized with antibodies against (6-4) photoproducts. Human fibroblasts transfected with DDB2-EGFP were UV-irradiated through the pores of polycarbonate filters and fixed 15 min after treatment. DNA is evidenced by the Hoechst reagent and the nuclei are shown with contrast images. **(B)** Representative cells demonstrating the defective translocation of DDB2-EGFP from undamaged nuclear areas to UV lesions after 18-h incubations with 75 μM formaldehyde. **(C)** Representative images illustrating that the 75-μM formaldehyde treatment impedes the redistribution of XPC-EGFP to UV lesions. **(D)** Reduced fluorescence intensity at UV lesion spots over the surrounding background in cells exposed for 18 h to 75 μM formaldehyde (N = 54; error bars, S.D.; *p<0.05, **p<0.01).

Inhibition of excision activity

Consistent with the compromised mobility of repair factors in the chromatin context, the low-dose (75 μM) formaldehyde treatment increased the sensitivity of human fibroblasts to the cytotoxic effect of UV light (**Figure 7A**). Next, the fibroblasts were transfected with a reporter vector (pGL3), carrying the *Photinus*

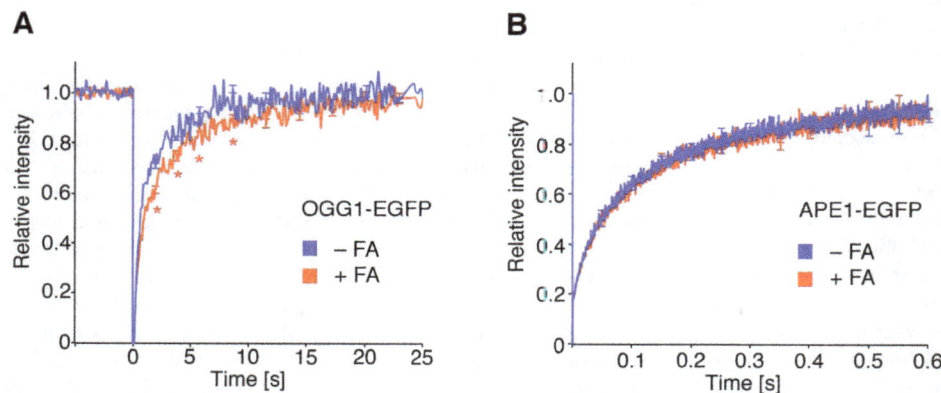

Figure 6. Formaldehyde-induced damage delays the nuclear trafficking of a DNA glycosylase. (A) Nuclear dynamics of OGG1, the DNA glycosylase that removes 8-oxo-dG, in human fibroblasts. Cells were transfected with the OGG1-EGFP construct, incubated for 18 h with 75 μM formaldehyde and subjected to FRAP analysis (N = 50; error bars, S.E.M.). The resulting fluorescence recovery curves were compared to those of untreated controls (*p<0.05). **(B)** FRAP studies (N = 50) demonstrating that the extremely fast movements of the APE1-EGFP fusion are not affected by the 75-μM formaldehyde treatment.

luciferase sequence, to determine the consequence of low-dose formaldehyde exposures on transcription and translation. The measurement of reporter luciferase activity in cell lysates showed that formaldehyde at concentrations of up to 125 μM exerted no inhibition on RNA or protein (luciferase) synthesis (**Figure 7B**). However, this reporter expression was suppressed by increasing the formaldehyde concentration to 1 mM, which leads to overt cytotoxicity.

Next, to monitor the NER pathway, pGL3 reporter vectors coding for the *Photinus* luciferase were UV-irradiated and introduced into fibroblasts together with an undamaged control coding for *Renilla* luciferase. DNA repair efficiency was assessed by measuring the *Photinus* luciferase activity in cell lysates, followed by normalization against the *Renilla* control. In this dual reporter assay, reactivation of UV-irradiated pGL3 is dependent on the ability of the NER system to remove UV lesions. The resulting *Photinus/Renilla* luciferase ratios revealed that NER activity is significantly reduced by formaldehyde exposure at 25-μM or higher (**Figure 7C**). This low-dose formaldehyde effect provides a proof for DNA repair inhibition as it reflects diminished reactivation of the UV-irradiated template rather than reduced transcription or translation.

The findings of **Figure 5** indicated that the inhibition of DNA repair is caused by an impaired nuclear trafficking of UV-DDB, resulting in reduced recognition of UV lesions during the NER process. This view was confirmed by the observation that NER inhibition could be reversed by co-transfection of the dual reporter system with a DDB2-EGFP expression vector (**Figure 7D**). Thus, by raising the cellular DDB2 level in formaldehyde-treated cells, it

was possible to restore the level of global-genome NER activity to that found in untreated cells. That the defective recognition function also jeopardizes the excision of chemically induced DNA damage is supported by host-cell reactivation assays where the pGL3 *Photinus* vector was damaged by cisplatin. Indeed, as a consequence of the inhibited removal of cisplatin-DNA adducts, expression of the cisplatin-damaged reporter was reduced by formaldehyde (**Figure 7E**). Finally, the induction of oxidative lesions, instead of UV or cisplatin adducts, in the same reporter vector demonstrated that also the hindrance of OGG1 mobility, observed in **Figure 6A**, translates to a significantly reduced BER efficiency as measured in host-cell reactivation assays (**Figure 7D**).

To confirm that an inhibition of DNA repair activity occurs in the chromosomal context, human fibroblasts were UV-irradiated after a 75-μM pretreatment with formaldehyde. The excision of UV lesions from genomic DNA was monitored by a slot-blot immunoassay taking advantage of antibodies against (6-4) pyrimidine-pyrimidone photoproducts (6-4PPs; **Figure 8A**) or cyclobutane pyrimidine dimers (CPDs). Among the UV lesions induced by sunlight, 6-4PPs are the ones removed most rapidly from human cells [40]. In fact, only ~25% of the initial amount of these 6-4PPs remained in the DNA of untreated control fibroblasts after a repair time of 3 h. In contrast, formaldehyde-exposed cells displayed a slower excision activity with only ~50% photoproduct repair during the same 3-h incubation period (**Figure 8B**). Similarly, the low-dose formaldehyde treatment slowed down the excision of CPDs in human fibroblasts (**Figure 8C**).

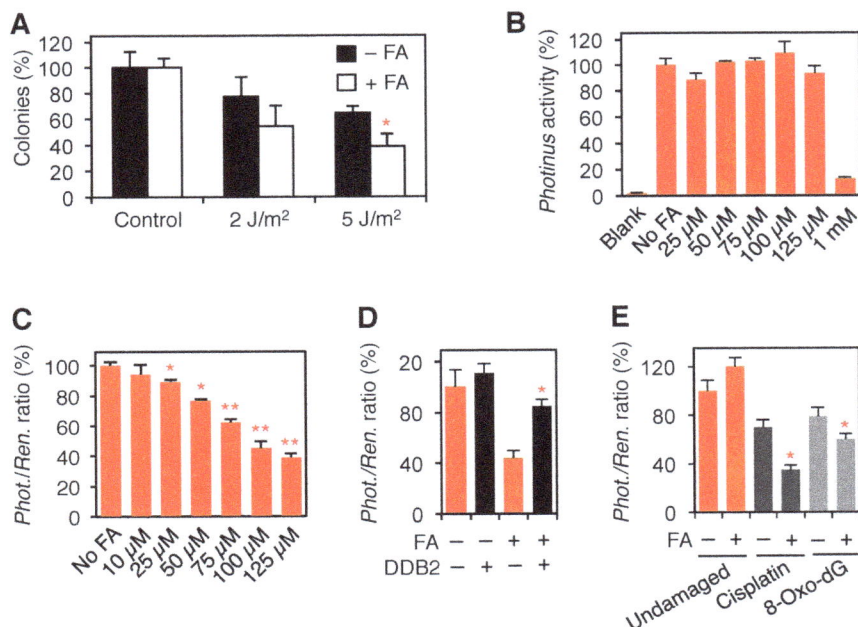

Figure 7. Inhibition of NER and BER activity by low-dose formaldehyde. (**A**) Colony-forming assay demonstrating that human fibroblasts exposed to formaldehyde (75 μM) are more sensitive to killing by UV radiation (2 and 5 J/m²) than the respective untreated controls (error bars, S.D.; n = 3, each measurement in triplicate). The asterisk (*p<0.05) denotes the significantly reduced colony formation ability. (**B**) Expression of *Photinus* luciferase in cells containing undamaged pGL3 and exposed (18 h) to formaldehyde. All values (N = 3; error bars, S.D.) are shown as a percentage of luciferase activity in untreated fibroblasts. Blank, untransfected cells. (**C**) Host-cell reactivation of UV-irradiated pGL3 reflecting NER activity in cells exposed (18 h) to formaldehyde. Values (N = 10; error bars, S.D.) are shown as a percentage of the *Photinus/Renilla* ratio in control cells. Asterisks, significant differences from the control (*p<0.05, **p<0.01). (**D**) Partial restoration of host-cell reactivation in 100-μM formaldehyde-treated cells overexpressing DDB2-EGFP (N = 5). Asterisk, significantly (*p<0.05) higher NER activity then controls without DDB2-EGFP. (**E**) Inhibition of host-cell reactivation of pGL3 containing cisplatin adducts or 8-oxo-dG lesions (N = 5). The asterisks (*p<0.05) denote significantly reduced DNA repair activity in formaldehyde-treated (75 μM, 18 h) cells.

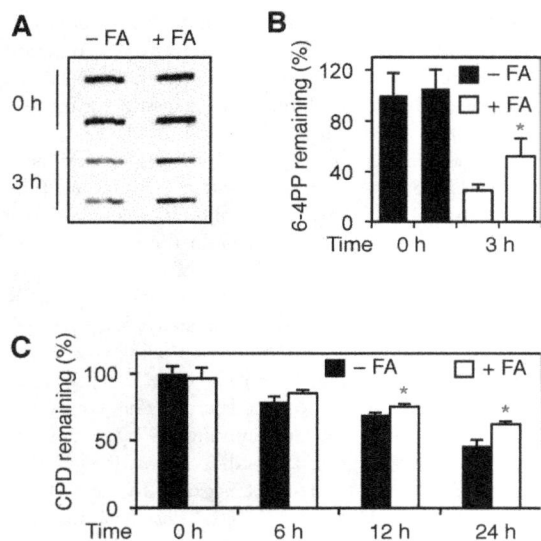

Figure 8. Inhibition of UV lesion removal from chromatin. (A) Untreated or formaldehyde-treated cells (75 μM, 18 h) were exposed to UV light (10 J/m^2) and collected immediately after irradiation or following 3-h repair incubations. Genomic DNA was isolated and analyzed for UV lesions using antibodies against 6-4PPs. **(B)** Quantification of three independent experiments demonstrating that 6-4PP excision is diminished by 75-μM formaldehyde exposure (error bar, S.D.). The asterisk (*p<0.05) denotes significantly reduced excision in formaldehyde-treated cells compared to the untreated control. **(C)** Quantification of three independent experiments demonstrating that CPD excision is also inhibited. The asterisks (*p<0.05) denote the significantly reduced excision in 75-μM formaldehyde-treated cells compared to untreated controls.

Discussion

The human skin is increasingly subjected to UV damage due to growing leisure times, the popularity of outdoor activities, frequent traveling to tropical areas and the use of artificial irradiation devices. Concomitantly, the DNA of the skin or mucous membranes is constantly attacked by exogenous or endogenous genotoxins. If not promptly excised by DNA repair machines, the resulting base lesions give rise to mutations, which in turn may cause cancer. However, DNA repair systems targeting such mutagenic lesions are prone to modulation by skin- or mucous membrane-penetrating xenobiotics. This is the first study that shows a direct effect of low-dose formaldehyde on the assembly of NER complexes. A previous report already suggested that formaldehyde delays DNA repair of UV lesions [23] but this earlier study was based on indirect measurements whereby transiently appearing single-stranded DNA breaks were taken as circumstantial evidence for ongoing excision repair. Another preceding report [22] revealed an inhibitory effect of formaldehyde on unscheduled DNA synthesis, reflecting DNA repair activity, but at considerably higher concentrations of > 100 μM.

One possible scenario to explain our findings is that formaldehyde-induced DPXs are themselves NER substrates that compete with UV lesions for being repaired. The processing of DPXs is not completely understood but recent studies concluded that the NER pathway plays only a marginal role in the removal of DPXs induced by low-dose formaldehyde [41]. Thus, the delayed NER

activity is unlikely to result from direct substrate competition between DPXs and UV lesions. As an alternative mechanism, we hypothesized that the formation of DPXs, mainly histone-DNA crosslinks, may slow down the molecular search for UV lesions by DNA damage sensors. It is believed that site-specific DNA-binding proteins, including DNA repair factors, locate their targets among the vast excess of non-target DNA by facilitated diffusion. This search process reduces the dimensionality of protein movements by "sliding" or "hopping" along DNA filaments [42]. In either case, the presence of covalently trapped histones or other chromatin proteins along the DNA path may interrupt such an effective search mode by facilitated diffusion, thereby restricting the rate by which DNA lesions are detected and channeled into DNA repair. This hypothesis was tested by *in situ* protein dynamics, thus demonstrating that formaldehyde slows down the recognition of DNA damage by UV-DDB and XPC. That this reduced damage recognition efficiency translates to slower repair has been confirmed by two different tests. In a host-cell reactivation assay, NER activity was inhibited by formaldehyde with a 50% reduction at concentrations of 75-100 μM. A NER inhibition was similarly observed by monitoring the removal of 6-4PPs and CPDs from the chromosomal DNA of human cells exposed to 75 μM formaldehyde.

Mechanistically, we found that the observed sequestration of UV-DDB in formaldehyde-damaged chromatin results from transient, non-covalent interactions that delay its movements during the search for DNA damage. Through protein-protein associations, UV-DDB bound to formaldehyde-damaged chromatin also impedes the function of the XPC complex, such that not only the excision of UV lesions but also the repair of chemically induced DNA adducts is diminished. We also found that the mobility of a representative DNA glycosylase is perturbed by low-dose formaldehyde and that the ensuing BER pathway is inhibited. Further dynamic chromatin transactions may be disturbed by a similar mechanism. For example, Rager et al. [9] recently reported that formaldehyde perturbs gene expression in the nasal epithelium.

In conclusion, the compromised DNA repair efficiency, reported in this study, raises the possibility that frequent exposures of the skin or mucous membranes to formaldehyde represents an unexpected additional risk factor for cutaneous cancer in combination with UV radiation or chemical carcinogens. In a broader perspective, the inhibition of DNA excision repair activity by formaldehyde implies that the carcinogenic endpoints of this highly reactive aldehyde may result, to a great extent, from the accumulation of DNA adducts and other mutagenic base lesions induced by constitutively occurring environmental carcinogens or endogenous genotoxic metabolites. In view of this adjuvant effect, we expect that a reduction of formaldehyde exposure in the general population would substantially reduce the overall cancer incidence.

Acknowledgments

We thank Dr. U. Camenisch and Dr. R. Müller for assistance in devising the FRAP assays.

Author Contributions

Conceived and designed the experiments: AL HN. Performed the experiments: FC RM JF. Analyzed the data: AL FC RM. Wrote the paper: AL HN.

References

1. Donaldson MR, Coldiron BM (2011) No end in sight: the skin cancer epidemic continues. Semin Cutan Med Surg 30:3–5.
2. Hoeijmakers JH (2009) DNA damage, aging and cancer. New Engl J Med 361:1475–1485.
3. Friedberg EC, Walker GC, Siede W, Wood RD, Schultz RA, et al. (2006) DNA Repair and Mutagenesis (ASM Press, Washington D.C).
4. Parsons JL, Dianov GL (2013) Co-ordination of base excision repair and genome stability. DNA Rep 12:326–333.
5. Maynard S, Schurmann SH, Harboe C, de Souza-Pinto NC, Bohr VA (2009) Base excision repair of oxidative DNA damage and association with cancer and aging. Carcinogenesis 30:2–10.
6. Flyvholm MA, Anderson P (1993) Identification of formaldehyde releasers and occurrence of formaldehyde and formaldehyde releasers in registered chemical products. Am J Ind Med 24:533–552.
7. Rumchev KB, Spickett JT, Bulsara MK, Phillips R, Stick SM (2002) Domestic exposure to formaldehyde significantly increases the risk of asthma in young children. Eur Respir J 20:403–408.
8. Zhang Y, Liu X, McHale C, Li R, Zhang L, et al. (2013) Bone marrow injury induced by oxidative stress in mice by inhalation exposure to formaldehyde. PLoS ONE 8:e74974.
9. Rager JE, Moeller BC, Doyle-Eisele M, Kracko D, Swenberg JA, et al. (2013) Formaldehyde and epigenetic alterations: microRNA changes in the nasal epithelium of nonhuman primates. Environ Health Perspect 121:339–344.
10. De Groot AC, Veenstra M (2010) Formaldehyde releasers in cosmetics in the USA and in Europe. Contact Dermatitis 62:221–224.
11. O'Connor PM, Fox BW (1989) Isolation and characterization of proteins cross-linked to DNA by the antitumor agent methylene dimethanesulfonate and its hydrolytic product formaldehyde. J Biol Chem 264:6391–6397.
12. Cohen-Hubal EA, Schlosser PM, Connoly RB, Kimbell J.S (1997) Comparison of inhaled formaldehyde dosimetry predictions with DNA-protein cross-link measurements in the rat nasal passages. Toxicol Appl Pharmacol 143:47–55.
13. Kerns WD, Pavkov KL, Donofrio DJ, Gralla EJ, Swenberg JA (1983) Carcinogenicity of formaldehyde in rats and mice after long-term inhalation exposure. Cancer Res 43:4382–4392.
14. Luce D, Gérin M, Leclerc A, Morcet JF, Brugère J, et al. (1993) Sinonasal cancer and occupational exposure to formaldehyde and other substances. Int J Cancer 53:224–331.
15. Vaughan TL, Stewart PA, Teschke K, Lynch CF, Swanson GM, et al. (2000) Occupational exposure to formaldehyde and wood dust and nasopharyngeal carcinoma. Occup Environ Med 57:376–384.
16. Bosetti C, McLaughlin JK, Tarone RE, Pira E, La Vecchia C (2008) Formaldehyde and cancer risk: a quantitative review of cohort studies through 2006. Ann Oncol 19:29–43.
17. National Toxicology Program, 12th Report on Carcinogens. 2011. Formaldehyde.
18. IARC Monographs on the Evaluation of Carcinogenic Risks to Humans (2012) A Review of Human Carcinogens: Chemical Agents and Related Occupations. WHO International Agency for Research on Cancer 100F:401–435.
19. Robbins JD, Norred W, Bathija A, Ulsamer AG (1984) Bioavailability in rabbits of formaldehyde from durable-press textiles. J Toxicol Environ Health 14:453–463.
20. Iverson OH. 1986. Formaldehyde and skin carcinogenesis. Environ Int 12:541–544.
21. Walrath J, Fraumeni JF (1983) Mortality patterns among balsamers. Int J Cancer 31:407–411.
22. Grafstrom RC, Fornace AJ, Autrup H, Lechner JF, Harris CC (1983) Formaldehyde damage to DNA and inhibition of DNA repair in human bronchial cells. Science 220:216–220.
23. Emri G, Schaefer D, Held B, Herbst C, Zieger W, et al. (2004) Low concentrations of formaldehyde induce DNA damage and delay DNA repair after UV irradiation in human skin cells. Exp Dermatol 13:305–315.
24. Scrima A, Konickova R, Czyzewski BK, Kawasaki Y, Jeffrey PD, et al. (2008) Structural basis of UV DNA-damage recognition by the DDB1-DDB2 complex. Cell 135:1213–1223.
25. Fei J, Kaczmarek N, Luch A, Glas A, Carell T, et al. (2011) Regulation of nucleotide excision repair by UV-DDB: prioritization of damage recognition to internucleosomal DNA. PLoS Biol 9:e1001183.
26. Mitchell D (2006) Revisiting the photochemistry of solar UVA in human skin. Proc Natl Acad Sci USA 103:13567–13568.
27. Spivak G, Hanawalt PC (2006) Host cell reactivation of plasmids containing oxidative DNA lesions in Cockayne syndrome but not in UV-sensitive syndrome fibroblasts. DNA Rep 5:13–22.
28. Maillard O, Solyom S, Naegeli H (2007) An aromatic sensor with aversion to damaged strands confers versatility to DNA repair. PLoS Biol 5:e79.
29. Houtsmuller AB, Vermeulen W (2001) Macromolecular dynamics in living cell nuclei revealed by fluorescence redistribution after photobleaching. Histochem Cell Biol 115:13–21.
30. Luijsterburg MS, Goedhart J, Moser J, Kool H, Geverts B, et al. (2007) Dynamic in vivo interaction of DDB2 E3 ubiquitin ligase with UV-damaged DNA is independent of damage recognition protein XPC. J Cell Sci 120:2706–2716.
31. Hoogstraten D, Bergink S, Verbiest VH, Luijsterburg MS, Geverts B, et al. (2008) Versatile DNA damage detection by the global genome nucleotide excision repair protein XPC. J Cell Sci 121:2850–2859.
32. Sprague BL, Pego RL, Stavreva DA, McNally JG (2004) Analysis of binding reactions by fluorescence recovery after photobleaching. Biophys J 86:3473–3495.
33. Payne A, Chu G (1994) Xeroderma pigmentosum group E binding factor recognizes a broad spectrum of DNA damage. Mutat Res 310:89–102.
34. Furuta T, Ueda T, Aune G, Sarasin A, Kraemer KH, et al. (2002) Transcription-coupled nucleotide excision repair as a determinant of cisplatin sensitivity of human cells. Cancer Res 62:4899–4902.
35. Szarvas T, Szatloczky E, Volford J, Trezl L, Tyihak E, et al. (1986) Determination of endogenous formaldehyde level in human blood and urine by dimedone-^{14}C radiometric method. J Radioanal Nucl Chem 106:357–367.
36. Lorenti Garcia C, Mechilli M, Proietti De Santis L, Schinoppi A, Kobos K, et al. (2009) Relationship between DNA lesions, DNA repair and chromosomal damage induced by acetaldehyde. Mutat Res 662:3–9.
37. Reardon JT, Cheng Y, Sancar A (2006) Repair of DNA-protein cross-links in mammalian cells. Cell Cycle 5:1366–1370.
38. Telford DJ, Stewart BW (1989) Micrococcal nuclease: its specificity and use for chromatin analysis. Int J Biochem 21:127–137.
39. Sugasawa K, Ng JM, Masutani C, Iwai S, van der Spek PJ, et al. (1998) Xeroderma pigmentosum group C protein complex is the initiator of global genome nucleotide excision repair. Mol Cell 2:223–232.
40. Hwang BJ, Ford JM, Hanawalt PC, Chu G (1999) Expression of the p48 xeroderma pigmentosum gene is p53-dependent and is involved in global genomic repair. Proc Natl Acad Sci USA 96:424–428.
41. de Graaf B, Clore A, McCullogh AK (2009) Cellular pathways for DNA repair and damage tolerance of formaldehyde-induced DNA-protein crosslinks. DNA Rep 8, 1207–1214.
42. Zharkov DO, Grollman AP (2005) The DNA trackwalkers: principles of lesion search and recognition by DNA glycosylases. Mutat Res 577:24–54.

13

Sporadic Premature Aging in a Japanese Monkey: A Primate Model for Progeria

Takao Oishi[1]*, Hiroo Imai[2], Yasuhiro Go[3,4], Masanori Imamura[2], Hirohisa Hirai[2], Masahiko Takada[1]

1 Systems Neuroscience Section, Primate Research Institute, Kyoto University, Inuyama, Japan, **2** Molecular Biology Section, Primate Research Institute, Kyoto University, Inuyama, Japan, **3** Department of Brain Sciences, Center for Novel Science Initiatives, National Institutes of Natural Sciences, Tokyo, Japan, **4** Department of Developmental Physiology, National Institute for Physiological Sciences, Okazaki, Japan

Abstract

In our institute, we have recently found a child Japanese monkey who is characterized by deep wrinkles of the skin and cataract of bilateral eyes. Numbers of analyses were performed to identify symptoms representing different aspects of aging. In this monkey, the cell cycle of fibroblasts at early passage was significantly extended as compared to a normal control. Moreover, both the appearance of senescent cells and the deficiency in DNA repair were observed. Also, pathological examination showed that this monkey has poikiloderma with superficial telangiectasia, and biochemical assay confirmed that levels of HbA1c and urinary hyaluronan were higher than those of other (child, adult, and aged) monkey groups. Of particular interest was that our MRI analysis revealed expansion of the cerebral sulci and lateral ventricles probably due to shrinkage of the cerebral cortex and the hippocampus. In addition, the conduction velocity of a peripheral sensory but not motor nerve was lower than in adult and child monkeys, and as low as in aged monkeys. However, we could not detect any individual-unique mutations of known genes responsible for major progeroid syndromes. The present results indicate that the monkey suffers from a kind of progeria that is not necessarily typical to human progeroid syndromes.

Editor: Hisao Nishijo, University of Toyama, Japan

Funding: The authors have no support or funding to report.

Competing Interests: The authors have declared that no competing interests exist.

* Email: oishi.takao.5e@kyoto-u.ac.jp

Introduction

Symptoms representing various aspects of aging at an early age are characteristic of progeria in humans. Progeria is a series of very rare genetic syndromes that are related to mutations in specific genes. For example, genes responsible for Werner syndrome (WS), Rothmund-Thomson syndrome (RTS), Bloom syndrome (BS), and Hutchinson-Gilford progeria syndrome (HGPS) are RECQ3 (WRN), RECQL4, RECQ2, and LMNA, respectively [1,2]. As summarized in Table 1, these syndromes are accompanied by common or separate physical abnormalities that can be ascribed to senile changes. An animal model is a powerful tool to investigate possible mechanisms underlying not only premature aging, but also normal aging. However, no useful animal models have so far been available. Although mice lacking the RECQL gene have been reported, their pathological changes are quite limited, probably due to a species difference between rodents and humans [3]. Thus, a primate model for progeria, which is taxonomically akin to humans and has a longer life-span than a rodent model, is needed to promote aging research. Recently, we have found a child Japanese monkey who displayed deep skin wrinkles and bilateral cataract at the age of one year. In the present study, we examined whether this monkey is actually progeroid or not from various viewpoints.

Methods

Animals and ethics statements

All animal experiments were planned and executed in strict concordance with Guide for the Care and Use of Laboratory Animals (8th Ed, 2011, The National Academies, USA) and Guidelines for Care and Use of Nonhuman Primates (Ver. 3, 2010, Primate Research Institute, Kyoto University). The protocol was approved by the Animal Welfare and Animal Care Committee, Primate Research Institute, Kyoto University (Permission No. 2012-115). All monkeys used in this study, weighing less than 10 kg, were kept in indoor individual cages (0.89 m×0.63 m×0.82 m) of the Primate Research Institute on a 12-h on/12-h off lighting schedule. In each cage, the monkeys were given *ad libitum* access to food and water, and a chain-hung wood block as a toy. No monkeys were sacrificed. Monkey N416, a female Japanese monkey (*Macaca fuscata*) who was supposed to be progeroid, was born in an outdoor cage (5 m×5 m×2.7 m, 6–7 monkeys per cage) and moved to an indoor individual cage at the age of one year and four months to avoid progress of cataract. Unfortunately, N416 suddenly died of bloat at the age of three years.

Table 1. Physical signs and symptoms characteristic of human progeroid syndromes and monkey N416.

Signs and symptoms	WS	RTS	HGPS	BS	N416
cataract	✓	✓			✓
abnormal glucose metabolism	✓				✓
urinary hyaluronan	✓		✓		✓
decline in proliferative potency of fibroblasts					
at early passage	✓				✓
increase in senescent fibroblast cells	(✓)	✓	✓	✓	✓
DNA repair deficiency	✓	✓		✓	✓
poikiloderma		✓		✓	✓
short statue, low bodyweight	✓	✓	✓		
facial proportion	✓	✓	✓		
loss of hair	✓	✓	✓		
facial rash		✓		✓	
intractable skin ulcers	✓				
prominent scalp vein			✓		
sun-sensitivity		✓		✓	
skeletal defects	✓	✓	✓		
abnormal dentation			✓		
characteristic voice	✓		✓	✓	
soft-tissue calcification	✓				
contracture of joints			✓		
atherosclerosis	✓		✓		
malignant tumors	✓	✓			
hypogonadism	✓	✓		✓	nt
immunodeficiency		✓		✓	nt
onset after adolescence	✓				
onset within one year of age		✓	✓		✓

BS: Bloom syndrome, HGPS: Hutchinson-Gilford progeria syndrome, nt: not tested, RTS: Rothmund-Thomson syndrome, WS: Werner syndrome, (✓): Controversial [10,11].

Measurements of cell proliferation

Cell proliferation was analyzed using fibroblast cells prepared from cultures of a tiny block of the ear skin as described previously [4]. To determine the duplication time of each cell, cultures were made using a medium of Amnio Max II Complete (Gibco BRL, USA) in a 60×15 mm petri dish for cell culture (FALCON, 353002, USA). After appropriate proliferation of cells in the primary dish culture, cells were harvested, and then cell suspension (25–30 cells/μl) was transferred to a 35×10 mm cell culture dish (Advanced glass bottom, Greiner Bio-one, 627965, Germany). One day later, recordings of the cell division were performed every five minutes with a time-lapse imaging system (Cell Observer, Carl Zeiss, Germany). The time of a cell cycle was figured out by chasing each cell division based on imaging data collected in 70-hr cultures: the time was calculated as an interval time from the first division to the second division of each cell. A total of 44 or 47 cells were analyzed in monkey N416 and a normal control, respectively. To measure the population growth activity, fibroblast cells derived from monkey N416 and normal infant and aged monkeys (all at passage 8) were seeded at 1×10^5 cells/well in gelatinized 6-well plates with DMED containing 15% FBS, 0.1 mM non-essential amino acids, 2 mM L-glutamine, 1 mM sodium pyruvate, 0.11 mM 2-mercaptoethanol, 100 U/ml penicillin, and 100 μg/ml streptomycin. The total cell numbers in three independent wells were counted every 3 days up to day 12.

Detection of senescent cells

Senescent cells were detected with ß-galactosidase assay using fibroblast cells prepared from monkey N416, a normal infant and a normal aged monkey. The number of passage of fibroblast cells derived from monkey N416 and the infant and the aged controls was 5, 7, and 5, respectively. Senescence-associated ß-galactosidase activity was detected by using Cellular Senescence Assay Kit (CBA-230, Cell Biolabs, USA). Cells were fixed and incubated with X-gal overnight. Five images were randomly captured (BZ-9000, Keyence, Japan) from each culture dish of the three monkeys, and the numbers of stained and unstained cells were counted.

Quantification of DNA repair

To test DNA repair deficiency in monkey N416, we quantified DNA damage by counting apurinic/apyrimidinic sites (AP sites), which are the target of base excision repair. DNA was freshly prepared with DNeasy Blood & Tissue Kit (69504, Qiagen, USA) from fibroblast cells of the three monkeys used for the senescent cell detection. Fibroblast cells from the infant control had been divided into normal and UV-irradiated (1 minute) groups. The AP sites were labeled with DNA Damage Quantification Kit (DK02, Dojindo, Japan) and then quantified (Flex Station 3, Molecular Devices, USA).

Pathological analyses

A small piece of skin tissue was taken from the upper arm of monkey N416 at her health check (1 year old), and histological specimens were prepared with hematoxylin-eosin (HE) staining. Immediately after her sudden death, the heart, lung, stomach, intestine, liver, kidney, and spleen were dissected and HE specimens were prepared.

X-ray CT and MRI

Images of the head and extremities of monkey N416 and a control monkey of the same age (2 years old) were obtained with Asteion TSX-021B (Toshiba Medical Systems). The monkeys were anesthetized with ketamine hydrochloride (5 mg/kg, i.m.) and medetomidine (100 µg/kg, i.m.), followed by head fixation with a stereotaxic apparatus [5]. Three-dimensional reconstruction was done with VirtualPlace (Aze).

T1-weighed MRIs of monkey N416, a child monkey of the same age (2 years old), and an aged monkey (28 years old) were taken with a 0.3 T apparatus (Vento, Hitachi). The monkeys were anesthetized with ketamine hydrochloride (5 mg/kg b.wt., i.m.) and medetomidine (100 µg/kg b.wt., i.m.), followed by head positioning with the stereotaxic apparatus [5].

Nerve conduction velocity

To assess the functional viability of the peripheral nervous system, the conduction velocity of sensory and motor nerves was measured from the ulnar nerve of monkey N416, in comparison with those in control Japanese monkeys including 5 child monkeys (Child; 2 years old), 5 adult monkeys (Adult; 5–8 years old), and 5 aged monkeys (Aged; 25–29 years old). Measurements were done with NeuroPak (Nihon Kohden, Tokyo). The monkeys were sedated with ketamine hydrochloride (2.5 mg/kg) and medetomidine (100 µg/kg), and anaesthetized with 2% sevoflurane during the measurement. The room temperature was kept at 20°C. To measure the sensory conduction velocity (SCV), pairs of recording surface electrodes were placed over the ulnar nerve 2–3 cm proximal to the elbow and 2–3 cm proximal to the wrist, and supramaximal stimuli were delivered through a pair of ring electrodes attached to the distal and proximal interphalangeal joints of the little finger. To measure the motor conduction velocity (MCV), recording surface electrodes were placed over the abductor digiti minimi and the ulnar nerve was stimulated 2–3 cm proximal to the elbow and 2–3 cm proximal to the wrist, with supramaximal amplitude.

Measurements of blood and urinary biomarkers

Blood and urine samples were obtained from monkey N416 (2 years and 6 months old), 5 child monkeys (age-matched), 5 adult monkeys (6–9 years old), and 5 aged monkeys (21–27 years old). Measurements of hemoglobin A1c (HbA1c; blood), hyaluronan (urine and serum), glucose (urine and serum), low-density lipoproteins (LDL; serum), high-density lipoproteins (HDL; serum), and triglyceride (TG; serum) were carried out by a commercial test company (Falco Biosystems).

Statistics

Student's t-test was applied for cell proliferation time comparison. One-way ANOVA and *post hoc* test (Bonferroni) were applied for analyses of the ratio of senescent cells, the number of AP sites, the levels of biomarkers, and the conduction velocity of peripheral nerves (Kaleidagraph 4.1, Hulinks, Japan). Linear discriminant analysis was also applied for the peripheral nerve conduction velocity.

DNA sequence

The sequences of candidate genes for progeroid syndromes, RECQ3 (Werner syndrome), RECQL4 (Rothmund-Thomson syndrome), RECQ2 (Bloom syndrome), and LMNA (Hutchinson-Gilford progeria syndrome), were determined by standard PCR and the direct sequence protocol with ExTaq DNA polymerase (Takara Bio Inc., Shiga, Japan) and BigDye Terminator v3.1 Cycle Sequencing Kit with a 3130 Genetic Analyzer (Applied Biosystems, CA, USA). Primers used in PCR are shown in Table S1. Furthermore, to identify possible genetic causes specifically occurring in monkey N416, broader candidate gene approaches were performed in 100 macaque monkeys including monkey N416 with Next Generation Sequencer (NGS) of HiSeq2000 (Illumina Inc., CA, USA). The candidates for progeroid-related genes were listed in Table S2. Genomic DNA samples were obtained from 100 macaque monkey blood samples including monkey N416. Each library for the NGS analysis was constructed by using the standard Illumina TruSeq DNA Library Prep protocol.

Results

Physical appearance

Two major appearances characteristic of monkey N416 were deep wrinkles of the skin and cataract on bilateral sides. The skin wrinkles have been eminent since she was a baby, while such wrinkles were not observed in her mother (Fig. 1A). As monkey N416 grew up, her wrinkles seemed to gradually decrease but still remained at the age of two years (Fig. 1B). Indeed, X-ray CT clearly revealed the wrinkled skin of this monkey (Fig. 1C; imaged later than the time when Fig. 1B was taken) and, as a control, the smooth skin of an age-matched normal monkey (Fig. 1D). There was no sign of ulceration of the skin, and hyperkeratosis was slightly observed in the extremities. Cataract occurred in one eye as early as monkey N416 was 9 months old, and progressed bilaterally as she grew around two years old (Fig. 1B). Based on the fact that this monkey usually reached for primate pellets in a box attached to her own cage and smelled them before eating, it appeared that her vision was still left but was only weak. When we examined her under anesthesia, we could not find out either rigidity or hyperflexibility of the limbs. No skeletal anomaly, osteoporosis, or dental crowding was found in monkey N416.

Fibroblast cell cycle

Our time-lapse video analysis at passage 5 confirmed that the time for fibroblast cell cycle was significantly longer in the model monkey N416 than in a normal control (Fig. 2A; N416 28.8 ± 6.8 hr, control 20.0 ± 4.6 hr, $t = -7.17$, $p < 0.01$). Population growth of fibroblast cells at passage 8 from the aged monkey reached a plateau at day 6. On the other hand, population growth of fibroblast cells from monkey N416 and the infant monkey continued up to day 12 (Fig. 2B).

Senescent cells

Cytochemical assay of ß-galactosidase activity revealed that the ratio of senescent (ß-galactosidase-positive) fibroblast cells was higher in the aged monkey than in the infant monkey. Their ratio in monkey N416 was significantly lower than in the aged monkey and higher than in the infant monkey (Fig. 2C; aged $19.6 \pm 2.8\%$,

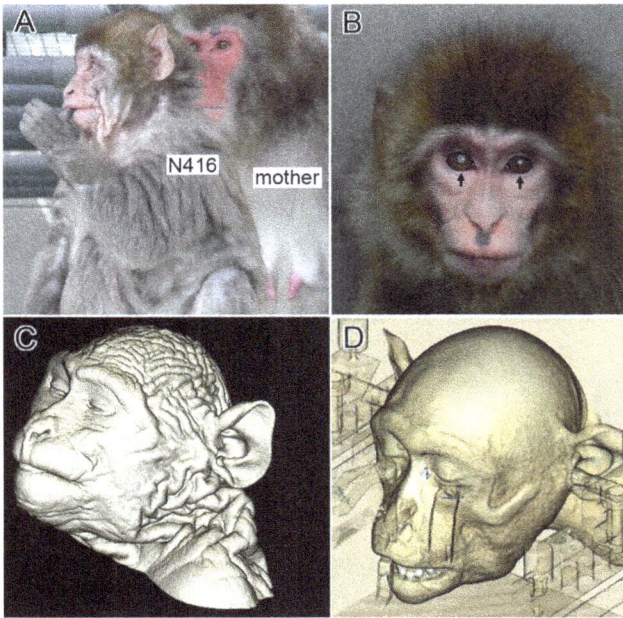

Figure 1. Monkey N416 resembled aged monkeys in appearance. A: Monkey N416 (10 months old) and her mother. Note that deep wrinkles of the skin are seen in monkey N416. B: Monkey N416 at the age of 1 year and 10 months. Cataract was first recognized in one eye when she was 10 months old and, progressed bilaterally within a few months (pointed to by arrows). The deep skin wrinkles become rather reduced as the body grows. C: Three-dimensionally-reconstructed CT image of monkey N416 (2 years and 5 months old). Note that the deep skin wrinkles clearly remain in both the scalp and the postcranial skin. D: Three-dimensionally-reconstructed CT image of an age-matched control. In this monkey, the skin is smooth. Semi-transparent objects are parts of the stereotaxic apparatus.

infant 7.0±2.2%, N416 14.5±1.5%, F(2,12)=41.9, p<0.001, one-way ANOVA; N416 vs. aged, N416 vs. infant, p<0.01, post hoc test).

DNA repair

Deficiency of DNA repair was estimated through AP site quantification of DNA extracted from fibroblast cells. One-way ANOVA and host hoc test revealed that numbers of AP sites were higher in cells from monkey N416 and UV-irradiated cells from the infant monkey than in cells from the aged and infant monkeys (Fig. 2D; aged 12.2±1.2, infant 14.6±1.7, UV-irradiated infant 22.3±1.4, N416 19.4±1.2, F(3,8)=33.2, p<0.001, one-way ANOVA; N416 vs. aged, UV-irradiated infant vs. aged, UV-irradiated infant vs. infant, p<0.01, N416 vs. infant, p<0.05, post hoc test).

Blood and urinary biomarkers

We measured levels of blood and urinary biomarkers in monkey N416 and compared these values with those in aged, adult, and child monkeys. HbA1c in monkey N416 was not different from those in the aged and child groups (Fig. 3A), but was significantly higher than that in the adult group (N416 4.8%, aged 4.1–4.5%, adult 3.8–4.2%, child 4.0–4.5%, F(3, 12)=7.642, p<0.01, one-way ANOVA; N416 vs. adult, p<0.01, post hoc test). Hyaluronan in urine in monkey N416 was not different from those in the aged and adult groups (Fig. 3B), but was significantly higher than that in

the child group (N416 590 µg/g creatinine, aged 148–582 µg/g, adult 72–333 µg/g, child 31–207 µg/g, F(3, 12)=5.938, p<0.05, one-way ANOVA; N416 vs. child, p<0.05, post hoc test). There was no significant difference in levels of hyaluronan in serum, glucose in serum and urine, LDL, HDL, or TG between monkey N416 and the other monkey groups (see Table S3).

Pathological features

Pathological examination of the skin showed slight acanthosis and slight hypertrophy in epidermis, atrophy in dermis, and normal feature in subcutis (Fig. 3C). Angiectasia in the superficial vascular plexus of dermis, and slight infiltration of monocular cells and eosinophils in the superficial perivascular region of dermis were also observed (Fig. 3D).

According to autopsy, slight hypertrophy of the heart was found, but no signs of atherosclerosis or malignant tumors were seen. Chronic pyelonephritis in the kidney was observed as another pathological change, except for congestion in the lung, the liver, and the stomach that was caused by bloat.

Brain structure

Shrinkage of the cerebral cortex and the hippocampus was markedly observed in monkey N416, although the size of her brain was not different from that of the control child monkey brain. Consequently, the cerebral sulci and lateral ventricles were largely expanded (Fig. 4A,B), as compared to an age-matched child monkey (Fig. 4C). The same feature was seen in an aged monkey (Fig. 4D).

Conduction velocity of peripheral nerves

The conduction velocity of peripheral sensory and motor nerves was measured from the ulnar nerve. The sensory conduction velocity (SCV) in monkey N416 was compared with that in aged, adult, and child monkeys; each value was 55.2, 53.4±5.7 (n=4), 78.5±3.5 (n=3), and 70.4±2.8 (n=4) ms, respectively (Fig. 5A). It was revealed that the SCV in monkey N416 and the aged group was significantly slower than in the adult and child groups (one-way ANOVA and post hoc test). Likewise, the motor conduction velocity (MCV) in monkey N416 was examined in comparison with that in the aged, adult, and child groups; each value was 62.1, 61.2±4.5 (n=5), 63.2±4.3 (n=5), and 57.0±13.8 (n=5) ms, respectively (Fig. 5B). Thus, the MCV in monkey N416 was not significantly different from that in the aged, adult, or child monkey group (one-way ANOVA). Scattergram of SCV and MCV shows that data points for the aged, adult, and child monkeys appeared to constitute three separate clusters, and the data point for monkey N416 was included within the cluster for the aged group (Fig. 5C).

DNA sequences of progeroid-related genes

The sequences of progeroid-related genes, RECQ3, RECQL4, RECQ2, and LMNA, were determined in monkey N416 and normal Japanese monkeys (n = 7, 41, 2, and 4, respectively) and a normal rhesus monkey, as referenced with those of rhesus monkey genomic sequence (rheMac2). Though there were five non-synonymous point mutations in recQL4 gene of monkey N416 with reference to rhesus sequences (exon3 183A>G, 387C>T, exon4 157G>A, 208T>C, 315C>T), such mutations were shared by 14 normal monkeys. Thus, these mutations were not symptom-specific. The sequences of RECQ3, RECQ2 and LMNA gene in monkey N416 were identical to that of rhesus monkey.

To further extend the DNA sequencing analysis, we adapted a large scale of DNA sequencing approaches using a next generation

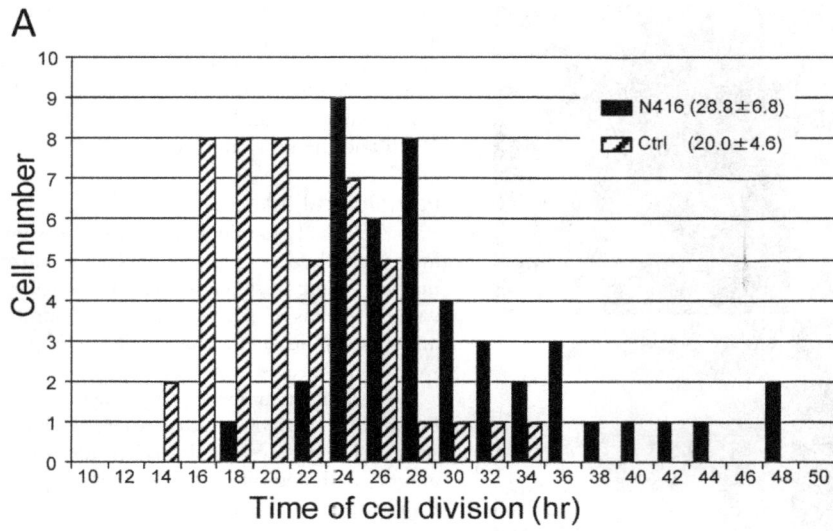

A

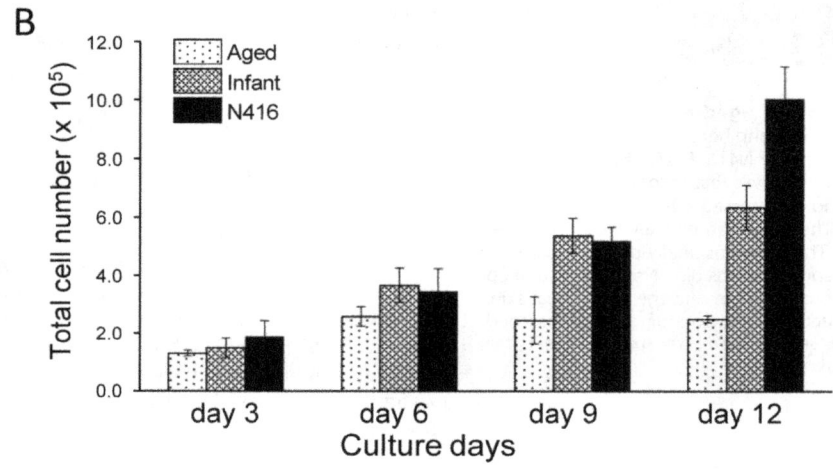

B

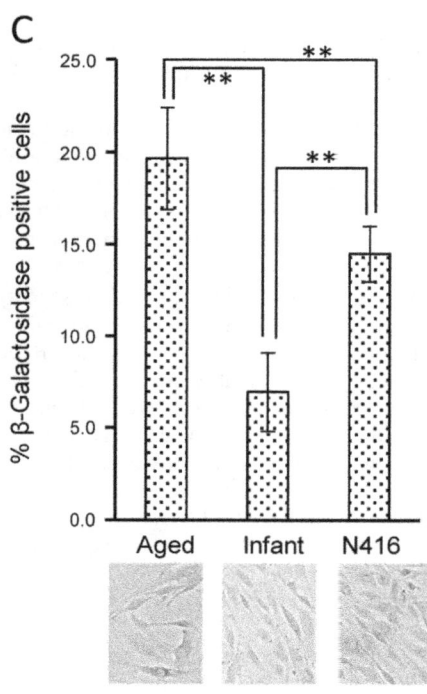

C

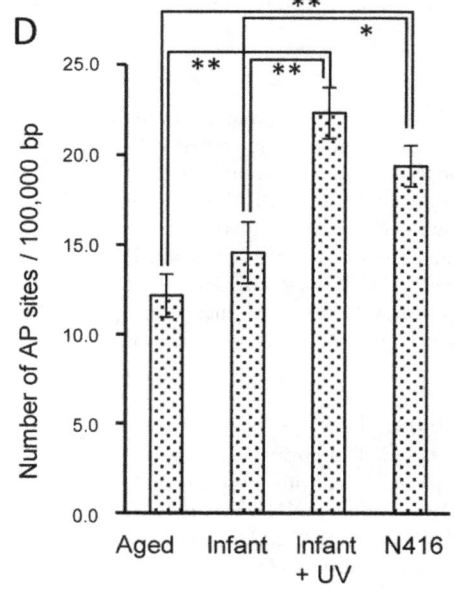

D

Figure 2. Fibroblast cells derived from monkey N416 resembled those derived from progeroid patients in cell biological features. A: Histograms of cell division intervals of fibroblasts derived from monkey N416 and a control monkey (Ctrl) at passage 5. Note that the time of cell division is significantly longer in monkey N416 (28.8 ± 6.6 hr; n = 44) than in the control (20.0 ± 4.6 hr, n = 47). B: Population growth of fibroblast cells derived from a normal aged monkey, a normal infant monkey, and monkey N416 at passage 8. C: Ratio of senescent-associated ß-galactosidase-positive fibroblast cells derived from monkey N416, in comparison with those derived from aged and infant monkeys (**: $p < 0.01$). Photographs are representative images. D: Number of apurinic/apyrimidinic (AP) sites per 100,000 bp in fibroblast cells derived from monkey N416, in comparison with those from aged and infant monkeys and, also, UV-irradiated cells from the infant monkey (*: $p < 0.05$, **: $p < 0.01$).

sequencer (NGS). Specifically, 21 candidates of progeroid-related genes were sequenced in 100 monkeys including monkey N416 (see Table S2). The results showed that two genes (NBN and DDB2) have N416-specific mutations out of the 100 monkeys. However, these mutations are synonymous substitutions in NBN genes and downstream mutations in DDB2 genes, respectively, suggesting that an impact of such mutations on phenotypic functions is thought to be very weak.

Discussion

In our institute, we found a child Japanese monkey (monkey N416) who exhibited deep skin wrinkles and bilateral cataract within one year after birth. The following abnormalities were also found in this monkey: retardation of fibroblast proliferation at passage 5, appearance of senescent cells, deficiency in DNA repair, impaired glucose metabolism, altered hyaluronan metabolism, poikiloderma, shrinkage of the cerebral cortex and the hippocampus, and decrease in the conduction velocity of peripheral sensory nerves. However, the growth of this monkey was not retarded

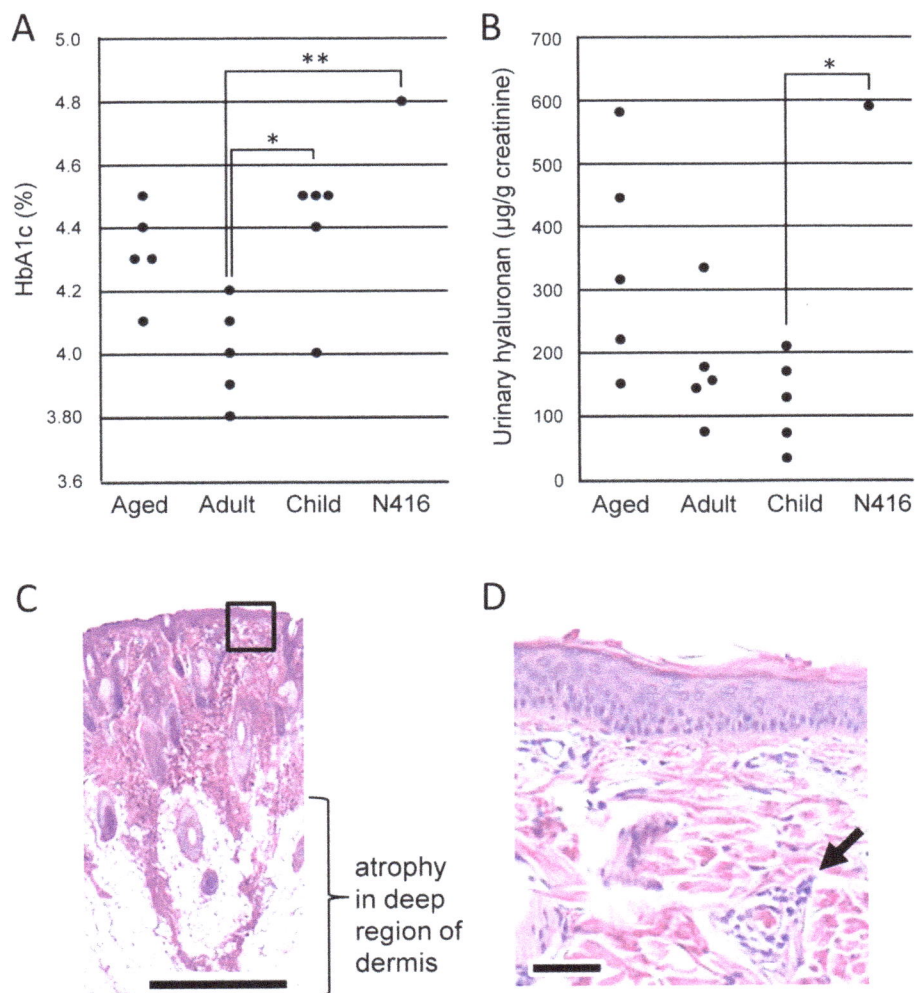

Figure 3. Monkey N416 resembled progeroid patients in metabolic and dermatological features. A,B: Levels of blood HbA1c (A) and urinary hyaluronan (B) in monkey N416 and the aged, adult, and child monkey groups. A significant difference between monkey N416 and the adult monkey group or between monkey N416 and the child monkey group is indicated (*: $p < 0.05$, **: $p < 0.01$). C,D: Poikiloderma with superficial telangiectasia of the skin in monkey N416. Atrophy in dermis is observed in C, while slight acanthosis in epidermis and expanded capillary in the superficial region of dermis (arrow) are seen in D (specified by the box in C). Bars, 1 mm in C and 100 μm in D.

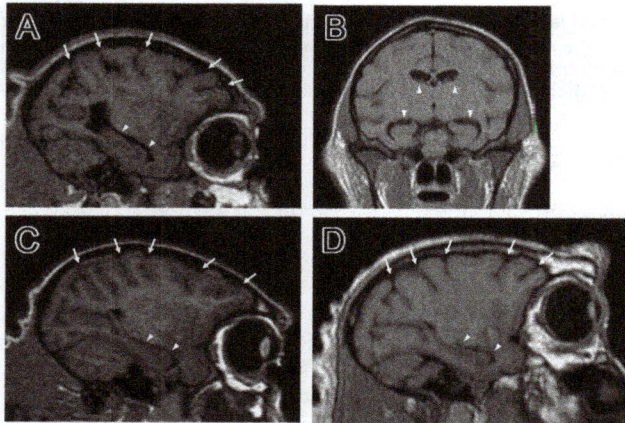

Figure 4. Shrinkage of the cerebral cortex and the hippocampus in monkey N416 was revealed with MRI. A,C,D: Parasagittal images of monkey N416 (1 year and 5 months old), an age-matched control, and an aged monkey (28 years old), respectively. Note that expansion of the cerebral sulci (arrows) in monkey N416 is as large as and expansion of the lateral ventricles (arrowheads) is larger than those in the aged monkey. This is probably ascribed to shrinkage of the cerebral cortex and the hippocampus. B: Coronal image of monkey N416. Seen is prominent expansion of the lateral ventricles (arrowheads).

before she was 3 years old. These findings suggest that monkey N416 may probably suffer from a certain type of progeria.

Cataract

Bilateral cataract in monkey N416 seemed severe. In an individual cage, this monkey frequently smelled food pellets. As normal monkeys do not smell ordinary food so frequently, such a behavior may result from weak vision caused by cataract.

Retardation of fibroblast proliferation

Retardation of fibroblast proliferation is characteristic of WS and HGPS [6,7]. Our cell-culture examination showed a prolonged duration for fibroblast proliferation at passage 5 in monkey N416 as compared to a normal control, at a single cell level. This suggests that this monkey is likely progeroid. At passages 8, on the other hand, the population growth activity in fibroblast cells derived from monkey N416 was as high as that from the infant monkey, and higher than that from the aged monkey. This discrepancy can be accounted for by postulating that the difference in the manner of fibroblast proliferation in these experiments may be attributed to the difference in the passage number: passage 5 cells used for single-cell level analysis vs. passage 8 cells used for population level analysis. To address this issue, further investigations, for example flow cytometry of cell cycle constitution in early and late cultures, are needed.

Appearance of senescent cells

Acidic ß-galactosidase is known to be a good hallmark of senescent cells [8]. Fibroblasts from HGPS patients express strong ß-galactosidase activity [9]. Though it was previously reported that oxydative stress did not increase senescence-associated ß-galactosidase activity in fibroblasts with WS [10], knockdown of RECQ2, RECQ3, and RECQL4 resulted in an increase in ß-galactosidase-positive fibroblasts [11]. Our cytochemical analysis showed that the ratio of ß-galactosidase-positive fibroblasts in monkey N416 was lower than in the aged monkey, and higher than in the infant

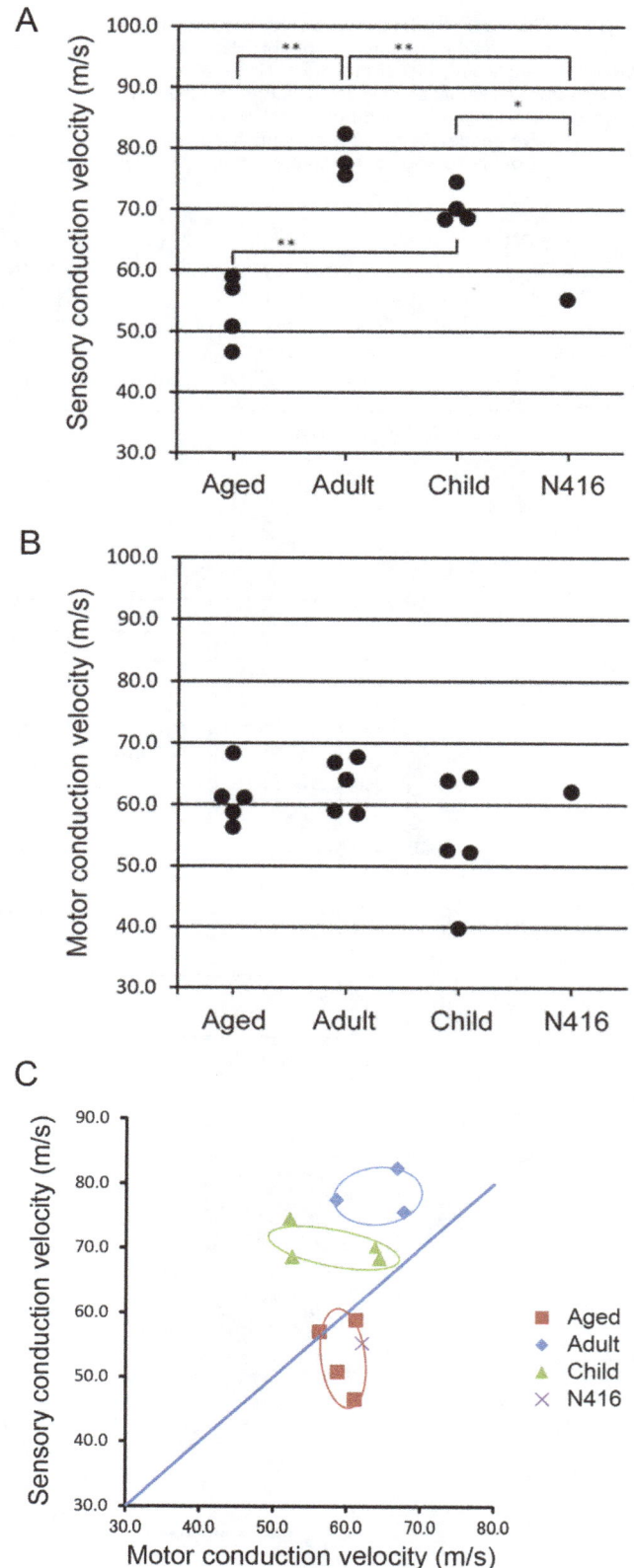

Figure 5. Conduction velocity of peripheral sensory but not motor nerve was slower in monkey N416. Measured from the ulnar nerve. A: Sensory conduction velocity (SCV). The SCV in monkey N416 and aged monkeys was slower than that in adult and child monkeys. **: $p < 0.01$, *: $p < 0.05$ (one-way ANOVA and Bonferroni *post*

hoc test). B: Motor conduction velocity (MCV). The MCV in monkey N416 was similar to that in aged, adult, and child monkeys. C: SCV versus MCV in individual monkeys. Ovals indicate 1.5 standard deviation limits. Note that each age group makes a cluster, and that the position for monkey N416 is located within the cluster for the aged monkey group.

monkey. Considering the difference in the number of passage of fibroblasts (N416 and aged 5, infant 7) and the age (infant 0 year, N416 3 years, aged 21 years), the ratio of senescent fibroblast cells in monkey N416 could be reasonable as a result of aging.

Deficiency in DNA repair

DNA repair deficiency is a common feature of HGPS [12] and other progeroid syndromes. RECQ3, RECQL4, and RECQ2 encode helicases, involved in DNA repair [1]. Our AP site assay have revealed that DNA repair does not simply deteriorate with the age, and that clear deficiency in DNA repair is observed in monkey N416. These findings strongly support that monkey N416 may suffer from a type of progeria.

Impaired glucose metabolism

It is well known that diabetes mellitus (DM) develops in most patients with WS [13]. Thus, we examined the stable DM marker, HbA1c. Levels of HbA1c in aged, adult, and child monkeys were as high as that in healthy rhesus (<4.7) [14] and long-tailed (4.4±0.1) [15] monkeys. However, HbA1c in monkey N416 was 4.8, estimated as "glucose impaired", and significantly higher than in adult monkeys. This suggests that monkey N416 exhibits glucose metabolism impairment in accordance with WS patients, though she has not developed DM yet.

Altered hyaluronan metabolism

Hyaluronan is a key molecule to shape skin characteristics. For example, the Shar-pei dog and human patients with similar abnormal deep skin wrinkles share impairment in hyaluronan metabolism. Skin wrinkles in the Shar-pei dog results from upregulation of hyaluronic acid synthase 2 [16]. Deep skin wrinkles seen in human patients are correlated with an elevated hyaluronan concentration in serum [17]. Also, the hyaluronan level in WS is relatively high in urine [18]. The urinary hyaluronan level in monkey N416 was significantly higher than in child monkeys and was located between human WS patients and healthy controls [18]. The urinary hyaluronan level increased with the age in Japanese monkeys like humans, though the level itself was lower [18].

Poikiloderma and other chronic pathological changes

Based on the present pathological findings on the skin (slight acanthosis in epidermis, atrophy in dermis, and normal feature in subcutis, angiectasia in the superficial vascular plexus of dermis, and slight cellular infiltration of monocular cell and eosinophils in the superficial perivascular region of dermis), it can be considered that monkey N416 has poikiloderma with superficial telangiectasia. As poikiloderma is a representative feature of RTS, and BS [1,19], this favors that such pathological features may be ascribable to progeria.

In addition, there were chronic pathological changes in monkey N416, such as slight cardiac hypertrophy and renal pyelonephritis. It should be emphasized here, however, that these are quite rare in normal child monkeys.

Nervous systems

Expansion of the cerebral sulci and lateral ventricles, due to shrinkage of the cerebral cortex and the hippocampus, is a prominent aging-related change in the central nervous system of humans and rhesus monkeys [20]. In the present study, our MRI analysis indicated that the same event occurred in monkey N416 as well as in an aged control, but not in a child control. It remains to be known whether there are any histological changes (i.e., neuronal degeneration/loss) or molecular events that may be caused by aging.

The conduction velocity of peripheral sensory and motor nerves decreases when demyelination progresses [21]. In our experiments, we analyzed the conduction velocity of the ulnar nerve in monkey N416 by comparing with those in aged, adult, and child monkeys. The sensory conduction velocity (SCV) was significantly lower in monkey N416 and the aged monkey group than in the adult and child monkey groups. The SCV in adult and child monkeys was similar to that reported in the baboon [22]. Thus, it is most likely that demyelination of the sensory nerve may occur not only in the aged monkeys, but also in monkey N416. By contrast, the motor conduction velocity (MCV) in monkey N416 was not different from that in aged, adult, or child monkeys. The MCV measured in our study was similar to that reported in rhesus and long-tailed monkeys [23,24]. Further investigations are needed to clarify whether demyelination of the sensory nerve may occur in monkey N416 and whether the discrepancy between the SCV and the MCV may be ascribed to demyelination.

Progeroid-related genes

No individual-unique mutations of known progeroid-related genes were found in monkey N416. Some point mutations with reference to rhesus sequences were shared with normal monkeys. Therefore, it can be considered that phenotypes of monkey N416 are not simply attributed to such point mutations. To survey candidate genes responsible for progeroid phenotypes observed in monkey N416, further intensive analyses of the whole genome, at least at the exome level, are indispensable.

Comparison with major progeroid syndromes

As described above, monkey N416 displayed eminent symptoms of premature aging, such as bilateral cataract, retardation of fibroblast proliferation, appearance of senescent cells, deficiency in DNA repair, impaired glucose metabolism, altered hyaluronan metabolism, and poikiloderma. However, we could not detect any individual-unique mutations of known genes responsible for major progeroid syndromes. Typical facial features of HGPS include a small jaw, loss of hair, and scalp veins [2], none of which were observed in monkey N416. Moreover, this monkey did not show growth impairment, sclerotic skin, or joint contractures, each of which is characteristic of HGPS [25]. RTS is characterized by juvenile cataract, skeletal abnormality, proportionate short stature, and poikiloderma [1]. Juvenile cataract and poikiloderma were the common feature in both monkey N416 and RTS. WS is the most popular progeroid syndrome, though its frequency is less than 1/ 100,000 [26]. Some symptoms in monkey N416 were consistent with those in WS, including bilateral cataract, increased urinary hyaluronan, increased HbA1c, and retarded cell proliferation [18]. However, many symptoms seen in WS emerge only after adolescence [13], whereas several symptoms appear as early as 1 year old in monkey N416. In addition, there was no sign of osteoporosis, ulceration or atrophy of the skin, or calcification of the Achilles tendon [27], suggesting that monkey N416 does not suffer from WS. Progeroid phenotypes manifested in monkey

N416 are similar to in some aspects, but rather different from those in such major progeroid syndromes in humans (see Table 1).

In conclusion, the present study is the first to report a naturally occurring primate model with sporadic symptoms of premature aging. Further investigations especially with genome analyses and cell cultures will help us understand normal aging as well as progeria in primates including humans.

Supporting Information

Table S1   Sequence of primers used in PCR.

Table S2   Candidates for progeroid-related genes. There were N416 specific mutations in two progeria-related genes (NBN and DDB2). In the NBN, heterozygous synonymous mutation was found at exon 4, positioning of chr8:92,628,839 (coordinated by rheMac2), and in the DDB2, homozygous mutation was found at the downstream region, positioning of chr14:24,879,090 (also coordinated by rheMac2).

Table S3   Levels of biomarkers in serum and urine, which showed no difference among N416 and age groups.

Acknowledgments

We are grateful to Mr. Y. Kamanaka, Ms. M. Morimoto, Ms. N. Suda-Hashimoto, and Mr. A. Yoshida for their excellent animal care and help for tests. We also thank V.D. A. Kaneko and V.D. A. Watanabe for their veterinary care, Ms. Y. Hirai, Ms. M. Hakukawa, V.D. A. Hirata, and Dr. K. Inoue for their special technical assistance. The monkey N416 was inspected by courtesy of National BioResource Project "Japanese Monkeys". We appreciate School of Medicine, Nihon University for permitting us to use Neuropak.

Author Contributions

Conceived and designed the experiments: TO HI YG HH MT. Performed the experiments: TO HI YG MI HH. Analyzed the data: TO HI YG MI HH. Contributed reagents/materials/analysis tools: TO HI YG MI HH. Wrote the paper: TO HI YG HH MT.

References

1. Mohaghegh P, Hickson ID (2002) Premature aging in RecQ helicase-deficient human syndromes. Int J Biochem Cell Biol 34: 1496–1501.

2. Pollex RL, Hegele RA (2004) Hutchinson-Gilford progeria syndrome. Clin Genet 66: 375–381.

3. Hoki Y, Araki R, Fujimori A, Ohhata T, Koseki H, et al. (2003) Growth retardation and skin abnormalities of the Recql4-deficient mouse. Hum Mol Genet 12: 2293–2299.

4. Hirai H, Hirai Y, Kawamoto Y, Endo H, Kimura J, et al. (2002) Cytogenetic differentiation of two sympatric tree shrew taxa found in the southern part of the Isthmus of Kra. Chrom Res 9: 313–327

5. Nishimura T, Oishi T, Suzuki J, Matsuda K, Takahashi T (2008) Development of the supralaryngeal vocal tract in Japanese macaques: implications for the evolution of the descent of the larynx. Am J Phys Anthropol 135: 182–194.

6. Epstein C, Martin G, Motulsky A (1965) Werner's syndrome; caricature of aging. A genetic model for the study of degenerative diseases. Trans Assoc Am Physicians 78: 73–81.

7. Danes BS (1971) Progeria: a cell culture study on aging. J Clin Invest 50: 2000–2003.

8. Dimri GP, Lee X, Basile G, Acosta M, Scott G, et al. (1995) A biomarker that identifies senescent human cells in culture and in aging skin *in vivo*. Proc Natl Acad Sci U S A 92: 9363–7.

9. Gordon LB, Cao K, Collins FS (2012) Progeria: translational insights from cell biology. J Cell Biol 199: 9–13.

10. de Magalhães JP, Migeot V, Mainfroid V, de Longueville F, Remacle J, et al. (2004) No increase in senescence-associated beta-galactosidase activity in Werner syndrome fibroblasts after exposure to H2O2. Ann N Y Acad Sci 1019: 375–8.

11. Lu H, Fang EF, Sykora P, Kulikowicz T, Zhang Y, et al. (2014) Senescence induced by RECQL4 dysfunction contributes to Rothmund-Thomson syndrome features in mice. Cell Death Dis 5:e1226.

12. Epstein J, Williams JR, Little JB (1973) Deficient DNA repair in human progeroid cells. Proc Natl Acad Sci U S A 70: 977–81.

13. Goto M (1997) Hierarchical deterioration of body systems in Werner's syndrome: Implications for normal ageing. Mechanisms of Ageing and Development 98: 239–254.

14. McTighe MS, Hansen BC, Ely JJ, Lee DR (2011) Determination of hemoglobin A1c and fasting blood glucose reference intervals in captive chimpanzees (Pan troglodytes). J Am Assoc Lab Anim Sci 50: 165–170.

15. Marigliano M, Casu A, Bertera S, Trucco M, Bottino R (2011) Hemoglobin A1C Percentage in Nonhuman Primates: A Useful Tool to Monitor Diabetes before and after Porcine Pancreatic Islet Xenotransplantation. J Transplant 2011: 965605.

16. Olsson M, Meadows J, Truvé K, Rosengren Pielberg G, Puppo F, et al. (2011) A novel unstable duplication upstream of HAS2 predisposes to a breed-defining skin phenotype and a periodic fever syndrome in Chinese Shar-Pei dogs. PLos Genetics 7: e1001332.

17. Ramsden CA, Bankier A, Brown TJ, Cowen PS, Frost GI, et al. (2000) A new disorder of hyaluronan metabolism associated with generalized folding and thickening of the skin. J Pediatr 136: 62–68.

18. Tanabe M, Goto M (2001) Elevation of serum hyaluronan level in Werner's syndrome. Gerontology 47: 77–81.

19. Arora H, Chacon AH, Choudhary S, McLeod MP, Meshkov L, et al. (2014) Bloom syndrome. Int J Dermatol 53: 798–802.

20. Peters A, Rosene DL (2003) In aging, is it gray or white? J Comp Neurol 462: 139–143.

21. McDonald WI (1963) The effects of experimental demyelination on conduction in peripheral nerve: a histological and electrophysiological study. II. Electrophysiological observations. Brain 86: 501–524.

22. Hopkins AP, Gilliatt RW (1971) Motor and sensory nerve conduction velocity in the baboon: normal values and changes during acrylamide neuropathy. J Neurol Neurosurg Psychiatry 34: 415–426.

23. Purser DA, Berrill KR, Majeed SK (1983) Effects of lead exposure on peripheral nerve in the cynomolgus monkey. Br J Ind Med 40: 402–412.

24. Weimer MB, Gutierrez A, Baskin GB, Borda JT, Veazey RS, et al. (2005) Serial electrophysiologic studies in rhesus monkeys with Krabbe disease. Muscle Nerve 32: 185–190.

25. Merideth MA, Gordon LB, Clauss S, Sachdev V, Smith AC, et al. (2008) Phenotype and course of Hutchinson-Gilford progeria syndrome. N Engl J Med 358: 592–604.

26. Goto M (2004) Clinical Aspects of Werner's Syndrome: Its Natural History and the Genetics of the Disease. In: M L, editor. Molecular Mechanisms of Werner's Syndrome. New York: Kluver Academic Plenum Publishers. pp. 1–11.

27. Takemoto M, Mori S, Kuzuya M, Yoshimoto S, Shimamoto A, et al. (2013) Diagnostic criteria for Werner syndrome based on Japanese nationwide epidemiological survey. Geriatr Gerontol Int 13: 475–481.

14

The Metagenomic Telescope

Balázs Szalkai[1][¶], **Ildikó Scheer**[2][¶], **Kinga Nagy**[2], **Beáta G. Vértessy**[2,3]*, **Vince Grolmusz**[1,4]*

1 PIT Bioinformatics Group, Eötvös University, Budapest, Hungary, **2** Laboratory of Genome Metabolism, Institute of Enzymology, Research Center for Natural Sciences, Hungarian Academy of Sciences, Budapest, Hungary, **3** Department of Applied Biotechnology and Food Sciences, Budapest University of Technology and Economics, Budapest, Hungary, **4** Uratim Ltd., Budapest, Hungary

Abstract

Next generation sequencing technologies led to the discovery of numerous new microbe species in diverse environmental samples. Some of the new species contain genes never encountered before. Some of these genes encode proteins with novel functions, and some of these genes encode proteins that perform some well-known function in a novel way. A tool, named the Metagenomic Telescope, is described here that applies artificial intelligence methods, and seems to be capable of identifying new protein functions even in the well-studied model organisms. As a proof-of-principle demonstration of the Metagenomic Telescope, we considered DNA repair enzymes in the present work. First we identified proteins in DNA repair in well-known organisms (i.e., proteins in base excision repair, nucleotide excision repair, mismatch repair and DNA break repair); next we applied multiple alignments and then built hidden Markov profiles for each protein separately, across well-researched organisms; next, using public depositories of metagenomes, originating from extreme environments, we identified DNA repair genes in the samples. While the phylogenetic classification of the metagenomic samples are not typically available, we hypothesized that some very special DNA repair strategies need to be applied in bacteria and Archaea living in those extreme circumstances. It is a difficult task to evaluate the results obtained from mostly unknown species; therefore we applied again the hidden Markov profiling: for the identified DNA repair genes in the extreme metagenomes, we prepared new hidden Markov profiles (for each genes separately, subsequent to a cluster analysis); and we searched for similarities to those profiles in model organisms. We have found well known DNA repair proteins, numerous proteins with unknown functions, and also proteins with known, but different functions in the model organisms.

Editor: Sebastian D. Fugmann, Chang Gung University, Taiwan

Funding: This work was supported by the Hungarian Scientific Research Fund (OTKA NK 84008, K109486), the Baross program of the New Hungary Development Plan (3DSTRUCT, OMFB-00266/2010 REG-KM-09-1-2009-0050), the Hungarian Academy of Sciences (TTK IF-28/2012), and the European Commission FP7 Biostruct-X project (contract number 283570). The funders had no role in study design, data collection and analysis, decision to publish, or preparation of the manuscript. Co-author Vince Grolmusz is employed by Uratim Ltd. Uratim Ltd provided support in the form of salary for author Vince Grolmusz, but did not have any additional role in the study design, data collection and analysis, decision to publish, or preparation of the manuscript. The specific role of the author is articulated in the 'author contributions' section.

* Email: vertessy.beata@ttk.mta.hu (BGV); grolmusz@pitgroup.org (VG)

¶ Joint first authors.

Introduction

The vast field of computer science, termed artificial intelligence (AI), offers powerful methods for distilling relevant information from large sets of data. Metagenomic databases have been increasingly used in the recent years to investigate the bacterial composition of samples taken from a variety of environments. To analyze and compare different genomic data, Hidden Markov Models [1] provide a useful methodology.

A Hidden Markov Model, applied to protein sequences, is basically a random amino acid sequence generator with multiple internal states, two of which are distinguished as START and STOP states. The generator starts from the START state. Until it arrives to the STOP state, it repeats the following two steps:

- it outputs a random amino acid, then
- it moves to a random new state (typically not in uniform distribution).

The role of the multiple internal states is that the probability distribution of the output amino acid and the distribution of the new state both depend on the current state. The model is named "hidden" because the internal states cannot be unambiguously determined by observing the output sequence.

HMMs are particularly useful because they can be trained by a set of input sequences to output similar sequences: if we have proteins of related functions, then we can build a Hidden Markov Model which will generate random amino acid sequences as output, similar to the ones used in training. It is even a more useful property of HMMs that if we take any amino acid sequence, denoted by w, our model can easily tell us the probability of generating exactly that sequence w as an output.

Consequently, if we have a HMM trained on a certain set of proteins, then the same HMM can assign higher scores (i.e., probabilities) to proteins *similar* to the training set, and lower scores (i.e., probabilities) to proteins *dissimilar* to the training set. Note that this scoring is usually not homogeneous as in the case of BLAST [2] and its clones [3]: in HMM models conservative subsequences are differentiated from those appearing in variable regions.

In the present work, we have applied HMM in a novel way to suggest and possibly discover still unknown protein functions in several well-studied model organisms. Starting from sequence alignments for proteins involved in DNA damage repair, we created Hidden Markov Models and used these models to search for similar genes in the metagenomic samples from different environments. Combining the original HMM with the genes found in the metagenomes, we created a second, more trained HMM that we used to interrogate proteomes of higher order model organisms. This search (termed as "Metagenomic Telescope" in the present study) generated numerous novel hits in the higher order organisms, containing proteins previously not yet described as closely similar to the DNA damage repair proteins. These results indicate the Metagenomic Telescope may be a powerful method for the identification of novel proteins in higher order model organisms.

Methods

First, we took some known *E. coli* and Archaean occurrences of a specific enzyme as listed in Table 1. We aligned these similar proteins using Clustal Omega [4]. The aligned sequences were then used to train a HMM with the `hmmbuild` utility of the HMMER3 package [5]. We term the resulting model as the "original HMM" (cf. Figure 1).

This "original HMM" was used twice: once in the direct projection to the model organisms (producing "original hits"), and second time for Projection 1 in the Telescope (here it represents the first step for producing "telescopic hits").

In the original projection, similarity scores are assigned to the protein sequences of the model organisms: the output of this single projection is the set of the highest scored proteins, using an inclusion threshold of E-value $\leq 10^{-6}$, found in the proteomes of the model organisms. We termed these highest scored proteins as "original hits" (this step is visualized on the upper panel of Figure 1).

For the application in the Telescope, we first extracted open reading frames from the metagenomes with the `getorf` application of EMBOSS [6], then applied the `hmmsearch` utility of HMMER3 [5] on the "original HMM" and the database of amino acid sequences extracted from each metagenome. The result of this search consists of hits in the metagenome and are referred to as "metagenome matches".

Three extreme metagenomes in the present study were accessed through the CAMERA portal [7].

Richmond Mine in Iron Mountain

CAMERA accession code: CAM_PROJ_AcidMine. The Iron Mountain, California mine was closed in the sixties. Later, the large, underground pyrite depositories became exposed to atmospheric oxygen and moisture, producing one of the most acidic mine drainages on Earth [8]. The metagenome consists of the data gained by sequencing samples from the thick, pink biofilm in this acidic and hot (42°C) environment, containing iron-oxidizing bacteria and other species.

Yellowstone Bison hot spring

CAMERA accession code: CAM_PROJ_BisonMetagenome. The Bison Pool environment is an alkaline hot spring in the Sentinel Meadow of Yellowstone National Park, situated in Wyoming, U.S. The samples were collected from sites with water temperature of 56°C through 92°C [9–11].

Phosphorus removing (EBPR) sludge community

CAMERA accession code: CAM_PROJ_EBPRSludge. The samples were taken from an enhanced biological phosphorus removal (EBPR) sludge community from the Thornside Sewage Treatment Plant in Brisbane, Queensland, Australia.

The metagenome matches (cf. Figure 1) were aligned and clustered using the OPTICS method [12]. The clusters were then used as inputs of `hmmbuild` [5], which yielded the "new HMMs". In other words, these models have been built on possible unknown DNA repair enzymes found in the metagenome. We then

Table 1. Protein families and proteomes used in the present study.

Protein families	Archaea proteomes in "original HMM"	Eukaryotic proteomes screened by HMM
dUTPase	Aeropyrum pernix	Saccharomyces cerevisae
uracil-DNA glycosylase (UNG)	Archeoglobus fulgidus	Arabidopsis thaliana
thymine-DNA glycosylase (TDG)	Halobacterium salinarum	Caenorhabditis elegans
Archaeal UDG	Haloferax volcanii	Drosophila melanogaster
NTHL1	Methanobacterium thermoautotrophicum	Danio rerio
OGG1	Methanococcus jannaschii	Gallus gallus
Rad50	Methanococcus maripaludis	Bos taurus
Mre11	Methanosarcina acetivorans	Canis lupus
	Pyrococcus abyssi	Mus musculus
	Pyrococcus furiosus	Sus scrofa
	Pyrococcus horikoshii	Rattus norvegicus
	Sulfolobus acidocaldarius	Homo sapiens
	Sulfolobus islandicus	
	Sulfolobus solfataricus	
	Thermococcus kodakaraensis	

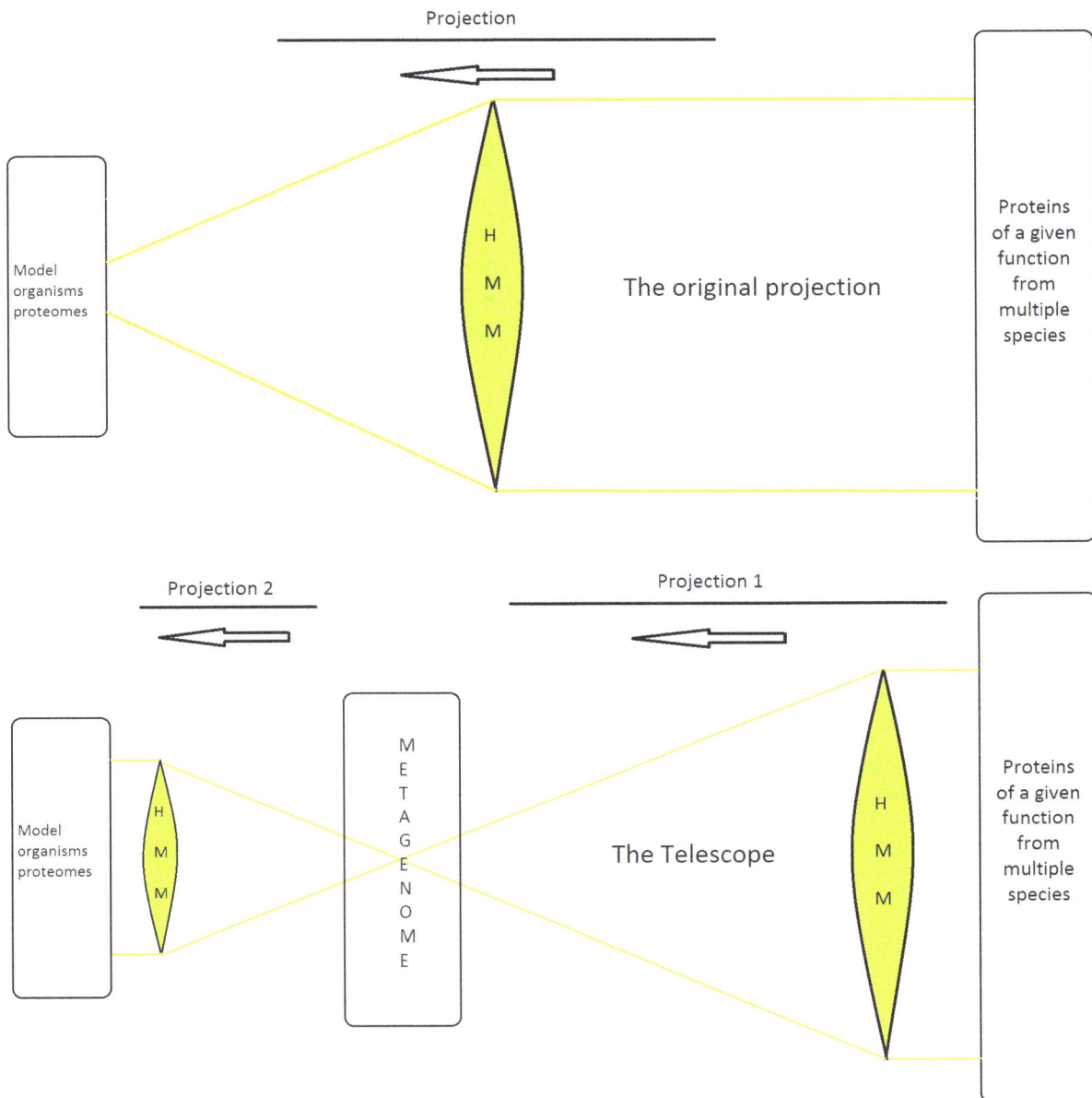

Figure 1. The original projection (upper panel) compared to the scheme of the Metagenomic Telescope (lower panel). Projection 1 discovers genes or proteins in the metagenome that probably have similar function as the well-known starting proteins in front of the objective lens on the right hand side. Projection 2 directly identifies these proteins within the proteomes of the model organisms (as a set of UniProt accession numbers).

performed the final step in the process pipeline, i.e., testing both the original and the new, telescopic HMMs on the proteomes of higher level organisms. As visualized on Figure 1, we compared the results of the projection on the upper panel and the projections of the lower panel. These organisms included *Arabidopsis thaliana, C. elegans* and *E. coli* as well as mouse, rat, human, and other model species. The flowchart of the application of the Metagenomic Telescope is shown on Figure 2. After the last projection (Projection 2 on Figure 1), the highest scoring proteins were selected, using again an inclusion threshold of E-value $\leq 10^{-6}$. These proteins were termed as "telescopic hits".

Our goal was to examine whether the possible new DNA repair enzymes found in the metagenomes could be used for finding new

DNA repair enzymes in the model organisms as well. This included comparison of the results of the searches with the original and the new models, respectively.

Results and Discussion

Design of the Metagenomic Telescope

The optical (refractive) telescope applies two projections: the first projection is done by the objective lens, the second by another lens called "the eyepiece": through the eyepiece one can see the enlarged image, generated by the objective.

Our Metagenomic Telescope also consists of two projections, each one is performed by applying HMMs. The key point is

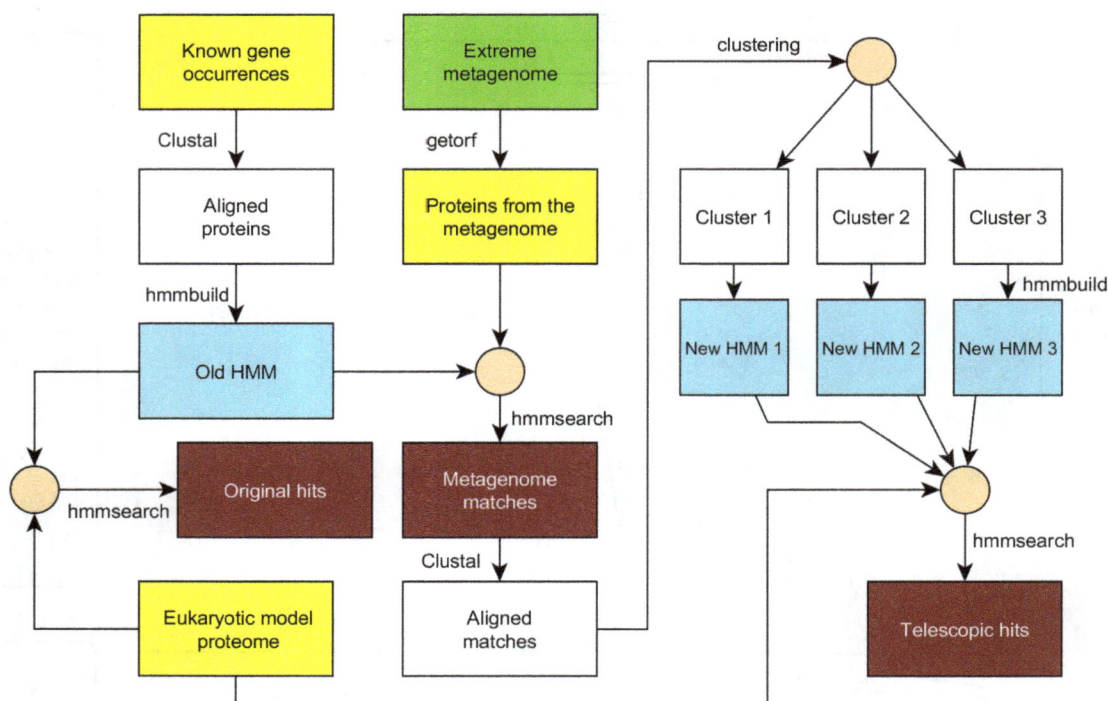

Figure 2. The flowchart of the Metagenomic Telescope applied to DNA repair enzymes.

making use of *metagenomes* in the projections: *first* we project *to* metagenomes, then we project *from* metagenomes. The lower panel of Figure 1 describes these two projections, together producing the "telescopic hits"; and compares these to a single HMM projection on the upper panel of Figure 1, producing the "original hits".

The starting point is a set of proteins of similar function or structure, taken from well–annotated organisms. This set is the teaching set for both the first HMM in the Metagenomic Telescope and the single HMM of the original projection.

In the original HMM or the original projection (upper panel of Figure 1), we use the HMM constructed in the step for finding similar protein sequences in model organisms: this is the only projection we use here. Using that HMM, similarity scores are assigned to the protein sequences of the model organisms. The output of this projection is the set of the highest scored proteins found in the proteomes of the model organisms (termed as "original hits").

In contrast, in the Metagenomic Telescope (lower panel of Figure 1), we apply two projections:

Projection 1 in the Telescope. Here we use the same HMM as in the original projection, but now we search for high-scored protein sequences in the metagenomes instead of proteins in the model organisms.

Projection 2 in the Telescope. The starting point is the highest scored proteins from the metagenome. After a suitable clustering, a new – second – HMM is built: its teaching set consists of these high scored proteins. Next, the proteomes of some model organisms are considered, and by this second HMM similarity scores are assigned to the protein sequences of the model organisms. The output of the second projection is the set of the highest scored proteins found in the proteomes of the model organisms (termed as "telescopic hits").

We believe that our telescope will facilitate the identification and annotation the functions of proteins in model organisms, since the diversity of well-chosen metagenomes is expected to help to assign new, still unknown functions to a number of proteins.

Proof of Concept: DNA Repair Enzymes

As a proof of concept, we applied the Metagenomic Telescope to DNA repair enzymes as the starting set of proteins, and metagenomes, found in extreme environments (acid mine leakage, a Yellowstone hot spring and a phosphorus removing sludge community), see Table 1. Our aim was to include metagenomes isolated from diverse extreme environmental sources, where chemical stress is present in addition to thermal effects.

The first discovered hyperthermophilic organism, *Sulfolobus acidocaldarius*, was found in Yellowstone National Park [13]. Nowadays, there are more than 90 known hyperthermophilic species, most of them are archaea, but there are some hyperthermophilic bacteria as well [13].

The metagenomes of two deep-sea hydrothermal vent chimneys, a black-smoker chimney called 4143-1 and a carbonate chimney from Lost City, were investigated in a survey [14]. The samples of these two chimneys are enriched in genes associated with mismatch repair (MMR) and homologous recombination repair [14]. These findings imply that these microorganisms have specific and extensive DNA repair systems to survive under the extreme environmental conditions (such as heavy metals, high concentrations of hydrogen sulfide, radionuclides and high temperature) present in their habitat [14].

(Hyper)thermophilic organisms are exposed to high temperatures, which may be expected to elevate the rate of spontaneous DNA mutations [13]. Interestingly, however, the genomic mutation rate of the hyperthermophilic archaeon *S. acidocaldarius* was found to be equal to that of mesophilic organisms [13]. It was also shown that base substitution rate in *S. acidocaldarius* is 10-

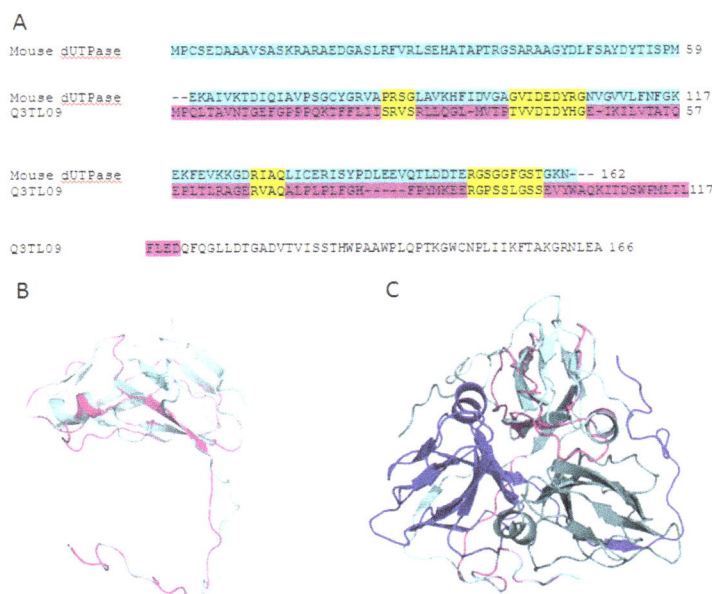

Figure 3. Identification of a novel dUTPase-like protein in the mouse proteome using the Metagenomic Telescope. Panel A shows an alignment (created by ClustalW) between the mouse dUTPase sequence (cyan) and the novel hit associated with the Uniprot accession number Q3TL09 (purple color indicates the part of this latter sequence that could be modeled in 3D using SwissModel or MUSTER). The conserved dUTPase motifs are shown in yellow. Panel B illustrates the structural alignment between human dUTPase (cyan) and the Q3TL09 modeled structure (purple) (at the subunit level). Panel C shows one of the models for Q3TL09 created by MUSTER software (purple), in this case the trimeric structure characteristic of dUTPases is shown (monomers are in shades of blue: cyan, royal blue and grayish blue). Protein structural models are shown in ribbon diagrams (PyMol).

fold lower compared to mesophilic organisms [13,15]. Surprisingly, *S. acidocaldarius* lacks all known bacterial mismatch repair genes. One explanation for this finding may involve a much better proofreading potency and insertion accuracy. Alternatively, a specific novel mismatch repair, distinct from the bacterial mutHLS model, can also account for that "normal" level of replication infidelity even in an extreme environment [15].

The structure of archaeon *Pyrococcus furiosus* proliferating cell nuclear antigen protein (PCNA) is an example of protein adaptation to increased temperature. The reduction of polar uncharged residues and elevated numbers of ion pairs likely contribute to increased stability of PCNA [16]. Besides, archaeal PCNAs are capable of self-loading onto DNA that can help higher DNA repair efficiency at a presumably increased DNA damage rate at extreme conditions [16].

Halophile and acidophile organisms live in relative high concentration of Na^+ and H^+, respectively [15]. These microorganisms cannot completely buffer against these ions, which can cause elevated stress to missense mutations carrying organisms. Still, the acidophile *S. acidocaldarius* has a 5-fold lower base substitution rate than the non-acidophile *T. thermophilus* [15], suggesting potent DNA repair systems in action.

The above data indicate that organisms living in extreme environments supposedly suffer more frequent DNA damage than organisms in ambient conditions, and to avoid drastic mutagenesis, they may contain specific and potent DNA repair mechanisms that are more efficient than that of other organisms. Therefore, it may be possible to identify new, more efficient DNA-repair enzymes in these extreme metagenomes. Certainly, there is a remarkable scientific interest in finding novel, more efficient enzymes in exotic species of the metagenomes mentioned. In addition, there is an even stronger interest in finding new functions

for already known enzymes and functions for proteins with unknown role in important eukaryotic model organisms, including *Homo sapiens*. Accordingly, we performed a second projection from the DNA-repair enzymes to several model organisms. Application of the Metagenomic Telescope resulted in an increased number of hits, as compared to the one-step original projection. The UniProt [17] accession numbers of these additional hits ("new telescopic hits"), appearing only among the telescopic hits, but not among the original hits are listed in Table S1, together with a short description as found in UniProt.

HMM projections starting with single domain protein families identify the relevant orthologues with few novel hits

Among the protein families involved in DNA damage recognition and repair selected for this present study, the trimeric dUTPase family — containing five well-conserved characteristic sequence motifs involved in building the active site — constitutes a well defined protein fold which can be also found in the family of prokaryotic dCTP deaminases [18]. In eukaryotes, however, to our present knowledge, this peculiar protein fold is exhibited only by dUTPases and no other proteins. Also, eukaryotic dUTPases are described as monogenic in the model eukaryotic organisms studied to date. dUTPases are responsible for hydrolysis of dUTP thereby preventing uracil incorporation into DNA and generating dUMP, the precursor for dTTP biosynthesis. These enzymes are essential to maintain genome integrity, and are found in all free-living organisms as well as in numerous DNA viruses as well as retroviruses [19]. Although it has been suggested that viral dUTPase sequences encode viral pseudo-proteases [20], later this suggestion was proved to be incorrect [21]; although some

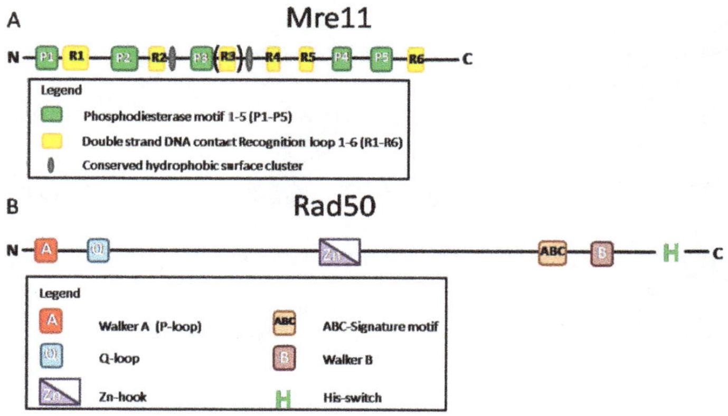

Figure 4. Schematic representation of Mre11 (Panel A) and Rad50 (Panel B) domains. (A) Mre11 has five phosphodiesterase motifs (green), 6 dsDNA recognition loop (yellow) and hydrophobic surface clusters (grey) (B) Rad50 has a bipartite ATPase domain: Walker A (red), Walker B (pale red), Q-loop (light blue), ABC-Signature motif (orange), histidine switch (green H) and has a Zinc-hook (purple).

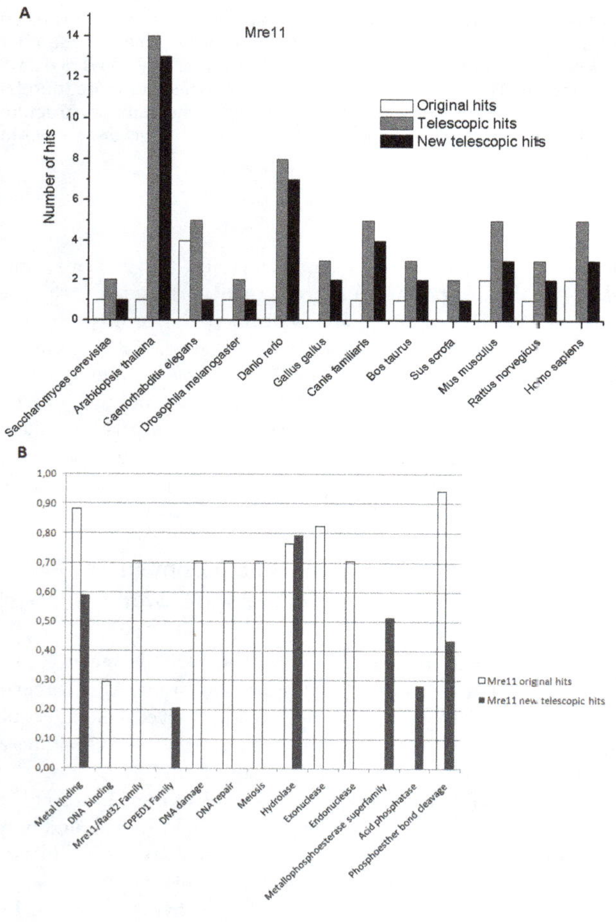

Figure 5. Original and telescopic hits for the Mre11 family. Panel A. Number of hits identified in the various eukaryotic model organisms after the original and the telescopic projections. Panel B. Distribution of genome ontology terms within the different hits. Note that new genome ontology classes can be observed in the telescopic hits.

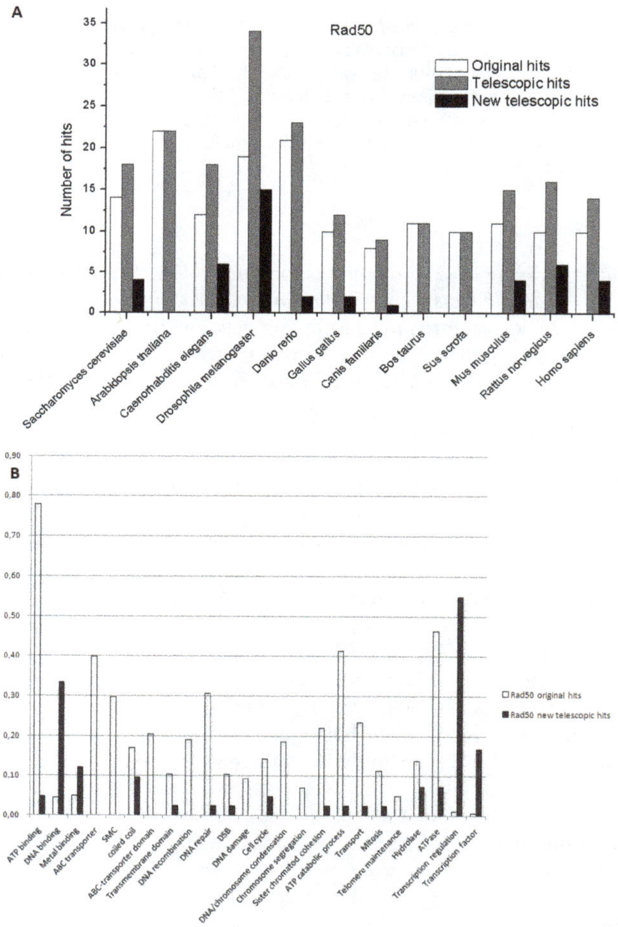

Figure 6. Original and telescopic hits for the Rad50 family. Panel A. Number of hits identified in the various eukaryotic model organisms after the original and the telescopic projections. Panel B. Distribution of genome ontology terms within the different hits. Note that new genome ontology classes can be observed in the telescopic hits.

database entries still contain the obsolete "pseudo-protease" annotation.

In accordance with the well-conserved character of this protein family, HMM searches did indeed find the orthologous dUTPase sequences, however, no novel protein could be found among the original hits. Still, among the telescopic hits, we found one novel hit in the mouse proteome (UniProt accession number Q3TL09). Although on the sequence level it showed rather low similarity to the authentic dUTPase sequence (identity 9%, similarity 23%), the sequence alignment indicates that out of the five characteristic dUTPase motifs, four can be identified in the sequence of this protein (Figure 3A). The actual functional relevance of this protein to dUTPases needs further experimental studies out of the scope of the present work. It was also of interest to investigate if the 3D structure of this protein may be similar to the dUTPase fold [22,23]. For such investigations, first we run the SwissModel software [24,25] by nominating the human dUTPase 3D structure (PDB ID 3EHW) [26–28] as the template. Results showed that the dUTPase fold can be adopted by this protein, however, the strength of this conclusion is somewhat weakened by the fact that the template was pre-defined and could strongly perturb the results.

Hence, we next used the MUSTER software [29] without any pre-defined template. This recently described software is based on an integrated use of protein profiling information and tries to fit a 3D structure from the Protein Data Bank on the sequence submitted. Results of the MUSTER-modeling showed that three slightly different 3D models could be created, and very interestingly, all of these used a dUTPase structure as the best-fitting model (Figures 3B,3C).

We conclude that for the dUTPase searches, the use of the telescopic HMM resulted in a promising finding. The newly found mouse protein, although with a very low level of sequence identity, may adopt the 3D structure of the antiparallel beta-sheeted jelly roll dUTPase-fold.

HMM models were also created for the numerous DNA-glycosylase families (listed in Table 1) that belong to either the alpha/beta superfamily of uracil-DNA glycosylases (UNG, TDG) or to the helix-turn-helix (HTH) superfamily of DNA glycosylases (NTH, NEI, OGG) [30]. These proteins, similarly to dUTPases, are also single domain proteins, with some N- or C-terminal extensions in several eukaryotic organisms. In several cases, eukaryotes encode different isoforms of DNA-glycosylases, dedicated to the different cellular compartments (nuclear vs. cytoplasmic). We found that while the original hits usually included the orthologues and their isoforms, the telescopic hits also included hits from the whole superfamily. For example, starting with the uracil-DNA glycosylase UNG, original hits showed the orthologous nuclear and mitochondrial isoforms of UNG, while telescopic hits included the closely related thymine-DNA glycosylases as well as SMUGs. Similarly, starting from any of the HTH superfamily DNA glycosylases, original hits were rather restricted to the different isoforms of the same proteins, while telescopic hits included proteins of the whole HTH superfamily. Hence, for the cases of the DNA-glycosylase families, the Metagenomic Telescope approach yielded new telescopic hits within the larger superfamily of these repair enzymes, but did not identify proteins within additional new families.

HMM projections of multi-domain proteins: telescopic hits suggest numerous novel associations

The Mre11 and Rad50 proteins play important roles in the repair of double-strand-DNA breaks. These two proteins are essential in both major pathways of double-stranded DNA break repair, in homologous recombination repair, as well as in non-homologous end-joining. Both Rad50 and MRE11 are multido-main proteins (cf., Figure 4). Rad50 has an ATPase globular domain and a highly lengthened coiled-coil domain connected together with a Zn-hook, whereas Mre11 contains a phosphodi-esterase core domain and several DNA-binding recognition loops.

Rad50 and Mre11 usually form a heterotetramer and this assembly is termed as the MRN complex. The MRN complex is crucial to (i) bridge DNA over short and long distances, (ii) DNA binding and processing, and (iii) activation of double strand break response and checkpoint signaling pathways [31]. Both Mre11 and Rad50 need a metal cofactor: manganese and magnesium, respectively [32]. Both of them can bind DNA. The dimerization of Rad50 is ATP dependent [33] and it belongs to the ABC-ATPase family [33]. Rad50 has a conserved "signature motif" that is needed for binding the γ-phosphate of ATP and is characteristic for ABC-ATPases [33]. The "signature motif" has a key role in the Rad50 dimer assembly [33]. Q-loop binds a magnesium ion [33]. The Walker A motif binds ATP and the Walker B motif hydrolyses it. The Walker A motif (also called P loop or phosphate-binding loop) forms the nucleotide binding site [32]. The D loop, which is a part of Walker B, binds one active magnesium ion and assists in dimerization [33]. The Mre11 binding site is on the coiled-coil region adjacent to the ABC domain [32]. Mre11 has five conserved phosphodiesterase motifs [32]. Conserved hydrophobic surface clusters are likely involved in macromolecular interaction sites [32]. The six DNA recognition loops (R1-R6) constitute a continuous DNA interaction surface [34]. All core DNA recognition loops are conserved in *S. pombe*, *S. cerevisiae* and *Xenopus*, except recognition loop 3 (R3) [34]. Rad50 and Mre11 homologs in *Escherichia coli* are termed SbcC and SbcD, respectively [35, 36].

The results of the application of the Metagenomic Telescope on these protein families are summarized on Figures 5 and 6 (for Mre11 and Rad50, respectively). In both figures, one panel (Figures 5A and 6A) shows the actual number of hits found in the original as well as in the telescopic projections in the model eukaryotic organisms. This representation provides a rather straightforward measure of the strength of the telescopic projection over the original projections. In some cases, the number of hits is just 1 (e.g., in the case for the original hits of Mre11 in several model organisms). In these cases, the hit was actually the *bona fide* Mre11 homologue in the given organism, and no additional "similar" proteins can be found. However, in the majority of cases, the number of hits is more than 1, and in these cases, in addition to the *bona fide* homologue that was always among the hits, additional proteins were also identified by the HMM projections.

The fact that the *bona fide* homologue is always identified indicates that the HMM projections are reliable. Nevertheless, these are the additional hits that may contain novel properties. It is easy to see for both Mre11 and Rad50 that the number of hits for a telescopic projection is never smaller than that for the corresponding original projection, on the contrary, these hits are quite frequently significantly more numerous. The additional hits, identified only in the telescopic projections are termed "new telescopic hits" on the respective panels in Figures 5A and 6A (cf. also Table S1).

To analyze the putative biological functions of the original and the new telescopic hits, in each cases we relied on the genome ontology classification categories, as provided in the UniProt database, and listed the different genome ontology definitions for each hit. The biological functions (genome ontology categories) found to be associated with most of the original hits are rather straightforward to assess. Accordingly, for both the Mre11 and

Table 2. New telescopic hits associated with a role in transcription regulation.

UniProt	Description	Organism
P18480	SWI/SNF chromatin-remodeling complex subunit SNF5	S.cer.
P29617	Homeobox protein prospero (May regulate transcription by binding to DNA)	D.mel.
P21519	Neurogenic protein mastermind	D.mel.
Q9VZY2	Myocardin-related transcription factor	D.mel.
Q24167	Protein similar (transcriptional regulator of the adaptive response to hypoxia)	D.mel.
Q9VPL6	Kismet (Hydrolase)	D.mel.
A8JNQ5	Ataxin-2 binding protein 1	D.mel.
Q8IQ98	PAR-domain protein 1	D.mel.
Q9VSK5	Grunge (Hydrolase)	D.mel.
P13002	Protein grainyhead/DNA-binding protein ELF-1/Transcription factor NTF-1	D.mel.
F1NZW0	Transcription initiation factor TFIID subunit 3 (Zinc-finger domain)	G.gal.
F1NY57	Uncharacterized protein (Contains: fork-head DNA-binding domain)	G.gal.
G3X8S4	Mediator of RNA polymerase II transcription subunit 15	M.mus.
Q8K4J6	Myocardin-like protein 1/Basic SAP coiled-coil transcription activator/ Myocardin-related transcription factor A	M.mus.
D4QGC2	Mastermind-2 (transcription coactivator activity -positive regulation of transcription from RNA polymerase II promoter)	M.mus.
G3V684	Positive cofactor 2/multiprotein complex/glutamine/Q-rich-associated protein (role in: stem cell maintenance)	R.nor.
F1LV40	Protein Mkl1 (negative regulation of cysteine-type endopeptidase activity involved in apoptotic process)	R.nor.
F1M7D7	Forkhead box protein P2	R.nor.
O14686	Histone-lysine N-methyltransferase 2D	Human
Q8IZL2	Mastermind-like protein 2	Human
Q0PRL4	Forkhead box P2	Human
Q96RN5	Mediator of RNA polymerase II transcription	Human

Using Rad50 sequences to build the starting HMM model, the Metagenomic Telescope approach identified the new telescopic hits listed in the table.

Rad50 families, we find that the functions listed (metal binding, DNA binding, DNA repair, etc) are already known to be associated with the Mre11 and Rad50 families.

Next, we compared the original and telescopic hits and found that the list of these properties is significantly enriched in the telescopic hits. Therefore, not only the number of hits was higher after using the telescopic HMMs, but also these hits were associated with additional functional properties (Figures 5B and 6B). The new telescopic hits were identified starting from different protein families involved in DNA repair. The criteria to affirm if any of these hits belong to e.g., families of Rad50 or Mre11 was to observe if these hits are listed in the UniProt database as belonging to the given protein family.

In order to evaluate the power of the Metagenomic Telescope method, we need to consider those genome ontology terms that show up only in the new telescopic hits. For the Mre11 family, such terms are the calcineurin-like phosphoesterase (CPPED1) family, the metallophosphoesterase family and the acid phosphatase biological function. While the latter two may be explained by the well-known characteristics of the Mre11 enzymatic action, the connection to the calcineurin-like phosphoesterase family seems to be novel. In this case, at least to our knowledge, the potentially similar characteristics of Mre11 and calcineurin-like phosphoesterases have not yet been addressed before. In the case of the hits

within the Rad50 family, the novel hits using the telescopic projections are even more evident. Perhaps the most intriguing result from these projections concerns the numerous occurrence of the "transcription regulation" and "transcription factor" genome ontology classes, which are evidently linked.

Table 2 presents these new telescopic hits, where we also listed the actual proteome within which the hits were identified. It is evident that these hits belong the different families involved in transcription regulation, each associated with its characteristic sequence motifs. Based on these findings, we suggest that Rad50-like proteins may also be involved not just in interacting with DNA but also interacting with the transcription process. It is known that e.g., DNA damage and repair occurs with higher frequency on transcriptionally active genomic segments since these are more accessible. Our present results may suggest that in addition to the less physical barrier in the actively transcribed genomic regions, Rad50-like proteins may also be involved in interacting with the transcription machinery in a more direct manner.

Our present method applied to the Mre11 and Rad50 protein families identified several new telescopic hits that are predicted to possess functional properties originally not found in Mre11 or Rad50. These proteins, according to the UniProt database, belong to protein families that still show some common traits with the Mre11 or Rad50 families with respect to catalytic action on nucleoside phosphates and/or nucleic acid binding. However, these newly found telescopic hits are not described in the UniProt database as members of the Mre11 or Rad50 families. We conclude that using the information within the metagenomes, the Metagenomic Telescope method leads to protein hits outside the protein families originally used as starting sequences, potentially facilitating search for proteins with additional functions.

References

1. Baum LE, Petrie T (1966) Statistical inference for probabilistic functions of finite state markov chains. Ann Math Statist 37: 1554–1563.
2. Altschul SF, Gish W, Miller W, Myers EW, Lipman DJ (1990) Basic local alignment search tool. J Mol Biol 215: 403–410.
3. Banky D, Szalkai B, Grolmusz V (2014) An intuitive graphical webserver for multiple-choice protein sequence search. Gene 539: 152–153.
4. Sievers F, Wilm A, Dineen D, Gibson TJ, Karplus K, et al. (2011) Fast, scalable generation of high-quality protein multiple sequence alignments using clustal omega. Mol Syst Biol 7: 539.
5. Eddy SR (2011) Accelerated profile HMM searches. PLoS Comput Biol 7: e1002195.
6. Rice P, Longden I, Bleasby A (2000) EMBOSS: the European Molecular Biology Open Software Suite. Trends Genet 16: 276–277.
7. Seshadri R, Kravitz SA, Smarr L, Gilna P, Frazier M (2007) CAMERA: a community resource for metagenomics. PLoS Biol 5: e75.
8. Baker BJ, Tyson GW, Webb RI, Flanagan J, Hugenholtz P, et al. (2006) Lineages of acidophilic archaea revealed by community genomic analysis. Science 314: 1933–1935.
9. Havig JR, Raymond J, Meyer-Dombard DR, Zolotova N, Shock EL (2011) Merging isotopes and community genomics in a siliceous sinter-depositing hot spring. Journal of Geophysical Research: Biogeosciences (2005–2012) 116.
10. Dick JM, Shock EL (2011) Calculation of the relative chemical stabilities of proteins as a function of temperature and redox chemistry in a hot spring. PLoS One 6: e22782.
11. Swingley WD, Meyer-Dombard DR, Shock EL, Alsop EB, Falenski HD, et al. (2012) Coordinating environmental genomics and geochemistry reveals metabolic transitions in a hot spring ecosystem. PLoS One 7: e38108.
12. Ankerst M, MBreunig M, Kriegel H, Sander J (1999) Optics: Ordering points to identify the clustering structure. In: Proc. ACM SIGMOD '99 Int. Conf. on Management of Data, Philadelphia PA.
13. van Wolferen M, Ajon M, Driessen AJM, Albers SV (2013) How hyperthermophiles adapt to change their lives: DNA exchange in extreme conditions. Extremophiles 17: 545–563.
14. Xie W, Wang F, Guo L, Chen Z, Sievert SM, et al. (2011) Comparative metagenomics of microbial communities inhabiting deep-sea hydrothermal vent chimneys with contrasting chemistries. ISME J 5: 414–426.
15. Drake JW (2009) Avoiding dangerous missense: thermophiles display especially low mutation rates. PLoS Genet 5: e1000520.
16. Winter JA, Bunting KA (2012) Rings in the extreme: PCNA interactions and adaptations in the archaea. Archaea 2012: 951010.
17. Consortium U (2010) The Universal Protein Resource (UniProt) in 2010. Nucleic Acids Res 38: D142–D148.
18. Vertessy BG, Persson R, Rosengren AM, Zeppezauer M, Nyman PO (1996) Specific derivatization of the active site tyrosine in dUTPase perturbs ligand binding to the active site. Biochem Biophys Res Commun 219: 294–300.
19. Vértessy BG, Tóth J (2009) Keeping uracil out of DNA: physiological role, structure and catalytic mechanism of dUTPases. Acc Chem Res 42: 97–106.

Supporting Information

Table S1 New telescopic hits identified in the present study.

Author Contributions

Conceived and designed the experiments: BGV VG. Performed the experiments: BS IS KN. Analyzed the data: BS IS KN BGV VG. Contributed reagents/materials/analysis tools: BS IS. Wrote the paper: BS IS BGV VG.

20. McClure MA, Johnson MS, Doolittle RF (1987) Relocation of a protease-like gene segment between two retroviruses. Proc Natl Acad Sci U S A 84: 2693–2697.
21. McGeoch DJ (1990) Protein sequence comparisons show that the 'pseudoproteases' encoded by poxviruses and certain retroviruses belong to the deoxyuridine triphosphatase family. Nucleic Acids Res 18: 4105–4110.
22. Fiser A, Vertessy BG (2000) Altered subunit communication in subfamilies of trimeric dutpases. Biochem Biophys Res Commun 279: 534–542.
23. Kovari J, Barabas O, Varga B, Bekesi A, Tolgyesi F, et al. (2008) Methylene substitution at the alpha-beta bridging position within the phosphate chain of dUDP profoundly perturbs ligand accommodation into the dUTPase active site. Proteins 71: 308–319.
24. Arnold K, Bordoli L, Kopp J, Schwede T (2006) The swiss-model workspace: a web-based environment for protein structure homology modelling. Bioinformatics 22: 195–201.
25. Kiefer F, Arnold K, Kunzli M, Bordoli L, Schwede T (2009) The swiss-model repository and associated resources. Nucleic Acids Res 37: D387–D392.
26. Nemeth-Pongracz V, Barabas O, Fuxreiter M, Simon I, Pichova I, et al. (2007) Flexible segments modulate co-folding of dutpase and nucleocapsid proteins. Nucleic Acids Res 35: 495–505.
27. Varga B, Barabas O, Takacs E, Nagy N, Nagy P, et al. (2008) Active site of mycobacterial dUTPase: structural characteristics and a built-in sensor. Biochem Biophys Res Commun 373: 8–13.
28. Varga B, Barabas O, Kovari J, Toth J, Hunyadi-Gulyas E, et al. (2007) Active site closure facilitates juxtaposition of reactant atoms for initiation of catalysis by human dutpase. FEBS Lett 581: 4783–4788.
29. Wu S, Zhang Y (2008) Muster: Improving protein sequence profile-profile alignments by using multiple sources of structure information. Proteins 72: 547–556.
30. Jacobs AL, Schar P (2012) DNA glycosylases: in DNA repair and beyond. Chromosoma 121: 1–20.
31. Williams GJ, Lees-Miller SP, Tainer JA (2010) Mre11-Rad50-Nbs1 conformations and the control of sensing, signaling, and effector responses at DNA double-strand breaks. DNA Repair (Amst) 9: 1299–1306.
32. Hopfner KP, Karcher A, Craig L, Woo TT, Carney JP, et al. (2001) Structural biochemistry and interaction architecture of the DNA double-strand break repair Mre11 nuclease and Rad50-ATPase. Cell 105: 473–485.
33. Hopfner KP, Karcher A, Shin DS, Craig L, Arthur LM, et al. (2000) Structural biology of Rad50 ATPase: ATP-driven conformational control in DNA double-strand break repair and the ABC-ATPase superfamily. Cell 101: 789–800.
34. Williams RS, Moncalian G, Williams JS, Yamada Y, Limbo O, et al. (2008) Mre11 dimers coordinate DNA end bridging and nuclease processing in double-strand-break repair. Cell 135: 97–109.
35. Delmas S, Shunburne L, Ngo HP, Allers T (2009) Mre11-Rad50 promotes rapid repair of DNA damage in the polyploid archaeon Haloferax volcanii by restraining homologous recombination. PLoS Genet 5: e1000552.
36. de Jager M, Trujillo KM, Sung P, Hopfner KP, Carney JP, et al. (2004) Differential arrangements of conserved building blocks among homologs of the Rad50/Mre11 DNA repair protein complex. J Mol Biol 339: 937–949.

15

The Hypothetical Protein 'All4779', and Not the Annotated 'Alr0088' and 'Alr7579' Proteins, Is the Major Typical Single-Stranded DNA Binding Protein of the Cyanobacterium, *Anabaena* sp. PCC7120

Anurag Kirti, Hema Rajaram*, Shree Kumar Apte

Molecular Biology Division, Bhabha Atomic Research Centre, Trombay, Mumbai, India

Abstract

Single-stranded DNA binding (SSB) proteins are essential for all DNA-dependent cellular processes. Typical SSB proteins have an N-terminal Oligonucleotide-Binding (OB) fold, a Proline/Glycine rich region, followed by a C-terminal acidic tail. In the genome of the heterocystous nitrogen-fixing cyanobacterium, *Anabaena* sp. strain PCC7120, *alr0088* and *alr7579* are annotated as coding for SSB, but are truncated and have only the OB-fold. *In silico* analysis of whole genome of *Anabaena* sp. strain PCC7120 revealed the presence of another ORF '*all4779*', annotated as a hypothetical protein, but having an N-terminal OB-fold, a P/G-rich region and a C-terminal acidic tail. Biochemical characterisation of all three purified recombinant proteins revealed that they exist either as monomer or dimer and bind ssDNA, but differently. The All4779 bound ssDNA in two binding modes i.e. (All4779)$_{35}$ and (All4779)$_{66}$ depending on salt concentration and with a binding affinity similar to that of *Escherichia coli* SSB. On the other hand, Alr0088 bound in a single binding mode of 50-mer and Alr7579 only to large stretches of ssDNA, suggesting that All4779, in all likelihood, is the major typical bacterial SSB in *Anabaena*. Overexpression of All4779 in *Anabaena* sp. strain PCC7120 led to enhancement of tolerance to DNA-damaging stresses, such as γ-rays, UV-irradiation, desiccation and mitomycinC exposure. The tolerance appears to be a consequence of reduced DNA damage or efficient DNA repair due to increased availability of All4779. The ORF *all4779* is proposed to be re-annotated as *Anabaena ssb* gene.

Editor: Sergey Korolev, Saint Louis University, United States of America

Funding: The authors have no support or funding to report.

Competing Interests: The authors have declared that no competing interests exist.

* E-mail: hemaraj@barc.gov.in

Introduction

Single-stranded DNA-binding proteins (SSB) are ubiquitous proteins found in all bacteria. The SSB proteins are characterised by their non-specific binding to single-stranded DNA (ssDNA) and active participation in the maintenance of genome integrity (DNA repair) as well as genetic information transfer (replication and transcription) [1]. A typical SSB monomer consists of (a) an N-terminal region having several conserved residues responsible for binding to ssDNA, tetramerisation and stabilization of monomer fold [2], and (b) a C-terminal region which displays a low sequence conservation except for the last few amino acids (known as acidic tail), and is responsible for protein-protein interactions and recruitment of DNA interactive proteins [3]. The highly conserved OB-fold has been extensively described for *Escherichia coli* SSB protein [1]. Deletion of C-terminus diminishes recruitment of other DNA interacting proteins [3], but enhances the affinity of N-terminus to ssDNA [4]. The spacer region between the N-terminal OB-fold and C-terminal acidic tail is rich in proline/glycine (P/G) residues and is thought to modulate the strength of DNA binding, possibly by distancing the highly negatively charged C-terminal end from the positively charged DNA binding N-terminal domain [5].

Nitrogen-fixing cyanobacteria, such as strains of *Anabaena* and *Nostoc* exhibit tolerance to a variety of abiotic stresses such as salinity, desiccation, heat and radiation [6,7], strongly indicative of a robust mechanism of DNA repair in these microbes [8]. Unfortunately, genes/proteins involved in DNA metabolism of cyanobacteria have not received adequate attention. The SSB protein is a key protein involved in all DNA related cellular activities. In the genomic database of *Anabaena* sp. strain PCC7120 (hereafter referred to as *Anabaena* 7120), two ORFs '*alr0088*' and '*alr7579*' are annotated as coding for SSB-like proteins (http://genome.microbedb.jp/cyanobase/Anabaena) and exhibit 28–30% homology at amino acid level with EcoSSB, and about 42% homologous to each other. However, BLAST search [9] of the amino acid sequence of these two proteins show that the protein sequence is terminated immediately after the N-terminal OB-fold region and have no region corresponding to either the P/G spacer or the C-terminal acidic tail. Since, C-terminal acidic tail is essential for interaction with other DNA replication/repair/recombination proteins [3], it seems unlikely that the proteins encoded by these two ORFs can perform all the functions of SSB proteins. However, this does not rule out that they are genuine SSBs, since the second SSB (SsbB) of naturally transformable bacteria, such as *Bacillus subtilis*, lacks the C-terminal acidic tail,

Table 1. Primers used and PCR amplicons generated in this study.

Primers	Nucleotide Sequence*	R.E.	Amplicon#
alr0088Fwd	5′ GGCCATATGAGCATTAACATTGTC 3′	NdeI	alr0088 ORF (0.35 kb)
alr0088 Rev	5′ GGCGGATCCTTAAAAATTTTCTGGTGC 3′	BamHI	
alr7579Fwd	5′ GGCCATATGAACTATATCAACAAA 3′	NdeI	alr7579 ORF (0.38 kb)
alr7579 Rev	5′ GGCGGATCCCTAGAAATTTGCGTTAGC 3′	BamHI	
all4779Fwd	5′ GGCCATATGAACAGCTGTGTTTTA 3′	NdeI	all4779 ORF (0.55 kb)
all4779 Rev	5′ GGCGGATCCTAAAATGGAATATCGTC 3′	BamHI	

*The restriction endonuclease (R.E.) site included in each primer is underlined and the corresponding R.E. indicated in the adjacent column.
#The amplicons generated with a given set of PCR primers are specified.

but functions as a SSB and is involved in genetic recombination [10,11]. Alr0088 and Alr7579 exhibit about 36–38% overall homology with BsSsbB. An *in silico* search for a SSB-like protein with the C-terminal region (i.e. the spacer region and acidic tail) in the genome of *Anabaena* 7120, revealed the ORF 'all4779'. The ORF encodes a 182 amino acid long protein with an N-terminal OB-fold, a P/G rich spacer region and a C-terminal acidic tail but has been annotated as a hypothetical protein possibly due to its lower homology (15–18%) with other bacterial SSBs. The homologs of the two truncated SSB-like proteins as well as hypothetical SSB-like protein of *Anabaena* 7120 are found across all cyanobacterial genomes (http://genome.microbedb.jp/cyanobase).

In the present work, we cloned, overexpressed, purified and biochemically characterised Alr0088, Alr7579 and All4779 proteins. All three proteins existed in monomeric and dimeric forms and showed differential binding to ssDNA. All4779 protein showed typical structural features, oligomerisation, ssDNA binding modes compared to *E. coli* SSB and conferred tolerance to DNA damage, upon overexpression in *Anabaena*, that identifies it as the major SSB of *Anabaena*.

Materials and Methods

Organism and Growth Conditions

E. coli cells were grown in Luria–Bertani (LB) medium at 37°C with shaking (150 rpm). When required antibiotics [34 µg chloramphenicol mL^{-1}(Cm$_{34}$), 50 µg kanamycin mL^{-1} (Kan$_{50}$), or 100 µg carbenicillin mL^{-1} (Cb$_{100}$)] were used in culture media. Axenic cultures of *Anabaena* 7120 were grown in BG-11 liquid

medium, pH 7.0 [12] in the absence of combined nitrogen (BG-11, N$^-$) under stationary conditions with continuous illumination (30 µE m^{-2} s^{-1}) at 27°C±2°C. Recombinant *Anabaena* strains were grown in the presence of 10 µg neomycin mL^{-1} (Nm$_{10}$) in BG-11 liquid media or with 25 µg neomycin mL^{-1} (Nm$_{25}$) on BG-11 agar plates. Growth was assessed in terms of chlorophyll *a* content per ml of culture as described earlier [13]. Cell survival was assessed in terms of colony forming units (cfu) by plating 100 µl of the culture on to BG-11, N$^-$ agar plates followed by incubation under illumination for 10 days as described earlier [14].

Three-day-old nitrogen-fixing *Anabaena* cultures were concentrated to a chlorophyll *a* density of 10 µg mL^{-1}, prior to subjecting them to one of the following stresses: (i) 6 kGy of ^{60}Co γ-rays at a dose rate of 4.5 kGy h^{-1}, or (ii) 6 days of desiccation or in humid chamber (control), or (iii) 0–4 µg mitomycinC (mitC) mL^{-1} for 30 min, or (iv) exposure to 0–1.5 kJ UV-B (280 nm) (dose rate 5 J m^{-2} sec^{-1}) for different duration. Survival in response to stress, and post-stress recovery were compared with unstressed/control cultures grown under illumination at 27°C±2°C as described earlier [14].

Generation of Plasmid Constructs for Overexpression of Proteins in *E. coli*

Different amplicons (*alr0088, alr7579, all4779*) were generated by PCR amplification of *Anabaena* 7120 genomic DNA (100 ng) using gene specific primers (as indicated in Table 1), 1 µM dNTPs and 1U Taq DNA polymerase in Taq buffer (Bangalore Genei, India). These amplicons were individually digested with *Nde*I and *Bam*HI restriction endonucleases and ligated to the expression

Table 2. Plasmids used in this study.

Plasmids	Description	Source/Reference
pET16b	Cb^r, expression vector	Novagen
pAM1956	Kan^r, promoterless vector with *gfpmut2* reporter gene	[19]
pBluescript (pBS)	Cb^r, cloning vector	Lab Collection
pFPN	Cb^r, Kan^r, integrative expression vector	[18]
pETalr0088	Cb^r, 0.35 kb *alr0088* gene cloned in pET16b at *Nde*I/*Bam*HI restriction sites	This study
pETalr7579	Cb^r, 0.38 kb *alr7579* gene cloned in pET16b at *Nde*I/*Bam*HI restriction sites	This study
pETall4779	Cb^r, 0.55 kb *all4779* gene cloned in pET16b at *Nde*I/*Bam*HI restriction sites	This study
pFPNall4779	Cb^r, Kan^r, 0.55 kb *all4779* gene cloned in pFPN at *Nde*I/*Bam*HI restriction sites	This study
pAMall4779	Kan^r, 0.81 kb *Xma*I-*Sal*I fragment from pFPNall4779 gene cloned in pAM1956	This study

Table 3. Bacterial strains used in this study.

Bacterial Strains	Description	Source/Reference
E. coli strains		
DH5α	F⁻ recA41 endA1 gyrA96 thi-1hsdR17 (rk⁻mk⁻) supE44 relA λ ΔlacU169	Lab Collection
HB101	F⁻ mcʳ Bmʳ rhsdS20(rB⁻ mB⁻) recA13 leuB6 ara-14 proA2 lacY1 galK2 xyl-5 mtl-1 rps (SmR) glnV44 λ⁻	Lab Collection
BL21(pLysS)	Cmʳ F⁻ ompT hSdSB (rB⁻ mB⁻) gal dcm pLysS (pLysS) (DE3)	Novagen
BL21(pLysS)(pETalr0088)	Cmʳ, Cbʳ, E. coli BL-21 cells harbouring the plasmid, pETalr0088	This study
BL21(pLysS)(pETalr7579)	Cmʳ, Cbʳ, E. coli BL-21 cells harbouring the plasmid, pETalr7579	This study
BL21(pLysS)(pETall4779)	Cmʳ, Cbʳ, E. coli BL-21 cells harbouring the plasmid, pETall4779	This study
Ec(pAMall4779)	Kanʳ, HB101 harbouring pAMall4779 plasmid	This study
HB101 (pRL623+ pRL443)	Donor strain carrying pRL623 (encoding methylase) and conjugal plasmid pRL443	(Wolk, C.P.)
Anabaena strains		
Anabaena 7120	Wild type strain	Lab Collection
AnpAM	Nmʳ, Anabaena 7120 harbouring the plasmid, pAM1956	[26]
Anall4779⁺	Nmʳ, Anabaena 7120 harbouring the plasmid, pAMall4779	This study

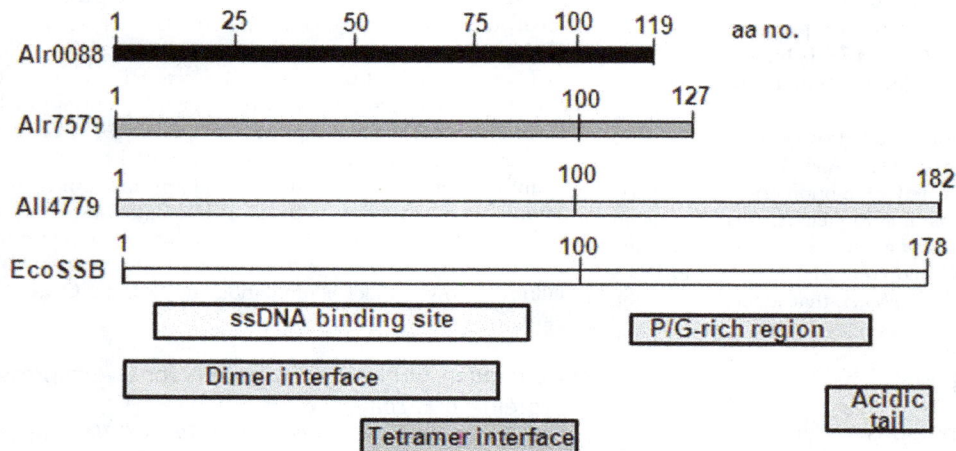

Figure 1. Bioinformatic analysis of Alr0088, Alr7579 and All4779 proteins of *Anabaena* **7120.** (A) Conserved Domain Database (CDD) analysis of *Anabaena* Alr0088, Alr7579 and All4779 proteins and *E. coli* SSB (EcoSSB) protein. The OB-fold corresponding ssDNA binding region, dimer and tetramer interfaces for all the proteins are indicated. (B) Comparison of homology between predicted amino acid sequence of All4779 and EcoSSB. The identical amino acids are indicated by letters and similar amino acids with a '+' sign. The proline (P) and glycine (G) residues beyond the OB fold are shown in larger font, while the acidic residues at the C-terminal end are in bold and underlined. The numbers on the left and right hand side correspond to amino acid residues.

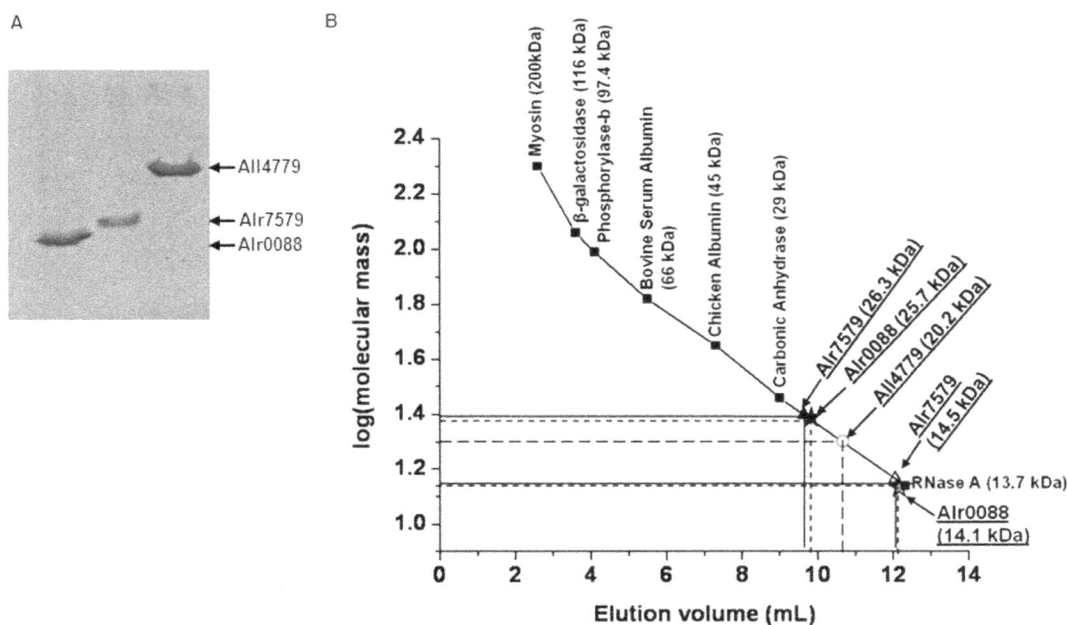

Figure 2. Molecular mass determination of purified native *Anabaena* proteins. (A) Ni-NTA affinity chromatography purified Alr0088, Alr7579 and All4779 proteins separated on 12% SDS-polyacrylamide gel followed by staining with Coomassie Brilliant Blue (CBB) G-250. The purified proteins are indicated by arrows. (B) Elution profile of purified native Alr0088, Alr7579 and All4779 proteins in gel filtration chromatography using Superdex HR200 matrix. A standard graph using the following standard proteins: [Myosin (200 kDa), β-galactosidase (116 kDa), Phosphorylase-b (97.4 kDa), Bovine Serum Albumin (66 kDa), Chicken Albumin (45 kDa), Carbonic Anhydrase (29 kDa) and RNaseA (13.7 kDa)] was drawn to calculate the molecular mass of the eluted native *Anabaena* proteins depending on their elution volume. The position of the eluted proteins has been depicted by 'star' and 'triangle' symbols. The vertical and horizontal lines from the two symbols indicate the elution volume and the corresponding log of molecular mass.

vector pET16b (Table 2), having His$_{10}$-tag at the 5′ end, at identical restriction sites. The resulting plasmid constructs were designated as pET*alr0088*, pET*alr7579* and pET*all4779* respectively (Table 2). DNA insert of all three plasmids were sequenced on both strands using Sanger's dideoxy method and were found to be completely identical to the corresponding nucleotide sequences available in the genomic database. The nucleotide sequences corresponding to *alr0088*, *alr7579* and *all4779* respectively were submitted to GenBank (GenBank Accession Nos. GU225949, GU225950, GU225951).

Overexpression and Purification of His-tagged Proteins

The plasmids pET*alr0088*, pET*alr7579* and pET*all4779* (Table 2) were transformed into *E. coli* BL21(pLysS) and transformants selected on LBCm$_{34}$Cb$_{100}$ plates (Table 3). Proteins were overexpressed from the respective logarithmic phase cultures of *E. coli* upon addition of 1 mM IPTG for 3 h at 37°C. The recombinant His-tagged proteins (Alr0088, Alr7579 and All4779) were extracted in lysis buffer (20 mM Tris-HCl, pH 8, 0.5 M NaCl, 5 mM imidazole and 0.1% TritonX-100) by sonication and purified by Ni-NTA affinity chromatography (Qiagen, Germany) using different concentrations of imidazole ranging from 10–1000 mM as described earlier. The proteins were eluted individually in 1 M imidazole fraction and visualised by electrophoretic separation on 14% SDS-polyacrylamide gel followed by staining with Coomassie Brilliant Blue (CBB) G-250. The proteins were quantified spectrophotometrically by measuring absorbance at 280 nm using 19480 M^{-1} cm^{-1}, 20970 M^{-1} cm^{-1} and 12950 M^{-1} cm^{-1} as the extinction coefficients, estimated using Expasy software (*web.expasy.org/protparam*) for Alr0088, Alr7579 and All4779 proteins respectively.

The purified recombinant native proteins were individually cross-linked using glutaraldehyde, as described earlier [15], followed by precipitation of protein with cold acetone. The pellet was air dried, solubilised in 1X Laemmli's buffer by heating at 80°C for 10 min, separated by SDS-PAGE and visualized by staining with CBB G-250.

Molecular mass determination of native purified proteins was carried out by gel filtration chromatography using Superdex HR200 column. The coumn was equilibrated with Tris-NaCl buffer and standard graph obtained using the following proteins: Myosin (200 kDa), β-galactosidase (116 kDa), Phosphorylase-b (97.4 kDa), Bovine Serum Albumin (66 kDa), Chicken Albumin (45 kDa), Carbonic Anhydrase (29 kDa) and RNase A (13.7 kDa). Molecular mass of the three *Anabaena* proteins was calculated from the standard graph on the basis of the elution volume. Presence of protein in different eluates/fractions was detected by measuring absorbance at 280 nm.

Electrophoretic Mobility Shift Assay (EMSA)

A 75-mer oligonucleotide (10 ng) was end-labelled with γ-^{32}P-ATP using Polynucleotide Kinase. The labelled oligo was incubated in the presence of specified concentrations of the Alr0088, Alr7579 and All4779 proteins in binding buffer [20 mM Tris-HCl, pH 8, 1 mM MgCl$_2$, 100 mM KCl, 8 mM Dithiothre-itol (DTT), 4% sucrose, 80 μg mL^{-1} Bovine Serum Albumin (BSA)] for 30 min at room temperature and electrophoretically separated subsequently on 6% non-denaturing polyacrylamide gel at 150 V for 2 h in 1X Tris-borate EDTA (TBE) buffer. Imaging of the radioactive gel was carried out using Phosphorimager Typhoon Trio Variable mode imager (Wipro-GE-HealthCare, USA).

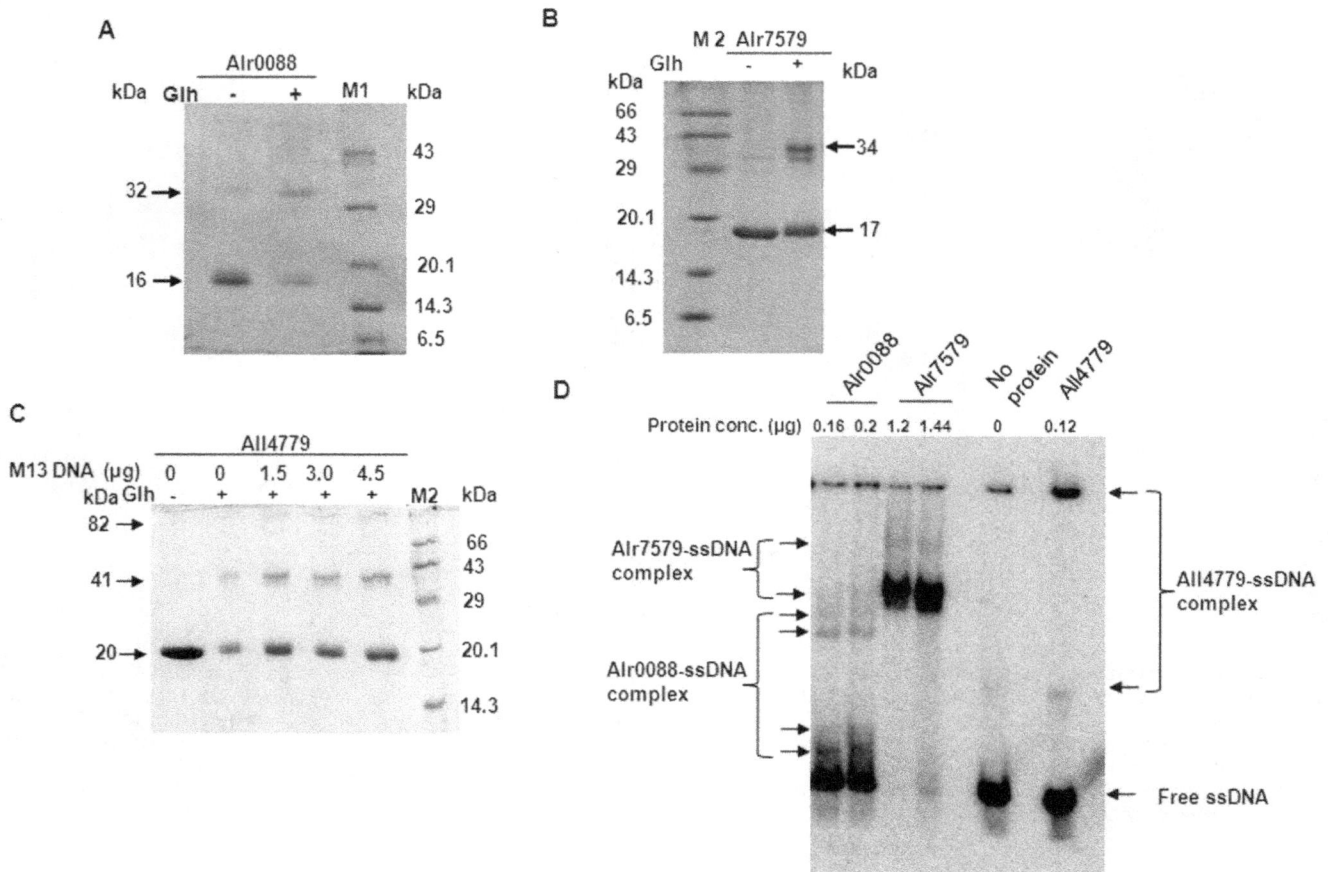

Figure 3. Glutarldehyde (Glh)-aided crosslinking of native purified *Anabaena* SSB-like proteins and their binding to ssDNA. The purified native *Anabaena* proteins (A) Alr0088, (B) Alr7579 and (C) All4779 were cross-linked with Glh in the presence or absence of M13 ssDNA as indicated. The proteins were electrophoretically separated on 12% SDS-polyacrylamide gel followed by staining with Coomassie Brilliant Blue (CBB) G-250. The molecular mass of the protein markers used (M1 and M2) are written to the immediate (right/left) of the marker lane. Different molecular forms of the native *Anabaena* proteins are indicated by the arrows. (D) Electrophoretic Mobility Shift Assay (EMSA) of a γ-^{32}P-ATP labeled 75-mer oligonucleotide in the presence of different concentrations of Alr0088, Alr7579 and All4779 proteins. Following *in solution* interaction, the assay mix was separated by 6% non-denaturing PAGE in 1X TBE and radioactive gel imaged using a phosphorimager. The free ssDNA substrate and the different ssDNA-protein complexes formed are indicated.

Fluorescence Measurements

All three proteins showed maximum excitation at 282 nm. The emission maxima were found to be 310, 340 and 335 nm respectively for Alr0088, Alr7579 and All4779 proteins in 20 mM Tris-HCl, pH 8, 1 mM EDTA buffer. The change in the intensity of the emitted fluorescence was measured in the presence of increasing concentration of ssDNA [poly(dT) or M13 ssDNA]. The graph of relative fluorescence (%) as a function of poly(dT) concentration was used to determine the binding constant for the individual proteins to ssDNA as described earlier [16]. The binding constants were calculated as the reciprocal of the concentration of poly(dT) at which 50% fluorescence compared to the initial 100% was detected. The graph depicting quenching expressed as the ratio of difference in fluorescence to initial fluorescence ($\Delta F/F_i$) as a function of the ratio of concentrations of poly(dT) and protein was used to determine the binding modes or occlusion site of the proteins as described earlier [17]. It corresponded to the $[nt]_{poly(dT)}/[Protein]$ value at the point of saturation of quenching of fluorescence. During titration, solutions were added from concentrated samples and correction for dilution was made as required. All fluorescence

measurements were performed with Jasco spectrofluorimeter FP6500 (Japan) using a quartz cuvette of 1 cm path length at room temperature.

Western Blotting and Immuno-detection

The purified All4779 protein was used to raise specific polyclonal antibody (anti-All4779 antibody) in rabbit. Proteins were extracted from three-day-old wild type and recombinant *Anabaena* cultures in Laemmli's buffer, separated by 14% SDS-PAGE followed by electroblotting on to nitrocellulose membrane. Immunodetection was carried out using the 1:5000 dilution of anti-All4779 antibody, followed by secondary anti-rabbit IgG antibody, coupled to alkaline phosphatase and colour development using NBT-BCIP.

Generation of Recombinant *Anabaena* Strains

The 0.55 kb *Nde*I-*Bam*HI fragment from pET*all4779* was ligated to pFPN vector (Table 2) [18] at the same restriction sites, resulting in the plasmid construct, pFPN*all4779*. The 0.81 kb *Sal*I–*Xma*I fragment from pFPN*all4779* was ligated to pAM1956 vector (Table 2) [19] digested with the identical restriction enzymes. The

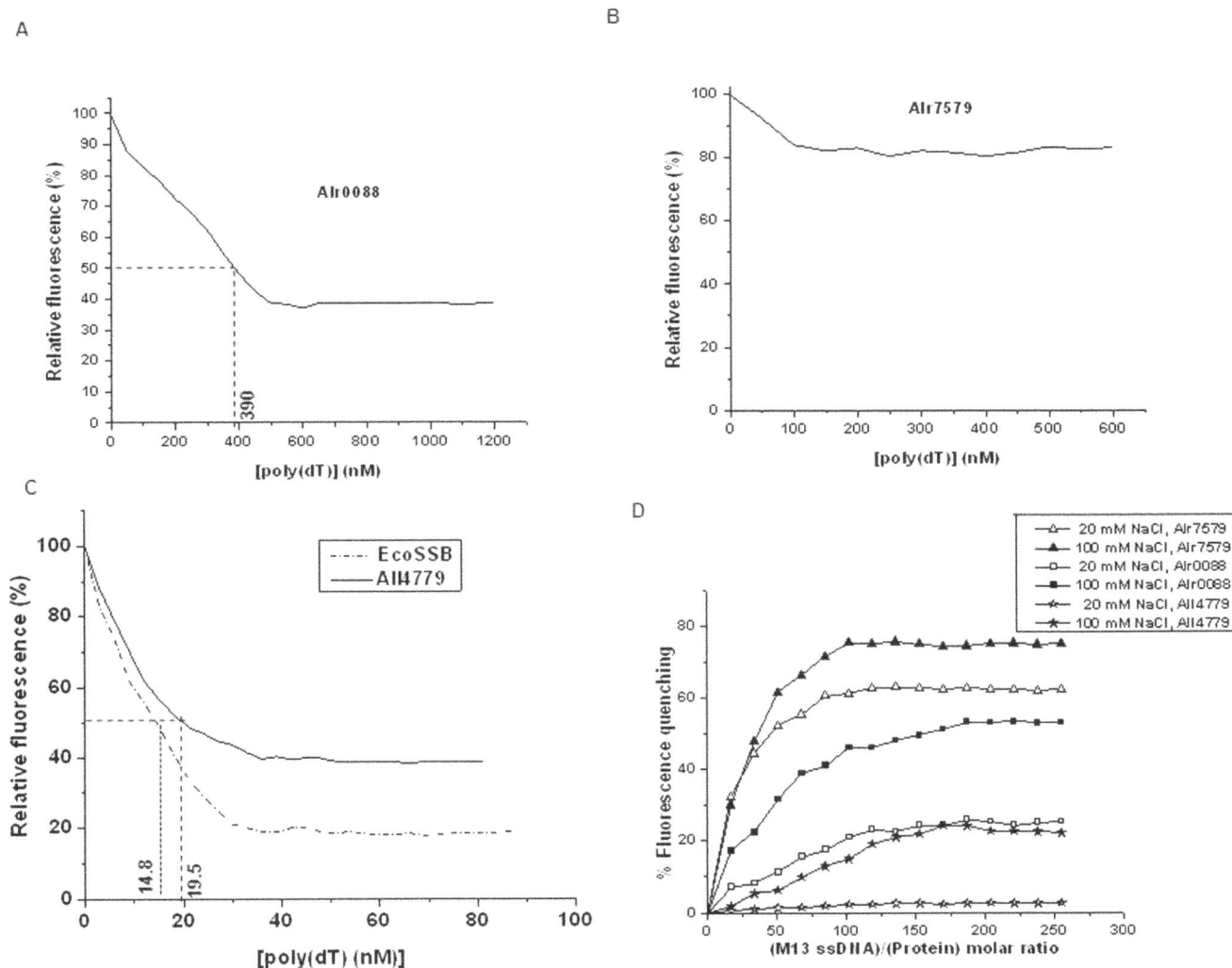

Figure 4. Relative quenching of fluorescence of native purified *Anabaena* SSB-like proteins and EcoSSB as a function of ssDNA concentration. (A–C) Quenching of fluorescence in 20 mM NaCl as a function of poly(dT) concentration of (A) Alr0088, (B) Alr7579 and (C) All4779 and purified EcoSSB (commercially available, Sigma) proteins represented as relative fluorescence, considering the observed fluorescence in the absence of any poly(dT) as 100%. The horizontal line designates the point on the graph wherein relative fluorescence is 50% and the corresponding vertical line indicates the concentration of poly(dT) at which it is achieved. Reciprocal of this concentration corresponds to the binding constant of the protein for poly(dT). (D) Percent fluorescence quenching of Alr0088, Alr7579 and All4779 proteins as a function of molar ratio of M13ssDNA and protein in the presence of 20 mM or 100 mM NaCl. The fluorescence quenching in the absence of M13ssDNA is considered as 0%.

resulting construct was designated as pAM*all4779* (Table 2). In this construct, the *gfpmut2* gene (coding for Green Fluorescent Protein, GFP) is co-transcribed with the upstream *all4779* gene from the P$_{psbA1}$ promoter, but the two transcripts are translated independently as described earlier [20]. Recombinant *Anabaena* strain overexpressing All4779 protein (An*all4779*$^{+}$) (Table 3) was generated by introducing the plasmid pAM*all4779* into *Anabaena* by conjugation as described earlier [20], and repeated selection on BG-11 agar plates containing 17 mM NaNO$_3$ (BG-11, N^{+}) and Neo$_{25}$, till completely segregated cells, uniformly expressing GFP, were obtained.

Results and Discussion

Bio-informatic Analysis of Alr0088, Alr7579 and All4779 Proteins

The *alr0088*, *alr7579* and *all4779* genes respectively encode 119, 127 and 182 amino acid long polypeptides (Figure 1A) with an estimated molecular mass of 13, 14 and 20 kDa. The prokaryotic SSBs are generally about 160–180 amino acids long, having a molecular mass of 17–18 kDa, except for SsbB of naturally transformable bacteria, which in case of BsSsbB is 113 amino acids long [11]. The SsbA protein of *B. subtilis* is 172 amino acids long, similar in size to EcoSSB [10]. Among the naturally non-transformable bacteria, the smallest known bacterial SSB are from the thermophilic bacteria, *Thermotoga maritima* (TmaSSB) and *T. neapolitana* (TneSSB) consisting of 141 and 142 amino acids respectively, having a single OB-fold domain and a C-terminal domain with the conserved DEPPF terminal amino acids [21].

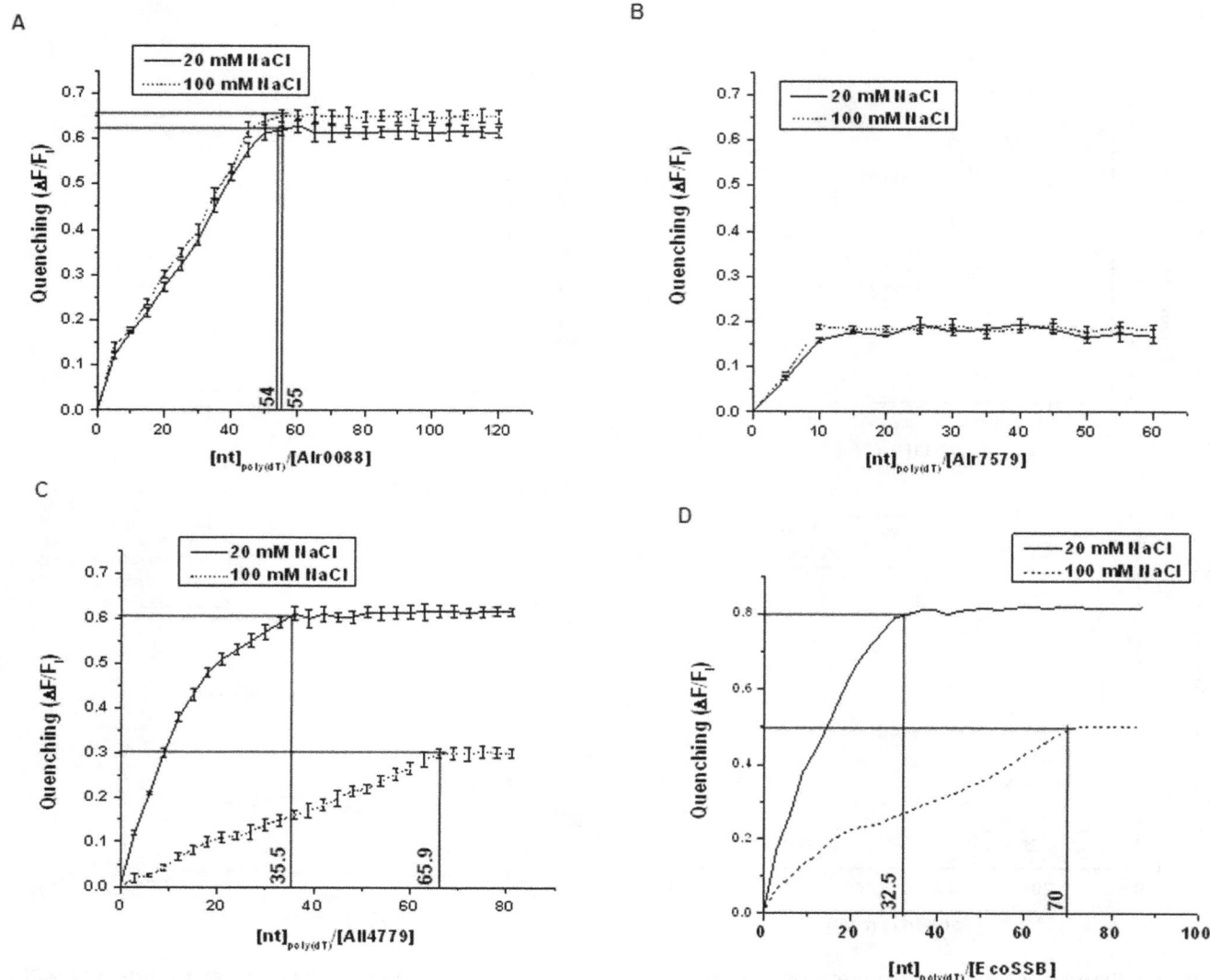

Figure 5. Quenching of fluorescence of native purified *Anabaena* SSB-like proteins compared with that of EcoSSB. The quenching of fluorescence of (A) Alr0088, (B) Alr7579, (C) All4779 and (D) EcoSSB in the presence of 20 mM or 100 mM NaCl was expressed as a ratio of change in fluorescence (ΔF) and initial fluorescence (F$_i$). The (ΔF/F$_i$) was expressed as a function of ratio of concentrations of poly(dT) and protein. The horizontal lines indicate the point of saturation and the vertical lines drawn from the point of saturation indicate the probable length of ssDNA bound by one molecular unit of the protein.

Among eukaryotes, the *Hs*mtSSB is 133 amino acids long and does not have the region corresponding to 1/3rd of the C-terminal region of EcoSSB [22]. Amino acid sequence analysis using Conserved Domain Database (CDD) [9] revealed the presence of a putative ssDNA-binding OB-fold domain and dimer/tetramer interface within N-terminal half in all three proteins similar to that in *E. coli* SSB (Figure 1A). All4779 additionally had a long C-terminal region similar to that observed for EcoSSB which comprised of a proline-rich region (19 residues) with two glycine residues, as compared to EcoSSB which is glycine rich (21 residues) and has 8 prolines in the corresponding region (Figure 1B). While multiple glycine residues allow flexibility in structure, multiple proline residues provide rigidity and kinks in the structure and thus no ordered structure results in gly-rich or pro-rich regions [23]. The proline-rich region of All4779 would also separate the positively charged N-terminal and the negatively

charged C-terminal regions, similar to that in EcoSSB [5]. The N-terminal region of All4779 exhibited 26% identical and 48% similar amino acid residues and a nearly identical acidic tail compared to EcoSSB (Figure 1B). In spite of having an N-terminal OB-fold, P/G rich region and a C-terminal acidic tail, the low homology of All4779 to other known bacterial SSB proteins may possibly account for it not being annotated earlier as SSB-like protein in the genome database of *Anabaena* 7120, unlike Alr0088 and Alr7579 which show a greater homology than All4779 in the OB-fold region.

Biochemical Characterisation of *Anabaena* Alr0088, Alr7579 and All4779 Proteins

The *Anabaena* Alr0088, Alr7579 and All4779 proteins overexpressed in *E. coli* BL21(pLysS) cells were purified to near homogeneity using Ni-NTA affinity chromatography (Figure 2A).

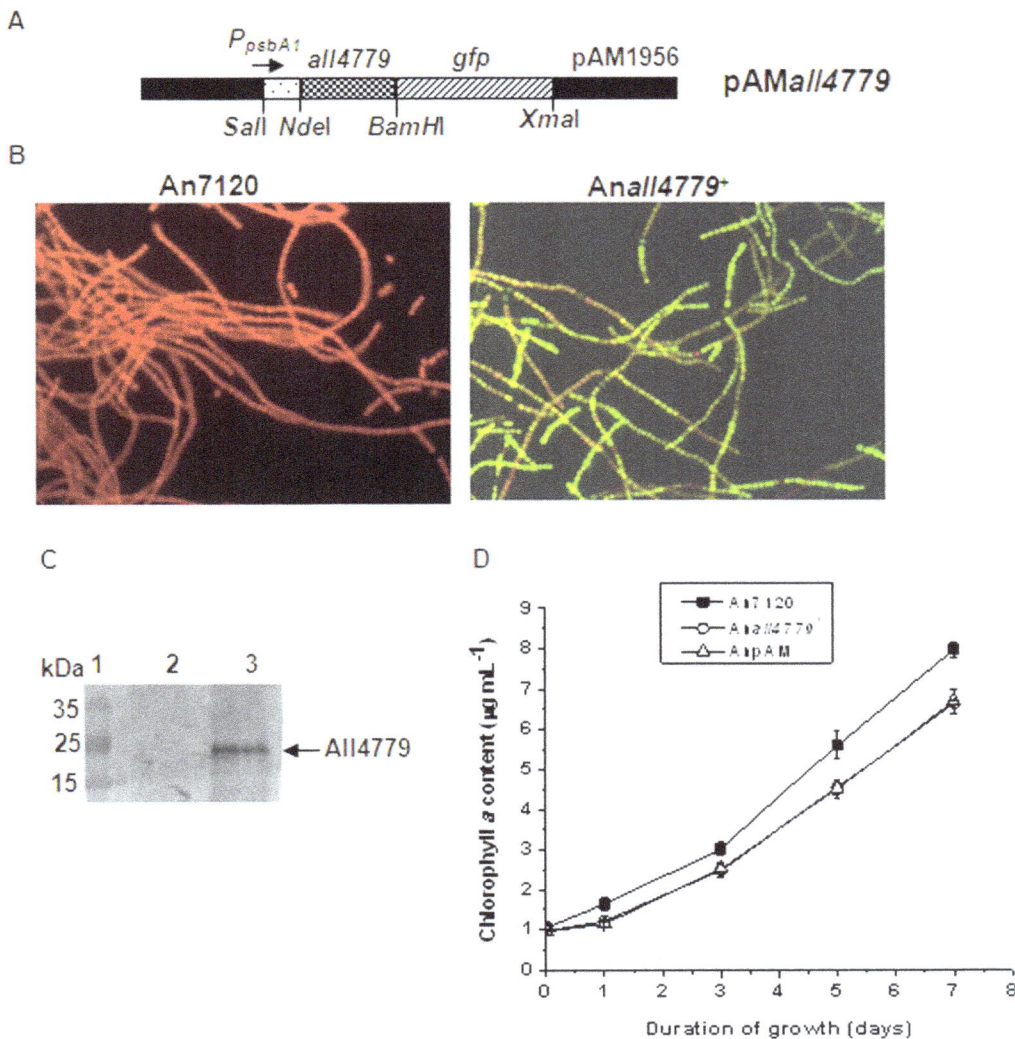

Figure 6. Construction of recombinant *Anabaena* strain overexpressing All4779 protein. (A) Schematic diagram of the plasmid construct, pAM*all4779* used for overexpression of All4779 protein in *Anabaena*. The different restriction enzymes used for cloning are indicated. (B) Fluorescence microphotograph (600X magnification) [using Hg-Arc lamp (excitation 470 nm, emission 508 nm)] of *Anabaena* 7120 [An7120] and recombinant strain, An*all4779*+, grown for 3 days in BG-11, N− media. (C) Protein extracts from *Anabaena* 7120 (lane 1) and An*all4779*+ (lane 2) were separated by 12% SDS-PAGE, followed by blotting on to nitrocellulose membrane and immunodetection of All4779 protein using anti-All4779 antibody. The cross-reacting All4779 protein is indicated by an arrow. Equal loading controls are shown below the blot. Other details were as described in legend to Figure 2. (D) Growth profile of wild type *Anabaena* 7120 and recombinant *Anabaena* strain, An*all4779*+ and AnpAM under nitrogen-fixing conditions over a period of 7 days. Growth was measured in terms of increase in chlorophyll *a* content. Recombinant strains were grown in presence of neomycin while wild type was grown without antibiotic.

Presence of dimer and tetramer interfaces in the amino acid sequence (Figure 1A) suggested possibility of formation of multimers by the protein. Alr0088 and Alr7579 were eluted in two distinct fractions and All4779 in a single fraction (Figure 2B) upon separation by gel filtration chromatography using Superdex HR200. On the basis of elution profile of standard proteins on the same matrix, the molecular mass of the different fractions was predicted as 14.1 kDa and 25.7 kDa for Alr0088, 14.5 kDa and 26.3 kDa for Alr7579 and 20.2 kDa for All4779 (Figure 2B). This indicated dimerisation of Alr0088 and Alr7579 proteins as against only the monomeric form detected for All4779 protein. Higher molecular forms of these proteins were not detected even at higher protein concentrations (data not shown). This did not conform to the bioinformatic prediction for the three proteins which indicate

the presence of dimeric and tetrameric interfaces (Figure 1A). Further probing of multimeric status was carried out by cross-linking the native proteins with glutaraldehyde followed by separation by SDS-PAGE. Upon cross-linking, the dimeric forms corresponding to 32 kDa for Alr0088 (Figure 3A), 34 kDa for Alr7579 (Figure 3B) and 41 kDa for All4779 (Figure 3C) were detected, with the levels of the dimeric form being lowest for All4779. This could be the reason for the inability to detect a higher molecular weight peak during gel-filtration chromatography for All4779 (Figure 2B). In the presence of M13 ssDNA the levels of the dimeric 41 kDa form as well as a probable tetrameric form of ~82 kDa increased (Figure 3C). This suggested that All4779 attains the native multimeric conformation preferably in the presence of ssDNA. In case of Alr0088 and Alr7579, no effect

A

B

C

D

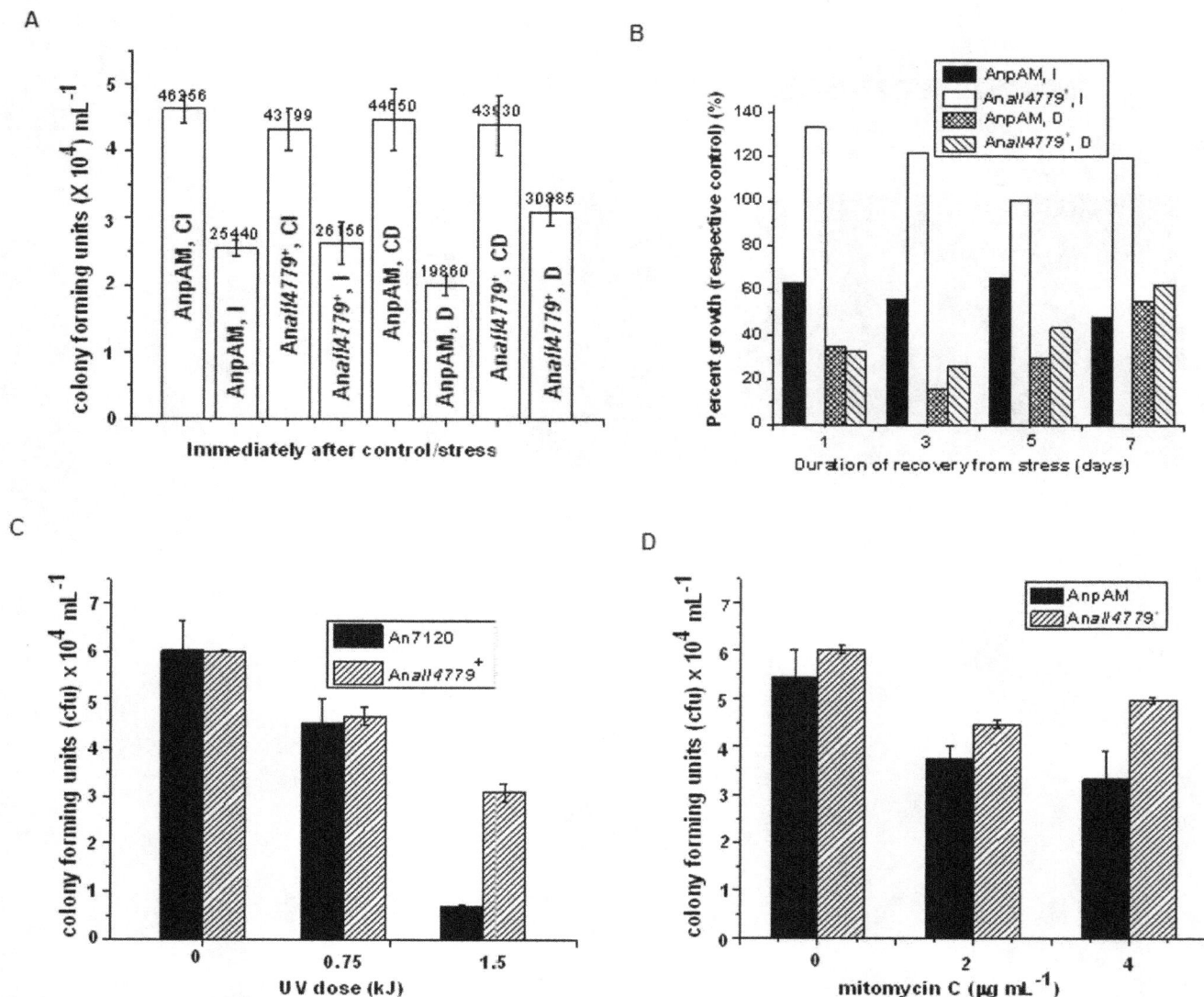

Figure 7. Effect of All4779 overexpression on the survival and tolerance of *Anabaena* **to DNA-damage inducing stresses.** (A and B) Three day-old cultures were concentrated to 10 µg chl*a* density mL^{-1} and exposed to 6 kGy of $^{6°}$Co γ-irradiation or to 6 days of desiccation. (A) Survival was measured in terms of colony forming units immediately after irradiation (I) or desiccation (D) and compared with the respective unirradiated control (CI) or undesiccated control (CD). (B) The stressed and control cultures were washed, inoculated in fresh BG-11, N$^-$, Neo$_{12.5}$ and allowed to recover under normal growth conditions for 7 days. Growth during post-irradiation/desiccation recovery was measured in terms of chlorophyll *a* content and expressed as percent of respective unirradiated/undesiccated controls. (C and D) Three-day-old cultures of recombinant strains AnpAM and An*all4779*$^+$ were concentrated to 10 µg chl*a* mL^{-1} density. (C) An 100 µl aliquot was spread on the corresponding BG-11, N$^-$, Neo$_{25}$ agar plates and exposed to UV-B (0–1.5 kJ) (D) Culture aliquots were exposed to mitomycinC (0–4 µg ml^{-1}) for 30 min in liquid media followed by plating 100 µl on BG-11, N$^-$ Neo$_{25}$ agar plate. Colonies were counted after 10 days of incubation at 27°±2°C with constant illumination.

on the levels of the dimeric form or generation of tetrameric form was observed even with ssDNA (data not shown). In general, bacterial SSB proteins function as tetramers [1], with the exception of thermophilic group of organisms (Thermus spp.) and the radioresistant microbe *Deinococcus radiodurans* [24] which function as dimer. However, the protomers of SSB of these organisms are twice the size of *E. coli* SSB and contain two OB-folds per monomer [24].

The DNA binding ability of the SSB-like proteins was assessed by Electrophoretic Mobility Shift Assay (EMSA) and fluorescence quenching techniques. Multiple shifts in the mobility of the 75-mer ss oligonucleotide was observed in the presence of Alr0088 (Fig. 3D). Increase in concentration beyond 0.2 µg Alr0088 did

not result in any further shifts in mobility (data not shown). Alr7579 decreased the mobility of the 75-mer oligo only when used at very high concentrations of 1.2–1.4 µg (Fig. 3D). Presence of Alr7579 also resulted in detection of multiple bands differing in their mobility, but majority of the complex formed, even when low concentrations of 0.12 µg of All4779 was used, was detected near the well (Fig. 3D). Based on this, the binding efficiency for ssDNA seems to be maximum for All4779, followed by Alr0088 and the least for Alr7579. The binding affinity for each of these proteins for ssDNA was calculated by fluorescence quenching technique using poly(dT) as the ssDNA substrate.

The relative fluorescence of native *Anabaena* proteins Alr0088, Alr7579 and All4779, was measured as a function of increasing concentration of poly(dT) at 20 mM NaCl. The relative fluorescence of (i) Alr0088 decreased to a maximum of 40% with ~450 nM poly(dT) (Figure 4A), (ii) Alr7579 showed less than 20% decrease (Figure 4B), and (iii) All4779 up to 40% of the initial fluorescence, but at much lower concentrations (~35 nM) of poly(dT) (Figure 4C). Based on this, the binding constant, as an average of three independent experiments, was calculated as $2.56 \pm 0.4 \times 10^6$ M^{-1} for Alr0088, $5.13 \pm 0.71 \times 10^7$ M^{-1} for All4779 and 6.76×10^7 M^{-1} for EcoSSB (Figure 4C), which was comparable to that reported for EcoSSB ($5.5 \pm 1.5 \times 10^7$ M^{-1}) [4]. In the absence of C-terminal acidic tail, the binding affinity for ssDNA has been shown to increase 10-fold in case of EcoSSB [4], as well as for HsmtSSB, which lacks the C-terminal tail, calculated as 4×10^8 M^{-1} [25]. However, the reverse was found to be true in case of Anabaena 7120, with Alr0088 which lacks the acidic tail, having 10-fold lower binding affinity than All4779. This could be due to the additional absence of the P/G-rich region as well in Alr0088.

The inability of Alr7579 to bind poly(dT) raised questions on whether the OB-fold, responsible for binding ssDNA [1] is active in Alr7579. To test this, a larger ssDNA, such as M13 ssDNA was used as a substrate. A 60–70% quenching of the fluorescence of Alr7579 was observed with the 7 kb M13 ssDNA, the efficiency being higher at high NaCl concentration (Figure 4D), which allows formation of a more compact structure of ssDNA. The quenching of fluorescence of Alr7579 was not observed with thermally denatured M13 ssDNA (data not shown). M13 ssDNA is known to form secondary structures [26], which are disrupted at higher temperature. This suggested that Alr7579 may be recognising secondary structures formed with long ssDNA, rather than short stretches of linear ssDNA. Both Alr0088 and All4779 also bound M13 ssDNA at high salt concentration, but with lower efficiency, the quenching of fluorescence being 40% and 18% respectively (Figure 4D). While low quenching of fluorescence of Alr0088 by M13 ssDNA was observed at low NaCl (Figure 4D), indicating low level interactions, no such interaction was observed for All4779 (Figure 4D).

In general, SSB proteins interact with ssDNA in multiple binding modes, differing in the number of OB-folds which interact with the ssDNA. In the (SSB)$_{35}$ mode, approximately 35 nucleotides of ssDNA interact with two subunits of the Ssb tetramer, while in (SSB)$_{65}$ mode, ~65 nucleotides of ssDNA wrap around all four subunits, which is more favoured at higher salt concentrations [25]. Based on the quenching of fluorescence (ΔF/F$_i$) of the three Anabaena proteins with poly(dT) at low (20 mM) NaCl and high (100 mM) NaCl concentrations, binding modes or occlusion size for each of the protein determined. A single binding mode of 54–55 nucleotides was estimated for Alr0088, which was independent of NaCl concentration (Figure 5A). No significant quenching of fluorescence of Alr7579 was observed at low or high concentrations of NaCl (Figure 5B), while two binding modes dependent on NaCl concentration was observed for All4779 (Figure 5C). The binding size was found to be 35.5 nucleotides at 20 mM NaCl and 65.9 nucleotides at 100 mM NaCl for All4779 (Figure 5C), and 32.5 and 70 nucleotides at 20 mM and 100 mM NaCl respectively for EcoSSB under identical experimental conditions, comparable to the (SSB)$_{35}$ and (SSB)$_{65}$ modes of binding, at low and high salt concentrations respectively, shown for EcoSSB [1,27]. Since, (SSB)$_{65}$ mode of binding requires the binding of ssDNA to the tetrameric form of SSB [27], and molecular form corresponding to a tetramer of All4779 was very low, the quenching of fluorescence of All4779 at higher NaCl was lower than that at lower NaCl (Figure 5C), as well as that observed with EcoSSB (Figure 5D).

Thus, though all the three proteins i.e. Alr0088, Alr7579 and All4779 bind ssDNA, their binding affinity and modes of binding are distinct and among these, the binding ability as well as binding modes of All4779 was quite similar to other known bacterial SSBs. The presence of (P/G)-rich spacer and a near identical C-terminal acidic tail, suggests that in Anabaena 7120, All4779 may also be performing in vivo functions similar to those carried out by the typical bacterial SSB proteins. Since, overexpression of bacterial SSBs are known to influence the repair of stress induced DNA damage [28,29], thereby enhancing tolerance to DNA damaging stresses, a similar role for All4779 was assessed in Anabaena 7120.

Physiological Role of All4779 Protein in Anabaena 7120

The All4779 protein was overexpressed in trans from the plasmid pAMall4779 (Table 2, Figure 6A) in the recombinant Anabaena strain, Anall4779$^+$ (Table 3). Due to growth under continuous illumination, the expression of the All4779 protein from the light-inducible psbA1 promoter was expected to be constitutive. Co-overexpression of the Green Fluorescent Protein (GFP), coded by gfpmut2 in the pAMall4779 plasmid, provided a handy tool to distinguish the fully segregated recombinant Anall4779$^+$ strain exhibiting green fluorescence, from the wild type Anabaena 7120 which exhibited red fluorescence upon excitation with λ_{470} light (Figure 6B). It also ensured expression of the upstream gene. The overexpression of All4779 in Anall4779$^+$ cells was indeed confirmed by immunodetection with anti-All4779 antibody (Figure 6C). Under normal growth conditions, the nitrogen-fixing cultures of Anall4779$^+$ grew marginally slower than the wild type Anabaena 7120 cultures, and at rates comparable to the recombinant Anabaena strain harbouring pAM1956 vector, AnpAM (Table 3) [30] (Figure 5D). This is possibly due to the presence of neomycin in the growth medium used for recombinant strain.

The effect of overexpression of All4779 on the ability of Anabaena 7120 to tolerate DNA damage inducing stresses was analysed in response to two distinct types of DNA damages i.e. (i) γ-irradiation and desiccation which cause single strand and double strand breaks, and (ii) UV-B and mitomycinC which cause formation of pyrimidine dimers and DNA adducts respectively. The empty vector control recombinant strain, AnpAM exhibited about 55% and 44% survival upon exposure to 6 kGy of ^{60}Co γ-rays or 6 days of desiccation respectively (Figure 7A). Upon constitutive overexpression of All4779 in Anall4779$^+$ cells, the survival increased to about 60% after exposure to 6 kGy of γ-rays and 70% after 6 days of desiccation (Figure 7A). The recovery of irradiated cultures of Anabaena, measured in terms of chlorophyll a content increased from about 50% to over 100% in cells overexpressing All4779 (Figure 7B) suggesting better tolerance to radiation. Such correlation was however, not found in post desiccation recovery, (Figure 7B), possibly due to additional stresses, such as osmotic stress experienced during desiccation followed by rehydration of these cells. Of the other two SSB-like proteins of Anabaena, overexpression of Alr0088 decreased the radiation tolerance of Anabaena, while that of Alr7579 had no effect [14]. This suggested that All4779 is the typical bacterial SSB of Anabaena, involved in the repair of single and double strand breaks in DNA, possibly as part of a larger DNA repair complex, which is yet to be identified.

Overexpression of All4779 was also beneficial in protection against stresses which caused formation of DNA adducts. The survival of AnpAM was about 75% and 11.8% respectively upon exposure to 0.75 and 1.5 kJ m^{-2} of UV-B irradiation, which increased to 77% and 50% respectively in Anall4779$^+$ cells, overexpressing All4779 protein (Figure 7C). The beneficial effect of the constitutive overexpression of All4779 was more pronounced when exposed to higher doses (1.5 kJ m^{-2}) of UV-B (Figure 7C), while at lower dose of (0.75 kJ m^{-2}), that of Alr0088 was more benficial [14]. AnpAM cells exhibited 50% survival upon exposure to $4 \text{ μg mitomycinC mL}^{-1}$ for 30 min, which increased to 85% upon overexpression of All4779 in An*all4779*$^+$ cells (Figure 7D), comparable to that observed upon overexpression of Alr7579, but lower than that with Alr0088 [14]. Thus, the presence of high levels of All4779 in *Anabaena* possibly decreased the net damage to DNA, both in terms of single and double stranded breaks as well as formation of DNA adducts, possibly by efficient repair of the damaged DNA. Overexpression of SSB has been shown to be beneficial by aiding DNA repair in *E. coli* cells [28].

Thus, All4779 is the major typical bacterial SSB of *Anabaena* 7120 in terms of structural domains, binding to ssDNA and physiological role in DNA repair. The genes coding for the two atypical truncated annotated SSB proteins, Alr0088 and Alr7579 may have arisen due to gene duplication as suggested for PriB, a dimeric protein with only OB-fold and capable of binding ssDNA [31] and may be involved in other functions such as replication and recombination. The unicellular cyanobacterium, *Synechocystis* PCC6803 has been shown to be naturally transformable with possible involvement of competence proteins, ComA (Slr0197) [32] and ComF (Slr0388) [33]. The orthologs of these genes are

also found in *Anabaena* 7120, annotated as *all3087* and *alr2926* respectively (http://genome.microbedb.jp/cyanobase/Anabaena), suggesting the possibility of *Anabaena* being also naturally transformable, though this needs to be ascertained. Thus, as has been observed in case of the naturally transformable *B. Subtilis*, the naturally C-terminal truncated BsSsbB, is involved in competence by protecting the incoming DNA [10,11], a similar role may also be associated with Alr0088 or/and Alr7579, both of which bear moderate homology to BsSsb, though this needs to be ascertained. The acidic tail, characteristic of most SSBs, has been shown to be the site of interaction with DNA repair proteins for *E. coli* SSB [3]. Owing to the presence of an acidic tail, All4779 upon overexpression offers better protection from DNA-damage when subjected to different DNA-damage-inducing stresses. Based on data presented, we propose that All4779 be re-annotated as the gene coding for typical single stranded DNA binding protein (SSB) and the corresponding ORF be annotated as the *ssb* gene of *Anabaena* 7120.

Author Contributions

Conceived and designed the experiments: AK HR. Performed the experiments: AK. Analyzed the data: AK HR. Contributed reagents/materials/analysis tools: AK HR. Wrote the paper: AK HR SKA.

References

1. Lohman TM, Ferrari ME (1994) *Escherichia coli* single-stranded DNA-binding protein: multiple DNA-binding modes and cooperativities. Annu. Rev. Biochem. 63: 527–570.
2. Carlini L, Curth U, Kindler B, Urbanke C, Porter RD (1998) Identification of amino acids stabilizing the tetramerization of the single stranded DNA binding protein from *Escherichia coli*. FEBS Lett. 430: 197–200.
3. Shereda RD, Kozlov AG, Lohman TM, Cox MM, Keck JL (2008) SSB as an organizer/mobilizer of genome maintenance complexes. Crit. Rev. Biochem. Mol. Biol. 43: 289–318.
4. Kozlov AG, Cox MM, Lohman TM (2010) Regulation of single-stranded DNA binding by the C termini of *Escherichia coli* single-stranded DNA binding (SSB) protein. J. Biol. Chem. 285: 17246–17252.
5. Eggington JM, Haruta N, Wood EA, Cox MM (2004) The single-stranded DNA- binding protein of *Deinococcus radiodurans*. BMC Microbiol. 4: 2doi 10.1186/1471-2180-4-2.
6. Apte SK (2001) Coping with salinity/water stress: Cyanobacteria show the way. Proc. Indian Natl. Acad. Sci (PINSA), B67: 285–310.
7. Singh H, Fernandes T, Apte SK (2010) Unusual radioresistance of nitrogen-fixing cultures of *Anabaena* strains. J. Biosci. 35: 427–434.
8. Singh H, Anurag K, Apte SK (2013) High radiation and desiccation tolerance of nitrogen-fixing cultures of the cyanobacterium *Anabaena* sp. strain PCC7120 emanates from genome/proteome repair capabilities. Photosynth. Res. 118: 71–81.
9. Marchler-Bauer A, Lu S, Anderson JB, Chitsaz F, Derbyshire MK et al. (2011) CDD: a Conserved Domain Database for the functional annotation of proteins. Nucleic Acids Res. 39: 225–229.
10. Yadav T, Carrasco B, Myers AR, George NP, Keck JL, et al., (2012) Genetic recombination in *Bacillus subtilis*: a division of labor between two single-strand DNA-binding proteins. Nuc. Acid Res. 40: 5546–5559.
11. Kidane D, Ayora S, Sweasy JB, Graumann PL, Alonso JC (2012) The cell pole: the site of cross talk between the DNA uptake and genetic recombination machinery. Crit. Rev. Biochem. Mol. Biol. 47: 531–555.
12. Castenholz RW (1988) Culturing methods for cyanobacteria. Methods Enzymol. 167: 68–93.
13. Mackinney G (1941) Absorption of light by chlorophyll solutions. J. Biol. Chem. 140: 315–322.
14. Kirti A, Rajaram H, Apte SK (2013) Characterization of two naturally truncated, Ssb-like proteins from the nitrogen-fixing cyanobacterium, *Anabaena* sp. PCC7120. Photosynth. Res. 118: 147–154.
15. Wadsworth RI, White MF (2001) Identification and properties of the crenarchaeal single-stranded DNA binding protein from *Sulfolobus solfataricus*. Nucleic Acids Res. 29: 914–920.
16. Molineux IJ, Pauli A, Gefter ML (1975) Physical studies of the interaction between the *Escherichia coli* DNA binding protein and nucleic acids. Nucleic Acids Res 2: 1821–1837.
17. Lohman TM, Overman LB (1985) Two binding modes in *Escherichia coli* single strand binding protein-single stranded DNA complexes. Modulation by NaCl concentration. J Biol. Chem. 260: 3594–3603.

18. Chaurasia AK, Parasnis A, Apte SK (2008) An integrative expression vector for strain improvement and environment applications of nitrogen-fixing cyanobacterium *Anabaena* sp. strain PCC7120. J. Microbiol. Methods. 73: 133–141.
19. Yoon HS, Golden JW (1998) Heterocyst pattern formation controlled by a diffusible peptide. Science. 282: 935–938.
20. Raghavan PS, Rajaram H, Apte SK (2011) Nitrogen status dependent oxidative stress tolerance conferred by overexpression of MnSOD and FeSOD proteins in *Anabaena* sp. strain PCC7120. Plant Mol. Biol. 77: 407–417.
21. Olszewski M, Grot A, Wojciechowski M, Nowak M, Mickiewicz M, et al. (2010) Charcterization of exceptionally thermostable single-stranded DNA-binding proteins from *Thermotoga maritima* and *Thermotoga neapolitana*. BMC Microbiol. 10: 260. doi.10.1186/1471-2180-10-260.
22. Curth U, Urbanke C, Greipel J, Gerberding H, Tiranti V, et al. (1994) Single-stranded-DNA binding proteins from human mitochondria and *Escherichia coli* have analogous physicochemical properties. Eur. J. Biochem. 221: 435–443.
23. Belts MJ, Russell RB (2003) Amino acid properties and consequences of substitutions. In Bioinformatics for Geneticists. Ed. Barnes MR and Gray IC. John Wiley and sons, 289–316.
24. Bernstein DA, Eggington JM, Killoran MP, Misic AM, Cox MM, et al. (2004) Crystal structure of the *Deinococcus radiodurans* single-stranded DNA-binding protein suggests a mechanism for coping with DNA damage. Proc. Natl. Acad. Sci. 101: 8575–8580.
25. Curth U, Genschel J, Urbanke C, Greipel J (1996) *In vitro* and *in vivo* function of the C-terminus of *Escherichia coli* single-stranded DNA binding protein. Nucleic Acids Res. 24: 2706–2711.
26. Reckmann B, Grosse F, Urbanke C, Frank R, Blocker H, et al. (1985) Analysis of secondary structures in M13mp8 (+) single-stranded DNA by the pausing of DNA Polymerase α. Eur. J. Biochem. 152: 633–643.
27. Bujalowski W, Lohman TM (1986) *Escherichia coli* single-strand binding protein forms multiple, distinct complexes with single-stranded DNA. Biochem. 25: 7799–7802.
28. Moreau PL (1987) Effects of overproduction of single-stranded DNA-binding protein on RecA protein-dependent processes in *Escherichia coli*. J. Mol. Biol. 194: 621–634.
29. Moreau PL (1988) Overproduction of single-stranded-DNA-binding protein specifically inhibits recombination of UV-irradiated bacteriophage DNA in *Escherichia coli*. J. Bacteriol. 170: 2493–2500.
30. Rajaram H, Apte SK (2010) Differential regulation of *groESL* operon expression in response to heat and light in *Anabaena*. Arch. Microbiol. 192: 729–738.
31. Ponomarev VA, Makarova KS, Aravind L, Koonin EV (2003) Gene duplication with displacement and rearrangement: origin of the bacterial replication protein PriB from the single-stranded DNA-binding protein, Ssb. J. Mol. Microbiol. Biotechnol. 5: 225–229.
32. Yura K, Toh H, Go M (1999) Putative mechanism of natural transformation as deduced form genome data. DNA Res. 6: 75–82.
33. Nakasugi K, Svenson CJ, Neilan BA (2006) The competence gene, *comF*, from *Synechocystis* sp. strain PCC6803 is involved in natural transformation, phototactic motility and piliation. Microbiol. 152:3623–3631.

DNA Binding Properties of the Actin-Related Protein Arp8 and Its Role in DNA Repair

Akihisa Osakabe²⁹, Yuichiro Takahashi¹⁹, Hirokazu Murakami¹, Kenji Otawa², Hiroaki Tachiwana²,
Yukako Oma¹, Hitoshi Nishijima³, Kei-ich Shibahara³, Hitoshi Kurumizaka²*, Masahiko Harata¹*

1 Laboratory of Molecular Biology, Graduate School of Agricultural Science, Tohoku University, Sendai, Japan, **2** Laboratory of Structural Biology, Graduate School of Advanced Science and Engineering, Waseda University, Tokyo, Japan, **3** Department of Integrated Genetics, National Institute of Genetics, Mishima, Japan

Abstract

Actin and actin-related proteins (Arps), which are members of the actin family, are essential components of many of these remodeling complexes. Actin, Arp4, Arp5, and Arp8 are found to be evolutionarily conserved components of the INO80 chromatin remodeling complex, which is involved in transcriptional regulation, DNA replication, and DNA repair. A recent report showed that Arp8 forms a module in the INO80 complex and this module can directly capture a nucleosome. In the present study, we showed that recombinant human Arp8 binds to DNAs, and preferentially binds to single-stranded DNA. Analysis of the binding of adenine nucleotides to Arp8 mutants suggested that the ATP-binding pocket, located in the evolutionarily conserved actin fold, plays a regulatory role in the binding of Arp8 to DNA. To determine the cellular function of Arp8, we derived tetracycline-inducible Arp8 knockout cells from a cultured human cell line. Analysis of results obtained after treating these cells with aphidicolin and camptothecin revealed that Arp8 is involved in DNA repair. Together with the previous observation that Arp8, but not $\hat{\gamma}$-H2AX, is indispensable for recruiting INO80 complex to DSB in human, results of our study suggest an individual role for Arp8 in DNA repair.

Editor: Fatah Kashanchi, George Mason University, United States of America

Funding: This work was supported by Grants-in-Aid for Scientific Research on Innovative Areas from the Japanese Society for the Promotion of Science (JSPS) and the Ministry of Education, Culture, Sports, Science and Technology (MEXT), and by the Human Frontier Science Program (RGP0017). H. Kurumizaka was also supported by the Waseda Research Institute for Science and Engineering. The funders had no role in study design, data collection and analysis, decision to publish, or preparation of the manuscript.

Competing Interests: The authors have declared that no competing interests exist.

* Email: kurumizaka@waseda.jp (HK); mharata@biochem.tohoku.ac.jp (MH)

⁹ These authors contributed equally to this work.

Introduction

Chromatin structure governs genome function, including transcription, DNA damage repair, and replication. The chromatin structure, in its default state, limits the accessibility of DNA binding factors. So, in order for gene expression and DNA repair to take place, chromatin must open up for these factors. Chromatin remodeling complexes are known to play a major role in chromatin opening. Consequently, their activity and recruitment to chromatin must be tightly regulated for exercising proper genome functioning. These remodeling complexes contain multiple regulatory subunits. Thus, to understand the epigenetic regulatory mechanisms of these complexes, it is imperative to know the properties of their regulatory subunits.

Several members of the actin family of proteins, which are evolutionarily conserved, are essential components of these chromatin remodeling complexes [1,2]. The actin family consists of conventional actin and other evolutionarily and structurally similar actin-related proteins (Arps). Although only a portion of actin is found in the nucleus, some of the Arps are predominantly localized in the nucleus. These nuclear Arps, in most cases together with actin, are known to be essential components of various chromatin modulating complexes. For example, the INO80 chromatin remodeling complex, which is evolutionarily conserved from yeast to man, have been reported to contain actin

and three Arps (Arp4, Arp5, and Arp8). Actin and Arps share the evolutionarily conserved actin fold, which contains the ATP-binding pocket at the center. A model has been proposed, wherein any structural change in the actin fold of actin or an Arp, occurred as a result of binding of an adenine nucleotide (ATP/ADP) to this ATP-binding pocket, contributes to the regulation of cellular functions of these proteins, including polymerization of actin, and also probably assembly of actin and Arps into chromatin remodeling complexes [1,3,4,5].

Two major roles have been proposed for the nuclear Arps in chromatin remodeling and histone modification complexes. First, Arps are responsible for recruiting the complexes to chromatin. Indeed, Arp4 and Arp8 have been shown to bind to core histones [6,7,8,9,10]. It has been shown that the yeast Arp8 binds to a 30 bp long DNA with low affinity (in the micromolar range), whereas the human Arp8 binds to the same 30-bp long DNA with about 3-fold less affinity [9]. Arp5 is also required for the recruitment of INO80 complex to chromatin, although direct binding of Arp5 to chromatin has not been detected so far (Chen et al., 2014; Shen et al., 2003). Second, it has been shown that nuclear Arps regulate the ATPase activity of the Snf2-type ATPase of the chromatin remodeling complexes (Matsuda et al., 2010; Wu et al., 2003; Wu et al., 2005). In yeast, Arp5 and Arp8 seem to regulate the ATPase activity of INO80 by different mechanisms.

Thus, the ATPase activity of INO80 lacking the Arp8 was not stimulated by DNA, but was simulated only by the nucleosome core particle, whereas the ATPase activity of INO80 lacking the Arp5 was stimulated by DNA, but was not stimulated by the nucleosome [11].

The INO80 complex binds to selected regions of the genome, including the 5′ and 3′ regions of the open reading frames of genes, and regulates gene expression [12,13]. In addition, the INO80 complex is recruited to double-strand breaks (DSBs) [14,15] and to stalled replication forks [16], and is involved in maintaining the genome integrity by promoting the repair processes and restarting the replication at the stalled fork. Both in budding yeast and human, the INO80 complex is required during the DSB repair for effective DNA end resection [14,15]. Since DNA end resection is an early event that take place during the homologous recombination (HR) repair process, it is believed that the INO80 complex assists the function of an endonuclease through the remodeling of nucleosomes proximal to DSB.

In humans, defects in the maintenance of genome stability can lead to cancer development and progression. Interestingly, by using a RNA interference assay, it was observed that among all the tested subunits only Arp8 was indispensable for recruiting the INO80 complex to DSB in human cells [17]. Thus, to understand the underlying molecular basis of multiple function of the INO80 complex in gene expression and genome integrity, it would be necessary to analyze the biochemical properties of human Arp8 and determine the phenotype of human cells lacking Arp8. In the present study, we purified and characterized the bacterially expressed human Arp8 and also established a tetracycline (tet)-inducible Arp8-knockout human cell line. We found that the purified human Arp8 possessed ssDNA-binding activity, and therefore, we subsequently examined its possible role in HR repair of human genome.

Results

Preparation of recombinant human Arp8 and its deletants

To analyze the biochemical properties of human Arp8, we purified recombinant human Arp8 after expressing it in *E. coli* cells as a His$_6$-tagged protein (Fig. 1A). During the purification process, the His$_6$ tag was removed by treating with PreScission protease to obtain purified recombinant Arp8 without the His$_6$ tag. A characteristic feature of Arps is the presence of specific insertions in the actin fold, a core structure common to both actin and Arps. The human Arp8 contains an extension in the N-terminal end. In addition, it has multiple insertions, of which insert IV is the largest (Fig. S1) [7,9,10,18]. To analyze the functions of the N-terminal extension and insertion IV, we created and purified an Arp8 deletion mutant lacking the N-terminal amino acids 1 to 38 (Arp8 Δ1-38; Fig. 1B and Fig. S1) and another Arp8 deletion mutant lacking the insertion IV (deletion of amino acids 403 to 463; Arp8 Δ403-463; Fig. 1B and Fig. S1).

Arp8 binds to double-stranded DNA

The N-terminal extension of the human Arp8 is abundant in basic amino acids. We previously analyzed this N-terminal sequence using a software and predicted that it might bind to DNA [19]. It was reported earlier that Arp8 associated with a synthetic 30-bp double-stranded DNA (dsDNA), albeit with very low affinity, and the N-terminal extension was not necessary for the association [9]. We expected that the human Arp8 would form stable complexes with DNAs and the N-terminal extension is involved in this DNA binding process. Gel shift analysis revealed

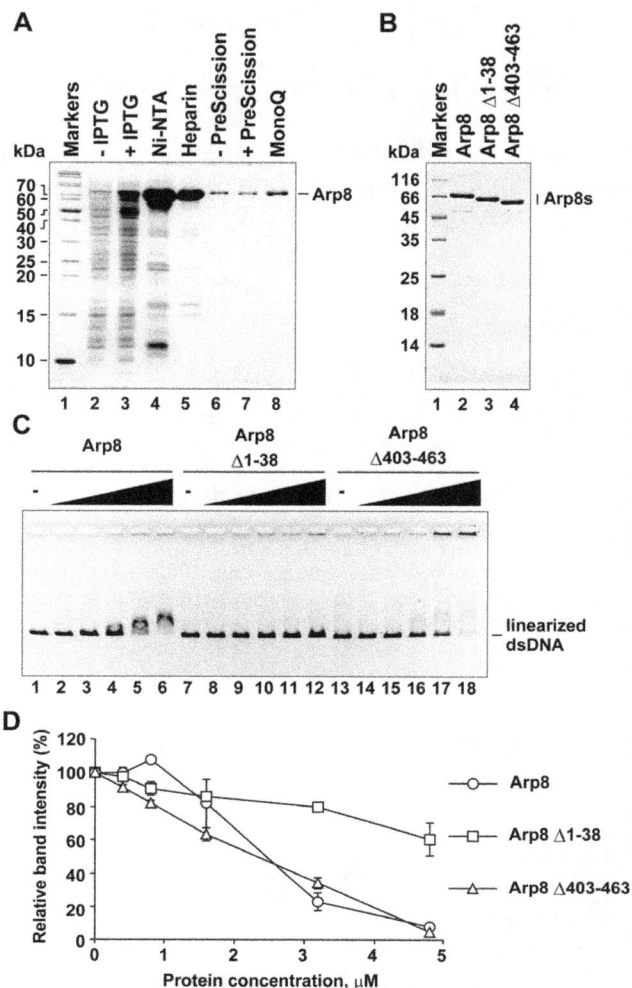

Figure 1. Double-stranded DNA binding activity of purified human Arp8 and its mutants. (A) Purification of human Arp8. Protein fractions from each purification step were analyzed by SDS-PAGE (gel was stained with Coomassie Brilliant Blue). Lane 1, molecular weight markers. Lanes 2 and 3, whole cell lysates before and after induction with IPTG, respectively. Lanes 4 and 5, peak fractions from Ni-NTA agarose and Heparin Sepharose columns, respectively. Lanes 6 and 7, Heparin Sepharose fraction before and after treatment with PreScission protease (removal of His6 tag). Lane 8, peak fraction from MonoQ column. (B) Purified wild-type and deletion mutants (deletants) of Arp8. Lane 1, molecular weight markers. Lanes 2-4, purified Arp8, Arp8 Δ1-38 (N-terminal deletion), and Arp8 Δ403-463 (insertion IV deletion), respectively. (C) dsDNA binding activities of Arp8 and its deletants. Bindings of Arp8 (lanes 2-6), Arp8 Δ1-38 (lanes 8-12), and Arp8 Δ403-463 (lanes 14-18) to linearized φX174 were examined at various protein concentrations: 0 μM (lanes 1, 7, and 13), 0.4 μM (lanes 2, 8, and 14), 0.8 μM (lanes 3, 9, and 15), 1.6 μM (lanes 4, 10, and 16), 3.2 μM (lanes 5, 11, and 17), and 4.8 μM (lanes 6, 12, and 18). (D)Intensity of the unbound DNA in each lane of panel C was quantified and then plotted as relative intensity (%) with respect to that of the unbound DNA from the control (no protein added control) lane.

that the full length Arp8 bound to a linearized plasmid DNA (Fig. 1C, lanes 1 to 6). Arp8 also bound to nicked-circular and closed-circular DNAs (Fig. S2). The DNA binding activity of Arp8 is consistent with the earlier observation that the INO80 complex lacking the Arp8 has only partial DNA binding activity, and that a recombinant complex consisting of the HSA domain of Ino80, actin, Arp4, and Arp8 binds to DNA [7,11].

In our analysis, the Arp8 mutant Arp8 Δ1–38, which lacked the N-terminal extension, was unable to cause any shift in the DNA mobility (Fig. 1C, lanes 7 to 12), indicating that this extension is required for the stable binding of Arp8 to DNA. We have also tested the DNA binging activity of the Arp8 Δ403–463 deletant (lacking insert IV), and found that this insertion is not required for the DNA binding (Fig. 1C, lanes 13 to 18). However, the positions of DNA bands shifted by the Arp8 Δ403–463 deletant (lacking insertion IV) differed from those shifted by the wild-type Arp8, and a portion of the DNA remained at the origin. Although it remains to seen whether this observed differences in shifts is due to an alteration in the properties of the bound protein, it is likely that the altered shift reflects some involvement of the insertion IV in forming a proper Arp8-DNA complex. These observations suggest that multiple regions, protruding from the actin fold of Arp8, might contribute to the formation and/or properties of the Arp8-DNA complex in different manners.

Arp8 binds to single-stranded DNA preferentially

During the HR repair process, the INO80 complex is involved in effective DNA end resection [14,20,21]. Therefore, we thought that Arp8 could also bind to single-stranded DNA (ssDNA), which is produced in the proximal regions of DSB by DNA end resection. Gel shift assay revealed that indeed Arp8 forms a stable complex with the ssDNA (Fig. 2, A and B). Remarkably, in the presence of same molar ratios of shorter and longer dsDNAs, the unbound fraction of ssDNA disappeared earlier than the unbound dsDNAs (Fig. 2A). These results suggest that Arp8 binds preferentially to ssDNA and that this property likely contributes to recruiting the INO80 complex to the DSB sites (see below).

As was found with the dsDNA, the Arp8 Δ403–463 deletant bound to ssDNA similarly as the full-length Arp8 (Fig. 2B, lanes 13 to 18). Although the Arp8 Δ1–38 mutant, which lacked the N-terminal extension, caused a shift in the DNA mobility (Fig. 2B, lanes 7 to 12), the apparent shift was less than that was observed for the dsDNA (Fig. 1C, lanes 7 to 12). These results suggest that the contribution of these protruding regions to the binding of ssDNA is probably similar, but not same, as their contribution to the binding of dsDNA.

During the HR process, DNA resection of the DSB generates ssDNA with 3'-overhangs and this gap between the ssDNA and dsDNA becomes the target of further DNA resection. Since the INO80 complex was reported to be involved in DNA resection, we tested a possibility that Arp8 has high affinity for DNA fragments with 3'-overhangs. The binding of Arp8 to 3'-overhang DNA was analyzed by a competitive gel shift assay in which the molar ratios of dsDNA and ssDNA were same (Fig. 2C). Results shown in Fig. 2C (plot below the gel panel) suggested that Arp8, which has a binding preference for ssDNA, binds to the 3'-overhang DNA with an apparent affinity that is in between its affinity for ssDNA and dsDNA). This result suggests that Arp8 binds to the gap between the ssDNA and dsDNA with an affinity that was less or similar to that of the ssDNA.

ATP affects the DNA binding activity of Arp8

It was reported earlier that ATP affects the dsDNA binding activity of the budding yeast INO80 complex *in vitro* [7]. Arp8, which shares the conserved ATP-binding pocket of the actin fold with actin, has been reported to have ATP binding activity [9,10]. We therefore examined the effect of ATP and ADP on the DNA binding activity of Arp8. Addition of ATP to the binding assay mixture significantly decreased the binding of Arp8 to linearized dsDNA (Fig. 3A, lanes 10 to 13) and circular dsDNA (Fig. S2, lanes 7 to 12). In contrast, addition of ADP only slightly decreased the binding of Arp8 to dsDNAs (Fig. 3A, lanes 14 to 17; and Fig. S2, lanes 13 to 18). The rabbit actin exhibited significantly low DNA binding activity under the same experimental condition (Fig. 3A, lanes 1 to 5). On the other hand, addition of ATP only slightly decreased the binding of Arp8 to ssDNA (Fig. 3B, lanes 10 to 13), whereas addition of ADP seemed to have little or no effect on the binding of Arp8 to ssDNA (Fig. 3B, lanes 14 to 17). These results suggested the possibility that ATP, but not ADP, may be a regulator of the DNA binding activity of Arp8.

To study the contribution of the ATP binding pocket of Arp8 in DNA binding, we designed and prepared three Arp8 mutants, namely Arp8 S55A T56A, Arp8 E266A, and Arp8 K288A S290A, by replacing the S55/T56, E266, or K288/S290 residue(s) with an Ala, respectively (Fig. 4A and Fig. S3). We chose these amino acids for the mutation analysis because they are expected to be positioned in or around the ATP-binding pocket of Arp8. In the absence of ATP, all three Arp8 mutants bound to the dsDNA (Fig. 4B, lanes 11 to 15 of upper panel, and lanes 1 to 5 and lanes 11 to 15 of lower panel, respectively) in the same manner as the wild-type Arp8 (Fig. 4B, lanes 1 to 5 of upper panel). However, in the presence of ATP, the DNA-binding activities of these Arp8 mutants were clearly different from that of the wild-type Arp8. Thus, addition of ATP significantly inhibited the DNA binding activity of the wild-type Arp8 (Fig. 4B, lanes 6 to 10), and consequently the amount of unbound DNA was increased in the presence of ATP (Fig. 4C, Arp8). In contrast, addition of ATP did not inhibit the DNA binding activities of the Arp8 S55A T56A and Arp8 K288A S290A mutants (Fig. 4B, lanes 16 to 20 of upper and lower panels, respectively), and as a result the amounts of free DNA hardly or slightly increased for these two mutants (Fig. 4B, Arp8 S55A T56A and Arp8 K288A S290A). However, addition of ATP only partially inhibited the DNA binding activity of the Arp8 E266A mutant (Fig. 4B, lanes 6 to 10 of lower panel, and Fig. 4C, Arp8 E266A). Taken together, these results suggest that the S55, T56, K288, and S290 residues of Arp8, all of which are located in the ATP binding pocket, might play some regulatory role in the binding of Arp8 to DNA in the presence of ATP.

Establishment of Arp8-overexpressing and Arp8-knockout human cell lines

So far, the function of Arp8 in mammalian cells has been analyzed by knocking down its expression by RNA interference. We have developed a conditional Arp8 knockout (KO) cell line from the parent human Nalm-6 B cell line (Fig. 5) (see also Materials and Methods). In the *ARP8*[−/−/transgene] cells, the expression of Arp8 is under the control of a tetracycline (tet)-repressible promoter. Western blot analysis showed that Arp8 was overexpressed in *ARP8*[−/−/transgene] cells in the absence of tetracycline (Fig. 6A, day 0). The amount of expressed Arp8 reduced gradually after the addition of tetracycline, and expression of Arp8 was not detectable after 8-day (Fig. 6A). Since our previous study indicated that this tet-repressible promoter shuts down within a day after the tetracycline addition [22], results obtained in this study suggest that Arp8 is a relatively stable protein. Overexpression of Arp8 did not have any apparent adverse effect on the cell growth (Fig. 6B, -tet versus WT). However, following the addition of tetracycline, when Arp8 became undetectable on Western blot, a significant decline in the growth of *ARP8*[−/−/transgene] cells, but no immediate cell death, was observed. This result suggests that Arp8 is required for normal cell growth.

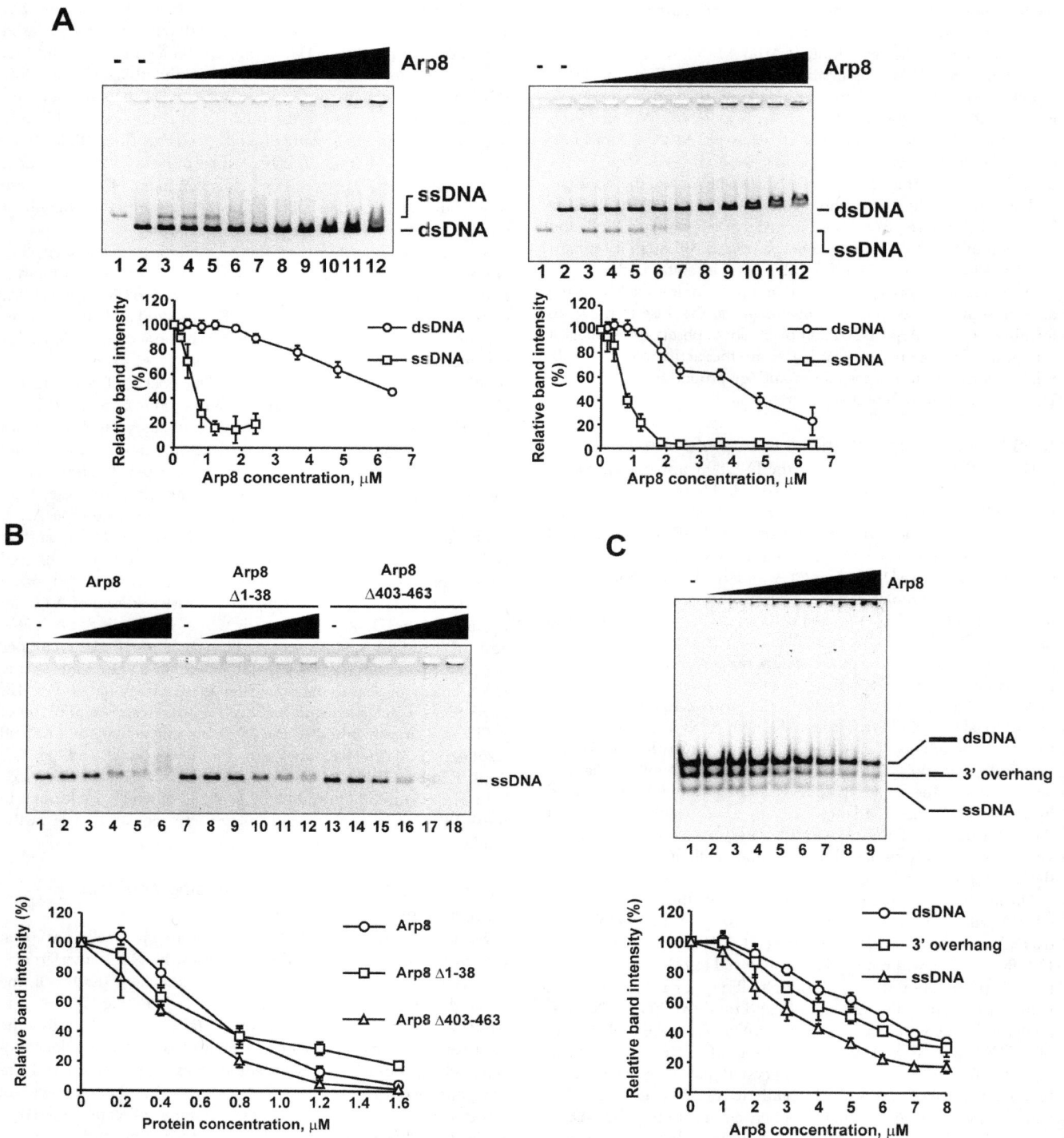

Figure 2. Binding of single-stranded DNA to Arp8. (A) Comparison of the binding of ssDNA (20 μM) and dsDNA (20 μM) to Arp8.
Lanes 1 and 2 contain only ssDNA and dsDNA, respectively, but no protein. Lanes 3, 4, 5, 6, 7, 8, 9, 10, 11, and 12 contain 0, 0.2, 0.4, 0.8, 1.2, 1.8, 2.4, 3.6, 4.8, and 6.4 μM of Arp8, respectively. Positions of protein-free dsDNA and ssDNA were shown. (B) Binding of ssDNA to Arp8 and its deletants. Bindings of Arp8 (lanes 2–6), Arp8 Δ1–38 (lanes 8–12), and Arp8 Δ403–463 (lanes 14–18) to linearized φX174 were examined at various protein concentrations: 0 μM (lanes 1, 7, and 13), 0.2 μM (lanes 2, 8, and 14), 0.4 μM (lanes 3, 9, and 15), 0.8 μM (lanes 4, 10, and 16), 1.6 μM (lanes 5, 11, and 17), and 3.2 μM (lanes 6, 12, and 18). (C) Competitive binding of Arp8 to dsDNA, 3'-overhang DNA, and ssDNA (3 μM each). Lanes 1–9 contain 0, 1, 2, 3, 4, 5, 6, 7, 8 μM of Arp8 protein, respectively. Positions of protein-free dsDNA, 3'-overhang, ssDNA are shown. Quantification of each gel: intensity of the unbound DNA in each lane was quantified and then plotted as relative intensity (%) with respect to that of the unbound DNA from the control (no protein added control) lane.

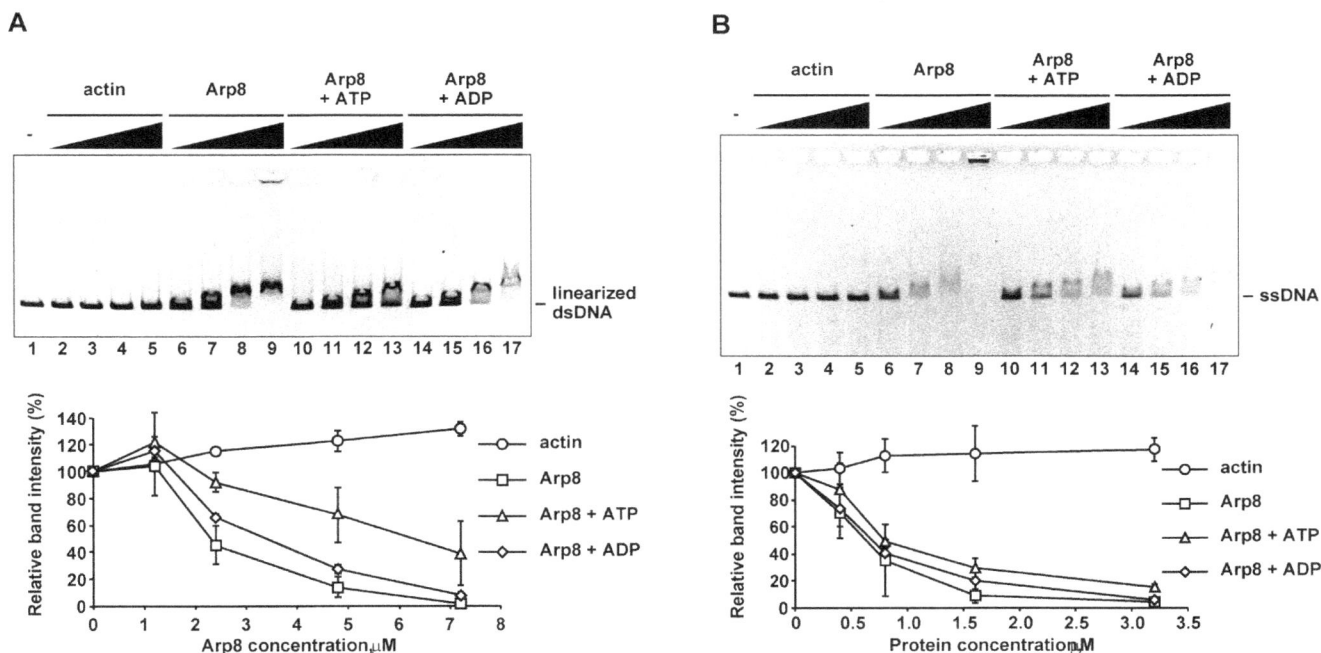

Figure 3. Effect of ATP on the binding of Arp8 to DNAs. Bindings of actin (lanes 2 to 5) and Arp8 (lanes 6 to 13) to dsDNA (A) and ssDNA (B) were examined in the absence (lanes 2 to 5 and 10 to 13) and presence (lanes 6 to 9) of 1 mM ATP, or in the presence of 1 mM ADP (lanes 14 to 17). Various concentrations of Arp8 (lanes 6–17) and actin (lanes 2–5) were used in this experiment. Concentrations of Arp8 used in panel (A) were: 1.2 μM (lanes 2, 6, 10, and 14), 2.4 μM (lanes 3, 7, 11, and 15), 4.8 μM (lanes 4, 8, 12, and 16), and 7.2 μM (lanes 5, 9, 13, and 17). Concentrations of Arp8 used in panel (B) were: 0.4 μM (lanes 2, 6, 10, and 14), 0.8 μM (lanes 3, 7, 11, and 15), 1.6 μM (lanes 4, 8, 12, and 16), and 3.2 μM (lanes 5, 9, 13, and 17). Lane 1 in both A and B: no protein added control. Intensity of the unbound DNA in each panel was quantified and plotted as before (see Fig. 2 legend).

DNA repair is impaired in Arp8-knockout cells

To test the possibility whether Arp8 may have any role in DNA repair, we treated wild-type, Arp8-overexpessed (Arp8 OE), and Arp8-knockout (Arp8 KO) cells with aphidicolin and camptothecin to induce DSBs in these cells. Although the repair process of aphidicolin- and camptothecin-induced DSBs are not closely connected to the HR-repair process of endonuclease-induced DSBs, there are some similarities between these two repair processes [23,24,25,26]. Fig. 6C shows the relative inhibition of increase in number of cells grown in the presence of aphidicolin as compared to that of the untreated cells. In this plot, 0% inhibition means that the cells grew in the presence of aphidicolin to the same extent as in the absence of the drug, and 100% inhibition means that no increase in cell number was observed in the presence of aphidicolin (see legend of Fig. 6C for further explanation). In wild-type cells, aphidicolin inhibited the cell growth by 50%. In Arp8 KO cells, the value was −15%, which meant that the cell number did not increase at all in the presence of aphidicolin, but instead decreased, possibly because of apoptosis. Thus, this induced inhibition in cell growth was most severe for the Arp8 KO cells than for the wild-type and Arp8 OE cells. This result suggests that knockout of Arp8 probably impairs DNA repair, which probably takes place via the HR or an HR-like repair process.

Camptothecin-induced repair was also analyzed in wild-type, Arp8 OE, and Arp8 KO cells (Fig. 6D). For this experiment, cells were treated with camptothecin for 1 h. Cells were then washed to remove camptothecin and were incubated for up to 8 h without the drug. Induced DSBs were analyzed by comparing the relative number of γ-H2AX foci. As shown, in all three cells the observed number of γ-H2AX foci was not significantly different at 0 h of the

drug free incubation (Fig. 6D, 0 h). However, the number of γ-H2AX foci in Arp8 KO cells, observed after 2 h and 8 h of drug free incubation, respectively, was significantly higher than the number of foci observed in the wild-type and Arp8 OE cells (Fig. 6D, 2 h and 8 h), suggesting that the repair of camptothecin-induced DSB is impaired in the absence of Arp8. Consistent with this observation, the camptothecin-induced inhibition in cell growth was most severe for the Arp8 KO cells than for the wild-type and Arp8 OE cells (Fig. 6E). The experiment to determine sensitivity to camptothecin was performed using Arp8 KO cells that were treated with tetracycline for eight days, and the results were normalized with respect to the Arp8 KO cells that were not exposed to camptothecin. These observations support the idea that Arp8 contributes to the progression of DNA repair which probably takes place via the HR or an HR-like repair process.

Discussion

Our study suggested that the binding of human Arp8 to both dsDNA and ssDNA is relatively stable. This is the first example demonstrating the binding of dsDNA and ssDNA to one of the components of the INO80 complex and also to a member of the actin family of proteins. Recently, three dimensional architecture of the budding yeast INO80 complex has been determined by cryo-electron microscopy [11]. Accordingly, the INO80 complex is organized in four modules (head-neck-body-foot architecture). Among these modules, Arp8 forms the foot module together with Arp4, another histone-binding Arp, and this module is called as the Arp8 module [11]. When the INO80 complex binds to chromatin, a nucleosome is captured and placed between the Arp8 module and the head module. As the link between the Arp8 module and body module is flexible, it allows the Arp8 module to

A

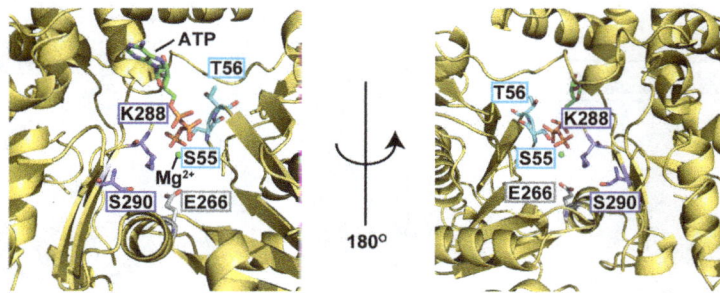

B

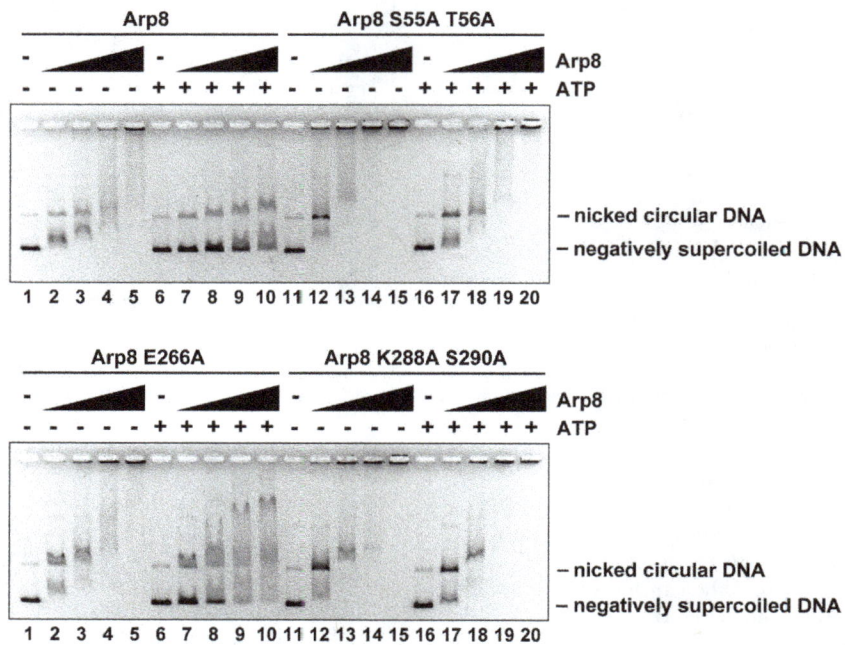

C

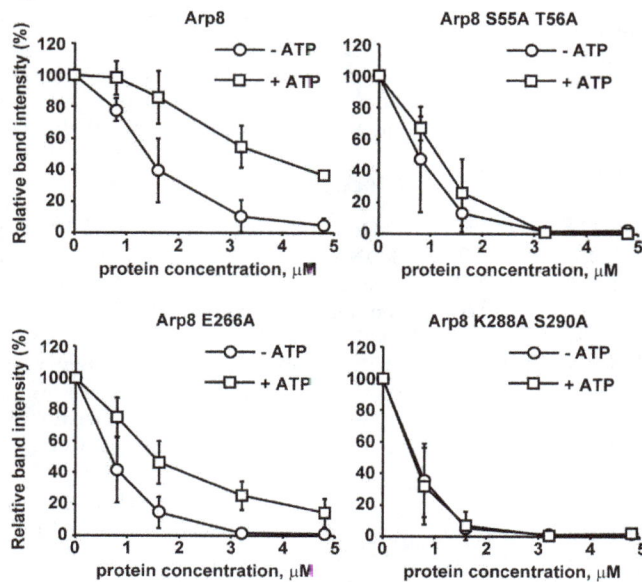

Figure 4. Binding of Arp8 mutants to DNAs. (A) Positions of mutations introduced in the ATP binding pocket of Arp8. The crystal structure of the ATP binding pocket of Arp8 was obtained from the Protein Database (PDB ID: 4FO0). (B) Binding of Arp8 and Arp8 mutants to supercoiled and nicked circular forms of φX174. Following concentrations of Arp8 (upper panel, lanes 1 to 10), Arp8 S55A T56A (upper panel, lanes 11 to 20), Arp8 E266A (lower panel, lanes 1 to 10), and Arp8 K288A S290A (lower panel, lanes 11 to 20) were used for this experiment: 0 μM (lanes 1, 6, 11 and 16), 0.8 μM (lanes 2, 7, 12 and 17), 1.6 μM (lanes 3, 8, 13 and 18), 3.2 μM (lanes 4, 9, 14 and 19), 4.8 μM (lanes 5, 10, 15 and 20). In lanes 6 to 10 and 16 to 20, 1 mM ATP was added to the reaction mixture. (C) Intensity of the unbound DNA in the absence or presence of ATP for each Arp8 protein (wild-type Arp8 and Arp8 mutants Arp8 S55A T56A, Arp8 E266A, and Arp8 K288A S290A) used in panel B was quantified and then the data was plotted as relative intensity (%) with respect to that of the unbound DNA from the control (no protein added control) lane.

fold back and stabilize the nucleosome [9,11]. It is thought that both Arp4 and Arp8 contribute to the folding back of the Arp8 module through their histone binding activities. In addition, it has been shown that actin in the Arp8 module is involved in associating the INO80 complex with the extranucleosomal DNA [27]. Thus, the binding of Arp8 to the DNA adjacent to a nucleosome, together with the histone binding activity of Arp8, is expected to stabilize the binding of INO80 complex to the nucleosome.

The INO80 complex lacking Arp8 showed a moderate, but not a complete, decrease in the DNA binding activity [11]. This observation suggested that some other component(s) of the complex might also have DNA binding activity. It is known that the ATPase activity of the INO80 complex is stimulated in the presence of DNA, and the DNA-stimulated ATPase activity is

completely abolished in the INO80 complex lacking Arp8 [11]. Interestingly, the INO80 complex lacking Arp8 is stimulated by nucleosome. Thus, our observations are consistent with the model that the DNA binding activity of Arp8 contributes to the DNA-stimulated INO80 functions. However, further analyses are necessary to clarify this issue fully.

The INO80 complex is recruited to double-strand breaks (DSBs) [14,15]. At these sites, γ-H2AX (phosphorylated-H2A in yeast and phosphorylated-H2AX in mammals) is accumulated as a signal for recruiting protein factors. In yeast, Arp4 and Nhp10 interact with γ-H2AX and are required for the recruitment of the INO80 complex. However, human Arp4 is not required for the recruitment of the complex to DSB, and Nhp10 is not conserved in human INO80 complex [17]. In yeast, DNA end resection is required for recruiting the INO80 complex [28]. Although the

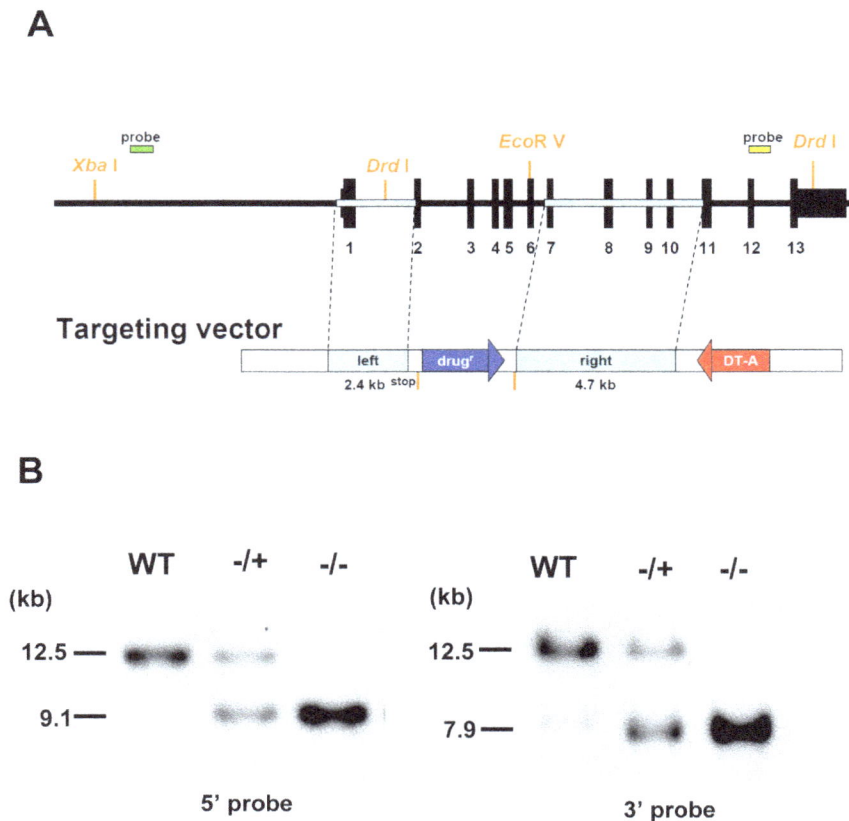

Figure 5. Generation of *ARP8* gene knockout in Nalm-6 cells. (A) Restriction map of the knockout construct used for the targeted disruption of the *ARP8* gene in Nalm-6 cells. Filled boxes (black) over the construct indicate positions of exons. This targeted construct was expected to disrupt five exons of *ARP8* and stop translation before the second exon. Positions of the 5' and 3' probes used for the Southern blot analysis and positions of the restriction enzyme sites are also indicated. (B) Restriction enzyme analysis of the targeted integration of the *ARP8* knockout construct in Nalm-6 cells. Genomic DNAs, prepared from the WT Nalm-6 cells and cells obtained after the first (+/−) and second (−/−) rounds of targeting, were digested with *XbaI/EcoRV* (left) or *DrdI* (right). Disruption of the *ARP8* gene was confirmed by Southern blot hybridization using the 5' (left panel) and 3' DNA probes (right panel) indicated in A.

A

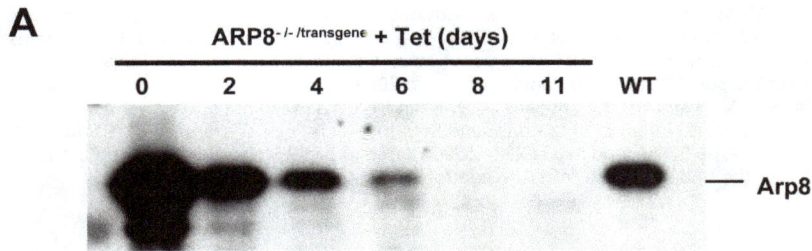

B

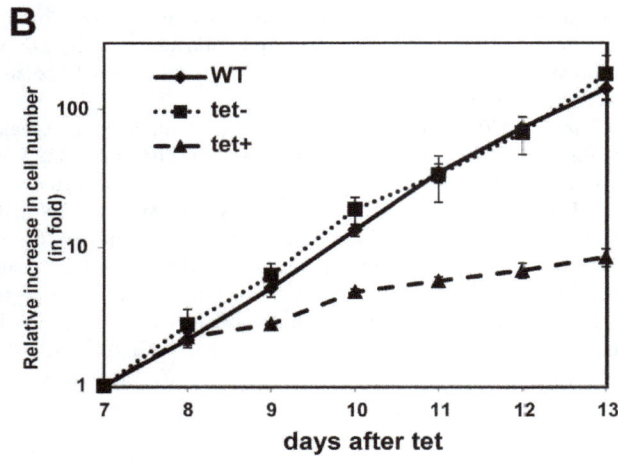

C

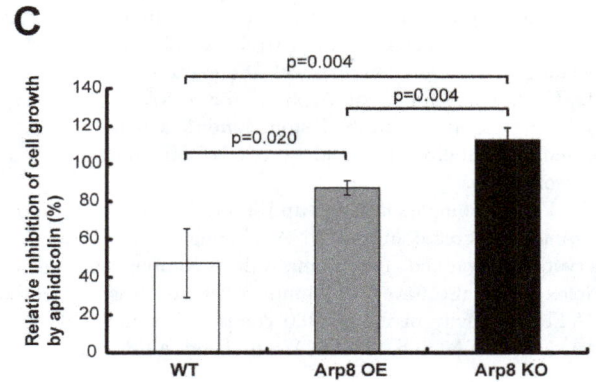

D

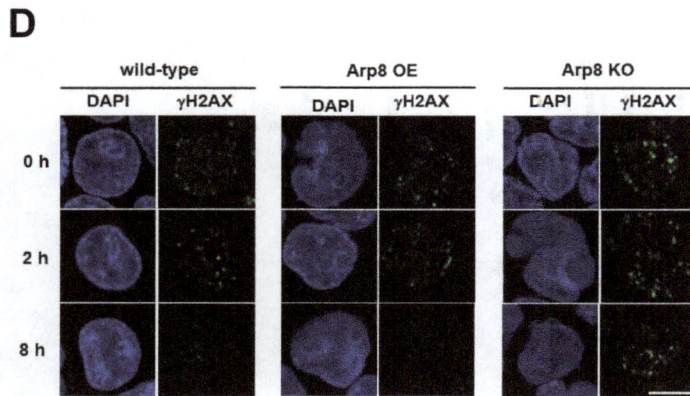

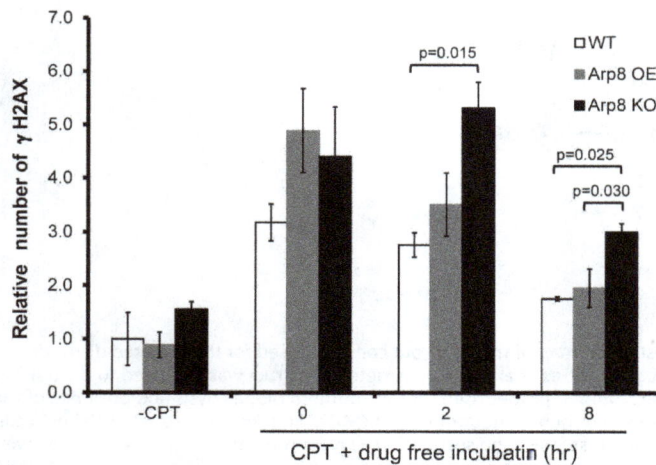

E

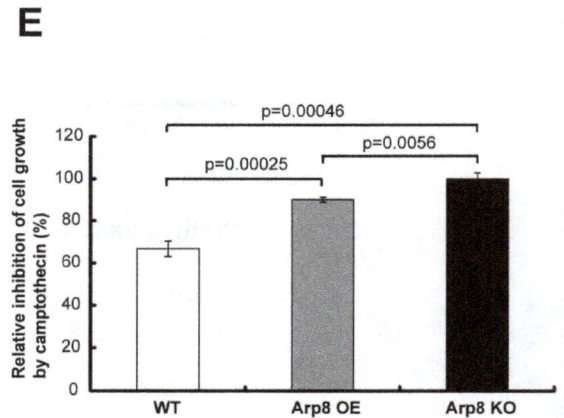

Figure 6. Characterization of Arp8-knockout cells. (A) Whole-cell extracts were prepared from same number of wild-type (WT) and $ARP8^{-/-/transgene}$ cells at the indicated times after the addition of 2 μg/ml of tet, and subsequently they were analyzed by Western blot using an anti-Arp8 antibody. (B) Representative growth curves for the WT and $ARP8^{-/-/transgene}$ cells with (tet+) or without (tet−) tetracycline treatment. Results shown are using cells from day 7 to day 13 after the addition of tetracycline. The number of living cells was counted after trypan blue staining and represented as fold increase in cell number. (C) Sensitivity of WT, Arp8 OE and Arp8-knockout (Arp8 KO; $ARP8^{-/-/transgene}$ cells cultured in the presence of tet for 8 days) cells to aphidicolin. Cells were cultured in the absence or presence of 0.25 μM aphidicolin for 48 h. The relative inhibition of increase in cell number by aphidicolin (%) was calculated as follows: [{(cell number at 48 h – cell number at 0 h) in the absence of aphidicolin – (cell number at 48 h – cell number at 0 h) in the presence of aphidicolin} ×100]/[(cell number at 48 h – cell number at 0 h) in the absence of aphidicolin]. If the cells did not grow at all in the presence of aphidicolin, then the (cell number at 48 h – cell number at 0 h) in the presence of aphidicolin becomes zero and the relative inhibition becomes 100%. However, sometimes in the presence of aphidicolin there were less number of cells at 48 h than at 0 h (instead of an increase in cell number), in which case the (cell number at 48 h – cell number at 0 h) becomes negative and the relative inhibition becomes more than 100%. (D) Comparison of γ-H2AX foci in wild-type, Arp8 OE, and Arp8 KO cells. The cells were treated with camptothecin (CPT) for 1 h, and after washing out the reagent, the cells were incubated without CPT for 2 h or 8 h. Immunostained γ-H2AX foci were observed under a fluorescence microscope, and the number of foci was counted using the ImageJ software. Plot below shows the number of γ-H2AX foci in the indicated cells relative to that in the camptothecin-untreated (-CPT) wild-type cells. (E) Wild-type, Arp8 OE, and Arp8 KO cells were treated with 1 μM camptothecin for 1 h. After washing out the reagent, the relative inhibition by camptothecin was shown as in C. Error bars indicate average mean ± SD (n = at least 3 independent experiments).

need for ssDNA in recruiting the human INO80 complex to DSB is not yet fully understood [16,17,28], it is worth noting that Arp8 is indispensable for recruiting the human INO80 complex to DSBs [17]. Therefore, it is likely that the ssDNA-binding activity of Arp8 contributes to the process of recruiting INO80 complex to DSBs.

Our results indicated that the DNA binding activity of Arp8 decreased in the presence of ATP. Previously, it was observed that ATP affects the intra-molecular interactions in budding yeast Arp4 [5]. It was also proposed that the binding of ATP to Arp4 promotes disassembly of Arp4 from the chromatin remodeling and modification complexes [3,5]. In this respect, Arp4 seems to have some similarity to actin, whose monomer-filament transition is regulated by ATP-binding. Since the INO80 complex contains multiple actin-family proteins, such as Arp4, Arp8, and actin as well [11,29], it is plausible that the binding of ATP to these actin family proteins regulates the function of the INO80 complex by affecting their intramolecular interactions. The actin fold consists of two major domains, and the relative configuration of these major two domains is shifted as a result of ATP binding [30,31,32]. Since the mutational analyses of the ATP binding pocket of Arp8 suggested that the binding of ATP to Arp8 play an important role in DNA binding (Fig. 4B), it is likely that the binding of ATP to Arp8 change the relative configuration of these two domains and thereby affect the DNA binding activity.

In both budding yeast and human, the INO80 complex is required for the efficient DNA end section during HR repair [14,15]. The INO80 complex is recruited to the DSB site in the early stage of the HR repair process. Importantly, disruption of the Arp8 gene in the budding yeast has caused a defect in HR repair [33]. Based on our results, we have proposed a model depicting how Arp8 might contribute to the function of the INO80 complex during the HR repair (Fig. 7). After DSB occurs, endonucleases start DNA end resection. Nucleosomes proximal to the DSB site pose obstacles for the DNA end resection by endonucleases, and this nucleosome barrier could halt the formation of ssDNA (Fig. 7, first row). The ssDNA binding activity of Arp8, together with the histone binding activities of Arp4 and Arp8, would facilitate the binding of INO80 complex to the nucleosomes flanking the ssDNA (Fig. 7, second row). This recruited INO80 complex could then evict or reposition the adjacent nucleosome, and this process may be required to overcome the nucleosome barrier in order for the DNA end resection to progress (Fig. 7, second row). After evicting or relocating the first nucleosome, the newly resected ssDNA adjacent to the next nucleosome barrier is targeted by the second INO80 complex (Fig. 7, third row). The first INO80 complex stays bound to the original position on the DNA through its own ssDNA binding activity, but without associating with

histones (Fig. 7, third row). This model is consistent with earlier observations that knockdown of Arp8 impairs RPA focus formation and that knockdown of Ino80, although affects an early event, is not necessary for the later stages of HR repair [15].

Alternatively, since Arp8 also has INO80 complex-independent function at least in mitotic chromosome segregation [34], DNA-binding activity of Arp8 might contribute to DNA repair independently of the INO80 complex. Further analyses would be necessary to clarify the mechanism of Arp8 in DNA repair.

It was previously shown that the DNA damage-induced stimulation of poly (ADP-ribose) polymerase led to a decrease in the ATP pool [35]. Since the DNA-binding ability of Arp8 is relatively high under an ATP-deprived condition (Fig. 2), this change in ATP concentration could also be involved in the function of Arp8 in DNA repair.

DNA damage repair is crucial for the maintenance of genome stability and cancer suppression. Therefore, defects in DNA repair could be relevant to human diseases. Recently, an association was found between the SNPs in the *INO80* gene and chronic kidney disease (CKD), an important public health problem with a genetic component [36]. Indeed, Zhou et al. (2012) have recently shown that inadequate DNA repair is relevant to CKD. Since Arp8 is essential for the proper functioning of the INO80 complex, its dysfunction in human cells would be expected to cause diseases. By using our tet-inducible Arp8-knockout cell line, we would be able to analyze the effects of Arp8 expression level on genome stability and would also be able to analyze functions of Arp8 mutants in the absence of endogenous Arp8. This system will provide further knowledge on the roles of Arp8 and INO80 complex in epigenetic regulations including DNA repair and transcription.

Materials and Methods

Expression and purification of proteins

Human Arp8 was overexpressed in *Escherichia coli* cells as an N-terminal hexahistidine (His6)-tagged protein. To achieve this, the full length wild-type (WT) or mutagenized Arp8 (Arp8 Δ1–38, Arp8 Δ403–463, Arp8 S55A T56A, Arp8 E266A, and Arp8 K288A S290A) coding cDNA fragment was ligated into the pET15b vector (Novagen), which harbors a His_6 tag and a PreScission protease-recognition sequence (GE Healthcare Bioscience) at the N-terminus. Primers used for the site-directed mutagenesis are shown in Table S1. Recombinant Arp8 was expressed in *E. coli* and purified as follows. Briefly, *E. coli* cells carrying the Arp8 expression plasmid, grown on ampicilin (100 μg/ml) and chloramphenicol (35 μg/ml) supplemented LB plates at 37°C, were used for inoculating 5 l growth medium (LB

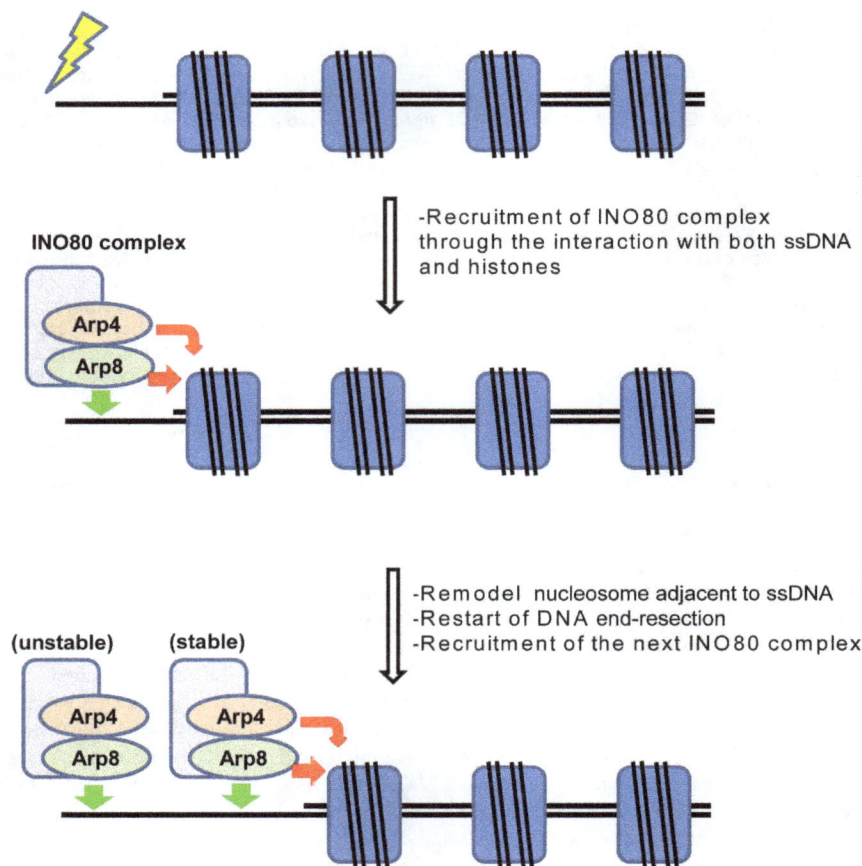

Figure 7. A schematic model depicting the role of Arp8 in the early stage of HR repair. Green and red arrows near the INO80 complex represent binding of Arp8 to ssDNA and binding of Arp4 and Arp8 to histones, respectively. See text for details.

containing 100 μg/ml ampicillin) and the culture was grown with shaking at 37°C. When the cell density reached an OD600 of 0.45–0.55, 1 mM isopropyl beta-D-1-thiogalacropyranoside (IPTG) was added to induce the expression of Arp8 and the culture was further incubated at 18°C for 15 h. Cells were harvested, resuspended in 30 ml of buffer A [50 mM Tris-HCl (pH 8.0), 0.7 M NaCl, and 10% glycerol] containing 1× Protease Inhibitor Cocktail (Nakalai Tasque) and disrupted by sonication. After removing cell debris by centrifugation, the lysate was mixed with 2 ml (50% slurry) of Ni-NTA beads. After packing the beads in an Econo-column (BioRad), the bound His6-tagged Arp8 was eluted by using a linear gradient of imidazole in 50 mM NaCl (pH 8.0), 0.1 M NaCl, and 10% glycerol. Fractions containing His6-Arp8 were identified by SDS-PAGE, and combined fractions were applied to a Heparin Sepharose column (GE Healthcare Biosciences). After washing the column with buffer B [20 mM Tris-HCl (pH 8.0), 0.1 M NaCl, 0.25 mM EDTA, 2 mM 2-mercaptoethanol and 10% glycerol], Arp8 was eluted using a linear gradient of 100 to 1,000 mM NaCl. Fractions containing Arp8 were collected and treated with PreScission protease (3 units/mg of protein) to remove the His6 tag. The resultant Arp8 was further purified using a MonoQ column (GE Healthcare Bioscience), from where the bound Arp8 was eluted with a gradient of 0 to 0.5 M NaCl in buffer C [20 mM Tris-HCl (pH 8.0), 0.25 mM EDTA, 2 mM 2-mercaptoethanol and 10% glycerol], and the eluted protein was dialyzed against buffer G [20 mM Tris-HCl (pH 8.0), 0.2 M NaCl, 0.25 mM EDTA,

2 mM 2-mercaptoethanol and 10% glycerol]. The concentration of purified Arp8 was determined by the Bradford method [37] with BSA as the standard protein.

DNA binding assay

Single-stranded φX174 viral (+) strand DNA was purchased from New England Biolabs. The linear dsDNAs were prepared by digesting the φX174 replicative form I DNA and pUC19 vector with the restriction enzyme *Pst*I. The DNAs were mixed with the purified Arp8 protein in 10 μl of reaction mixture containing 20 mM HEPES-NaOH (pH 7.5), 1 mM DTT, 0.1 mg/ml bovine serum albumin, 1 mM MgCl₂, 160 mM NaCl, 8% glycerol, and with or without 1 mM ATP. The reaction mixture was incubated at 37°C for 15 min, and was then analyzed by 0.8% agarose gel electrophoresis in 1xTAE buffer (electrophoresis was carried out at 3.0 V/cm for 2 h). The DNA bands were visualized by ethidium bromide staining.

To perform the competitive DNA-binding assay, we prepared double-stranded DNA (dsDNA), 3′-overhang DNA, and single-stranded DNA (ssDNA) as described previously (MacKay C et al., 2010). We used oligonucleotides a3, a3-cp, and c described earlier [38], whose nucleotide sequences were as follows: a3, 5′- CCTCG ATCCT ACCAA CCAGA TGACG CGCTG CTACG TGCTA CCGGA AGTCG; a3-cp, 5′- CGACT TCCGG TAGCA CGTAG CAGCG CGTCA ACTGG TTGGT AGGAT CGAGG; and c, 5′- GCCTA GAGTG CAGTT CGTGG CGAGC. To prepare dsDNA and 3′-overhang DNA, the oligonucleotide pairs a3 and a3-cp, and a3 and c, respectively, were annealed, and the annealed DNAs were then purified by polyacrylamide gel electrophoresis. The oligonucleotide a3 was used as the ssDNA. A mixture containing 3 μM each of the dsDNA, 3′-overhang DNA, and ssDNA was incubated with the indicated amounts of Arp8 at 37°C for 15 min in 10 μl of 20 mM HEPES-NaOH buffer (pH 7.5) supplemented with 250 mM NaCl, 1 mM MgCl$_2$, 0.1 mg/ml bovine serum albumin, and 1 mM DTT. The protein-DNA complexes were separated by electrophoresis on 10% polyacrylamide gel in 0.5xTBE buffer (at 6.25 V/cm for 150 min), and were visualized after staining with SYBR Gold (Invitrogen).

Cell culture and cell viability analyses

Nalm-6 cells were cultured at 37°C in Roswell Park Memorial Institute medium containing GlutaMAX-I (Invitrogen) supplemented with 10% fetal calf serum, penicillin, and streptomycin as described earlier (Ono et al., 2009). To suppress the expression of the tetracycline (tet)-responsive Arp8 transgene, tetracycline (Sigma) was added to the culture medium to a final concentration of 2 μg/ml. The number of viable cells after the addition of aphidicolin (Wako) or camptothecin (Wako) was counted by the trypan blue dye exclusion assay.

Establishment of Arp8-deficient cells by using a tetracycline (Tet)-regulated gene depletion (Tet-Off) system

Tet-Off Arp8 cells were established following a protocol described previously [22,39]. In brief, the full length cDNA of *ARP8* was cloned into the pTRE-IRES-neo vector to yield a tetracycline-regulated Arp8 expression plasmid. The left (2.4 kb) and right (4.2 kb) arms of the targeting vector were respectively amplified by genomic PCR. The left arm and the right arm contained exon 1 and exons 7–10, respectively (see Fig. 5A). These fragments were used to generate the final targeting plasmids, pTARGET-*ARP8*-His and pTARGET-*ARP8*-Puro. Gene targeting and screening of purposive clones were performed as described previously [22,39]. The disruption of both alleles of *ARP8* gene was confirmed by Southern blot analysis using probes shown in Fig. 5A and the results are shown in Fig. 5B. Established Arp8 Tet-Off cells were confirmed by Western blot analysis using an anti-Arp8 antibody.

Indirect immunofluorescence staining and Western blot analysis

Cells were fixed in 4% paraformaldehyde in phosphate-buffer saline. To visualize γ-H2AX foci, an anti-γ-H2AX antibody (Milipore) was used for immunostaining, and the bound antibody was detected using a fluorescent-conjugated secondary anti-mouse antibody. Cellular DNA was stained with 1 mg/ml of DAPI, and

fluorescence of bound DAPI and γ-H2AX was observed under a confocal laser scanning microscope (FV1000, Olympus). The number of γ-H2AX foci was counted by using the Image J software. Western blot analysis was performed using an anti-Arp8 antibody [34] or an anti-H3 antibody (Abcam ab1791). An anti-IgG conjugated to horseradish peroxidase (Promega) was used as the secondary antibody, and ECL Western blotting detection reagents (GE Healthcare) were used for the detection of bound antibodies [40].

Supporting Information

Figure S1 Schematic diagrams of full-length and deletion mutants of Arp8. The N-terminal extension and insertions are shown in red and light blue, respectively.

Figure S2 Binding of Arp8 to supercoiled and nicked circular forms of φX174 in the presence of ATP or ADP. Binding of Arp8 was examined in the absence (lanes 1 to 6) or presence of 1 mM ATP (lanes 7 to 12), and in the presence of 1 mM ADP (lanes 13 to 18) as well. Concentrations of Arp8 used were: 0 μM (lanes 1, 7, and 13), 0.4 μM (lanes 2, 8, and 14), 0.8 μM (lanes 3, 9, and 15), 1.6 μM (lanes 4, 10, and 16), 3.2 μM (lanes 5, 11, and 17), and 4.8 μM (lanes 6, 12, and 18). Intensity of the unbound DNA band in each lane was quantified and plotted as relative intensity (%) with respect to the intensity of the unbound DNA from the control (no protein added) lane.

Figure S3 SDS-PAGE analysis of purified wild-type and ATP binding pocket mutants of Arp8. Lane 1: molecular weight markers. Lane 2: wild-type Arp8, Lane 3: Arp8 S55A T56A, Lane 4: Arp8 E266A, and Lane 5: Arp8 K288A S290A.

Table S1 Oligonucleosides for cytodirected mutagenesis.

Acknowledgments

We thank Mrs. Y. Wakata at NIG for establishing the tetracycline (tet)-inducible Arp8-knockout cells. This work was supported by Grants-in-Aid for Scientific Research on Innovative Areas and the Human Frontier Science Program (RGP0017). H. Kurumizaka was also supported by the Waseda Research Institute for Science and Engineering.

Author Contributions

Conceived and designed the experiments: YO KS HK MH. Performed the experiments: AO YT HM KO HT YO HN. Analyzed the data: AO YT YO HK MH. Wrote the paper: AO HK MH.

References

1. Oma Y, Harata M (2011) Actin-related proteins localized in the nucleus: from discovery to novel roles in nuclear organization. Nucleus 2: 38–46.
2. Dion V, Shimada K, Gasser SM (2010) Actin-related proteins in the nucleus: life beyond chromatin remodelers. Curr Opin Cell Biol 22: 383–391.
3. Kast DJ, Dominguez R (2011) Arp you ready for actin in the nucleus? EMBO J 30: 2097–2098.
4. Boyer LA, Peterson CL (2000) Actin-related proteins (Arps): conformational switches for chromatin-remodeling machines? Bioessays 22: 666–672.
5. Sunada R, Gorzer I, Oma Y, Yoshida T, Suka N, et al. (2005) The nuclear actin-related protein Act3p/Arp4p is involved in the dynamics of chromatin-modulating complexes. Yeast 22: 753–768.
6. Harata M, Oma Y, Mizuno S, Jiang YW, Stillman DJ, et al. (1999) The nuclear actin-related protein of Saccharomyces cerevisiae, Act3p/Arp4, interacts with core histones. Mol Biol Cell 10: 2595–2605.
7. Shen X, Ranallo R, Choi E, Wu C (2003) Involvement of actin-related proteins in ATP-dependent chromatin remodeling. Mol Cell 12: 147–155.
8. Nishimoto N, Watanabe M, Watanabe S, Sugimoto N, Yugawa T, et al. (2012) Heterocomplex Formation by Arp4 and beta-Actin Involved in Integrity of the Brg1 Chromatin Remodeling Complex. J Cell Sci.
9. Gerhold CB, Winkler DD, Lakomek K, Seifert FU, Fenn S, et al. (2012) Structure of Actin-related protein 8 and its contribution to nucleosome binding. Nucleic Acids Res 40: 11036–11046.
10. Saravanan M, Wuerges J, Bose D, McCormack EA, Cook NJ, et al. (2012) Interactions between the nucleosome histone core and Arp8 in the INO80 chromatin remodeling complex. Proc Natl Acad Sci U S A 109: 20883–20888.
11. Tosi A, Haas C, Herzog F, Gilmozzi A, Berninghausen O, et al. (2013) Structure and subunit topology of the INO80 chromatin remodeler and its nucleosome complex. Cell 154: 1207–1219.

12. Cai Y, Jin J, Yao T, Gottschalk AJ, Swanson SK, et al. (2007) YY1 functions with INO80 to activate transcription. Nat Struct Mol Biol 14: 872–874.

13. Bhatia S, Pawar H, Dasari V, Mishra RK, Chandrashekaran S, et al. (2010) Chromatin remodeling protein INO80 has a role in regulation of homeotic gene expression in Drosophila. Genes Cells 15: 725–735.

14. van Attikum H, Fritsch O, Hohn B, Gasser SM (2004) Recruitment of the INO80 complex by H2A phosphorylation links ATP-dependent chromatin remodeling with DNA double-strand break repair. Cell 119: 777–788.

15. Gospodinov A, Vaissiere T, Krastev DB, Legube G, Anachkova B, et al. (2011) Mammalian Ino80 mediates double-strand break repair through its role in DNA end strand resection. Mol Cell Biol 31: 4735–4745.

16. Shimada K, Oma Y, Schleker T, Kugou K, Ohta K, et al. (2008) Ino80 chromatin remodeling complex promotes recovery of stalled replication forks. Curr Biol 18: 566–575.

17. Kashiwaba S, Kitahashi K, Watanabe T, Onoda F, Ohtsu M, et al. (2010) The mammalian INO80 complex is recruited to DNA damage sites in an ARP8 dependent manner. Biochem Biophys Res Commun 402: 619–625.

18. Muller J, Oma Y, Vallar L, Friederich E, Poch O, et al. (2005) Sequence and comparative genomic analysis of actin-related proteins. Mol Biol Cell 16: 5736–5748.

19. Kumar M, Gromiha MM, Raghava GP (2007) Identification of DNA-binding proteins using support vector machines and evolutionary profiles. BMC Bioinformatics 8: 463.

20. van Attikum H, Fritsch O, Gasser SM (2007) Distinct roles for SWR1 and INO80 chromatin remodeling complexes at chromosomal double-strand breaks. EMBO J 26: 4113–4125.

21. Chambers AL, Downs JA (2012) The RSC and INO80 chromatin-remodeling complexes in DNA double-strand break repair. Prog Mol Biol Transl Sci 110: 229–261.

22. Ono T, Nishijima H, Adachi N, Iiizumi S, Morohoshi A, et al. (2009) Generation of tetracycline-inducible conditional gene knockout cells in a human Nalm-6 cell line. J Biotechnol 141: 1–7.

23. Rothkamm K, Kruger I, Thompson LH, Lobrich M (2003) Pathways of DNA double-strand break repair during the mammalian cell cycle. Mol Cell Biol 23: 5706–5715.

24. Saleh-Gohari N, Bryant HE, Schultz N, Parker KM, Cassel TN, et al. (2005) Spontaneous homologous recombination is induced by collapsed replication forks that are caused by endogenous DNA single-strand breaks. Mol Cell Biol 25: 7158–7169.

25. Yonetani Y, Hochegger H, Sonoda E, Shinya S, Yoshikawa H et al. (2005) Differential and collaborative actions of Rad51 paralog proteins in cellular response to DNA damage. Nucleic Acids Res 33: 4544–4552.

26. Arnaudeau C, Lundin C, Helleday T (2001) DNA double-strand breaks associated with replication forks are predominantly repaired by homologous recombination involving an exchange mechanism in mammalian cells. J Mol Biol 307: 1235–1245.

27. Kapoor P, Chen M, Winkler DD, Luger K, Shen X (2013) Evidence for monomeric actin function in INO80 chromatin remodeling. Nat Struct Mol Biol 20: 426–432.

28. Bennett G, Papamichos-Chronakis M, Peterson CL (2013) DNA repair choice defines a common pathway for recruitment of chromatin regulators. Nat Commun 4: 2084.

29. Fenn S, Breitsprecher D, Gerhold CB, Witte G, Faix J, et al. (2011) Structural biochemistry of nuclear actin-related proteins 4 and 8 reveals their interaction with actin. EMBO J 30: 2153–2166.

30. Nolen BJ, Pollard TD (2007) Insights into the influence of nucleotides on actin family proteins from seven structures of Arp2/3 complex. Mol Cell 26: 449–457.

31. Oda T, Iwasa M, Aihara T, Maeda Y, Narita A (2009) The nature of the globular- to fibrous-actin transition. Nature 457: 441–445.

32. De La Cruz EM, Mandinova A, Steinmetz MO, Stoffler D, Aebi U, et al. (2000) Polymerization and structure of nucleotide-free actin filaments. J Mol Biol 295: 517–526.

33. Kawashima S, Ogiwara H, Tada S, Harata M, Wintersberger U, et al. (2007) The INO80 complex is required for damage-induced recombination. Biochem Biophys Res Commun 355: 835–841.

34. Aoyama N, Oka A, Kitayama K, Kurumizaka H, Harata M (2008) The actin-related protein hArp8 accumulates on the mitotic chromosomes and functions in chromosome alignment. Exp Cell Res 314: 859–868.

35. Berger NA (1985) Poly(ADP-ribose) in the cellular response to DNA damage. Radiat Res 101: 4–15.

36. Pattaro C, Kottgen A, Teumer A, Garnaas M, Boger CA, et al. (2012) Genome-wide association and functional follow-up reveals new loci for kidney function. PLoS Genet 8: e1002584.

37. Bradford MM (1976) A rapid and sensitive method for the quantitation of microgram quantities of protein utilizing the principle of protein-dye binding. Anal Biochem 72: 248–254.

38. MacKay C, Declais AC, Lundin C, Agostinho A, Deans AJ, et al. (2010) Identification of KIAA1018/FAN1, a DNA repair nuclease recruited to DNA damage by monoubiquitinated FANCD2. Cell 142: 65–76.

39. Nishijima H, Yasunari T, Nakayama T, Adachi N, Shibahara K (2009) Improved applications of the tetracycline-regulated gene depletion system. Biosci Trends 3: 161–167.

40. Kimura H, Hayashi-Takanaka Y, Goto Y, Takizawa N, Nozaki N (2008) The organization of histone H3 modifications as revealed by a panel of specific monoclonal antibodies. Cell Struct Funct 33: 61–73.

Maintenance of Sex-Related Genes and the Co-Occurrence of Both Mating Types in *Verticillium dahliae*

Dylan P. G. Short[1,9]**, Suraj Gurung**[1,9]**, Xiaoping Hu**[2]**, Patrik Inderbitzin**[1]**, Krishna V. Subbarao**[1]*

1 Department of Plant Pathology, University of California Davis, Salinas, CA, United States of America, **2** State Key Laboratory of Crop Stress Biology for Arid Areas and College of Plant Protection, Northwest A&F University, Yangling, Shaanxi, China

Abstract

Verticillium dahliae is a cosmopolitan, soilborne fungus that causes a significant wilt disease on a wide variety of plant hosts including economically important crops, ornamentals, and timber species. Clonal expansion through asexual reproduction plays a vital role in recurring plant epidemics caused by this pathogen. The recent discovery of recombination between clonal lineages and preliminary investigations of the meiotic gene inventory of *V. dahliae* suggest that cryptic sex appears to be rare in this species. Here we expanded on previous findings on the sexual nature of *V. dahliae*. Only 1% of isolates in a global collection of 1120 phytopathogenic *V. dahliae* isolates contained the *MAT1-1* idiomorph, whereas 99% contained *MAT1-2*. Nine unique multilocus microsatellite types comprised isolates of both mating types, eight of which were collected from the same substrate at the same time. Orthologs of 88 previously characterized sex-related genes from fungal model systems in the Ascoymycota were identified in the genome of *V. dahliae*, out of 93 genes investigated. Results of RT-PCR experiments using both mating types revealed that 10 arbitrarily chosen sex-related genes, including *MAT1-1-1* and *MAT1-2-1*, were constitutively expressed in *V. dahliae* cultures grown under laboratory conditions. Ratios of non-synonymous (amino-acid altering) to synonymous (silent) substitutions in *V. dahliae MAT1-1-1* and *MAT1-2-1* sequences were indistinguishable from the ratios observed in the *MAT* genes of sexual fungi in the *Pezizomycotina*. Patterns consistent with strong purifying selection were also observed in 18 other arbitrarily chosen *V. dahliae* sex-related genes, relative to the patterns in orthologs from fungi with known sexual stages. This study builds upon recent findings from other laboratories and mounts further evidence for an ancestral or cryptic sexual stage in *V. dahliae*.

Editor: Stefanie Pöggeler, Georg-August-University of Göttingen Institute of Microbiology & Genetics, Germany

Funding: Funding for this study was provided by USDA-NIFA-SCRI grant no. 2010-51181-21069 and the California Leafy Greens Research Board. The funders had no role in study design, data collection and analysis, decision to publish, or preparation of the manuscript.

Competing Interests: The authors have declared that no competing interests exist.

* Email: kvsubbarao@ucdavis.edu

9 These authors contributed equally to this work.

Introduction

Sexual reproduction is thought [1] to act as a mechanism to combine fit alleles from different individuals, and to break apart locally disadvantageous allele combinations under dynamic selection pressures [2]. While sexual reproduction may in theory be costly and disrupt favorable gene combinations, experimental evidence has suggested that sex in fungi increases the rate of adaptation to new environments [3]. Prior to molecular techniques, the formation of sexual structures and spores was the primary evidence of sex in fungi. It is now evident that sex in many taxa is rare, unpredictable and elusive. For many fungi, the only documented sexual structures are formed on certain media and/or growth conditions *in vitro* [4,5]. Some putatively asexual plant pathogens have been found to sexually reproduce in nature only in specific ecological conditions and geographic locales, such as near the center of origin of the species [6].

Advances in genetic markers and population biology have led to significant advances in the discovery of rare or cryptic sexual stages in fungi [7]. Populations of many species that lack obvious sexual stages in nature nevertheless have been found to harbor molecular patterns of sexuality based on investigations of mating type frequencies, population structure, multilocus linkage disequilibrium [8,9] and computer simulations [6]. Additionally, bioinformatic surveys of complete genomes, have been used to infer sexuality based on the meiotic gene inventory [10–13]. Advances in genomics have enabled the unprecedented implementation of these approaches to investigate sexuality in fungi. Many seemingly asexual fungi have retained the genes required for the sexual "machinery", including many that are important to the fields of agriculture and medicine [12,14–16].

Verticillium is a small genus of phytopathogenic fungi that causes billions of dollars in agricultural losses annually [17]. *Verticillium dahliae* is a cosmopolitan, soilborne plant pathogen that causes an economically significant wilt disease. It is known for its extremely wide host range [18] and its ability to survive in soils as dormant resting structures for many years [17,19]. Historically, *V. dahliae* has been considered strictly asexual because it has failed to form sexual structures under the laboratory conditions tested. Vegetative anastomosis, the fusion of growing hyphae under laboratory conditions, has been reported [20,21], and several vegetative compatibility groups (VCGs) have been classified. Deep sequencing of all known VCGs of *V. dahliae* has revealed that VCGs are strongly correlated to clonal lineages [22], but has also revealed that putative sexual recombination between clonal lineages has occurred rarely [23].

Sexual compatibility and fruiting body formation in heterothallic fungi in the Ascoymycota is determined by a variety of sex-related gene pathways. Of primary importance are the two idiomorphs of the *MAT* locus, which differ in gene content and are the master regulators of sexual recombination in the Ascoymycota [24]. One idiomorph contains a critical gene that encodes an α domain (*MAT1-1-1*), while the other contains a critical gene that encodes a DNA-binding domain of the high-mobility group (HMG) type (*MAT1-2-1*) [25]. Isolates with either of the idiomorphs are referred to as *MAT1-1* or *MAT1-2* [26]. *Verticillium dahliae* is considered heterothallic because both idiomorphs are known to exist [27], and only one idiomorph has been observed in any one isolate.

Previous sequences of the α and HMG domains of *V. dahliae MAT* genes showed high amino acid conservation with other fungi in the subphylum *Pezizomycotina* [27,28]. Mating type frequencies in *V. dahliae* have been reported in multiple studies as skewed [23,28,29]. Although previous studies have reported mating type distributions skewed towards *MAT1-2* in *V. dahliae*, they have not clearly stated whether both mating types are sympatric in nature, that is, whether isolates of opposite mating coexist in nature. It is also unknown whether genetically identical multilocus microsatellite types contain both *MAT* idiomorphs, a condition which has previously been interpreted as unequivocal evidence for sexual recombination [11].

In addition to the presence of both mating types, other molecular signatures suggestive of sex have been reported in *V. dahliae*. Multilocus linkage equilibrium has been reported in collections of *V. dahliae* [30], although clonal expansion is of primary importance in pathogen reproduction and dissemination within regions where this pathogen is a severe problem in agriculture [31]. However, even in species with known sexual stages, signatures of clonality can predominate in multilocus data sets [32]. Gene trees with incongruent topologies may be a robust indicator of meiotic recombination when they occur within a strongly supported phylogenetic species [33,34]. Gene trees with incongruent topologies were previously reported in *V. dahliae* based on sequences of the protein coding genes *actin* (*ACT*), *elongation factor 1-alpha* (*EF*), *glyceraldehyde-3-phosphate dehydrogenase* (*GPD*), and *tryptophan synthase* (*TS*) [35]. The strongest evidence yet of recombination between lineages of *V. dahliae* was based on over 20,000 single nucleotide polymorphisms (SNPs) [23].

Genomic investigations of *V. dahliae* have also provided some evidence of sexuality in *V. dahliae*. For example, a single homolog of the gene encoding the DNA methyltranferase (DMT) *RID* exists in *V. dahliae* reference strain Ls 17, a gene which was first characterized as part of the Repeat-Induced Point (RIP) machinery in *N. crassa* [36]. Patterns consistent with RIP-like mutation were subsequently discovered in the *V. dahliae* genome in multiple long interspersed element (LINE)-like and long terminal repeat (LTR) retroelement sequences [37] and other transposons [38]. Furthermore, preliminary explorations of the meiotic gene inventory have revealed the presence of genes known to function in sex-related pathways in other fungal systems [23].

Comparative population genomics of *V. dahliae* has significantly advanced the understanding of the molecular basis of races, as well as the existence of inter-Kingdom horizontal gene transfer [39], and has also led some researchers to posit chromosomal

reshuffling (genomic rearrangements and chromosomal length polymorphisms, despite a high degree of sequence conservation) as the sole mechanism for generating the diversity observed within *V. dahliae* [40]. Significant chromosomal rearrangements are expected to interfere with meiosis [41], so it is reasonable to expect sex to be impossible between isolates with extreme karyotypic polymorphisms [40,42].

It has been postulated that a detailed understanding of the genes required for the initiation and completion of meiosis in sexual fungi, that it should be possible to understand the molecular mechanisms that control sexual compatibility and to determine which of these genes are missing or nonfunctional in asexual fungi [43]. In fact, imperfect functioning of mating type genes and other sexual factors such as pheromone receptors have been hypothesized in *V. dahliae* [27]. In the context of exploring the functionality of sex-related genes (and not merely the existence of pseudogenes), Reverse transcriptase-PCR has been used to show that both mating type genes are expressed in fungi, for which no known sexual stage has been documented [44], while other studies have demonstrated pheromone receptor and precursor gene expression in other putatively asexual fungi [45]. To date, the expression of *MAT* genes and other sex-related genes in *V. dahliae* has never been investigated.

Evolutionary theory predicts that if amino acid-altering genetic mutations occur in genes or domains of critical function and result in lower fitness, they will be purged from populations through purifying selection [46]. Conversely, selection acting on mutations in non-essential genes or domains is "relaxed", and thus accumulation of amino acid-altering mutations is more likely in such regions. Calculations of the Ka/Ks ratios in a set of amino acid sequences can thus be used to estimate an evolutionary history of both positive and purifying selection at each amino-acid site. Strong purifying selection in 9,471 core eukaryotic genes was previously reported in the genomes of several isolates of *V. dahliae* [40]. Whether sex-related genes in the *V. dahliae* genome are similarly conserved, compared to related sexual fungi, is currently unknown.

The goals of this study were to: 1) characterize the mating types of *V. dahliae* from a large collection of phytopathogenic isolates; 2) determine whether isolates of opposite mating types are present concurrently in the same habitat; 3) determine whether genetically identical multilocus microsatellite types contain both *MAT* idiomorphs; 3) determine if the complete genome sequence of *V. dahliae* strain Ls 17 contains orthologs of fungal sex-related genes; 4) test whether such genes are constitutively expressed in both mating types under laboratory conditions; and 5) estimate the extent of positive (relaxed) and purifying selection in a subset of sex-related genes in *V. dahliae*, relative to fungi with known sexual stages.

Results and Discussion

Molecular assays to identify *Verticillium* species, *MAT* type, and multilocus microsatellite types

All isolates used in this study were identified as the phylogenetic species *V. dahliae sensu stricto* [35]. The frequency of *MAT* idiomorphs was extremely skewed towards an overabundance of *MAT1-2* (Table S1). The *MAT1-1* idiomorph was only observed in 1% (12/1120) of isolates characterized. The *MAT1-1* isolates comprised eight isolates from commercial spinach seed lots from Washington State, USA, two isolates from a commercial artichoke field in California, and one isolate each from two commercial tomato field in CA (Table S1).

Complete multilocus microsatellite types were generated for 941 isolates; all 12 *MAT1-1* isolates had different MLMTs, whereas 410 different MLMTs were observed for *MAT1-2* isolates. Thus, after clone correction, 97% (410/422) isolates were *MAT1-2*. Nine of the *MAT1-1* MLMTs were identical to MLMTs of one or more *MAT1-2* isolates (Table 1). Of the nine MLMTs that comprised both mating types, three of them were found to have overlapping ecological niches. That is, they were collected at the same time from the same location and were isolated from the same substrate (artichoke, spinach seed, and tomato) (Table S1). The presence of multilocus genotypes common to both mating types has been interpreted as evidence of sexual recombination [11,47]. However, this interpretation assumes no homoplasy, and assumes that isolates of opposite mating types did not acquire the same alleles at the thirteen loci independently through mutation.

Verticillium genome queries and ortholog searches

Out of 93 sex-related genes considered, 88 were found in the *V. dahliae* genome (Table 2). The five genes not found in *V. dahliae* genome searches were the *N. crassa* accessions NCU09793, NCU04329, which are DNA helicase and repair proteins, respectively, and *S. cerivisae* accessions YIL072W, YGL033W, and YGL183C, which correspond to *HOP1*, *HOP2* and *MND1*. Since no orthologs to *HOP1*, *HOP2* or *MND1* were found among any of the Sordariomycetes in the FUNGIPath database, including the sexual fungi *Neurospora crassa*, *Podospora anserina*, and *Nectria haematococca*, it is reasonable to speculate that these three genes are not required for a fully functional sexual cycle for taxa in this group.

SELECTON analyses of positive and purifying selection in sex-related genes of V. dahliae

Selective pressures were estimated in 20 *V. dahliae* genes, including *MAT1-1-1* and *MAT1-2-1*. The subset of 20 genes chosen for SELECTON analysis were distributed in the *V. dahliae* genome on chromosomes 1, 2, 3, 4, 5, 7 and 8. No codons under positive selection were detected in either *MAT1-1-1* or *MAT1-2-1* or any of the other 18 genes using the M8 model (Figure 1A, Figure 2A). However, using the MEC model, positive selection was detected in 12/20 genes investigated (Figure 1B, Figure 2B, Figure S1). Likelihood ratio tests between the MEC and M8a models revealed that in all cases, the AIC score of the MEC model was lower than the M8a model.

Using the MEC model, *Verticillium dahliae MAT1-1-1* contained 12% of codons under positive selection and 34% of codons under strong purifying selection (Table 3); *V. dahliae MAT1-2-1* contained 9% of codons under positive selection and 35% of codons under strong purifying selection (Table 4). Of the 21 codons under positive selection in *V. dahliae MAT1-1-1*, only 3 were within the highly conserved α domain (Figure 1B); similarly of the 51 codons under positive selection in *V. dahliae MAT1-2-1*, only 1 was within the highly conserved HMG domain (Figure 2B). When only sequences from sexual fungi were considered, *MAT1-1-1* codons under positive and purifying selection ranged from 9–15% and 33–43% respectively (Table 3), whereas *MAT1-2-1* codons under positive and purifying selection ranged from 12–22% and 21–30% respectively (Table 4). Thus, the extent and type of selection estimated for *V. dahliae MAT* genes were comparable to the estimates for *MAT* genes from sexual fungi. Interestingly, *MAT1-1-1* from the putatively asexual *P. fulva* contained the highest relative numbers of codons under positive selection and the lowest under strong purifying selection (Table 3); however, *P. fulva MAT1-2-1* Ka/Ks estimates were similar to sexual fungi (Table 4).

Table 1. Ecological characteristics of multilocus microsatellite types that comprised isolates of both mating types.

MLMT	Alleles for 13-locus MLMT [a]	MAT1-1 (n) [b]	MAT1-2 (n) [c]	MAT1-1 plant hosts	MAT1-2 plant hosts	MAT1-1 origins	MAT1-2 origins
1	366.315.369.333.329.577.361.350.367.373.392.334.317	1	2	Tomato	Lettuce	CA, USA	CA, USA
2	372.299.369.301.263.521.333.330.367.289.332.246.277	1	1	Spinach seed	Spinach seed	WA, USA	Netherlands
3	372.303.369.301.263.521.333.330.367.283.332.246.277	1	2	Spinach seed	Spinach seed	WA, USA	WA, USA
4	372.303.369.301.263.521.333.330.387.295.332.246.277	1	5	Spinach seed	Olive, Spinach seed	WA, USA	Denmark, Italy, WA, USA
5	378.299.369.301.263.521.333.330.387.283.332.246.277	1	45	Tomato	Cotton, Spinach seed, Tomato	CA, USA	Chile, CA and WA USA
6	378.299.369.301.263.521.333.330.387.301.332.246.277	1	18	Spinach seed	Spinach seed	WA, USA	WA, USA
7	378.315.376.301.263.513.361.330.367.301.332.246.277	1	1	Spinach seed	Spinach seed	WA, USA	WA, USA
8	384.299.369.301.263.521.333.330.367.289.332.246.277	1	12	Spinach seed	Spinach seed	WA, USA	WA, USA
9	384.299.376.301.263.545.333.330.367.295.401.250.277	1	11	Artichoke	Artichoke, Lettuce	CA, USA	CA, USA

[a] "0" indicates no amplification at locus; alleles are presented in the order: VD2.VD1.VD9.VD11.VD92.VD69.VD12.VD27.VD73.VD8.VD10.VD3. [b], [c] Total number of *MAT1-1*, *MAT1-2* isolates for each microsatellite type.

Table 2. *Verticillium dahliae* orthologs of *Neurospora crassa, Saccharomyces cerevisiae, Podospora anserina* sex-related genes.

Gene annotation/putative function	V. dahliae accession	Synonym	Other accession	Annotated fungal species
Meiosis				
Double-strand DNA breaks formation and processing				
Meiotic recombination protein REC12	VDAG_09359	SPO11	NCU01120	*Neurospora crassa*
Meiotic recombination protein REC4	VDAG_07486	SKI8	NCU03517	*Neurospora crassa*
DEAD/DEAH box DNA helicase MER3	NA		NCU09793	*Neurospora crassa*
Splicing factor 3B subunit 4	VDAG_08454		NCU04182	*Neurospora crassa*
Double-strand break repair protein MUS23	VDAG_07631		NCU08730	*Neurospora crassa*
DNA repair protein RAD50	VDAG_06865	USV6	NCU00901	*Neurospora crassa*
DNA repair protein of the MRE11 complex	NA		NCU04329	*Neurospora crassa*
Single strand invasion				
DNA repair protein RAD51	VDAG_08796	MEI3	NCU02741	*Neurospora crassa*
DNA repair and recombination protein RAD52	VDAG_00265	MUS11	NCU04275	*Neurospora crassa*
DNA repair and recombination protein RAD54	VDAG_02310		NCU11255	*Neurospora crassa*
Replication factor-A protein1	VDAG_08650	RPA1	NCU03606	*Neurospora crassa*
Replication factor-A protein 2	VDAG_10269		NCU07717	*Neurospora crassa*
Strand exchange protein RAD55p	VDAG_00585		NCU08806	*Neurospora crassa*
DNA-repair protein XRCC3	VDAG_07164		NCU01771	*Neurospora crassa*
DNA damage checkpoint				
Genome integrity checkpoint protein	VDAG_05896		NCU00274	*Neurospora crassa*
Cell cycle checkpoint protein RAD17	VDAG_03081		NCU00517	*Neurospora crassa*
Proteins involved in crossing over				
DNA mismatch repair protein	VDAG_07693		NCU05385	*Neurospora crassa*
DNA mismatch repair protein MUTS	VDAG_02856	MSH4	NCU10895	*Neurospora crassa*
DNA mismatch repair MUTS family	VDAG_08845	MSH5	NCU09384	*Neurospora crassa*
ATP-dependent helicase SGS1	VDAG_04304	MUS19	NCU08598	*Neurospora crassa*
Meiosis specific protein	VDAG_05193		NCU10836	*Neurospora crassa*
DNA repair protein RAD16	VDAG_01793	MUS38	NCU07440	*Neurospora crassa*
DNA repair protein RAD13	VDAG_00986		NCU07498	*Neurospora crassa*
Synaptonemal complex				
Histone H2A.Z	VDAG_07626		NCU05347	*Neurospora crassa*
Structural maintenance of chromosome: SMC protein	VDAG_01776		NCU09065	*Neurospora crassa*
Structural maintenance of chromosome: SMC protein	VDAG_09439		NCU02402	*Neurospora crassa*
Exodeoxyribonuclease	VDAG_02157		NCU06089	*Neurospora crassa*
Casein kinase I	VDAG_02638		NCU00685	*Neurospora crassa*
Nucleotide excision repair protein RAD23	VDAG_09770	RAD23	NCU07542	*Neurospora crassa*
ATP-dependent DNA helicase SRS2	VDAG_01559	MUS50	NCU04733	*Neurospora crassa*
Mismatch repair proteins				
DNA mismatch repair protein MSH2	VDAG_02253	MSH2	NCU02230	*Neurospora crassa*
DNA mismatch repair protein MSH3	VDAG_04229	MSH3	NCU08115	*Neurospora crassa*
DNA mismatch repair protein MSH6	VDAG_01192	MSH6	NCU08135	*Neurospora crassa*
DNA mismatch repair protein PMS1	VDAG_09041		NCU08020	*Neurospora crassa*
DNA mismatch repair protein MUTL	VDAG_08805		NCU09373	*Neurospora crassa*
Resolution of recombination intermediates				
Protein involved in DNA repair and recombination	VDAG_05488		NCU04047	*Neurospora crassa*
Crossover junction endonuclease MUS81	VDAG_03195	MUS81	NCU07457	*Neurospora crassa*
GIY-YIG catalytic domain containing protein	VDAG_09308		NCU01236	*Neurospora crassa*
DNA topoisomerase	VDAG_04479		NCU09118	*Neurospora crassa*
DNA topoisomerase	VDAG_00604		NCU06338	*Neurospora crassa*

Table 2. Cont.

Gene annotation/putative function	*V. dahliae* accession	Synonym	Other accession	Annotated fungal species
DNA topoisomerase	VDAG_06518		NCU00081	*Neurospora crassa*
Non-homologous end joining				
Ku70 protein	VDAG_10247	MUS51	NCU08290	*Neurospora crassa*
Ku80 protein	VDAG_06524	MUS52	NCU00077	*Neurospora crassa*
Other				
Protein required for meiotic recombination	VDAG_07839		NCU04415	*Neurospora crassa*
Repeat-induced point mutation gene	VDAG_05093	RID	NCU02034	*Neurospora crassa*
Synaptonemal complex protein HOP1	NA		YIL072W	*Saccharomyces cerevisiae*
Interhomolog meiotic recombination HOP2	NA		YGL033W	*Saccharomyces cerevisiae*
Interhomolog meiotic recombination MND1	NA		YGL183C	*Saccharomyces cerevisiae*
Cohesion				
Adherin				
Subunit of cohesin loading factor	VDAG_00695		NCU05250	*Neurospora crassa*
Chromosome cohesion				
Cohesin complex subunit	VDAG_04575		NCU01323	*Neurospora crassa*
Chromosome segregation protein SUDA	VDAG_06558		NCU07554	*Neurospora crassa*
Cohesin complex subunit required for sister chromatid cohesion	VDAG_08327		NCU01247	*Neurospora crassa*
Double-strand-break repair protein RAD21	VDAG_08702	RAD21	NCU03291	*Neurospora crassa*
Rec8 protein	VDAG_02664	REC8	NCU03190	*Neurospora crassa*
Protein required for establishment and maintenance of sister chromatid cohesion	VDAG_03579	V-SNARE	NCU00242	*Neurospora crassa*
Separin				
Separin	VDAG_05810		NCU00205	*Neurospora crassa*
Condensins				
Nuclear condensin complex subunit Smc2	VDAG_00648		NCU07679	*Neurospora crassa*
Nuclear condensin complex subunit Smc4	VDAG_10489		NCU09063	*Neurospora crassa*
Condensin	VDAG_09545		NCU09297	*Neurospora crassa*
Condensin subunit Cnd3	VDAG_06322		NCU06216	*Neurospora crassa*
Chromosome segregation				
Spindle pole body component alp14	VDAG_10219		NCU04535	*Neurospora crassa*
HEC/Ndc80p family protein	VDAG_10087		NCU03899	*Neurospora crassa*
Chromosome segregation protein	VDAG_09035		NCU07984	*Neurospora crassa*
Swi3 domain-containing protein	VDAG_04932		NCU01858	*Neurospora crassa*
Carboxy-terminal kinesin 2	VDAG_09024		NCU04581	*Neurospora crassa*
Tubulin alpha chain	VDAG_04060		NCU09132	*Neurospora crassa*
Tubulin gamma chain	VDAG_01827	TBG	NCU03954	*Neurospora crassa*
Tubulin alpha chain	VDAG_04060	TBA2	NCU09468	*Neurospora crassa*
Anaphase-promoting complex				
Anaphase-promoting complex/cyclosome subunit APC1	VDAG_09956		NCU05901	*Neurospora crassa*
Anaphase-promoting complex protein	VDAG_02447		NCU01963	*Neurospora crassa*
Anaphase-promoting complex subunit CUT9	VDAG_01327		NCU01377	*Neurospora crassa*
WD repeat-containing protein slp1	VDAG_06090		NCU02616	*Neurospora crassa*
Anaphase-promoting complex subunit 8	VDAG_08529		NCU01174	*Neurospora crassa*
Nuclear protein BIMA	VDAG_05870		NCU00213	*Neurospora crassa*
Anaphase-promoting complex subunit 10	VDAG_07093		NCU08731	*Neurospora crassa*
WD repeat-containing protein SRW1	VDAG_04583		NCU01269	*Neurospora crassa*
Meiosis-specific APC/C activator protein AMA1	VDAG_01235		NCU01572	*Neurospora crassa*
Transcription factor and gene regulation				
Meiosis-specific transcription factor	VDAG_00592		NCU09915	*Neurospora crassa*

Table 2. Cont.

Gene annotation/putative function	V. dahliae accession	Synonym	Other accession	Annotated fungal species
Histone-lysine N-methyltransferase	VDAG_10394		NCU06266	Neurospora crassa
Ankyrin repeat protein	VDAG_06433		NCU00388	Neurospora crassa
SNF2 family ATP-dependent chromatin-remodeling factor SNF21	VDAG_06547		NCU06488	Neurospora crassa
Signal transduction				
Calcium/Calmodulin-dependent protein kinase	VDAG_04474		NCU09123	Neurospora crassa
Protein kinase GSK3	VDAG_08431		NCU04185	Neurospora crassa
Serine/Threonine-protein kinase RIM15	VDAG_03223		NCU07378	Neurospora crassa
Pheromone proteins essential for mating				
Pheromone processing	VDAG_05762	STE23	YLR389C	Saccharomyces cerevisiae
Peptide pheromone maturation	VDAG_06292	RCE1	YMR274C	Saccharomyces cerevisiae
Pheromone processing	VDAG_09962	AFC1	YJR117W	Saccharomyces cerevisiae
Protein processing	VDAG_00116	KEX1	YGL203C	Saccharomyces cerevisiae
Pheromone receptor	VDAG_05622	PRE2	Pa_4_1380	Podospora anserina
Farnesyltransferase subunit beta	VDAG_05598	RAM1	Pa_4_7760	Podospora anserina
Putative ABC transporter expressed in the mitochondrial inner membrane	VDAG_01200	STE6	Pa_5_11640	Podospora anserina

In addition to the *MAT* genes, Ka/Ks patterns were investigated in 18 other sex-related genes (Table 5, Table S2). The percentage of codons in *V. dahliae* genes under positive and strong purifying selection ranged from 0–5% and 35–62%, respectively. Six genes, *KEX1*, *MEI3*, *RAD21*, *RAD54*, *STE23*, and *V-SNARE* contained no codons under positive selection using either the M8 or MEC model (Table 5, Figure S1).

Expression of sex-related genes based on RT-PCR

RT-PCR using RNA from both mating types of *V. dahliae* successfully amplified all 10 sex-related genes investigated (Figure 3). As expected, RNA from *MAT1-1-1* and *MAT1-2-1* only amplified from the strain that carried the respective *MAT1-1* and *MAT1-2* idiomorph (Figure 3). DNAse was used to treat extracted RNA, and no amplification was observed in reactions with reverse transcriptase omitted, indicating that DNA contamination was not present in the reactions (gels not shown). Since fungal isolates were cultured independently, it appears that *V. dahliae* expressed these genes during vegetative growth on PDA in the absence of a compatible culture of opposite mating type.

Conclusions

The overabundance of *MAT1-2* in *V. dahliae* has been reported on multiple scales, from heavily sampled single agricultural fields to larger scales such as countries. This phenomenon may be partly explained by clonal expansion of certain successful, highly fit genotypes which do not require sexual reproduction to complete the disease cycle [23,31], unlike some other plant pathogens. Nevertheless, in two field sites in coastal California and in commercial spinach seed lots from WA, identical multilocus microsatellite types comprising both mating types were found, indicating at the very least, that both *MAT1-1* and *MAT1-2* co-occur in some niches currently.

The sample of *V. dahliae* characterized in the current study was biased toward virulent, phytopathogenic isolates collected from diseased plant tissue, because most were isolated from plants with visible wilt symptoms in agricultural settings. This raises the hypothesis that *V. dahliae MAT1-2* may be associated with higher

virulence on some, if not all hosts, which is a phenomenon that has been reported in other fungal systems [48–51]. Preliminary data on the virulence of isolates from both idiomorphs originally collected from tomato suggest that *MAT1-2* isolates are significantly more virulent than *MAT1-1* isolates (Subbarao, unpublished data). A more comprehensive analysis of the virulence of the two idiomorphs is required to confirm these results, however with more experiments and by investigating the mating-type structure in populations of non-pathogenic, endophytic *V. dahliae* [52,53].

Although the current study clearly documents patterns of purifying selective pressures in protein coding regions of the sex-related *V. dahliae* genes investigated, it is possible that there are mutations in non-coding, regulatory regions of the genome that affect the level, timing or location of sex-related gene expression and therefore hinder the sexual cycle. Furthermore, it is possible that genes that were originally associated with sexual reproduction in ancestral populations have evolved new functions, and this is the reason they are being maintained under selection. Yet, it has been previously supposed that the presence of the majority, if not all, of the meiosis-specific genes in the genome of a microorganism is the "strongest indicator" that genes are maintained for meiosis and sex, even if it is rare [10]. The *V. dahliae* genome is clearly replete with orthologs to genes known for their roles in pathways associated with the sexual cycle. Further, the SELECTON analyses provide evidence that sex-related genes are not in the process of becoming pseudogenes.

The production of actual sexual structures *in vitro* currently remains a mystery in *V. dahliae*, possibly due to the lack of research into the growth medium content requirements, such as nutrient (i.e. carbon) content and pH, which are highly variable for sexual fungi in the *Pezizomycotina* [5,16]. Nevertheless, the genomic evidence presented in the current study, taken together with previous studies of population structure and recombination [23], is compelling and could be reasonably interpreted as evidence of an ancestral or rare sexual cycle in this predominantly asexual species.

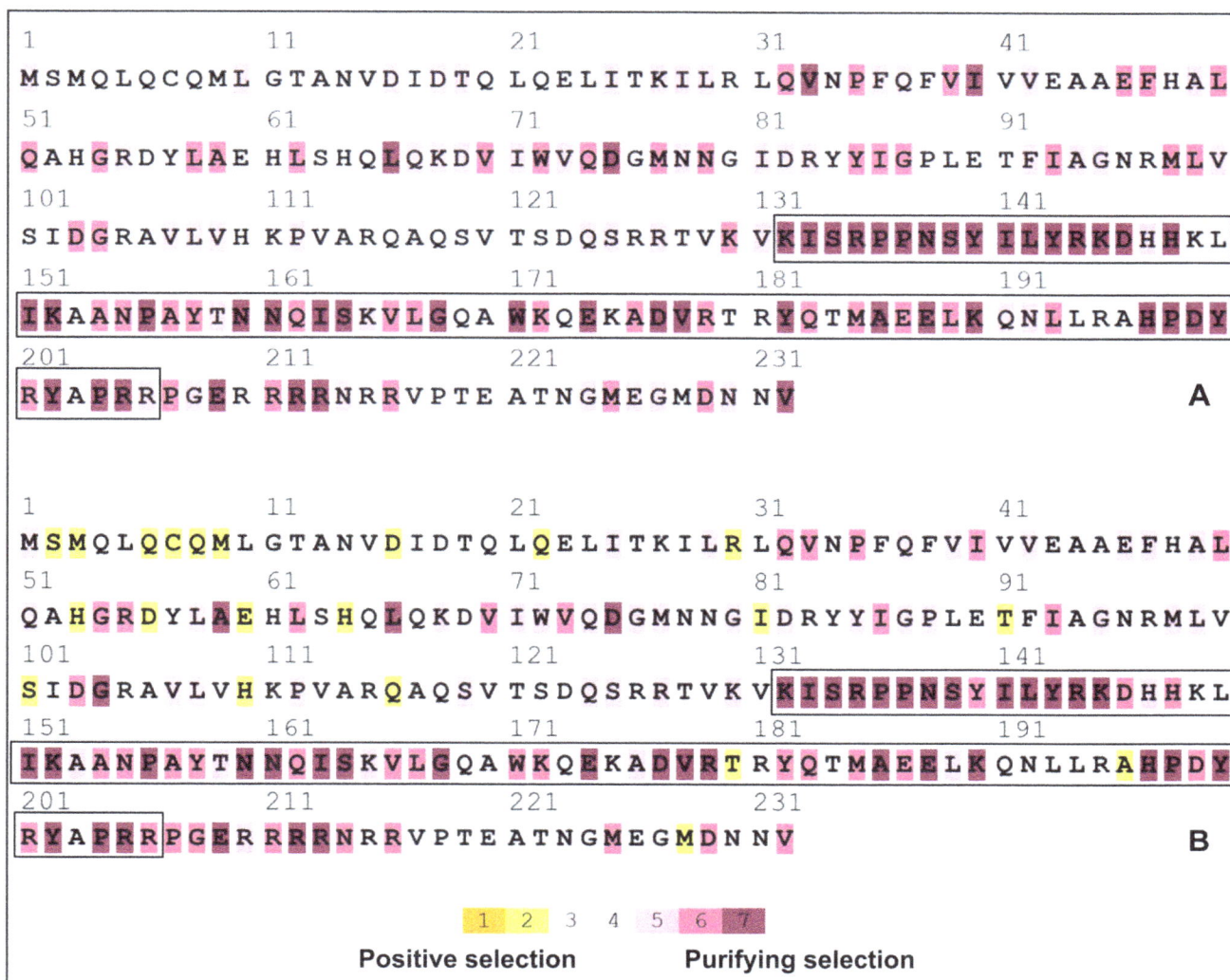

Figure 1. Color-coded results of SELECTON analyses of *Verticillium dahliae MAT1-1-1,* **compared to sequences from nine different sexual fungi in the** *Pezizomycotina.* Shades of yellow (colors 1 and 2) indicate a Ka/Ks ratio>1 (positive selection), and shades of purple (colors 3 through 7) indicate a Ka/Ks ratio<1 (purifying selection); A) results from the M8 model; B) results of the MEC model; amino acid sequence of the α domain is indicated by black border.

Materials and Methods

Fungal culture maintenance and DNA extraction for *MAT* characterization

In this study, 1120 isolates of *V. dahliae*, collected from 10 different countries, were characterized for mating type (Table S1). No specific permissions were required for isolating *Verticillium* from any of the regions in the current study. The field isolations did not involve endangered or protected species. Importation of *Verticillium* cultures was performed under the appropriate USDA-APHIS permits (P526P-11-02218, P526P-11-02476, P526P-11-02806). *Verticillium* cultures were originally cultivated on semi-selective NP–10 medium [54], and then single-conidium purified and transferred to potato dextrose agar (PDA). Cultures were stored long-term as spore suspensions in 25% glycerol at −20°C. Mycelia for DNA extraction were grown in 250 ml Erlenmeyer flasks containing 50 ml potato dextrose broth (PDB). Each flask was inoculated with a piece of PDA culture with an approximate surface area of 1 cm^2. Mycelia from PDB were harvested after

10 days, washed with sterile distilled water, dried using paper towels, lyophilized, and ground to a fine powder using a high-speed mixer mill (Model MM301; Retsch Inc., Newtown, PA). Genomic DNA of each isolate was extracted using a FastDNA Kit (MP Biomedicals LLC, Solon, OH) following the manufacturer's instructions. A Nano Drop (Model ND–1000, Thermo Scientific Inc., Waltham, MA) was used to quantify DNA extractions, which were diluted to 10 ng/μl, and stored in a freezer at −20°C until needed for PCR assays.

Molecular assays to identify *Verticillium* species and mating type

All isolates used in this study were identified as *V. dahliae* using a *Verticillium* species-specific multiplex as previously described [35]. Mating types were determined for 1120 *V. dahliae* isolates PCR assay with the previously developed primers Alf3 (CGATCGCGA-TATCGGCAAGG), MAT11r (CAGTCAGATCCAACCTG-CTGGCC), HMG21f (CGGCCGCCCAATTCGTACATCC)

Figure 2. Color-coded results of SELECTON analyses of *Verticillium dahliae MAT1-2-1*, **compared to sequences from nine different sexual fungi in the** *Pezizomycotina.* Shades of yellow (colors 1 and 2) indicate a Ka/Ks ratio>1 (positive selection) and shades of purple (colors 3 through 7) indicate a Ka/Ks ratio<1 (purifying selection); A) results from the M8 model; B) results of the MEC model; amino acid sequence of the HMG domain is indicated by black border.

Table 3. Comparison of codons under positive (relaxed) and purifying selection in *MAT1-1-1*, in a variety of fungi in the subphylum *Pezizomycotina* using the MEC model.

Fungal taxon	*MAT1-1 −1* accession no.	Transcript length (codons)	Codons under positive selection	Codons under strong purifying selection
Verticillium dahliae[1]	NCBI GenBank AB505215	421	51 (12%)	146 (34%)
Aspergillus fumigatus	NCBI GenBank AY898660	369	45 (12%)	125 (33%)
Aspergillus nidulans[2]	ANID_02755	362	45 (12%)	129 (35%)
Cochliobolus heterostrophus	NCBI GenBank X68399	384	46 (12%)	129 (33%)
Eupenicillium crustaceum[2]	NCBI GenBank FR729897	343	34 (9%)	121 (35%)
Fusarium graminearum[2]	FGSG_08892	345	53 (15%)	126 (36%)
Fusarium verticillioides	FVEG_02491	383	54 (14%)	129 (33%)
Histoplasma capsulatum	HCAG_09679	305	34 (11%)	107 (35%)
Nectria heamatococca	NCH17696	214	20 (9%)	92 (43%)
Penicillium chrysogenum	PC_255945071	342	34 (9%)	119 (34%)
Sclerotinia sclerotiorum	SS1G_04004	258	35 (13%)	91 (35%)
Passalora fulva[3]	DQ659350	358	60 (16%)	98 (27%)

[1]SELECTON results for the putatively asexual fungus *V. dahliae* were calculated by analyzing a *MAT1-1-1* codon sequence alignment including sequences from all other fungi listed except *P. fulva*. Results for the ten species *A. fumigatus – S. sclerotiorum* were calculated using a codon alignment of only these ten species.
[2]Homothallic fungus.
[3]Results for the putatively asexual fungus *P. fulva* were calculated by analyzing a *MAT1-2-1* codon sequence alignment including sequences from all other fungi listed except *V. dahliae*.

and MAT21r (CATGCCTTCCATGCCATTAGTAGCC). These primers amplify a ~600-bp fragment from *MAT1–1–1* isolates and a ~300-bp fragment from *MAT1–2–1* isolates, as previously described [29,35,37]. PCR assays to characterize mating types were performed in 25 µl reactions using GoTaq Green Mastermix (Promega, Madison, WI). All PCR assays in this study were performed in a PTC-100 Peltier Thermal cycler (MJ Research, Inc., Waterman, MA). For mating type multiplex PCR, the following thermal profile was used: 2 min initial denaturation at 94°C, 35 cycles of 10 sec at 94°C, 20 sec at 57°C, and 1 min at 72°C, followed by a final extension of 7 min at 72°C. PCR amplicons were stained with 5 µl SyberGold (Invitrogen Life Technologies, Carlsbad, CA), and aliquots were loaded in a 1.5% (wt/vol) agarose gel and run for 120 min at 75 V in 0.5% TBE buffer [55]. A 100–bp DNA ladder (Invitrogen Life Technologies, Carlsbad, CA) was

Table 4. Comparison of codons under positive (relaxed) and purifying selection in *MAT1-2-1*, in a variety of fungi in the subphylum *Pezizomycotina* using the MEC model.

Fungal taxon	*MAT1-2-1* accession no.	Transcript length (codons)	Codons under positive selection	Codons under strong purifying selection
Verticillium dahliae[1]	VDAG_02444	232	21 (9%)	81 (35%)
Chaetomium globosum	CHGG_03580	342	74 (22%)	101 (30%)
Aspergillus nidulans[2]	ANID_04734	318	70 (22%)	95 (30%)
Colletotrichum graminicola	GLRG_04643	238	42 (18%)	76 (32%)
Fusarium graminearum[2]	FGSG_08893	253	52 (21%)	76 (30%)
Fusarium sacchari	NCBI GenBank JF776855	227	48 (21%)	69 (30%)
Magnaporthe grisea	MG_02978	437	52 (12%)	150 (34%)
Ophiostoma novo-ulmi	NCBI GenBank FJ959052	183	33 (18%)	59 (32%)
Podospora anserina	Pa_1_20590	582	74 (13%)	124 (21%)
Penicillium chrysogenum	NCBI GenBank AM904545	303	64 (21%)	91 (30%)
Trichoderma ressei	TRI14830	241	46 (19%)	56 (23%)
Passalora fulva[3]	DQ659351	384	45 (11%)	133 (34%)

[1]SELECTON results for the putatively asexual fungus *V. dahliae* were calculated by analyzing a *MAT1-2-1* codon sequence alignment including sequences from all other fungi listed except *P. fulva*. Results for the ten species *C. globosum – T. reseei* were calculated using a codon alignment of only these ten species.
[2]Homothallic fungus.
[3]Results for the putatively asexual fungus *P. fulva* were calculated by analyzing a *MAT1-2-1* codon sequence alignment including sequences from all other fungi listed except *V. dahliae*.

Table 5. Comparison of codons under positive (relaxed) and purifying selection in 18 sex-related genes in *Verticillium dahliae* using the MEC model.

V. dahliae accession[1]	Locus	Transcript length (codons)	Codons under positive selection	Codons under strong purifying selection
VDAG_00116	KEX1	384	0 (0%)	154 (40%)
VDAG_08796	MEI3	354	0 (0%)	142 (40%)
VDAG_02856	MSH4	843	40 (3%)	313 (37%)
VDAG_08845	MSH5	863	8 (1%)	340 (39%)
VDAG_01559	MUS50	1166	5 (<1%)	462 (40%)
VDAG_01559	MUTL	704	2 (<1%)	281 (40%)
VDAG_08702	RAD21	530	0 (0%)	258 (49%)
VDAG_02310	RAD54	651	0 (0%)	261 (40%)
VDAG_05598	RAM1	469	4 (<1%)	185 (39%)
VDAG_06292	RCE1	304	1 (<1%)	122 (40%)
VDAG_02664	REC8	452	33 (2%)	281 (62%)
VDAG_01783	RID	957	66 (5%)	343 (36%)
VDAG_07486	SKI8	336	1 (<1%)	135 (40%)
VDAG_09359	SPO11	425	38 (3%)	149 (35%)
VDAG_05762	STE23	941	0 (0%)	377 (40%)
VDAG_06443	STE24	300	21 (1%)	107 (36%)
VDAG_01200	STE6	1416	23 (1%)	526 (37%)
VDAG_03579	V-SNARE	128	0 (0%)	51 (40%)

[1]Fungal taxa and gene accessions used to estimate selective pressures in *V. dahliae* genes are provided in Table S2. Color-coded SELECTON results for each gene are provided in Figure S1.

included in each gel and a transilluminator (Ultra-Violet Products, Ltd., Upland, CA) was used to visualize PCR products.

Multilocus microsatellite genotyping

Thirteen previously developed microsatellite loci were used in this study: VD1, VD2, VD3, VD8, VD9, VD10, VD11, VD12, VD27, VD69, VD73, VD92 and VD97 [56] which were developed using the *V. dahliae* strain Ls 17 complete genome sequence [30,37]. For all microsatellite loci, PCR was performed in 20 µl total volumes containing 4 µl of sterile, distilled water, 2 µl of 10 ng/µl genomic DNA, 2 µl each of 10 µM reverse and forward primer, and 12.5 µl of GoTaq Green PCR mix (Promega Inc., Madison, WI). Published thermocycling parameters were used as previously described [30]. PCR amplicons labeled with up to four fluorophores FAM, HEX, ROX and TAMRA (Invitrogen, Carlsbad, CA) were pooled [57]. One µl of the pooled amplicons was then combined with Hi-Di formamide and 0.3 µl of LIZ–500 size standard and separated on an ABI 3100 capillary electrophoresis genetic analyzer (Applied Biosystems, Carlsbad, CA) at the University of California-Davis DNA Sequencing Facility, Davis CA. The peaks in were scored using the GeneMarker software (SoftGenetics, State College, PA).

To assess reliability of microsatellite allele calls using capillary electrophoresis [58], 192 microsatellite amplicons representative of all 13 loci were arbitrarily selected for DNA sequencing using unlabeled forward and reverse primers. Amplicons from *V. dahliae* strain Ls 17 were also generated and compared to the results reported from the same strain in previous studies [30,56]. Different amplicon sizes at each locus were considered unique. Alleles were compiled across loci into multilocus microsatellite types (MLMTs).

Verticillium genome queries and ortholog searches

The FUNGIpath ortholog database was queried using a panel of 93 genes that have been characterized for functions related to sexual reproduction in the fungal model systems *Neurospora crassa*, *Saccharomyces cerevisiae*, and *Podospora anserina*. The set of 93 genes comprised the two mating type genes *MAT1-1-1* and *MAT1-2-1*, 81 previously described *Neurospora crassa* genes associated with meiosis [36,59,60] which were retrieved from the *Neurospora* Genome Database [61,62], four previously described *Saccharomyces cerevisiae* pheromone-related genes *STE23*, *RCE1*, *AFC1*, *KEX1* [63], which were retrieved from the *Saccharomyces* Genome Database [24,64], and three *Podospora anserina* pheromone-related genes *PRE2*, *RAM1*, *STE6* [63] which were retrieved from the *Podospora anserina* Genome Database [25,65]. Since *V. dahliae* is heterothallic and the sequenced strain contains only *MAT1-2-1*, a sequence of *V. dahliae MAT1-1-1* was obtained through National Center of Bioinformatics (NCBI) GenBank, Accession AB505215 [27]. Finally, three additional *Saccharomyces cerevisiae* genes broadly associated with meiosis in eukaryotes (*HOP1*, *HOP2*, and *MND1*) [10] were queried against the FUNGIpath database. For FUNGIpath ortholog database searches, either gene accession ids. or amino acid sequences were used as input [41]. In this way, *V. dahliae* genes were verified as orthologous to genes from sexual fungi. Ortholog gene accession ids. from other fungi in the *Pezizomycotina* were noted and downloaded from the respective genome databases for subsequent analyses.

Primer design

After identifying orthologs to sex-related genes in the genome of *V. dahliae*, coding sequences of *MAT* genes and eight other genes associated with meiosis in other systems were arbitrarily chosen

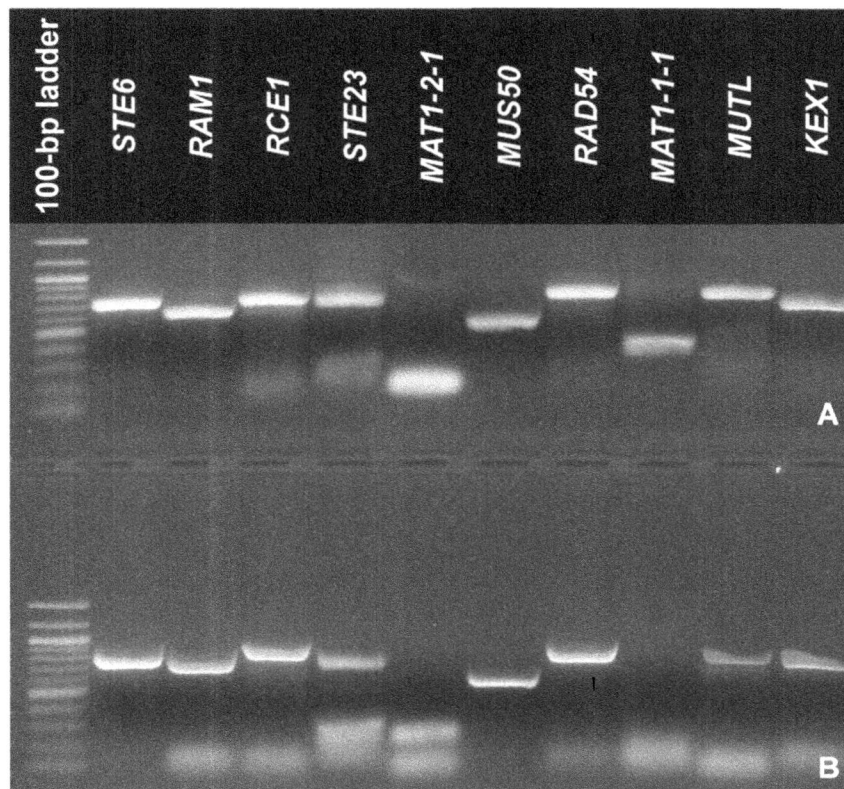

Figure 3. Reverse-transcriptase PCR results of 10 *Verticillium dahliae* orthologs of genes associated with the sexual cycle in model fungal systems; gene names are provided for each lane; A) RT-PCR results from *V. dahliae* strain 58 (*MAT1-1*). B) RT-PCR results from *V. dahliae* strain Ls 17 (*MAT1-2*);

and downloaded from the Broad Institute website [41]. Forward and reverse primers were designed to amplify ~500 to 1000-bp targets within coding sequences for 8 of the genes, whereas the previously described primers Alf3-MAT11r and HMG21f-MAT21r [29] were used to amplify *MAT1-1-1* and *MAT1-2-1*, respectively (Table 6).

RNA Extraction and RT-PCR

The two *V. dahliae* isolates 58 (*MAT1-1-1*) and Ls 17 (*MAT1-2-1*) were grown on PDA. For each culture, after ten days, 3 ml of sterile distilled water was poured onto the culture surface and spread with a plate spreader. One ml of the resulting conidia and hyphal suspensions was transferred to a 47 mm nitrocellulose membrane (0.45 μm pore size; Whatman, Maidstone, England) overlaid on a PDA plate. Cultures were maintained in the dark at 25°C. After 10 days, the nitrocellulose membranes covered with fungal tissue were harvested with sterilized forceps and ground to a fine powder in liquid nitrogen using a mortar and pestle. Total RNA was extracted from 100 mg of the ground powder using TRIzol Reagent (Life Technologies, Carlsbad, CA) following the manufacturer's protocol. Total RNA extracts were treated with TURBO DNase (Life Technologies, Carlsbad, CA) following the manufacturer's protocol, in order to degrade genomic DNA.

Reverse-transcriptase PCR (RT-PCR) was performed using a SuperScript III OneStep RT-PCR system with Platinum Taq DNA polymerase (Life Technologies, Carlsbad, CA) following the manufacturer's protocol. For RT-PCR the following thermal profile was used: a cDNA synthesis cycle of 30 min at 55°C, an

initial denaturation of 94°C for 2 minutes, 40 cycles of 94°C for 15 sec, 55°C for 30 sec, and 68°C for 1 min, followed by a final extension of 68°C for 5 min. Separate reactions including ten micromolar concentrations of forward and reverse primers for each and every locus described above were performed. For a positive control, RT-PCR was performed with the primers AaDTr (CTGGATGGAGACGTAGAAGGC) and Df (CTCGATGCT-CAAGCAGTACAT), which target *ACT* (VDAG_08445). Amplicons were visualized as above.

To verify the absence of genomic DNA in both of the RNA preparations, SuperScript III/RT Platinum Taq mix was omitted from PCR assays, and instead, two units of Platinum Taq DNA polymerase (Life Technologies, Carlsbad, CA) were used in reactions using the primers AaDTr and Df, in accordance with the manufacturer's instructions.

SELECTON analyses of positive and purifying selection in *MAT1-1-1*, *MAT1-2-1*, and other sex-related genes of *V. dahliae*

To test the hypothesis that *V. dahliae* mating type and meiosis-associated genes are being maintained under strong purifying selection, ratios of non-synonymous (amino-acid altering) to synonymous (silent) substitutions in *V. dahliae* genes (relative to sexual fungi) were calculated through the SELECTON server [66,67]. All *MAT* genes used in this study were either identified directly through the FUNGIPath database, or were obtained through NCBI GenBank and verified as orthologs to either

Table 6. Primers used to amplify *V. dahliae* sex-related genes with RT-PCR.

Gene name	*V. dahliae* accession[1]	Fw primer 5'–3'	Rv primer 5'–3'
MAT1-1-1	NA [1]	CGATCGCGATATCGGCAAGG	CAGTCAGATCCAACCTGCTGGCC
MAT1-2-1	VDAG_02444	GCAATGTCAGATGCTCGGTA	CTGCGAGATAATCACGACCA
STE6	VDAG_01200	GCAAACTTCTCACCCTCTGC	CAGGTCGTCTCCCACTTTGT
MUS50	VDAG_01559	CGACCTTATCGGCGATCTAC	CTCTCTTCTGGGTCGACAGG
RAD54	VDAG_02310	GCAAACGAGCTTGTCAAGTG	GGTTGCAGAGCTTCTTGAGG
RAM1	VDAG_05598	GCTTCTACGCCAGCAGACAC	GTCGACTTCACCGCCATAC
STE23	VDAG_05762	ACAGGTTCTCGTCACCATCC	GGACATGGTGTCAATGATCG
RCE1	VDAG_06292	ACAGAGGAGCTGCTTTTTCG	TCCACCACGCTTCTTGAACT
MUTL	VDAG_08805	AAGGCTCTACCGCCAATTTT	TCATCGTTTCGTCTGCTCTG
MSH5	VDAG_08845	CGGGACATTTACCGATGAAC	TCCTCAGCATCCCTCAGTCT

[1]The genome of *V. dahliae* strain Ls 17 contains only *MAT1-2-1*. *MAT1-1-1* sequence obtained from NCBI GenBank.

MAT1-1-1 or *MAT1-2-1* using the ortholog search function in the FUNGIPath.

Additionally, ortholog search results from the FUNGIpath database from taxa within the subphylum *Pezizomycotina* were downloaded for 18 arbitrarily chosen, previously characterized genes associated with meiosis (Table S2), which represented a subset of the aforementioned 93 genes. Unaligned nucleotide sequences of *V. dahliae* orthologs and sequences identified through the FUNGIpath database [68] from at least nine other *Pezizomycotina* fungi were used as input to the SELECTON server, to provide the recommended number of sequences.

Selection pressure was estimated in the following 20 sex-related *V. dahliae* genes: *MAT1–1–1* and *MAT1–2–1* [28,69]; the *RID* gene [36]; the nine *N. crassa* meiosis–specific genes *SPO11*, *SKI8*, *MUTL*, *RAD54*, *MSH4*, *MSH5*, *MUS50*, *RAD21*, and *REC8* [13]; the *N. crassa* gene *V-SNARE*, required for establishment and maintenance of sister chromatid cohesion [60]; and finally, seven *P. anserina* genes encoding pheromones, receptors, and genes related to pheromone biogenesis *STE24*, *RAM1*, *RCE1*, *KEX1*, *STE23*, *STE6*, and *PRE2* [63].

Nonsynonymous to synonymous substitution ratios (Ka/Ks) of *V. dahliae* genes were calculated using the SELECTON server [66,67], based on alignments of *V. dahliae* genes with sequences from the following fungi with known sexual stages: *Aspergillus fumigatus* [70], *Aspergillus nidulans* [71], *Botrytis cinerea* [72,73], *Chaetomium globosum* [74], *Colletotrichum graminicola* [75,76], *Epichloë festucae* [77], *Eupenicillium crustaceum* [78], *Fusarium graminearum* [79], *Histoplasma capsulatum* [80], *Mcgnaporthe oryzae* [81], *Neurospora crassa* [61], *Nectria haematococca* [82], *Ophiostoma novo-ulmi* (NCBI GenBank ADB96163), *Penicillium chrysogenum* [83], *Podospora anserina* [65], *Sclerotinia sclerotiorum* [73], *Trichoderma reesei* [84] and *Zymoseptoria tritici* [85]. For each of the 20 *V. dahliae* genes analyzed, sequences from different taxa were used as input, based on availability. Transcript sequences of the relevant genes from fungal taxa were obtained from multiple sources, and accession numbers of fungal gene sequences are provided in Table 2 and Table S2.

Codon alignments were generated by the SELECTON server and for each codon, the Ka/Ks ratio was estimated using a Bayesian approach. SELECTON results for each codon were reported on a scale of 1–7, with scores of one or two indicating positive selection, and scores of six or seven indicating strong purifying selection. For comparative purposes, two evolutionary models with positive selection enabled were used in the analyses,

namely the M8 model [86,87] and the mechanistic–empirical combination (MEC) model [88]. SELECTON implements several codon models, each of which assumes different biological assumptions. The MEC model takes into account the differences between different amino-acid replacement probabilities. For analyses with the MEC model, eight categories for the distribution, a JTT empirical amino-acid matrix, and a high precision level were used. In cases where positive selection sites were detected using the MEC model, a likelihood ratio test between the results of the MEC model and the M8a (null) model was performed, by comparing Akaike Information Content (AIC) scores [89].

Estimates of selection in genes may be influenced by the choice of taxa used in the codon alignment. Therefore, for comparative purposes of the two mating type genes, Ka/Ks ratios within each of the other *MAT1-1-1* and *MAT1-2-1* sequences from other species were also calculated as above. For these analyses, the *V. dahliae* sequence was removed from the set of nucleotide sequences, and each sequence from every fungal species was considered independently as the query sequence. Thus, the Ka/Ks ratios of *MAT* loci were calculated for several sexual fungi, relative to the same set of taxa used to estimate selective pressures in *V. dahliae* mating type genes. Lastly, the Ka/Ks ratios in *MAT1-1-1* and *MAT1-2-1* from *Passalora fulvum*, a putatively asexual species, were calculated in comparison with the same set of sexual fungi used in the analyses of *V. dahliae* genes.

Supporting Information

Figure S1 Color-coded results of SELECTON analyses of 18 *Verticillium dahliae* sex-related genes, compared to sequences from nine different sexual fungi in the *Pezizomycotina*.

Table S1 *V. dahliae* isolates used in this study along with country of origin, location, plant host, and mating types, as determined by PCR assays.

Table S2 List of fungal gene sequence accessions and results from SELECTON analyses of *Verticillium dahliae* genes associated with meiosis in model systems.

Acknowledgments

The authors are grateful for the fine technical assistance of Rosa Marchebout.

Author Contributions

Conceived and designed the experiments: DPGS SG XH PI KVS. Performed the experiments: DPGS SG XH. Analyzed the data: DPGS SG XH PI. Contributed reagents/materials/analysis tools: KVS. Contributed to the writing of the manuscript: DPGS SG XH PI KVS.

References

1. Weismann A (1889) The significance of sexual reproduction in the theory of natural selection. In: Poulton EB, Schönland S, Shipley AE, editors. Essays upon heredity and kindred biological problems. Oxford: Clarendon. pp. 251–332.
2. Otto SP (2009) The evolutionary enigma of sex. Am Nat 174: S1-S14.
3. Goddard MR, Godfray HCJ, Burt A (2005) Sex increases the efficacy of natural selection in experimental yeast populations. Nature 434: 636–640.
4. O'Gorman CM, Fuller HT, Dyer PS (2008) Discovery of a sexual cycle in the opportunistic fungal pathogen Aspergillus fumigatus. Nature 457: 471–474.
5. Short DPG, O'Donnell K, Thrane U, Nielsen KF, Zhang N, et al. (2013) Phylogenetic relationships among members of the Fusarium solani species complex in human infections and the descriptions of F. keratoplasticum sp. nov. and F. petroliphilum stat. nov. Fungal Genet Biol 53: 59–70.
6. Saleh D, Xu P, Shen Y, Li C, Adreit H, et al. (2012) Sex at the origin: an Asian population of the rice blast fungus Magnaporthe oryzae reproduces sexually. Mol Ecol 21: 1330–1344.
7. Dyer PS, O'Gorman CM (2011) A fungal sexual revolution: Aspergillus and Penicillium show the way. Curr Opin Microbiol 14: 649–654.
8. Milgroom MG (1996) Recombination and the multilocus structure of fungal populations. Annu Rev Phytopathol 34: 457–477.
9. Tibayrenc M, Kjellberg F, Arnaud J, Oury B, Brenière SF, et al. (1991) Are eukaryotic microorganisms clonal or sexual? A population genetics vantage. P Natl Acad Sci 88: 5129–5133.
10. Schurko AM, Logsdon JM (2008) Using a meiosis detection toolkit to investigate ancient asexual' "scandals" and the evolution of sex. Bioessays 30: 579–589.
11. Henk DA, Shahar-Golan R, Devi KR, Boyce KJ, Zhan N, et al. (2012) Clonality despite sex: the evolution of host-associated sexual neighborhoods in the pathogenic fungus Penicillium marneffei. PLoS Pathog 8: e1002851.
12. Ropars J, Dupont J, Fontanillas E, De La Vega RCR, Malagnac F, et al. (2012) Sex in cheese: evidence for sexuality in the fungus Penicillium roqueforti. PLoS ONE 7: e49665.
13. Ramesh MA, Malik S-B, Logsdon Jr JM (2005) A phylogenomic inventory of meiotic genes: evidence for sex in Giardia and an early eukaryotic origin of meiosis. Curr Biol 15: 185–191.
14. Butler G (2010) Fungal sex and pathogenesis. Clin Microbiol Rev 23: 140–159.
15. Dyer PS, Paoletti M, Archer DB (2003) Genomics reveals sexual secrets of Aspergillus. Microbiol 149: 2301–2303.
16. Dyer PS, O'Gorman CM (2012) Sexual development and cryptic sexuality in fungi: insights from Aspergillus species. FEMS Microbiol Rev 36: 165–192.
17. Pegg GF, Brady BL (2002) Verticillium Wilts. Wallingford: CABI Publishing. 522 p.
18. Inderbitzin P, Subbarao KV (2014) Verticillium systematics and evolution: how confusion impedes Verticillium wilt management and how to resolve it. Phytopathology 104: 564–574.
19. Wilhelm S (1955) Longevity of the Verticillium wilt fungus in the laboratory and field. Phytopathology 45: 180–181.
20. Joaquim TR, Rowe RC (1991) Vegetative compatibility and virulence of strains of Verticillium dahliae from soil and potato plants. Phytopathology 81: 552–558.
21. Puhalla J, Hummel M (1983) Vegetative compatibility groups within Verticillium dahliae. Phytopathology 73: 1.
22. Milgroom MG, del Mar Jimenez-Gasco M, Concepcion O-G, Jiminéz-Diaz (2013) New insights on the phylogenetic relationships between strains of Verticillium dahliae. In: Koopmann B, Tiedemann Av, editors. 11th International Verticillium Symposium - Göttingen 2013. Braunschweig: DPG-Verlag.
23. Milgroom MG, del Mar Jiménez-Gasco M, García CO, Drott MT, Jiménez-Díaz RM (2014) Recombination between clonal lineages of the asexual fungus Verticillium dahliae detected by genotyping by sequencing. PLoS ONE 9: e106740.
24. Metzenberg RL, Glass NL (1990) Mating type and mating strategies in Neurospora. Bioessays 12: 53–59.
25. Debuchy R, Turgeon BG (2006) Mating-Type Structure, Evolution, and Function in Euascomycetes. In: U K, Fischer R, editors. The Mycota Volume I: Growth, Differentiation and Sexuality Berlin: Springer-Verlag. 293–323.
26. Turgeon BG, Yoder O (2000) Proposed nomenclature for mating type genes of filamentous ascomycetes. Fungal Genet Biol 31: 1–5.
27. Usami T, Itoh M, Amemiya Y (2009) Asexual fungus Verticillium dahliae is potentially heterothallic. J Gen Plant Pathol 75: 422–427.
28. Usami T, Itoh M, Amemiya Y (2008) Mating type gene MAT1-2-1 is common among Japanese isolates of Verticillium dahliae. Physiol Mol Plant Pathol 73: 133–137.
29. Inderbitzin P, Davis RM, Bostock RM, Subbarao KV (2011) The ascomycete Verticillium longisporum is a hybrid and a plant pathogen with an expanded host range. PLoS ONE 6: e18260.
30. Atallah ZK, Maruthachalam K, Toit Ld, Koike ST, Michael Davis R, et al. (2010) Population analyses of the vascular plant pathogen Verticillium dahliae detect recombination and transcontinental gene flow. Fungal Genet Biol 47: 416–422.
31. Gurung S, Short DPG, Atallah Z, Subbarao KV (2014) Clonal expansion of Verticillium dahliae in lettuce. Phytopathology 104: 641–649.
32. Short DPG, Kerry O, Geiser DM (2014) Clonality, recombination, and hybridization in the plumbing-inhabiting human pathogen Fusarium keratoplasticum inferred from multilocus sequence typing. BMC Evol Biol 14: 91.
33. Smith JM, Smith N (1998) Detecting recombination from gene trees. Mol Biol Evol 15: 590–599.
34. Taylor JW, Jacobson DJ, Kroken S, Kasuga T, Geiser DM, et al. (2000) Phylogenetic species recognition and species concepts in fungi. Fungal Genet Biol 31: 21–32.
35. Inderbitzin P, Bostock RM, Davis RM, Usami T, Platt HW, et al. (2011) Phylogenetics and taxonomy of the fungal vascular wilt pathogen Verticillium, with the descriptions of five new species. PLoS ONE 6: e28341.
36. Freitag M, Williams RL, Kothe GO, Selker EU (2002) A cytosine methyltransferase homologue is essential for repeat-induced point mutation in Neurospora crassa. P Natl Acad Sci USA 99: 8802–8807.
37. Klosterman SJ, Subbarao KV, Kang S, Veronese P, Gold SE, et al. (2011) Comparative genomics yields insights into niche adaptation of plant vascular wilt pathogens. PLoS Pathog 7: e1002137.
38. Amyotte SG, Tan X, Pennerman K, del Mar Jimenez-Gasco M, Klosterman SJ, et al. (2012) Transposable elements in phytopathogenic Verticillium spp.: insights into genome evolution and inter-and intra-specific diversification. BMC Genomics 13: 314.
39. de Jonge R, van Esse HP, Maruthachalam K, Bolton MD, Santhanam P, et al. (2012) Tomato immune receptor Ve1 recognizes effector of multiple fungal pathogens uncovered by genome and RNA sequencing. P Natl Acad Sci USA 109: 5110–5115.
40. de Jonge R, Bolton MD, Kombrink A, van den Berg GC, Yadeta KA, et al. (2013) Extensive chromosomal reshuffling drives evolution of virulence in an asexual pathogen. Genome Res 23: 1271–1282.
41. Kistler HC, Miao VP (1992) New modes of genetic change in filamentous fungi. Annu Rev Phytopathol 30: 131–153.
42. Inderbitzin P, Thomma BP, Klosterman SJ, Subbarao KV (2014) Verticillium alfalfae and V. dahliae, Agents of Verticillium Wilt Diseases. In: Dean R, Kole C, Lichens-Park A, editors. Genomics of Plant-Associated Fungi and Oomycetes: Dicot Pathogens. Berlin: Springer-Verlag.
43. Turgeon BG, Sharon A, Wirsel S, Yamaguchi K, Christiansen SK, et al. (1995) Structure and function of mating type genes in Cochliobolus spp. and asexual fungi. Can J Bot 73: 778–783.
44. Yun S-H, Arie T, Kaneko I, Yoder OC, Turgeon BG (2000) Molecular organization of mating type loci in heterothallic, homothallic, and asexual Gibberella/Fusarium species. Fungal Genet Biol 31: 7–20.
45. Paoletti M, Rydholm C, Schwier EU, Anderson MJ, Szakacs G, et al. (2005) Evidence for sexuality in the opportunistic fungal pathogen Aspergillus fumigatus. Curr Biol 15: 1242–1248.
46. Massingham T, Goldman N (2005) Detecting amino acid sites under positive selection and purifying selection. Genetics 169: 1753–1762.
47. Berbee ML, Payne BP, Zhang G, Roberts RG, Turgeon BG (2003) Shared ITS DNA substitutions in isolates of opposite mating type reveal a recombining history for three presumed asexual species in the filamentous ascomycete genus Alternaria. Mycological Research 107: 169–182.
48. Zhan J, Torriani SF, McDonald BA (2007) Significant difference in pathogenicity between MAT1-1 and MAT1-2 isolates in the wheat pathogen Mycosphaerella graminicola. Fungal Genet Biol 44: 339–346.
49. Clarke D, Woodlee G, McClelland C, Seymour T, Wickes B (2001) The Cryptococcus neoformans STE11α gene is similar to other fungal mitogen-activated protein kinase kinase kinase (MAPKKK) genes but is mating type specific. Mol Microbio 40: 200–213.
50. Kwon-Chung KJ, Edman JC, Wickes BL (1992) Genetic association of mating types and virulence in Cryptococcus neoformans. Infect Immun 60: 602–605.
51. Lockhart SR, Wu W, Radke JB, Zhao R, Soll DR (2005) Increased virulence and competitive advantage of a/α over a/a or α/α offspring conserves the mating system of Candida albicans. Genetics 169: 1883–1890.
52. Malcolm GM, Kuldau GA, Gugino BK, Jiménez-Gasco MdM (2013) Hidden host plant associations of soilborne fungal pathogens: an ecological perspective. Phytopathology 103: 538–544.
53. Shittu HO, Castroverde DC, Nazar RN, Robb J (2009) Plant-endophyte interplay protects tomato against a virulent Verticillium. Planta 229: 415–426.
54. Kabir Z, Bhat R, Subbarao K (2004) Comparison of media for recovery of Verticillium dahliae from soil. Plant Dis 88: 49–55.

55. Gurung S, Short D, Adhikari T (2013) Global population structure and migration patterns suggest significant population differentiation among isolates of *Pyrenophora tritici-repentis*. Fungal Genet Biol 52: 32–41.

56. Almany GR, De Arruda MP, Arthofer W, Atallah Z, Beissinger SR, et al. (2009) Permanent genetic resources added to molecular ecology resources database 1 May 2009-31 July 2009. Mol Ecol Resour 9: 1460–1466.

57. Maruthachalam K, Atallah ZK, Vallad GE, Klosterman SJ, Hayes RJ, et al. (2010) Molecular variation among isolates of *Verticillium dahliae* and polymerase chain reaction-based differentiation of races. Phytopathology 100: 1222–1230.

58. Pasqualotto AC, Denning DW, Anderson MJ (2007) A cautionary tale: lack of consistency in allele sizes between two laboratories for a published multilocus microsatellite typing system. J Clin Microbiol 45: 522–528.

59. Borkovich KA, Alex LA, Yarden O, Freitag M, Turner GE, et al. (2004) Lessons from the genome sequence of *Neurospora crassa*: tracing the path from genomic blueprint to multicellular organism. Microbiol Mol Biol Rev 68: 1–108.

60. Ropars J, Dupont J, Fontanillas E, Rodríguez de la Vega RC, Malagnac F, et al. (2012) Sex in cheese: evidence for sexuality in the fungus *Penicillium roqueforti*. PLoS ONE 7: e49665.

61. Galagan JE, Calvo SE, Borkovich KA, Selker EU, Read ND, et al. (2003) The genome sequence of the filamentous fungus *Neurospora crassa*. Nature 422: 859–868.

62. Clutterbuck AJ (1996) Parasexual recombination in fungi. J Genet 75: 281–286.

63. Bidard F, Aït Benkhali J, Coppin E, Imbeaud S, Grognet P, et al. (2011) Genome-wide gene expression profiling of fertilization competent mycelium in opposite mating types in the heterothallic fungus *Podospora anserina*. PLoS ONE 6: e21476.

64. Cherry JM, Hong EL, Amundsen C, Balakrishnan R, Binkley G, et al. (2012) *Saccharomyces* Genome Database: the genomics resource of budding yeast. Nucleic Acids Res 40: D700-D705.

65. Espagne E, Lespinet O, Malagnac F, Da Silva C, Jaillon O, et al. (2008) The genome sequence of the model ascomycete fungus *Podospora anserina*. Genome Biol 9: R77.

66. Doron-Faigenboim A, Stern A, Mayrose I, Bacharach E, Pupko T (2005) Selecton: a server for detecting evolutionary forces at a single amino-acid site. Bioinform 21: 2101–2103.

67. Stern A, Doron-Faigenboim A, Erez E, Martz E, Bacharach E, et al. (2007) Selecton 2007: advanced models for detecting positive and purifying selection using a Bayesian inference approach. Nucleic Acids Res 35: W506-W511.

68. Grossetête S, Labedan B, Lespinet O (2010) FUNGIpath: a tool to assess fungal metabolic pathways predicted by orthology. BMC Genomics 11: 31.

69. Usami T, Itoh M, Amemiya Y (2009) Asexual fungus *Verticillium dahliae* is potentially heterothallic. Journal of General Plant Pathology 75: 422–427.

70. Nierman WC, Pain A, Anderson MJ, Wortman, Jr., Kim HS, et al. (2006) Genomic sequence of the pathogenic and allergenic filamentous fungus *Aspergillus fumigatus*. Nature 439: 502–502.

71. Galagan JE, Calvo SE, Cuomo C, Ma L-J, Wortman JR, et al. (2005) Sequencing of *Aspergillus nidulans* and comparative analysis with *A. fumigatus* and *A. oryzae*. Nature 438: 1105–1115.

72. Staats M, van Kan JAL (2012) Genome update of *Botrytis cinerea* strains B05.10 and T4. Eukaryot Cell 11: 1413–1414.

73. Amselem J, Cuomo CA, van Kan JAL, Viaud M, Benito EP, et al. (2011) Genomic analysis of the necrotrophic fungal pathogens *Sclerotinia sclerotiorum* and *Botrytis cinerea*. PLoS Genet 7: e1002230.

74. *Chaetomium globosum* Database. Available: http://www.broadinstitute.org/annotation/genome/chaetomium_globosum.2/Home.html. Accessed 1 May 2014.

75. O'Connell RJ, Thon MR, Hacquard S, Amyotte SG, Kleemann J, et al. (2012) Lifestyle transitions in plant pathogenic *Colletotrichum* fungi deciphered by genome and transcriptome analyses. Nature Genetics 44: 1060–1065.

76. Ohm RA, Feau N, Henrissat B, Schoch CL, Horwitz BA, et al. (2012) Diverse lifestyles and strategies of plant pathogenesis encoded in the genomes of eighteen *Dothideomycetes* fungi. PLoS Pathog 8: e1003037.

77. Schardl C, Moore N, Zhao P, Arnaoudova E, Bullock C, et al. (2012) Genome sequence of *Epichloë festucae*. In: Young CA, Aiken GE, McCulley RL, Strickland JR, Schardl CL, editors. Epichloae, endophytes of cool season grasses: implications, utilization and biology; Proceedings of the 7th International Symposium on Fungal Endophytes of Grasses, Lexington, Kentucky, USA, 28 June to 1 July 2010. Lexington: Samuel Roberts Noble Foundation. pp. 59–64.

78. Pöggeler S, O'Gorman CM, Hoff B, Kück U (2011) Molecular organization of the mating-type loci in the homothallic Ascomycete *Eupenicillium crustaceum*. Fungal Biol 115: 615–624.

79. Cuomo CA, Gueldener U, Xu JR, Trail F, Turgeon BG, et al. (2007) The *Fusarium graminearum* genome reveals a link between localized polymorphism and pathogen specialization. Science 317: 1400–1402.

80. *Histoplasma capsulatum* Database. Available: http://www.broadinstitute.org/annotation/genome/histoplasma_capsulatum/MultiHome.html. Accessed 1 May 2014.

81. Dean RA, Talbot NJ, Ebbole DJ, Farman ML, Mitchell TK, et al. (2005) The genome sequence of the rice blast fungus *Magnaporthe grisea*. Nature 434: 980–986.

82. Coleman JJ, Rounsley SD, Rodriguez-Carres M, Kuo A, Wasmann CC, et al. (2009) The genome of *Nectria haematococca*: contribution of supernumerary chromosomes to gene expansion. PLoS Genet 5: e1000618.

83. van den Berg MA, Albang R, Albermann K, Badger JH, Daran J-M, et al. (2008) Genome sequencing and analysis of the filamentous fungus *Penicillium chrysogenum*. Nat Biotechnol 26: 1161–1168.

84. Martinez D, Berka RM, Henrissat B, Saloheimo M, Arvas M, et al. (2008) Genome sequencing and analysis of the biomass-degrading fungus *Trichoderma reesei* (syn. *Hypocrea jecorina*). Nat Biotechnol 26: 1193–1193.

85. Goodwin SB, Ben M'Barek S, Dhillon B, Wittenberg AHJ, Crane CF, et al. (2011) Finished genome of the fungal wheat pathogen *Mycosphaerella graminicola* reveals dispensome structure, chromosome plasticity, and stealth pathogenesis. PLoS Genet 7: e1002070.

86. Swanson WJ, Nielsen R, Yang Q (2003) Pervasive adaptive evolution in mammalian fertilization proteins. Mol Biol and Evol 20: 18–20.

87. Wong WS, Nielsen R (2004) Detecting selection in noncoding regions of nucleotide sequences. Genetics 167: 949–958.

88. Doron-Faigenboim A, Pupko T (2007) A combined empirical and mechanistic codon model. Mol Biol Evol 24: 388–397.

89. Yang Z, Nielsen R, Goldman N, Pedersen A-MK (2000) Codon-substitution models for heterogeneous selection pressure at amino acid sites. Genetics 155: 431–449.

Gemcitabine Induces Poly (ADP-Ribose) Polymerase-1 (PARP-1) Degradation through Autophagy in Pancreatic Cancer

Yufeng Wang[1], Yasuhiro Kuramitsu[1]*, Kazuhiro Tokuda[1], Byron Baron[1], Takao Kitagawa[1], Junko Akada[1], Shin-ichiro Maehara[2], Yoshihiko Maehara[2], Kazuyuki Nakamura[1,3]

1 Department of Biochemistry and Functional Proteomics, Yamaguchi University Graduate School of Medicine, Ube, Yamaguchi, Japan, 2 Department of Surgery and Science, Graduate School of Medical Science, Kyusyu University, Fukuokashi, Fukuoka, Japan, 3 Centre of Clinical Laboratories in Tokuyama Medical Association Hospital, Shunan, Japan

Abstract

Poly (ADP-ribose) polymerase-1 (PARP-1) and autophagy play increasingly important roles in DNA damage repair and cell death. Gemcitabine (GEM) remains the first-line chemotherapeutic drug for pancreatic cancer (PC). However, little is known about the relationship between PARP-1 expression and autophagy in response to GEM. Here we demonstrate that GEM induces DNA-damage response and degradation of mono-ADP ribosylated PARP-1 through the autophagy pathway in PC cells, which is rescued by inhibiting autophagy. Hypoxia and serum starvation inhibit autophagic activity due to abrogated GEM-induced mono-ADP-ribosylated PARP-1 degradation. Activation of extracellular regulated protein kinases (ERK) induced by serum starvation shows differences in intracellular localization as well as modulation of autophagy and PARP-1 degradation in GEM-sensitive KLM1 and -resistant KLM1-R cells. Our study has revealed a novel role of autophagy in PARP-1 degradation in response to GEM, and the different impacts of MEK/ERK signaling pathway on autophagy between GEM-sensitive and -resistant PC cells.

Editor: Shaida A. Andrabi, Johns Hopkins University, United States of America

Funding: The work was supported by grant no. 24501352, www.jsps.go.jp/g-grantsinaid. The funder had no role in study design, data collection and analysis, decision to publish, or preparation of the manuscript.

Competing Interests: The authors have declared that no competing interests exist.

* Email: climates@yamaguchi-u.ac.jp

Introduction

Gemcitabine (GEM) is currently the standard treatment for advanced and metastatic pancreatic cancer (PC) in both adjuvant and palliative settings, but resistance to GEM has been a big problem as its response rate has been reduced to <20% [1–4]. GEM can inhibit DNA synthesis by targeting ribonucleotide reductase, leading to its inclusion into cellular DNA, causing DNA replication errors [5,6]. A previous study has reported that GEM-induced DNA replication stress, stalled replication forks and triggered checkpoint signaling pathways [7]. Inhibition of checkpoint kinase 1 (Chk1) with chemical inhibitors induced sensitization of PC cells in response to GEM [8,9]. Moreover mismatch repair-deficient HCT116 cells are more sensitive *in-vitro* to GEM-mediated radiosensitization [8]. Although the evidence has shown the relationship between DNA repair and sensitization of cells to GEM, the mechanisms responsible for the repair of GEM-induced DNA damage are not clearly understood.

Autophagy is a cellular pathway involved in the routine turnover of proteins or intracellular organelles with close connections to human disease and physiology [10]. Autophagic dysfunction is associated with cancer, neurodegeneration, microbial infection and as well as resistance of cancer cells to anticancer therapy [11,12]. GEM induced autophagy in Panc-1 and MiaPaCa-2 cells, and inhibition of autophagy by 3-methyladenine (3-ME) or vacuole membrane protein 1 knockdown decreased apoptosis in gemcitabine-treated cells [13]. Therefore this evidence indicates that autophagy may play an essential role in apoptosis of PC cells in response to GEM.

Poly (ADP-ribose) polymerase-1 (PARP-1) plays critical roles in many molecular and cellular processes, including DNA damage repair, genome stability, transcription and apoptosis [14]. PARP1 is involved in the repair of both single-stranded DNA (ssDNA) and double-strand DNA (dsDNA) breaks by binding with DNA ends and/or interacting with DNA repair proteins, example (Ataxia Telangiectasia Mutated) ATM and Ku subunits [15–18]. Inhibition of PARP-1 enhances the cytotoxicity of DNA-damaging agents and radiation *in-vitro* [19]. PARP-1 inhibitors have been reported as potential chemotherapeutic drugs for BRCA1/ BRCA2-deficient breast cancer and lung cancer [20–21]. Thus it is necessary to assess the changes of PARP-1 responsible for GEM-induced DNA damage in PC.

In the present study, we demonstrate that microtubule-associated protein 1A/1B-light chain 3 (LC3), a key factor of autophagosome formation, is down-regulated in KLM1-R compared to KLM1 cells. GEM induced a DNA damage response and autophagy in KLM1 and KLM1-R cells and down-regulated PARP-1 expression. Inhibition of autophagy blocked GEM-induced degradation of mono-ADP ribosylated PARP-1.

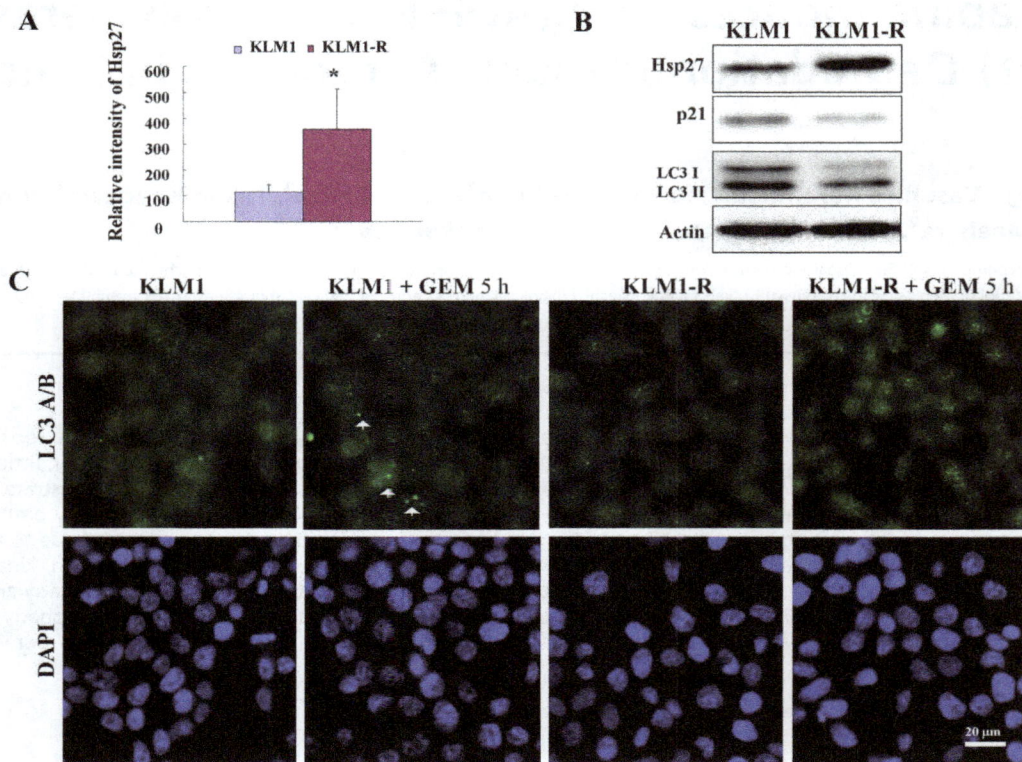

Figure 1. GEM induces autophagy in PC cells. (A) The expression of Hsp27 was tested by western blot and the relative intensity was measured by student-*t* test (n = 3) (B) KLM1 and KLM1-R cells were lysed and resolved by SDS-PAGE and probed with specific antibodies. Actin was used to normalize the loading levels of protein. (C) The indicated cells were treated with 100 µg/mL of GEM for 5 h and the expression of LC3A/B and formation of LC3-positive autophagosomes were examined by confocal microscopy. LC3A/B: green and DAPI: blue. Arrows indicate the autophagosome. Scale bar, 20 µm.

The MEK/ERK signaling pathway showed a different effect on autophagy and GEM-induced PARP-1 degradation between KLM1 and KLM1-R cells. Thus we highlight new insight regarding the autophagy pathway in regulating PARP-1 degradation in PC cells.

Material and Methods

Materials

U0126 (9903S) and wortmannin (9951S) were purchased from Cell Signaling Technology. The antibodies specific for p-ERK (sc-7383), ERK (sc-94200), p21 (sc-65595), Hsp27 (sc-13132), PARP-1 (sc-8007 for western blot and sc-1562 for confirmation and immunofluorescence), CtIP (sc-3970) and actin (sc-1616) were purchased from Santa Cruz Biotechnology. The antibodies specific for LC3A/B (4108S), SIRT6 (12486), caspase-3 (9665) and PI3KCIII (4263S) were purchased from Cell Signaling Technology. The antibodies specific for Bcl2 (B3170), Ulk1 (SAB4200106) and Beclin1 (B6061) were purchased from Sigma. The antibodies specific for AMPKα1 (07–350) was purchased from Millipore.

Cell culture

All the cell lines used in this study were previously published cell lines that were provided to us as a gift. Human pancreatic cell line KLM1 (ID: TKG0490) was cloned and established by Dr. Kobari M from the Institute of Department, Aging and Cancer, Tohoku University (Sendai, Japan) in 1996 [39]. GEM-sensitive KLM1 and -

resistant KLM1-R human pancreatic cancer cell lines were generously provided as a gift by the Department of Surgery and Science at Kyushu University Graduate School of Medical Science. KLM1-R has been established exposing KLM1 cells to GEM in previous studies [40–41]. These cells were cultured in Roswell Park Memorial Institute 1640 medium (RPMI 1640; GIBCO, 05918), supplemented with 10% heat-inactivated fetal bovine serum (FBS; GIBCO, 26140–079), and 2 mM L-glutamine and incubated at 37°C in a humidified incubator containing 5% CO_2.

Transient transfection

KLM1 cells were seeded and incubate at 37°C in a CO_2 incubator until the cells are 70% confluent. The cells were transfected with validated human LC3B siRNA (sc-43390, Santa Cruz Biotechnology) or control siRNA (sc-37007, Santa Cruz Biotechnology) by following a siRNA Transfection Protocol (Santa Cruz Biotechnology).

Western blotting

The cells were lysed with lysis buffer ((1% NP-40, 1 mM sodium vanadate, 1 mM PMSF, 50 mM Tris, 10 mM NaF, 10 mM EDTA, 165 mM NaCl, 10 µg/mL leupeptin, and 10 µg/mL aprotinin) on ice for 1 h. Cell lysates were then centrifuged at 15,000×g for 20 min at 4°C. The supernatant was collected and the protein concentration was determined by Lowry assay. Equal amounts of protein (20 µg) were resolved by 5–20% SDS-polyacrylamide gel and then transferred onto PVDF membrane (Immobilon-P; Millipore, Bedford, MA). The membrane was incubated with the appropriate primary antibody at 4°C overnight. Then, the membrane was washed and incubated with a horseradish peroxidase (HRP)-conjugated secondary antibody for 1 h at room temperature. The immunoblots were visualized with a chemiluminescence reagent (Immunostar, Wako). All of experiments were repeated for three times.

Immunofluorescence

Cells were cultured on 15 mm round coverslips in 12 well plates at a density of 1×10^5 cells per well. Cells on the coverslips were fixed using fresh 3.7% paraformaldehyde in phosphate buffered saline (PBS) for 30 min when they reached 70–80% confluency. Samples were then washed with PBS, followed by permeabilization with 0.1% Triton X-100 for 15 min. After washing with PBS they were incubated in blocking solution (1% goat serum or 1% donkey serum in PBS with 0.1% Tween 20) for 1 h at room temperature. Cells were treated with a primary antibody in blocking solution overnight at 4°C. After incubation with primary antibody, cells were rinsed with PBS with 0.1% Tween 20 (PBS-T) and incubated with a secondary antibody for 1 h at room temperature. After washing with PBS-T, their nuclei were counter-stained with 1.43 μM DAPI (4,6'-diamidino-2-phenylindole) for 5 minutes. Coverslips were washed with PBS-T, then mounted face-down onto microscope slides with Fluoromount (Diagnostic BioSystems, Pleasanton, CA, USA). Confocal images were obtained using Nikon Plan Apo 60X/1.40 objective, BZ-9000 series (BIOREVO) and BZ-II Viewer software (Keyence, Osaka, Japan) by an operator who was unaware of the experimental condition. All parameters were

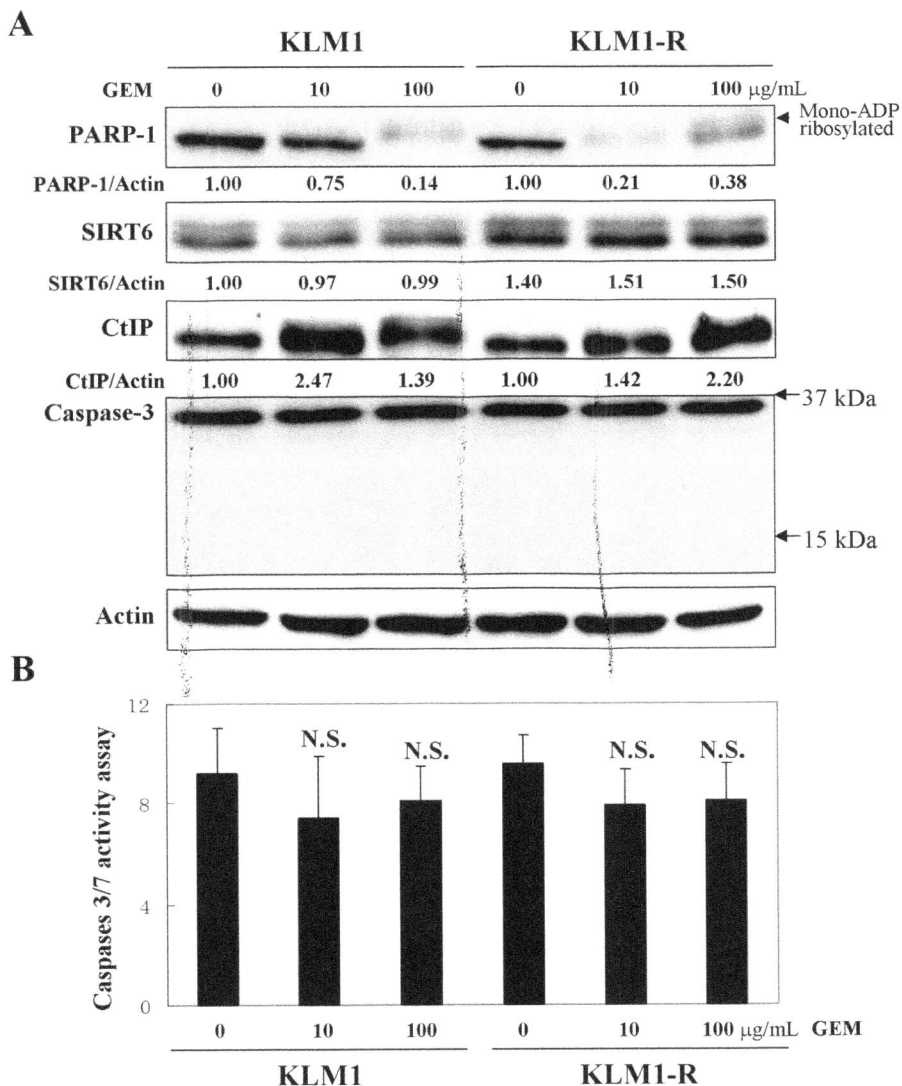

Figure 2. GEM down-regulates mono-ADP ribosylated PARP-1 in a caspase-independent manner. (A) KLM1 and KLM1-R cells were treated by GEM with the indicated concentrations for 24 h. Cell lysates were resolved in SDS-PAGE and probed with specific antibodies. The expression of PARP-1 was confirmed repeatedly by a distinct PARP-1 antibody described in Materials. An arrow head indicated the mono-ADP ribosylated form of PARP-1. Arrows indicated the position area of cleaved caspase-3. (B) The indicated cells were treated as in (A) and then stained using a caspases 3/7 assay kit (A). Caspase 3/7 activity was tested and measured by confocal microscopy and Image J. N.S., non significant. Error bars, SD.

kept constant within each experiment. Digital images were analyzed and the average intensity was measured using Image J. software.

Apoptosis assay

An appropriate number of cells was plated and treated for 24 h. Cells were stained by using Apo-BrdU *In Situ* DNA fragmentation Assay kit (80101, Biovision, Inc.) (data not shown) or Caspases 3/7 assay kit (12D51, ImmunoChemistry Technologies, LLC.). These experiments were performed strictly following the instructions of the relative protocols.

Results

Gemcitabine (GEM) induces autophagy in PC cells

Two PC cancer cell lines GEM-sensitiive KLM1 and - resistant KLM1-R, were used in this study. These cell lines are defined by their expression of heat shock protein 27 (Hsp27) (Fig. 1 A and B), which has been reported as a potential marker for PC-resistant to GEM [22–24]. Moreover the expression of p21 was shown to be reduced in KLM1-R compared to KLM1

cells (Fig. 1 B), indicating the different phenotypes of cell cycle between them. We then investigated autophagic activity in KLM1 and KLM1-R cells, which was determined by the expression of LC3 [25]. We demonstrated that both LC3-I and II were down-regulated in KLM1-R compared to KLM1 cells (Fig. 1 B). Moreover, down-regulation of AMP-activated protein kinase A1 (AMPKα1) and unc-51-like kinase 1 (Ulk1) were shown, unlike phosphatidylinositol 3- kinase (PI3K CIII) or Coiled-coil myosin-like BCL2-interacting protein (Beclin-1), in KLM1-R compared to KLM1 cells (Fig. S1 A and B), indicating that the reduction of autophagic activity in GEM-resistant KLM1-R cells may be related to the down-regulation of AMPKα1 and/or Ulk1 expression. To determine the effect of autophagy induced by GEM, cells were treated with GEM for 5 hours (h) and then observed by immunofluorescent microscopy using anti-LC3 antibody staining. In this experimental setting, we demonstrated that the LC3 II spots were increased after the cells were exposed to GEM (Fig. 1 C). These data suggested that GEM induces autophagy in PC cells.

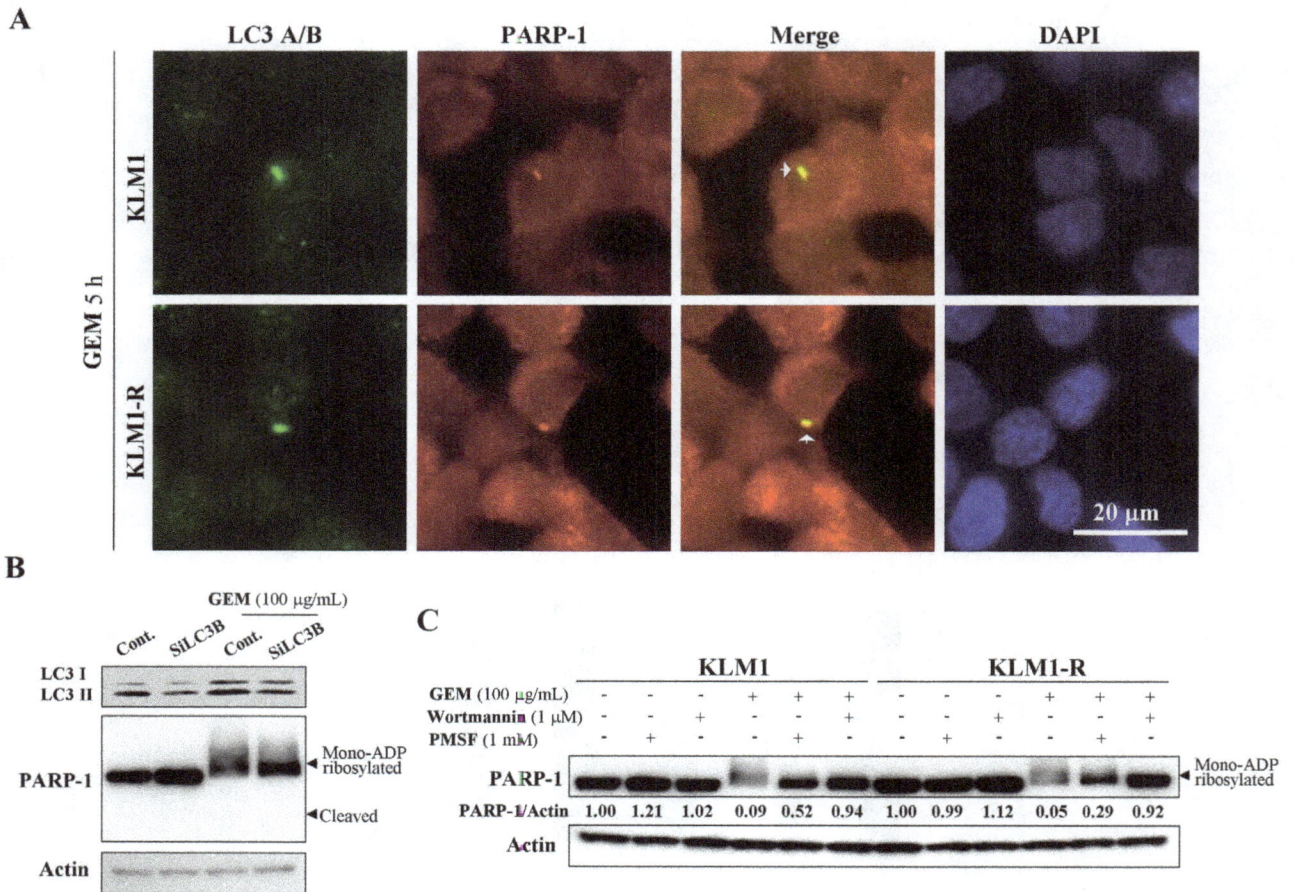

Figure 3. GEM suppresses mono-ADP ribosylated PARP-1 expression via autophagy. (A) KLM1 and KLM1-R cells were treated with 100 μg/mL of GEM for 5 h. Treated cells were stained with specific antibodies against LC3A/B and PARP-1 and then detected by confocal microscopy. LC3A/B: green, PARP-1: red and DAPI: blue. Scale bar, 20 μm. Arrows indicate the yellow staining of co-localizations between autophagosome and PARP-1. (B) KLM1 cells were exposed to GEM for 24 h after LC3B knockdown. (C) KLM1 and KLM1-R cells were exposed to GEM for 24 h in the present or absent of either PMSF or wortmannin at the indicated concentrations. Cell lysates were resolved by SDS-PAGE and probed with specific antibodies against to PARP-1. The arrow head indicates the mono-ADP ribosylated form of PARP-1. The expression of PARP-1 was confirmed repeatedly by a distinct PARP-1 antibody described in Materials.

A

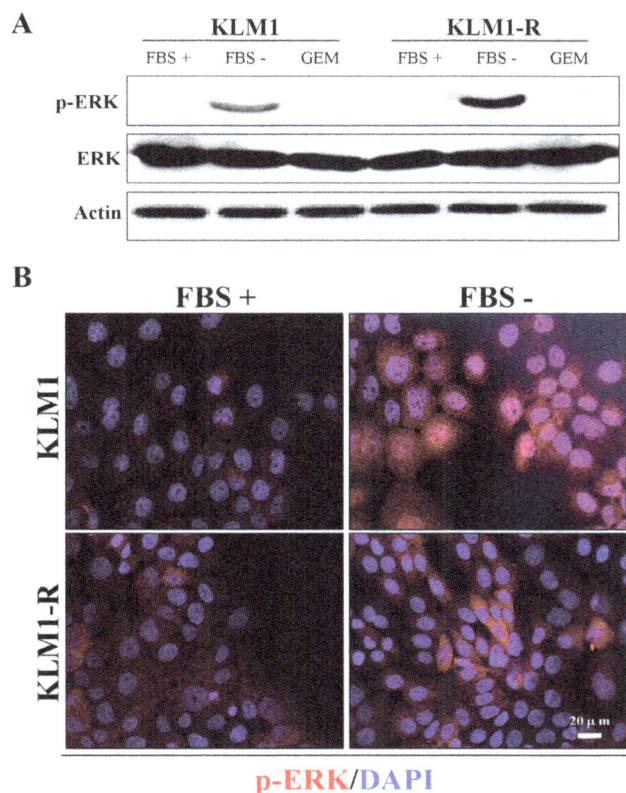

Figure 4. Serum starvation induces activation and different localization of extracellular ERK between KLM1 and KLM1-R cells. (A) KLM1 and KLM1-R cells were cultured in medium with or without FBS or exposed to 10 μg/mL of GEM for 24 h. Cell lysates were resolved by SDS-PAGE and probed with specific antibodies against p-ERK and ERK. (B) and (C) The indicated cells were stained with specific antibodies against p-ERK, Hsp27 and LC3A/B after cells were cultured in medium with or without FBS for 24 h. DAPI: blue and p-ERK: red in (B) and LC3A/B: green and Hsp27: red in (C). Scale bar, 20 μm.

GEM specifically down-regulates mono-ADP ribosylated PARP-1 in a caspase-independent manner

We further tested the effect of GEM on PARP-1 expression in PC cells using a western blot analysis. KLM1 and KLM1-R showed a remarkable reduction of PARP-1 when cells were exposed to 10 or 100 μg/mL of GEM for 24 h (Fig. 2 A). Interestingly, it was observed that the reduced bands of PARP-1 in the GEM-induced panels showed a slight increase in molecular weight than in the untreated panels. Thus this suggested that GEM specifically induced the down-regulation of PARP-1 which was mono-ADP ribosylated (Fig. 2 A). Zhiyong Mao *et al.* had defined the upper band of PARP-1 as the mono-ADP ribosylated form at lysine residue 521 which was induced by sirtuin 6 (SIRT6) [26]. SIRT6 is a mammalian homolog of the yeast Sir2 deacetylase and involved in cytokine production and migration of PC [27]. Moreover, KLM1-R showed a higher sensitivity to GEM-induced down-regulation of PARP-1 than KLM1 cells. Ten μg/mL of GEM was enough to significantly reduce much of the PARP-1 expression in KLM1-R compared to KLM1 cells (Fig. 2 A). To find the explanation, we then compared the expression of SIRT6 between KLM1 and KLM1-R cells. Expectedly SIRT6 showed stronger expression (approximately 1.5-fold) in KLM1-R than KLM1

cells (Fig. 2 A). This indicated that the efficiency of GEM on the down-regulation of mono-ADP ribosylated PARP-1 may depend on the expression of SIRT6. We also observed that GEM induced the overexpression of CtBP-interacting protein (CtIP) in both KLM1 and KLM1-R cells (Fig 2 A). Because the caspase family protease cleaves the death substrate PARP-1 to a specific 85 kDa form observed during apoptosis [28], we next investigated whether caspase-3/7 were related to the PARP-1 down-regulation herein. We revealed that the level of caspase-3/7 activity as well as apoptosis showed no differences between KLM1 and KLM1-R cells even when cells underwent GEM treatment for 24 h as shown by western blotting (Fig. 2 A) and caspase-3/7 activity assay (Fig. 2 B). These data suggested that GEM specifically down-regulated mono-ADP ribosylated PARP-1 in a caspase-independent manner.

GEM suppresses the expression of mono-ADP ribosylated PARP-1 through the autophagy degradation pathway

As GEM induced autophagy and the mono-ADP ribosylated PARP-1 down-regulation in a caspase-independent manner, we investigated if mono-ADP ribosylated PARP-1 could be directly degraded by autophagy. KLM1 and KLM1-R cells were stained with both anti-LC and anti-PARP-1 antibody after cells were exposed to GEM for 5 h and then examined by immunofluorescent microscopy. GEM-induced autophagosome formation was found to co-localize with PARP-1 in both KLM1 and KLM1-R cells (Fig. 3 A), indicating the relationship between autophagy and PARP-1. To test GEM-induced down-regulation of mono-ADP ribosylated PARP-1 through autophagy, LC3B siRNA, wortmannin (a PI3K inhibitor) and PMSF (a vacuolar protease inhibitor) were used to inhibit autophagy of cells in response to GEM. GEM induced PARP-1 mono-ADP ribosylation and down-regulation in KLM1, and the down-regulation of PARP-1 was reversed by LC3B knockdown (Figure 3 B). This reduction of PARP-1 expression by GEM in both KLM1 and KLM1-R cells were also rescued by treatment with either wortmannin or PMSF (Fig. 3 C). Together, these data demonstrated that GEM-induced down-regulation of mono-ADP ribosylated PARP-1 was mediated by the autophagy degradation pathway.

Serum starvation induces activation and different localization of extracellular signal-regulated kinase (ERK) in KLM1 and KLM1-R cells

To determine the effect of serum starvation on ERK activity and autophagy, cells were cultured in fresh medium with or without FBS for 24 hours and examined by western blot and immunofluorescent microscopy. We demonstrated that ERK was activated by serum starvation in KLM1 and KLM1-R cells and that GEM has no effect on ERK activity (Fig. 4 A). Next we examined the intracellular localization of phospho-ERK by immunofluorescence in KLM1 and KLM1-R cells. Interestingly, a remarkable difference in intracellular localization of p-ERK was shown between them. Under serum starvation, p-ERK was partially translocated into the nucleus in KLM1 cells; on the contrary, activated ERK was present solely in the cytoplasm in KLM1-R cells (Fig. 4 B). These results suggested that serum starvation induced activation of ERK in both of KLM1 and KLM1-R, but resulted in a difference in intracellular localization between them.

A

B

C

D

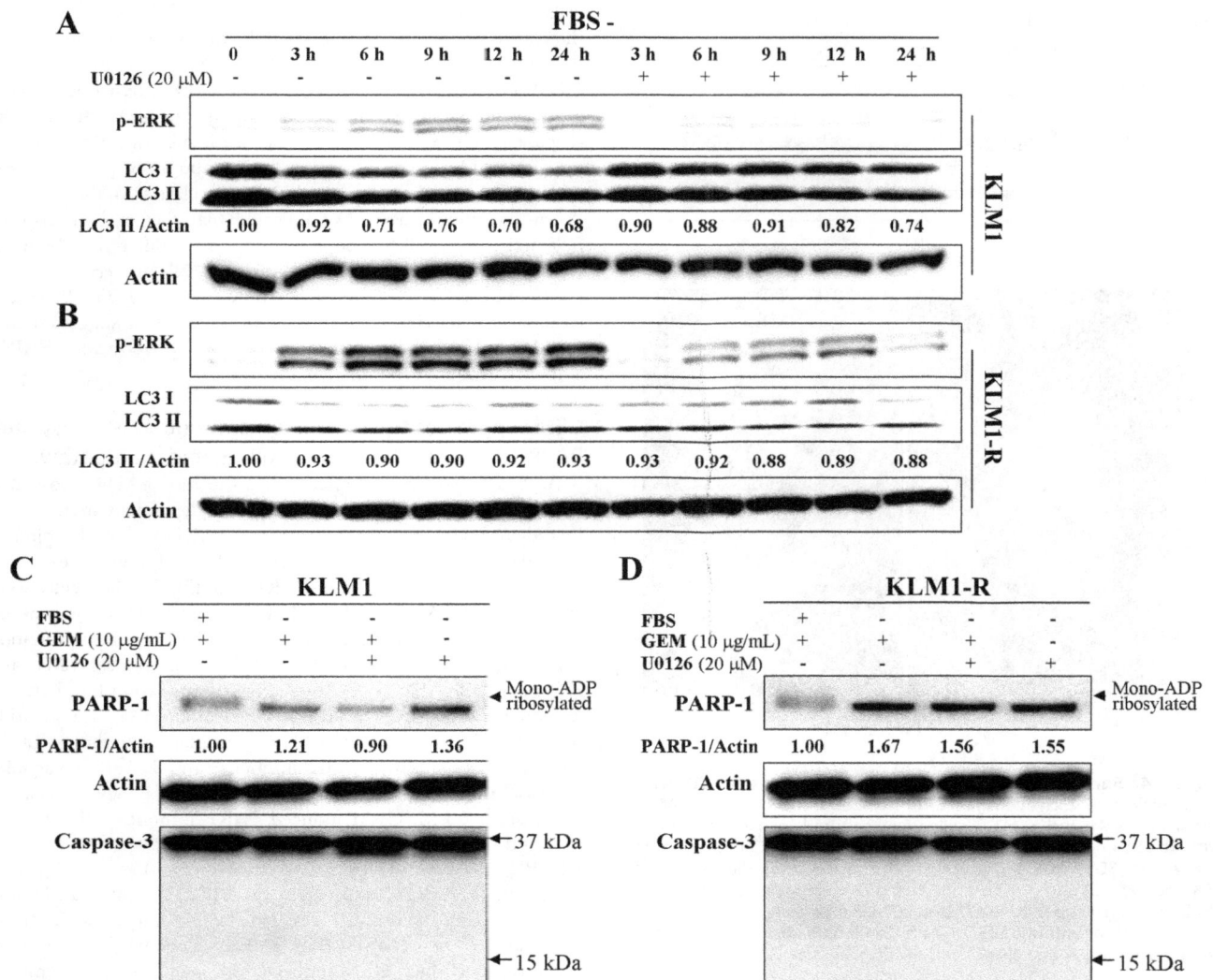

Figure 5. Serum starvation suppresses GEM-induced PARP-1 degradation through inhibition of autophagy via the ERK signaling pathway. (A) KLM1 and KLM1-R cells were exposed to 10 µg/mL of GEM in the presence or absence of 20 µM of U0126 for the indicated time courses. Cell lysates were resolved by SDS-PAGE and probed with specific antibodies against p-ERK and LC3A/B. (B) and (C) KLM1 and KLM1-R cells were cultured in the medium with or without FBS and meanwhile exposed to either or both GEM and U0126 at the indicated concentration. Cell lysates were resolved by SDS-PAGE and probed with specific antibodies. The arrow head indicates the mono-ADP ribosylated form of PARP-1. Arrows indicate the position area of cleaved caspase-3. The expression of PARP-1 was confirmed repeatedly by a distinct PARP-1 antibody described in Materials.

Serum starvation suppresses GEM-induced PARP-1 degradation by inhibiting autophagy via the ERK signaling pathway

We next examined the autophagic activity after serum deprivation in KLM1 and KLM1-R cells in combination with an extracellular signal–regulated (ERK) kinase (MEK) inhibitor, U0126, to assess the effects of the MEK/ERK pathway on autophagy. We demonstrated that the expression of LC3 was reduced by serum starvation over a time course of 24 h in KLM1 and KLM1-R cells (Fig. 5 A and B) and rescued by U0126 in KLM1 (Fig. 5 A) but much less in KLM1-R (Fig. 5 B). The activity of ERK still showed an increasing trend in KLM1-R when treated with U0126 (Fig. 5 B), indicating a tolerance to U0126 in KLM1-R compared to KLM1 cells. This data indicated that serum starvation-induced autophagy inhibition was mediated by

the MEK/ERK signaling pathway. Under serum starvation, U0126 had no effect on the expression of B-cell leukemia/lymphoma 2 (Bcl2) in KLM1 and KLM1-R cells and the reduction of p21 was delayed by treatment with U0126 in KLM1 but not in KLM1-R cells (Fig. S2 A and B), suggesting that the MEK inhibitor had different efficacy on the cell cycle progression between KLM1 and KLM1-R cells. Because the MEK inhibitor showed different efficacy on the modulation of autophagy between KLM1 and KLM1-R cells under serum starvation, we investigated its efficacy on the PARP-1 degradation in response to GEM. Indeed, GEM-induced PARP-1 degradation was inhibited by serum starvation in both KLM1 and KLM1-R cells in a caspase-independent manner and was reversed only in KLM1 cells by U0126 (Fig. 5 C and D). Moreover, the treatment by U0126 alone had no influence on PARP-1 expression. This data suggested that serum starvation suppresses autophagy and

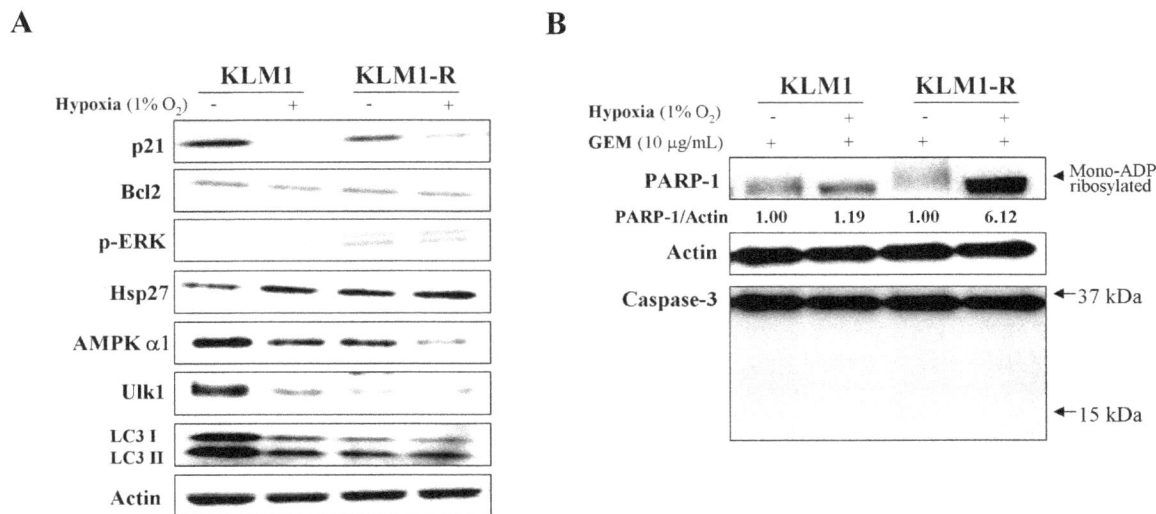

Figure 6. Hypoxia suppresses autophagy and GEM-induced PARP-1 degradation. (A) KLM1 and KLM1-R cells were cultured in normal conditions or 1% O$_2$ hypoxia for 24 hours, and then cell lysates were resolved by SDS-PAGE and probed with specific antibodies. (B) The indicated cells were cultured in normal conditions or 1% O$_2$ hypoxia together with 10 µg/mL of GEM for 24 hours. Cell lysates were resolved by SDS-PAGE and probed with specific antibodies. An arrow head indicates the mono-ADP ribosylated form of PARP-1. Arrows indicate the position area of cleaved caspase-3. The expression of PARP-1 was confirmed repeatedly by a distinct PARP-1 antibody described in Materials.

GEM-induced PARP-1 degradation through activation of the ERK signaling pathway, with KLM1 and KLM1-R cells showing different sensitivities to the MEK inhibitor U0126.

Hypoxia suppresses autophagy and GEM-induced PARP-1 degradation

Hypoxia leads to cell cycle arrest via decreased p21 synthesis [29]. We confirmed that the expression of p21 was down-regulated and Hsp27 was up-regulated by 1% O$_2$ hypoxia for 24 hours in KLM1 and KLM1-R cells; however, hypoxia had no influence on the expression of phospho-ERK and Bcl2 (Fig. 6 A). Under hypoxic condition, both AMPKα1 and Ulk1 expression were down-regulated in KLM1 and KLM1-R cells and the expression of LC3 in KLM1 was down to the same level as in untreated KLM1-R cells (Fig. 6 A), indicating that hypoxia induced phenotypic change in KLM1 leading to a KLM1-R-like condition and inhibition of autophagy. Thus we tested if hypoxia inhibited GEM-induced PARP-1 degradation. Western blot analysis demonstrated that GEM-induced PARP-1 degradation was remarkably abolished by hypoxia in a caspase-independent manner in both KLM1 and KLM1-R cells (Fig. 6 B). These results indicated that hypoxia suppresses GEM-induced PARP-1 degradation by reducing autophagic activity.

Discussion

Autophagy can be elevated by GEM in the treatment of PC cells [13]. PC cells showed to be more sensitive to the cytotoxic effect of GEM when this was combined with cannabinoids via reactive oxygen species (ROS)-mediated autophagic cell death [30]. In this study, we demonstrated that autophagic activity was reduced in GEM-resistant KLM1-R compared to -sensitive KLM1 cells. Therefore, reactivation of autophagy might be a useful strategy for resensitising PC to GEM. However little is known about the role of autophagy in DNA damage response induced by GEM. There is important evidence that autophagy is associated with the processing of double-strand breaks (DSBs) and cell death in

response to DNA damage in yeast through degradation of acetylated recombination protein Sae2 (human CtIP) [31]. Inhibition/ablation of histone deacetylases (HDACs) induces autophagy and acetylation of a number of DNA damage response (DDR) proteins, including Sea2 and Exo1 [31,32]. Here we show that GEM induces a DNA damage response (observed through CtIP overexpression by western blot) and mono-ADP ribosylated PARP-1 degradation leading to increased autophagy. Thus autophagy involved in DNA damage repair may be through the control of PARP-1 degradation rather than CtIP (yeast Sae2) in human PC. However, whether the mono-ADP ribosylation of PARP-1 is necessary for autophagy degradation and how this specific degradation is implemented in response to GEM should be clarified in further study.

Ablation of PARP-1 does not interfere with DSBs repair, but delays reactivation of stalled replication forks [33]. Therefore, GEM-induced degradation of PARP-1 may contribute to GEM-stalled replication forks. DSB response factors ATM, Mre11, and Rad50 are required for cell survival after replication fork stalling in response to GEM-induced DNA damage [34]. Restart of stalled replication forks and repair of collapsed replication forks require RAD51 activity and RAD51-mediated homologous recombination (HR) pathway, respectively [35]. Moreover, we demonstrate that CtIP is overexpressed in response to GEM in KLM1 and KLM1-R cells (Fig. 2 A). Taken together, these results suggest that DSBs repair is required for survival of cells as well as PC cells after GEM-induced DNA damage. CtIP was shown to be up-regulated in both KLM1 and KLM1-R induced by GEM, but its stability seems stronger in KLM1-R, particularly over a concentration rage of 10–100 µg/mL GEM, which expresses a higher level of SIRT6 compared to KLM1 cells (Fig. 2 A). SIRT6 promotes DNA stability and activation and stabilization of CtIP by deacetylation [31,35], indicating a possible mechanism for the lower sensitivity of KLM1-R to GEM compared to KLM1 cells. Moreover, recent studies suggest that PARP inhibitors are particularly lethal to cells deficient in homologous recombination (HR) proteins through deregulation of error-prone non-homologous end joining [36].

Likewise, therefore, combination of a kind of HR inhibitor with GEM (as a PARP-1 suppressor) may be a potential therapeutic strategy for PC. However, there is a major limitation because serum starvation and hypoxia (mimicking tumor microenvironments *in-vivo*) inhibit GEM-induced PARP-1 degradation by reducing autophagic activity (Fig. 5 and 6). Thus, the preferred candidate for combination therapy with GEM should not only inhibit the components of DSBs but also promote autophagy, for example a histone deacetylases inhibitor, namely, valproic acid [37]. Further studies are needed to test whether this inhibitor could enhance PC cell death in response to GEM.

The MEK inhibitor U0126 shows different effects on the level of autophagy and PARP-1 degradation in response to GEM between KLM1 and KLM1-R cells, indicating that possibly limitations exist on the therapeutic strategy for targeting the EGFR/Ras/ERK pathway in PC. It is presumed that the observed differences depend on one of three possibilities: 1) the differences in intracellular localization of activated ERK between KLM1 and KLM1-R cells (Fig. 4 B); 2) an unknown feedback signaling pathway for ERK reactivation in response to the MEK inhibitor (Fig. 5 B) [37,38]; 3) serum-induced down-regulation of ULK1 in KLM1-R cells leading to autophagy not controlled by ERK (Fig. S1 B).

In this study, we reveal new highlights that GEM functions as a suppressor of PARP-1 by promoting the autophagy degradation pathway and put forward the related suggestions about the desired characteristics of possible candidates for combination therapy with GEM for PC.

References

1. Heinemann V, Boeck S, Hinke A, Labianca R, Louvet C (2008) Meta-analysis of randomized trials: evaluation of benefit from gemcitabine-based combination chemotherapy applied in advanced pancreatic cancer. BMC Cancer 8: 82.
2. El-Rayes BF, Philip PA (2003) A review of systemic therapy for advanced pancreatic cancer. Clin Adv Hematol Oncol 1: 430–4.
3. Wheatley SP, McNeish IA (2005) Survivin: a protein with dual roles in mitosis and apoptosis. Int Rev Cytol 247: 35–88.
4. Moore MJ, Goldstein D, Hamm J, Figer A, Hecht JR, et al. (2007) Erlotinib plus gemcitabine compared with gemcitabine alone in patients with advanced pancreatic cancer: a phase III trial of the National Cancer Institute of Canada Clinical Trials Group. J Clin Oncol 25: 1960–6.
5. Baker CH, Banzon J, Bollinger JM, Stubbe J, Samano V, et al. (1991) 2′-Deoxy-2′-methylenecytidine and 2′-deoxy-2′,2′-difluorocytidine 5′-diphosphates: potent mechanism-based inhibitors of ribonucleotide reductase. J Med Chem 34: 1879–84.
6. Huang P, Chubb S, Hertel LW, Grindey GB, Plunkett W (1991) Action of 2′,2′-difluorodeoxycytidine on DNA synthesis. Cancer Res 51: 6110–7.
7. Karnitz LM, Flatten KS, Wagner JM, Loegering D, Hackbarth JS, et al. (2005) Gemcitabine-induced activation of checkpoint signaling pathways that affect tumor cell survival. Mol Pharmacol 68: 1636–44.
8. Matthews DJ, Yakes FM, Chen J, Tadano M, Bornheim L, et al. (2007) Pharmacological abrogation of S-phase checkpoint enhances the anti-tumor activity of gemcitabine in vivo. Cell Cycle 6: 104–10.
9. Parsels LA, Morgan MA, Tanska DM, Parsels JD, Palmer BD, et al. (2009) Gemcitabine sensitization by checkpoint kinase 1 inhibition correlates with inhibition of a Rad51 DNA damage response in pancreatic cancer cells. Mol Cancer Ther 8: 45–54.
10. Meijer AJ, Dubbelhuis PF (2004) Amino acid signalling and the integration of metabolism. Biochem Biophys Res Commun 313: 397–403.
11. Mizushima N, Levine B, Cuervo AM, Klionsky DJ (2008) Autophagy fights disease through cellular self-digestion. Nature 451: 1069–75.
12. Chen S, Rehman SK, Zhang W, Wen A, Yao L, Zhang J (2010) Autophagy is a therapeutic target in anticancer drug resistance. Biochim Biophys Acta 1806: 220–9.
13. Pardo R, Lo Ré A, Archange C, Ropolo A, Papademetrio DL, et al. (2010) Gemcitabine induces the VMP1-mediated autophagy pathway to promote apoptotic death in human pancreatic cancer cells. Pancreatology 10: 19–26.
14. Kim MY, Zhang T, Kraus WL (2005) Poly(ADP-ribosyl)ation by PARP-1: 'PAR-laying' NAD+ into a nuclear signal. Genes Dev 19: 1951–67.
15. Godon C, Cordelières FP, Biard D, Giocanti N, Mégnin-Chanet F, et al. (2008) PARP inhibition versus PARP-1 silencing: different outcomes in terms of single-strand break repair and radiation susceptibility. Nucleic Acids Res 36: 4454–64.
16. Schultz N, Lopez E, Saleh-Gohari N, Helleday T (2003) Poly(ADP-ribose) polymerase (PARP-1) has a controlling role in homologous recombination. Nucleic Acids Res 31: 4959–64.
17. Wang M, Wu W, Wu W, Rosidi B, Zhang L, et al. (2006) PARP-1 and Ku compete for repair of DNA double strand breaks by distinct NHEJ pathways. Nucleic Acids Res 34: 6170–82.
18. Aguilar-Quesada R, Muñoz-Gámez JA, Martín-Oliva D, Peralta A, Valenzuela MT, et al. (2007) Interaction between ATM and PARP-1 in response to DNA damage and sensitization of ATM deficient cells through PARP inhibition. BMC Mol Biol 8: 29.
19. Calabrese CR, Almassy R, Barton S, Batey MA, Calvert AH, et al. (2004) Anticancer chemosensitization and radiosensitization by the novel poly(ADP-ribose) polymerase-1 inhibitor AG14361. J Natl Cancer Inst 96: 56–67.
20. De Soto JA, Deng CX (2006) PARP-1 inhibitors: are they the long-sought genetically specific drugs for BRCA1/2-associated breast cancers? Int J Med Sci 3: 117–23.
21. Lee YR, Yu DS, Liang YC, Huang KF, Chou SJ, et al. (2013) New approaches of PARP-1 inhibitors in human lung cancer cells and cancer stem-like cells by some selected anthraquinone-derived small molecules. PLoS One 8: e56284.
22. Mori-Iwamoto S, Kuramitsu Y, Ryozawa S, Mikuria K, Fujimoto M, et al. (2007) Proteomics finding heat shock protein 27 as a biomarker for resistance of pancreatic cancer cells to gemcitabine. Int J Oncol 31: 1345–50.
23. Taba K, Kuramitsu Y, Ryozawa S, Yoshida K, Tanaka T, et al. (2010) Heat-shock protein 27 is phosphorylated in gemcitabine-resistant pancreatic cancer cells. Anticancer Res 30: 2539–43.
24. Mori-Iwamoto S, Kuramitsu Y, Ryozawa S, Taba K, Fujimoto M, et al. (2008) A proteomic profiling of gemcitabine resistance in pancreatic cancer cell lines. Mol Med Rep 1: 429–34.
25. Kabeya Y, Mizushima N, Ueno T, Yamamoto A, Kirisako T, et al. (2000) LC3, a mammalian homologue of yeast Apg8p, is localized in autophagosome membranes after processing. EMBO J 19: 5720–8.
26. Mao Z, Hine C, Tian X, Van Meter M, Au M, et al. (2011) SIRT6 promotes DNA repair under stress by activating PARP1. Science 332: 1443–6.
27. Bauer I, Grozio A, Lasiglè D, Basile G, Sturla L, et al. (2012) The NAD+-dependent histone deacetylase SIRT6 promotes cytokine production and migration in pancreatic cancer cells by regulating Ca2+ responses. J Biol Chem 287: 40924–37.
28. Lazebnik YA, Kaufmann SH, Desnoyers S, Poirier GG, Earnshaw WC (1994) Cleavage of poly (ADP-ribose) polymerase by a proteinase with properties like ICE. Nature 371: 346–7.
29. Mizuno S, Bogaard HJ, Voelkel NF, Umeda Y, Kadowaki M, et al. (2009) Hypoxia regulates human lung fibroblast proliferation via p53-dependent and -independent pathways. Respir Res 6: 10:17.

Supporting Information

Figure S1 (A) KLM1 and KLM1-R cells were lysed and resolved in SDS-PAGE and probed with specific antibodies. Actin was used to normalize the loading levels of protein. (B) KLM1 and KLM1-R cells were cultured in medium with or without FBS or exposed to 10 μg/mL of GEM for 24 h. Cell lysates were resolved in SDS-PAGE and probed with specific antibodies.

Figure S2 KLM1 (A) and KLM1-R (B) cells were exposed to 10 μg/mL of GEM in present or absent of 20 μM of U0126 for the indicated time courses. Cell lysates were resolved in SDS-PAGE and probed with specific antibodies against to p21 and Bcl2. The relative intensities of western blot were measured and shown in this figure.

Acknowledgments

We thank Ikeda E. and Cui D. to let us use a hypoxia incubator.

Author Contributions

Conceived and designed the experiments: YW YK. Performed the experiments: YW YK KT TK. Analyzed the data: YW BB JA SM YM KN. Contributed reagents/materials/analysis tools: JA YW YK BB KN. Wrote the paper: YW YK BB KN.

30. Vara D, Salazar M, Olea-Herrero N, Guzmán M, Velasco G, Díaz-Laviada I (2011) Anti-tumoral action of cannabinoids on hepatocellular carcinoma: role of AMPK-dependent activation of autophagy. Cell Death Differ 18: 1099–111.

31. Robert T, Vanoli F, Chiolo I, Shubassi G, Bernstein KA, et al. (2011) HDACs link the DNA damage response, processing of double-strand breaks and autophagy. Nature 471: 74–9.

32. Botrugno OA, Robert T, Vanoli F, Foiani M, Minucci S (2012) Molecular pathways: old drugs define new pathways: non-histone acetylation at the crossroads of the DNA damage response and autophagy. Clin Cancer Res 18: 2436–42.

33. Yang YG, Cortes U, Patnaik S, Jasin M, Wang ZQ (2004) Ablation of PARP-1 does not interfere with the repair of DNA double-strand breaks, but compromises the reactivation of stalled replication forks. Oncogene 23: 3872–82.

34. Ewald B, Sampath D, Plunkett W (2008) ATM and the Mre11-Rad50-Nbs1 complex respond to nucleoside analogue-induced stalled replication forks and contribute to drug resistance. Cancer Res 68: 7947–55.

35. Kaidi A, Weinert BT, Choudhary C, Jackson SP (2010) Human SIRT6 promotes DNA end resection through CtIP deacetylation. Science 329: 1348–53.

36. Patel AG, Sarkaria JN, Kaufmann SH (2011) Nonhomologous end joining drives poly(ADP-ribose) polymerase (PARP) inhibitor lethality in homologous recombination-deficient cells. Proc Natl Acad Sci U S A 108: 3406–11.

37. Dai P, Xiong WC, Mei L (2006) Erbin inhibits RAF activation by disrupting the sur-8-Ras-Raf complex. J Biol Chem 281: 927–33.

38. Shi M, Zhao M, Hu M, Liu D, Cao H, et al. (2012) β2-AR-induced Her2 transactivation mediated by Erbin confers protection from apoptosis in cardiomyocytes. Int J Cardiol [Epub ahead of print].

39. Kimura Y, Kobari M, Yusa T, Sunamura M, Kimura M, et al. (1996) Establishment of an experimental liver metastasis model by intraportal injection of a newly derived human pancreatic cancer cell line (KLM-1). Int J Pancreatol 20: 43–50.

40. Maehara S, Tanaka S, Shimada M, Shirabe K, Saito Y, et al. (2004) Selenoprotein P, as a predictor for evaluating gemcitabine resistance in human pancreatic cancer cells. Int J Cancer 112: 184–9.

41. Iwasaki I, Sugiyama H, Kanazawa S, Hemmi H. (2002) Establishment of cisplatin-resistant variants of human neuroblastoma cell lines, TGW and GOTO, and their drug cross-resistance profiles. Cancer Chemother Pharmacol 49: 438–44.

Regulation of 53BP1 Protein Stability by RNF8 and RNF168 Is Important for Efficient DNA Double-Strand Break Repair

Yiheng Hu[1], Chao Wang[1], Kun Huang[1], Fen Xia[2], Jeffrey D. Parvin[1]*, Neelima Mondal[3]*

1 Department of Biomedical Informatics, The Ohio State University, Columbus, Ohio, United States of America, 2 Department of Radiation Oncology, The Ohio State University, Columbus, Ohio, United States of America, 3 School of Life Sciences, Jawaharlal Nehru University, New Delhi, India

Abstract

53BP1 regulates DNA double-strand break (DSB) repair. In functional assays for specific DSB repair pathways, we found that 53BP1 was important in the conservative non-homologous end-joining (C-NHEJ) pathway, and this activity was dependent upon RNF8 and RNF168. We observed that 53BP1 protein was diffusely abundant in nuclei, and upon ionizing radiation, 53BP1 was everywhere degraded except at DNA damage sites. Depletion of RNF8 or RNF168 blocked the degradation of the diffusely localized nuclear 53BP1, and ionizing radiation induced foci (IRIF) did not form. Furthermore, when 53BP1 degradation was inhibited, a subset of 53BP1 was bound to DNA damage sites but bulk, unbound 53BP1 remained in the nucleoplasm, and localization of its downstream effector RIF1 at DSBs was abolished. Our data suggest a novel mechanism for responding to DSB that upon ionizing radiation, 53BP1 was divided into two populations, ensuring functional DSB repair: damage site-bound 53BP1 whose binding signal is known to be generated by RNF8 and RNF168; and unbound bulk 53BP1 whose ensuing degradation is regulated by RNF8 and RNF168.

Editor: Michael Shing-Yan Huen, The University of Hong Kong, Hong Kong

Funding: CA141090 National Cancer Institute (JDP and KH) and DST-PURSE and UGC resource networking grant (NM). The funders had no role in study design, data collection and analysis, decision to publish, or preparation of the manuscript.

Competing Interests: The authors have declared that no competing interests exist.

* Email: Jeffrey.Parvin@osumc.edu (JDP); nmondal@mail.jnu.ac.in (NM)

Introduction

DNA double-strand break (DSB) repair involves two major pathways: homologous recombination (HR) and nonhomologous end-joining (NHEJ). HR has a major homology-directed repair (HDR) pathway, which is a relatively precise form of repair and a minor subpathway called single-strand annealing (SSA), which causes DNA resection until homology at repair junctions is revealed [1]. To date, two types of end-joining systems are defined in the NHEJ: the major one is the conservative-NHEJ (C-NHEJ), which is predominantly associated with precise joining of DSB ends without altering the DNA sequence [2]. The alternative pathway for NHEJ (Alt-NHEJ) is highly mutagenic since it catalyzes DNA resection and utilizes imperfect microhomology for end-joining partners and thus resulting in deletions at repair junctions [3].

53BP1 is known to promote the repair of DSBs by NHEJ [4–9]. 53BP1 deletion in mouse results in a severe defect in class-switch recombination, a process dependent on NHEJ and associated with increased DNA end resection at the IgH locus [10–12]. Loss of 53BP1 restores homologous recombination in BRCA1-deficient murine cells, indicating that 53BP1 inhibits DNA resection in DSB repair, by the regulation of the downstream effector RIF1 to control 5′ end resection [13–15]. Since it is known that 53BP1 directly regulates efficient total NHEJ repair events in mammalian

cells [16] and 53BP1 inhibits end resection, it is predictable that 53BP1 might favor the C-NHEJ pathway over Alt-NHEJ.

Upon DNA DSB induction, a cascade of protein modification and relocalization is triggered: phosphorylation of H2AX (γ-H2AX) results in the recruitment of downstream factors, such as the E3 ubiquitin ligases RNF8 and RNF168, leading to the formation of K63-linked polyubiquitin chains on histones at DSBs. This ubiquitination cascade regulated by RNF8 and RNF168 is responsible for the localization of repair mediators, including BRCA1 and 53BP1 to the DNA damage sites [17–27]. Localization of 53BP1 to DSBs involves its recognition of H2A ubiquitinated on Lys-15 (H2AK15ub), the latter being a product of RNF168 via its ubiquitination-dependent recruitment (UDR) motif binding to K63-linked ubiquitination on chromatin. 53BP1 binding to the chromatin at the damage sites also requires dimethylation of histone H4 on lysine 20 (H4K20me2) via the 53BP1 tandem Tudor domain [28–31] plus a K63-linked ubiquitination of the 53BP1 protein at lysine 1268 by RNF168 [32].

In this study, we identified that 53BP1 acts specifically to promote conservative-NHEJ in a RNF8- and RNF168-dependent manner. We found that RNF8 and RNF168 not only mark histones at the break site to create a 53BP1 binding site, but these ubiquitin ligases also regulate the proteasome-mediated degradation of 53BP1. Failure to degrade 53BP1 protein not bound to

DSBs leads to mislocalization of a downstream factor RIF1, thus impairing DSB repair.

Materials and Methods

Antibodies and reagents

We used the following primary antibodies: anti-53BP1 (Santa Cruz, H-300, for immunoblot and immunofluorescence), anti-RAD51 (Santa Cruz, H-92), anti-γ-H2AX (Millipore, clone JBW301, for immunoblot and Immunofluorescence), anti-RIF1 (Santa Cruz, N-20, for immunoblot and Immunofluorescence), anti-RNF8 (Abnova), anti-RNF168 (Abcam), anti-α-tubulin (Sigma), anti-H4 (Millipore), anti-β-actin (Cell Signaling), anti-HA (purified from mouse ascites fluid), and anti-RHA (purified from rabbit serum). MG132 (Enzo life Sciences, dissolved in DMSO, treated 30 min prior to irradiation), caffeine (Sigma, dissolved in ddH$_2$O, treated 1 h prior to irradiation), cycloheximide (Fluka Analytical, dissolved in ethanol, treated 15 min prior to irradiation).

HR and end-joining assays

HDR, SSA and Alt-NHEJ were performed as previously described [33–36]. The HeLa-derived cell lines that stably integrate the HDR and SSA recombination substrates have been described [35]. The Alt-NHEJ substrate [33] was stably integrated into the HeLa genome. For the DSB repair assays, the appropriate cell line was transfected with siRNAs and, if necessary for the experiment, an expression plasmid on day 1. On day 3 the cells were re-transfected plus a plasmid expressing the I-SceI endonuclease, which initiates a DSB lesion. The amount of repair activity was determined by counting the percentage of GFP-positive cells using flow cytometry.

The C-NHEJ assay utilized quantitative real-time PCR and was carried out as described in [37] with the following modification. The genomic DNA isolated 3 days after transfection of the I-SceI plasmid was treated with the restriction enzyme XhoI and purified by Qiagen PCR purification kit before real-time PCR was applied. RPS17 probe (Hs00734303_g1, Applied Biosystems) was used as an internal control and quantitative $\Delta\Delta C_T$ method was used to analyze the data.

RNA interference and plasmids

We used the following siRNAs produced by Sigma: siControl targeting the luciferase gene: 5′-CGUACGCGGAAUACUU-CGA-3′ [34]; si53BP1: 5′-GAAGGACGGAGUACUAAUA-3′ [38]; si53BP1-2 starting at nucleotide 6051: 5′-UACUUGGU-CUUACUGGUUU-3′; siRNF8: 5′-GGACAAUUAUGGACAA-CAA-3′ [39]; siRNF168: 5′-GGCGAAGAGCGAUGGAGGA-3′ [38]; siLigase IV: 5′-AGGAAGUAUUCUCAGGAAUUA-3′ [38]; siBRCA1: 5′-GCUCCUCUCACUCUUCAGU-3′ [34]; siBRCA2: 5′-UAAAUUUGGACAUAAGGAGUCCUCC-3′ [34]. HA-tagged wild-type 53BP1 expression plasmid was a kind gift from Kuniyoshi Iwabuchi (Kanazawa Medical University). I-SceI expression plasmid has been previously described [34] and was a kind gift from Maria Jasin (Memorial Sloan-Kettering Cancer Institute). The total siRNA amount was adjusted to be the same in each sample by adding siControl. All RNAi transfections were carried out using Oligofectamine (Life Technologies) and plasmid transfections were using Lipofectamine2000 (Life Technologies).

Preparation of whole cell lysates

Whole cell extracts (if not otherwise indicated) were prepared by lysing cells in cell extraction buffer (50 mM Tris, pH 7.9, 300 mM NaCl, 0.5% Nonidet-40, 1 mM EDTA, 5% glycerol, 1 mM phenylmethylsulfonyl fluoride, 1 mM dithiothreitol, 1X complete protease inhibitor cocktail from Sigma). Alternatively, when indicated, the whole cell lysates were either prepared by direct boiling in sodium dodecyl sulfate (SDS) containing buffer (2% SDS in phosphate-buffered saline; PBS) and sonicated five times using a Fisher Scientific probe sonicator for 10 s pulses each at 45% amplitude; or cells were lysed and boiled in urea containing buffer (8 M urea and 2% SDS in PBS), and sonicated as above.

Immunofluorescence microscopy

Cells were fixed with cold 4% paraformaldehyde for 15 min and permeabilized with cold 70% ethanol for 5 min before blocking in 8% bovine serum albumin/PBS for 1 h. Primary antibodies were diluted at 1:500 (the rabbit γ-H2AX antibody was used at 1:1000 dilution) for incubation at room temperature for 2 h. Cells were washed with PBS and stained with secondary antibodies. DAPI was then added at 1: 10,000 for 5 min to stain the nucleus. For experiments in which the cells were extracted with NP40 prior to fixation, the extraction buffer (same as the cell extraction buffer above) was applied to cells that had been washed in PBS, followed by the above protocol for fixation and staining. Images were viewed and acquired using the 60X oil objective lens with a Zeiss Axiovert 200 M microscope. Fluorescence signals for the same indicated protein were captured on the same day with the same exposing times for all samples using software AxioVision 4.8.

Image analysis

For the immunofluorescence images, relative 53BP1 signal within the nuclei in each sample was assessed using ImageJ software (NIH). Briefly, 30 or 100 nuclei were assessed, depending on the different cell line. Mean intensity gray values of 53BP1 signal at foci regions in each nucleus in irradiated samples were scored and normalized according to the mean intensity of diffuse 53BP1 signal in unirradiated sample.

Alternatively, quantitative image analysis of 53BP1 protein signal in the nuclei was carried out using an algorithm in MATLAB R2013a. The program code is available on request. Briefly, 80–100 cells in each sample were randomly selected. DAPI stain was used to segment the outlines of nuclei and masked on the 53BP1 stain to measure the 53BP1 signal within the nuclei mask area. The distribution of all individual pixel intensity values of 53BP1 signal (X axis) in the 80–100 nuclei was then plotted against the total number of pixels (Y axis) in each sample and compared among the samples. Distribution of pixel intensity in each cohort in the 80–100 nuclei was examined using histogram with the x-axis being for intensities of 0 to 4096, (binsize = 1).

Statistical analysis

Data were objectively compared between different groups for each sample using unpaired and two-tailed Student's t test (*, **, and *** represent $p<0.05$, $p<0.01$, and $p<0.001$, respectively).

Results

53BP1 functions in conservative-NHEJ dependent on RNF8 and RNF168

53BP1 regulates DNA double-strand break (DSB) repair in the NHEJ pathway, but its specific function is unclear. To explore the role of 53BP1 in a specific pathway of DSB, we used cell lines that contain integrated into their genomes recombination substrates that specifically probe the conservative-NHEJ (C-NHEJ), alternative-NHEJ (Alt-NHEJ), homology directed repair (HDR), and single-strand annealing (SSA) repair pathways [33–35,37] (diagrammed in Figure 1A and C–E, *right*). The general strategy is to deplete by siRNA transfection 53BP1 or another factor, followed by transfection of a plasmid that expresses the rare-cutting restriction endonuclease I-SceI, which simulates a DSB at a specific site. 293/HW1 cells [37] contain a DNA substrate with two neighboring I-SceI sites in the genome for which repair by the C-NHEJ pathway can be measured by the precise joining of the DNA ends following I-SceI expression (Figure 1A, *right*). The concentration of DNAs repaired by C-NHEJ was measured by real-time PCR using an oligonucleotide probe that spans the break site. Depletion of ligase IV, which is known to affect the NHEJ repair frequency, reduced the C-NHEJ repair to 18% relative to the control siRNA. To test 53BP1 in the C-NHEJ pathway, 53BP1 was depleted in 293/HW1 cells, and we found that repair efficiency decreased to approximately 41% relative to the control siRNA (Figure 1A). Transfection of another siRNA targeting the *TP53BP1* 3′ untranslated region (3′UTR) sequence to deplete endogenous mRNA resulted in a decrease of C-NHEJ to 70% relative to the control. This siRNA reproducibly yielded less inhibition of the C-NHEJ than did the siRNA targeting the 53BP1 coding region. When this 3′-UTR specific siRNA was co-transfected with a plasmid expressing wild-type 53BP1 resistant to this siRNA, C-NHEJ repair efficiency was restored to 98% relative to the control, demonstrating the specificity of the siRNA depletions (Figure 1B). This observation of the role of 53BP1 in C-NHEJ is consistent with observations that 53BP1 promotes fusions of deprotected telomeres via C-NHEJ repair of the telomeres during G1 phase [40].

Depletion of BRCA1, a protein that regulates multiple DSB repair pathways, reduced C-NHEJ repair efficiency to about 49%. Co-depletion of 53BP1 and BRCA1 depressed the ratio further to about 19%, indicating that in this repair pathway BRCA1 and 53BP1 were not antagonistic, but rather each functioned to independently stimulate C-NHEJ.

The E3 ubiquitin ligases RNF8 and RNF168 have been demonstrated to be required for 53BP1 and BRCA1 localization at DSBs in an ubiquitination-dependent manner [17]. We depleted RNF8 or RNF168 by siRNA in 293/HW1 cells, resulting in a decrease in the C-NHEJ to 14% and 31%, respectively (lanes 6, 7). Co-depletion of both RNF8 and RNF168 did not have any additive effect compared to either single depletion (lane 10), indicating the epistatic role of RNF8 with RNF168 in the C-NHEJ pathway. We then tested whether RNF8 or RNF168 also regulates 53BP1 function in the C-NHEJ process. Co-depletion of RNF8 or RNF168 with 53BP1 had no additive effect relative to single depletion, consistent with the concept that RNF8 and RNF168 function in the same NHEJ pathway as 53BP1. These results along with prior observations [19,20,22] indicate that RNF8 and RNF168 are epistatic with 53BP1, which functions in C-NHEJ in a RNF8/RNF168-dependent manner.

We next tested 53BP1 function in the Alt-NHEJ pathway using a cell line, HeLa-EJ2, which has integrated in its genome a recombination substrate that is repaired by Alt-NHEJ to generate a functional GFP gene [33] (Figure 1C, *right*). BRCA1 depletion caused a decrease to 55% relative to the control siRNA, consistent with the literature [41]. Depletion of 53BP1, RNF8, or RNF168 each had minimal effect on Alt-NHEJ, which was not statistically significant (Figure 1C). We had anticipated that depletion of 53BP1 would enhance Alt-NHEJ function since the Alt-NHEJ pathway depends on resection of DNA ends, an activity thought to be inhibited by 53PB1. We suggest that the major product of this assay depends on resection of a short stretch of DNA, ~25 bp, and was not inhibited by 53BP1. Co-depletion of 53BP1 with RNF8 or

RNF168 had little impact on Alt-NHEJ though co-depletion of both RNF8 and RNF168 did impair the Alt-NHEJ repair efficiency compared to control (lanes 8–10). This last result suggests that RNF8 and RNF168 have redundant function in the Alt-NHEJ pathway, but it is independent of 53BP1.

We compared 53BP1 to BRCA1 function in homologous recombination, which has a major pathway of HDR and the minor SSA pathway. We utilized HeLa-DR cells and HeLa-SA cells to conduct the HDR and SSA assays, respectively. Repair by each pathway is measured by the conversion of cells to GFP-positive (Figure 1D and E, *right*). Depletion of 53BP1 had no effect in HDR (Figure 1D) but increased SSA (Figure 1E). The increase in SSA activity in 53BP1-depleted cells probably reflected relief from 53BP1-mediated inhibition of DNA resection needed for the SSA pathway. BRCA1 depletion affected both homologous recombination pathways (Figure 1D, E). Co-depletion of 53BP1 partially rescued the deficit caused by single depletion of BRCA1 in HDR or SSA, and this is consistent with the literature [42]. RNF8 and RNF168 depletions resulted in a statistically significant decrease in HDR but not in SSA. Co-depletion of both RNF8 and RNF168 had an additive effect in HDR and caused a decrease of repair efficiency. Co-depletion of 53BP1 and either RNF8 or RNF168 did not show any additive effect in both assays. BRCA2, which suppresses SSA [35,43], was used as a negative control (Figure 1E). Depletions of protein by siRNAs were confirmed by immunoblot (Figure 1F).

In summary for Figure 1, we investigated the role of 53BP1 in DNA DSB repair pathways and identified that it functions positively in C-NHEJ pathway, has no function in either HDR or Alt-NHEJ, and 53BP1 suppresses SSA. 53BP1 functions in the same C-NHEJ pathway as RNF8 and RNF168, which are known to regulate 53BP1 localization to sites of DSBs. Combined with prior studies, these results suggest that 53BP1 positively regulates C-NHEJ pathway in a RNF8- and RNF168-dependent manner.

53BP1 is destabilized upon irradiation damage

53BP1 forms ionizing radiation induced foci (IRIF) in response to DNA damage, and the protein has highest abundance during G1 phase, a stage in the cell cycle associated with NHEJ activity [44]. To investigate 53BP1 protein dynamics in response to irradiation-induced DNA damage, we evaluated changes in 53BP1 protein bulk level four hours post-irradiation (10 Gy). Surprisingly, the 53BP1 protein level decreased markedly compared to the non-irradiated control (Figure 2A, lanes 1–6). In this experiment we were careful to extract all of the 53BP1 protein in the HeLa cells by including 2% SDS in the lysis solution followed by a thorough sonication and heating at 100°C. Thus, the absence of 53BP1 protein in the irradiated cell lysates was not due to compartmentalization of the protein into an insoluble fraction. In contrast to 53BP1, the protein abundance of the homologous recombination factor RAD51 did not change upon irradiation. The DSB damage signal sensor γ-H2AX was used as a positive control and histone H4 served as the loading control. We found that 53BP1 protein levels decreased to very low concentration as early as 15 minutes following ionizing radiation and were restored after 24 hours (Figure 2A, lanes 2–7). In order to rule out that the detection of degradation of 53BP1 protein was specific for the antibody, we also tested for degradation when using HA-tagged 53BP1 protein expressed from a transiently transfected plasmid. Upon irradiation of 293T cells transfected with a HA-53BP1 expression plasmid and lysis in SDS containing buffer, detection of the HA-53BP1 via its epitope tag also revealed a substantial decrease in protein levels 4 h post-irradiation (Figure 2B, lanes 1, 2). Detection of the HA-53BP1 protein by the anti-HA antibody was specific for

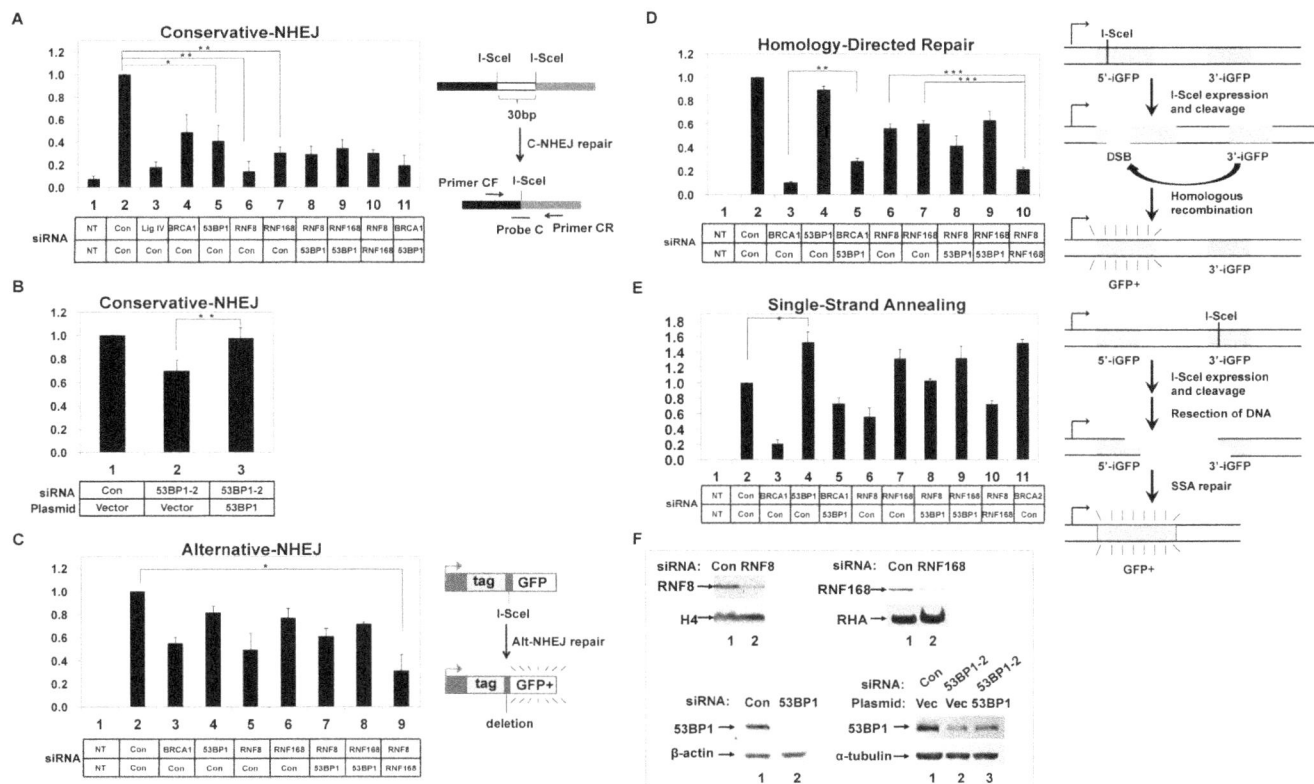

Figure 1. 53BP1 function in conservative-NHEJ is dependent on RNF8 and RNF168. The recombination substrates are diagrammed on the right with details described previously [33,35,37]. **A.** 293/HW1 cells transfected with indicated siRNAs (bottom grid) followed by transfection of the I-SceI expression plasmid to induce DSB. After 3 days, the repair efficiency was measured by applying quantitative real-time PCR on extracted DNA, represented by the percentage on the Y axis. In each experiment, the yield of conservatively repaired DNA was normalized relative to the result from the control siRNA transfection. Results (+/− SEM) are from three independent experiments. NT indicates no transfection of the I-SceI expressing plasmid. **B.** same as in panel A except that siRNA targeting the 53BP1 3′UTR was transfected in combination with the wild-type 53BP1 expression plasmid or an empty vector, as indicated. **C–E.** cells were subjected to two rounds of transfections as in A and the percentages of GFP-positive cells were determined by flow cytometry. In each experiment, the percentage of GFP-positive cells from control siRNA transfections was set equal to 1, and the fraction of GFP-positive cells was determined relative to the control siRNA to measure Alt-NHEJ, HDR, and SSA, respectively. **F.** immunoblots show the depletion of indicated protein by RNAi interference, or the expression of 53BP1 protein by plasmid transfection. H4, RHA, β-actin or α-tubulin were loading controls.

transfected samples (Figure 2B, lane 3, 4). Furthermore, in two non-cancer cell lines, human retinal pigment epithelium (RPE) cells and normal mammary epithelium MCF10A cells, the endogenous 53BP1 protein levels in 2% SDS lysates were diminished 4 h post-IR (Figure 2B, lane 6, 8), consistent with the results from the HeLa and 293T cell lines. To rule out any potential artifact due to the incomplete extraction of 53BP1 protein from chromatin, urea containing buffer (8 M urea and 2% SDS in phosphate-buffered saline) was used as a reliable means for the complete extraction of the chromatin-bound protein followed by a thorough sonication and heating at 100°C. 53BP1 protein extracted by this method showed a greatly diminished level 4 h post-IR in HeLa cells, compared to no irradiation (Figure 2B, lane 9, 10), further confirming the above results. The level of repair protein RAD51 did not change upon irradiation. γ-H2AX was a positive damage sensor and histone H4 was used as a loading control in 293T, RPE, MCF10A and HeLa cell lysates (Figure 2B).

Since 53BP1 is a DSB repair protein, we were surprised to observe that its protein abundance sharply decreased as early as 15 minutes post DNA damage. It is well known that 53BP1 protein forms prominent IRIF after ionizing radiation, and this

seemed to contradict the results from post-irradiation whole cell lysates. To confirm the immunoblot results, we evaluated 53BP1 IRIF 4 h post-IR in HeLa cells and then utilized image analysis to measure the mean intensity of 53BP1 signal at IRIF regions in the nuclei. In non-irradiated cells, 53BP1 protein was diffusely abundant in nuclei, whereas upon ionizing radiation, 53BP1 was everywhere diminished except at DNA damage sites (Figure 2C). Staining 53BP1 IRIF in RPE cells or MCF10A cells, showed the same observation in comparison with non-irradiated cells (Figure S1 in File S1). We quantified the relative mean fluorescence intensity of the 53BP1 signal at foci regions from 100 irradiated nuclei in HeLa cells and 30 irradiated nuclei in human RPE cells or MCF10A cells by using ImageJ software. Similarly, we measured the 53BP1 mean intensity throughout the non-irradiated nuclei as 53BP1 diffusely stains the nuclei. The comparison between irradiated and non-irradiated nuclei showed no difference in the mean intensity of 53BP1 signal nucleoplasm before IR and at IRIF regions in the three cell lines (Figure 2D). This result together with observation in Figure 2C and Figure S1 in File S1 suggests that 53BP1 foci appearance upon IR is due in large part to the degradation of diffuse 53BP1 in the nucleoplasm at undamaged regions.

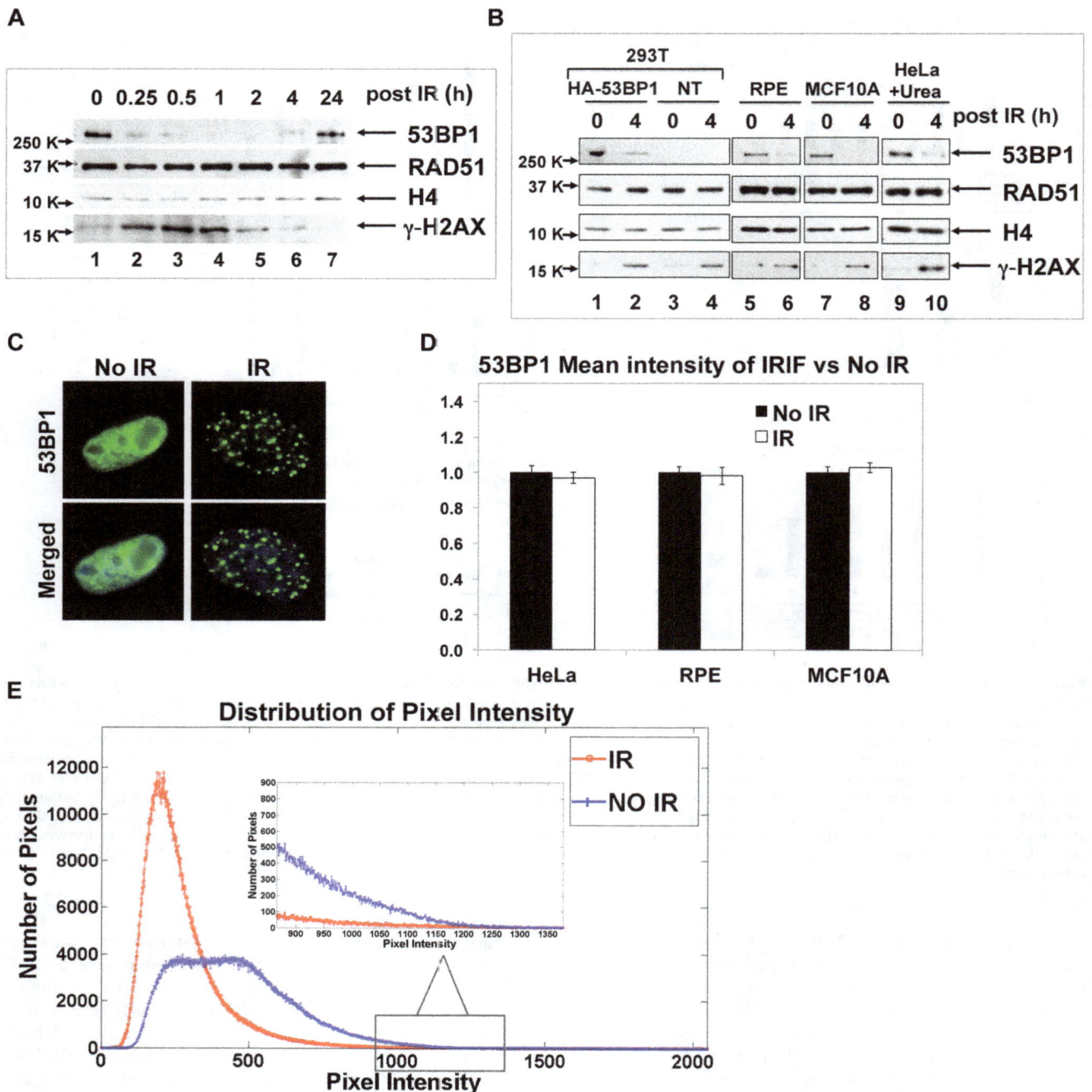

Figure 2. 53BP1 protein abundance decreases upon irradiation. A. HeLa cells were subjected to 10 Gy X-rays and total cell lysates were prepared in 2% SDS containing buffer at the indicated time points. Immunoblots of indicated protein were shown. H4 was a loading control. The positions of the molecular mass markers in kDa (K) are indicated at the *left*. B. HA-53BP1 plasmid was transfected into 293T cells, and 48 h post-transfection, cells were irradiated and extracted in SDS containing buffer; NT indicates no transfection (lanes 1–4). Retinal pigment epithelia cells (RPE; lanes 5, 6) and normal mammary epithelial cells (MCF10A; lanes 7, 8) were exposed to 10 Gy X-rays and total cell lysates were prepared in SDS containing buffer 4 h after irradiation. HeLa cells were exposed to 10 Gy irradiation similarly and total cell lysates were extracted 4 h post-IR in buffer containing 2% SDS and 8 M urea (lanes 9, 10). Immunoblots were developed using anti-HA antibody to detect HA-tagged 53BP1 protein in 293T cells (lane 1–4), and antibody specific for 53BP1 to recognize endogenous 53BP1 in RPE, MCF10A and HeLa cells (lane 5–10). C. HeLa cells were subjected to immunofluorescence microscopy 4 h post-irradiation (10 Gy). Cells were stained for 53BP1 (green; *top*) and merged with DAPI stain of DNA (blue; *bottom*). D. Similar to the immunofluorescence microscopy experiment in panel C, RPE and MCF10A cells were stained for 53BP1 protein before and after irradiation. For the immunofluorescence images, 100 nuclei were analyzed in HeLa cells, and 30 nuclei were analyzed in RPE or MCF10A cells using ImageJ software. Mean intensity of 53BP1 signal at foci regions in each nucleus in irradiation sample were scored and normalized according to the mean intensity of diffuse 53BP1 signal in no irradiation sample. Results (mean +/− SEM) were shown for three cell lines. E. The distribution of pixel intensities of 53BP1 nuclear stain in HeLa cells, as in panel C, was plotted for No IR (blue) and IR (red). Results were shown from 80–100 nuclei in each sample. Pixel intensities within the range of 850–1400 were shown in the graph inset.

Using the above image analysis method, we have assessed 53BP1 signal at damage sites represented by the mean intensity of fluorescence signal at IRIF, which is computed by the total intensity of pixels based on an area unit. There is a potential for bias in this approach in defining the regions for measuring the mean pixel intensity. We thus developed an unbiased method to assess the intensities of 53BP1 signal at all individual pixels within the nuclei (Figure 2E). 80–100 nuclei were outlined by DAPI stain, and the DAPI stained areas were analyzed for 53BP1 signal in irradiated or non-irradiated samples in HeLa cells using an algorithm in MATLAB R2013a. We plotted the distribution of all individual pixel intensities in the nuclei, and observed that for the irradiated nuclei, most of the pixels were in the very low intensity group (red trace in Figure 2E) as would be expected for foci, whereas the 53BP1 stain in the non-irradiated nuclei was more evenly distributed for pixel intensity and with fewer low intensity pixels but more high intensity pixels (blue trace in Figure 2E). Even for the foci in the irradiated sample with only a few pixels with very high intensity (representing bright dots of small area), the diffuse 53BP1 stain in non-irradiated cells was as intense (Figure 2E, inset). If the appearance of 53BP1 IRIF resulted from the movement of protein from the nucleoplasm to damage sites, then we would have anticipated that post IR a small number of pixels would gain intensity of 53BP1 stain above the baseline. Such an expectation was the opposite of what was observed: following IR most pixels had a decrease in 53BP1 staining intensity and foci did not have higher intensity stain when compared to the unirradiated nucleus. This result was consistent with data shown in Figure 2D and clearly suggests that the appearance of 53BP1 foci was due in large part to its degradation in the nucleoplasm. We also measured γ-H2AX signal by image analysis in both conditions (No IR versus IR, data not shown) and the protein level increased upon irradiation, consistent with observations by immunoblotting (Figure 2A). Though the appearance of 53BP1 at IRIF suggests recruitment from the nucleoplasm to DNA damage sites, these results presented here suggested instead that 53BP1 binding to damage sites renders it resistant to the universal degradation after ionizing radiation.

Next we tested if the decrease in 53BP1 protein abundance in response to irradiation in HeLa cells was due to proteasome-dependent degradation. Inclusion in medium of MG132, the proteasome inhibitor, blocked the decrease in 53BP1 protein concentration upon irradiation whereas inclusion of caffeine in the medium, an ATM inhibitor, did not. This result indicated that protein degradation, but not loss of ATM-dependent phosphorylation of 53BP1, caused the decrease of 53BP1 level after irradiation (Figure 3A). 53BP1 immunofluorescence following irradiation in HeLa cells was then analyzed in the presence of MG132, which resulted in the same diffuse 53BP1 pattern as observed in the absence of irradiation (Figure 3B). Image analysis confirmed that the IRIF do not have higher intensity 53BP1 stain than did the unirradiated sample or the IR plus MG132 sample (Figure 3C), indicating that 53BP1 protein was degraded upon irradiation.

Depletion of RNF8 or RNF168 blocks 53BP1 degradation upon irradiation

RNF8 and RNF168 are E3 ubiquitin ligases that mediate the conjugation of ubiquitin multimers on histone H2A via the degradation-independent lysine-63 side-chain of ubiquitin and via this activity recruit other proteins, such as 53BP1 and BRCA1, to the sites of DNA damage [17,19–22,26]. We tested the possibility that these two enzymes are involved in 53BP1 protein degradation. Indeed, depletion of RNF8 or of RNF168 from HeLa cells

and following irradiation-induced DNA damage, 53BP1 degradation was blocked (Figure 4A, lanes 6, 8, 10). Depletion of BRCA1, another E3 ubiquitin ligase involved in the DNA damage response, did not affect the 53BP1 protein level, indicating RNF8/RNF168 had a specific role in the control of 53BP1 protein levels upon irradiation (Figure 4A, lane 4). Consistent with the model that 53BP1 is degraded dependent on RNF8 and RNF168, the 53BP1 protein remained diffusely localized in the nucleus in HeLa cells in which these factors were depleted (Figure 4B). The distribution of pixel intensity in immunofluorescence images showed that depletion of RNF8 and/or RNF168 had a similar distribution pattern as no irradiation, though with more intense pixels (Figure 4C), suggesting that RNF8 and RNF168 regulate proteasome-dependent degradation and IRIF formation of 53BP1 in response to irradiation-induced DNA damage.

53BP1 turnover is accelerated upon irradiation damage

Since 53BP1 protein abundance changed following irradiation, we speculated that irradiation shortened the 53BP1 protein half-life. Using MCF7 (Figure 5) or HeLa cells (Figure S2 in File S1), we blocked new protein synthesis by the addition of cycloheximide to tissue culture media and made whole cell lysates in 2% SDS containing buffer. In the absence of IR, protein levels were stable (Figure 5A, lanes 1–3). By contrast, following IR, 53BP1 turnover was apparent as early as 30 min post-IR, and the protein level decreased to 4% 4 hours post IR (Figure 5A, B). By contrast, RAD51 half-life was not affected by IR (Figure 5A). The results together implicated that ionizing radiation accelerates 53BP1 protein turnover via the proteasome-dependent pathway.

53BP1 is protected from degradation at damage sites

Previous reports had suggested that 53BP1 is recruited to damage sites in an RNF8/RNF168 dependent manner [19,22]. We speculated that after 53BP1 stably localizes and binds to DSBs at chromatin at time points as early as four minutes post-irradiation [22], degradation of 53BP1 rather than the protein movement to the sites of DSBs, leads to prominent 53BP1 focus formation. To differentiate between movement of the protein versus degradation of bulk 53BP1, we irradiated HeLa cells and after foci formed we blocked proteasome-mediated degradation. If 53BP1 protein relocated within the nucleus, then foci would remain even if the proteasome were blocked. If, on the other hand, the 53BP1 bound to the DNA damage site was stabilized then the foci would be surrounded by diffuse 53BP1 as new protein was synthesized. MG132 was added to the cells 1 hour post-irradiation and a series of times points were analyzed to observe the 53BP1 foci at the damage sites. (Refer to time-line in Figure 6A, *top*.) In the absence of the MG132, 53BP1 containing IRIF were apparent at all post-IR time-points analyzed. By comparison, in the cells in which MG132 was added to the medium one hour post-IR, foci were still apparent, but these IRIF were in the presence of diffuse 53BP1 stain at late time points (Figure 6A, *bottom*). As an indication of the diffuse 53BP1 localization, in the presence of MG132 the nucleoli become apparent as holes in the diffuse pattern. The results from this experiment were quantified in Figure 6B and show that MG132 treated cells primarily had diffuse nuclear 53BP1 stain or diffuse stain with foci, suggesting that degradation event in the nucleoplasm happened prior to MG132 treatment, and the diffuse stain was due to the appearance of newly synthesized 53BP1. When MG132 was added to the cells prior to IR, no foci were apparent (Figure 3B). These results are most consistent with a model in which 53BP1 is continuously synthesized at a high rate, and post-IR it is rapidly degraded via the ubiquitin-proteasome system (Figure 5A, B). Only at sites of

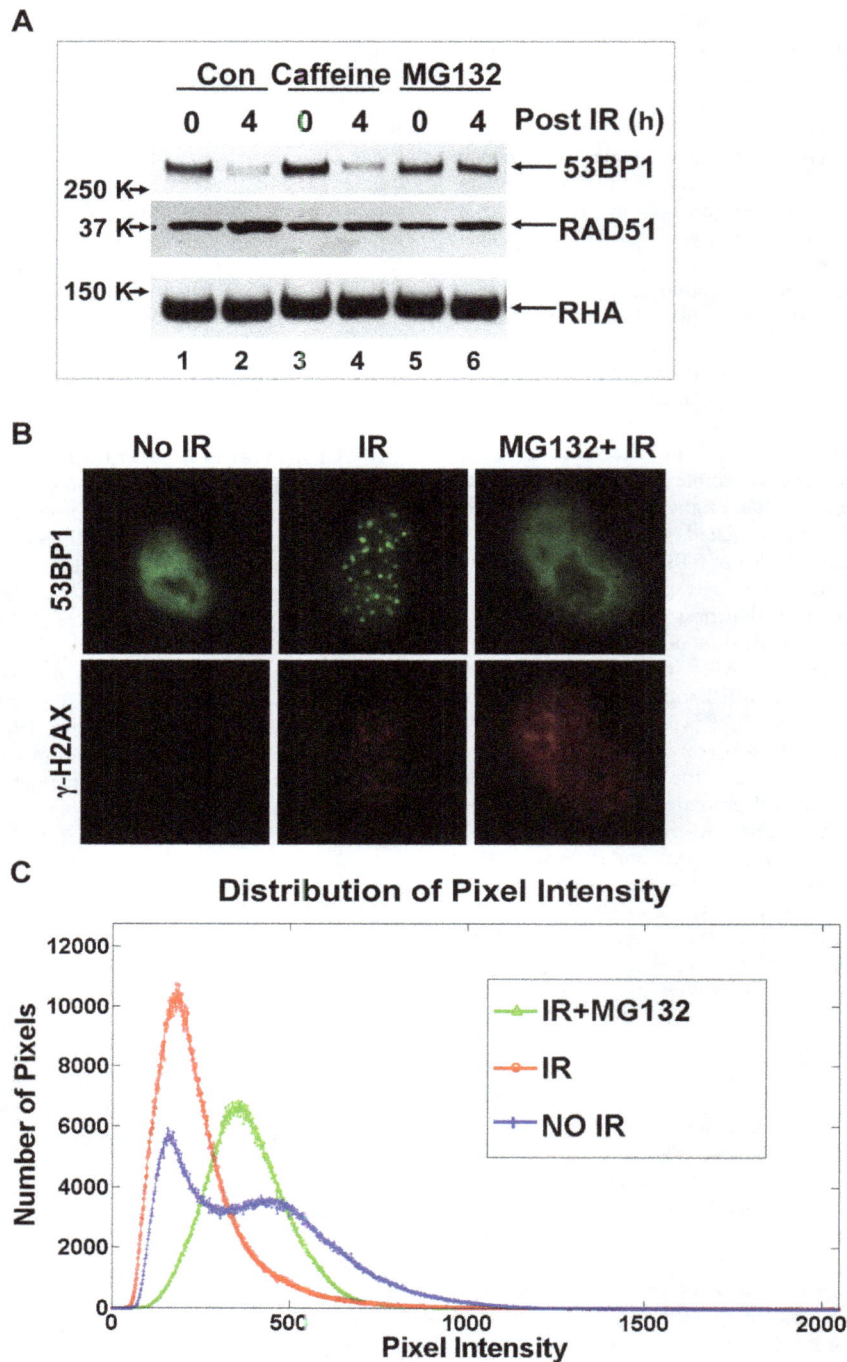

Figure 3. 53BP1 protein content decreases upon irradiation is due to protein degradation. A. Before subjection to 10 Gy X-rays, HeLa cells were treated with medium alone (Con; lanes 1, 2), caffeine (10 mM; lane 3, 4) or with MG132 (20 μM; lane 5, 6). 4 h post irradiation, cell lysates were prepared for immunoblots. **B.** HeLa cells were treated with or without MG132 as in panel A, and immunofluorescence microscopy was applied after 4 h irradiation (10 Gy) as in Figure 2C. **C.** Image analysis of cells as shown in panel B was done as in Figure 2E. The distribution of pixel intensity was plotted for each sample.

DNA damage is it protected from ubiquitin-dependent degradation. Immunoblot results of 53BP1 protein from experiment in Figure 6A, *top* were consistent with the notion that 53BP1 accumulated to high levels when in the presence of MG132 (Figure 6C). By comparison, the concentration of the downstream effector RIF1 did not change following DNA damage (Figure 6C). Together, these data suggest that following ionizing radiation, 53BP1 is rapidly synthesized and rapidly degraded except when bound to repair sites.

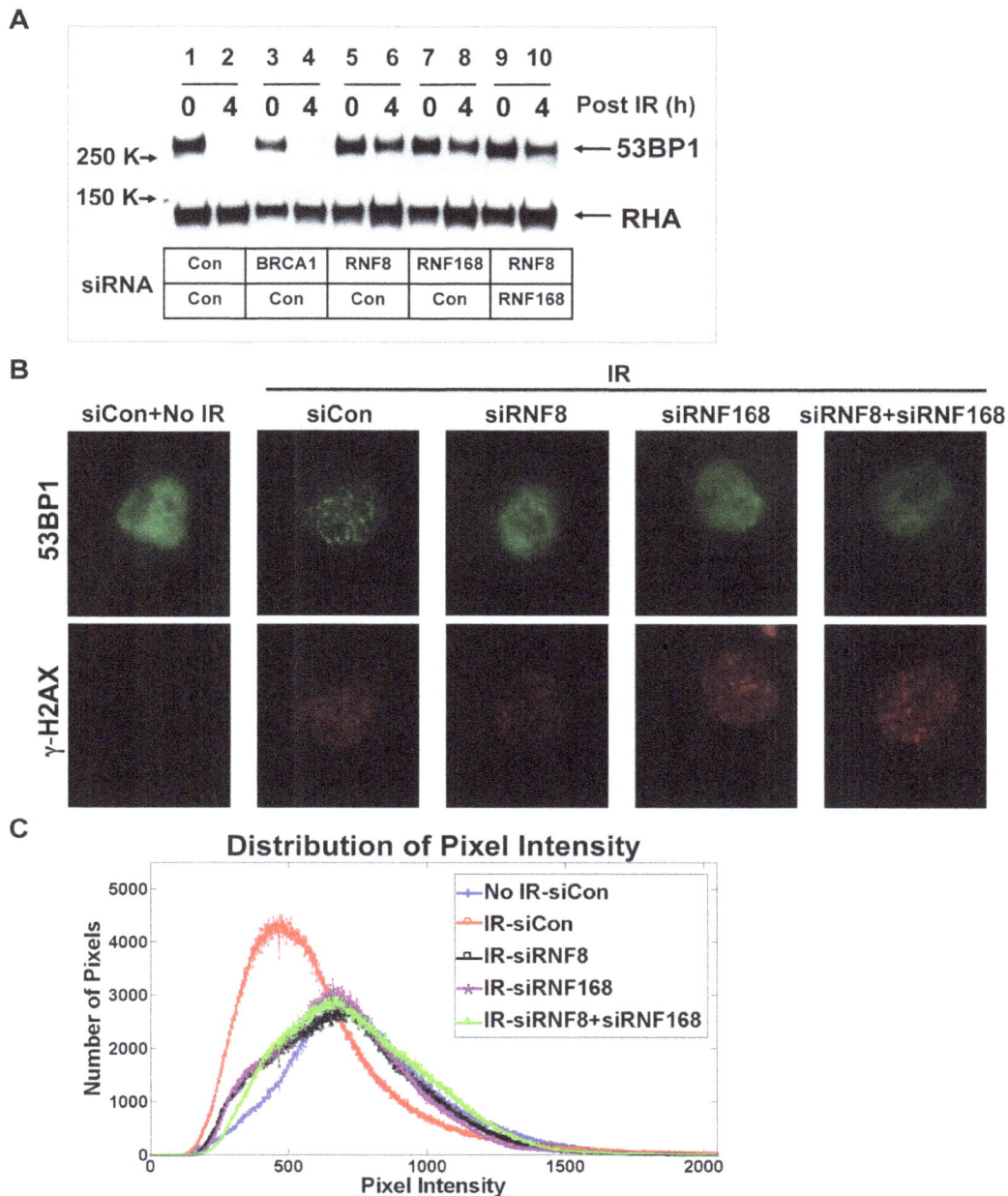

Figure 4. 53BP1 degradation upon irradiation is regulated by RNF8 and RNF168. A. HeLa cells transfected with two different siRNAs (indicated in the grid) were treated with 10 Gy X-rays. 4 hours post-IR cell lysates were prepared for 53BP1 immunoblot analysis. RHA served as a loading control. **B.** Immunofluorescence microscopy analysis of cells from panel A were stained for 53BP1 (green) and γ-H2AX (red). **C.** distribution of pixel intensity was analyzed from microscopic images in panel B.

53BP1 stability is important for RIF1 recruitment

RIF1 is the only known DNA damage repair factor that requires 53BP1 for its recruitment to damage sites [45], and which indirectly depends on RNF8 and RNF168 [46]. We tested if inhibition of 53BP1 protein degradation affects RIF1 association with IRIF. In contrast to 53BP1, irradiation did not affect RIF1 protein levels detected from immunoblots of lysates from HeLa cells (Figure 7B, lane 2). Inclusion of MG132 abolished RIF1 association with IRIF in HeLa cells (Figure 7A) without changing its protein level (Figure 7B, lane 4). Similarly, caffeine or RNF8

and/or RNF168 depletion did not affect RIF1 abundance (Figure 7B, lane 3, 5–7 and [45]).

To test whether the inhibition of RIF1 IRIF in the presence of MG132 was directly associated with the failure to degrade the unbound bulk 53BP1, but not due to the impaired ubiquitin-dependent DSB signaling, we modified the immunofluorescence staining protocol to include a detergent extraction step prior to fixation. Following IR in the presence of MG132 in HeLa cells, the unbound bulk 53BP1 protein within the nucleus was removed by using cell extraction buffer containing 0.5% NP40 and 300 mM NaCl, and then cells were fixed and stained as usual.

A

B

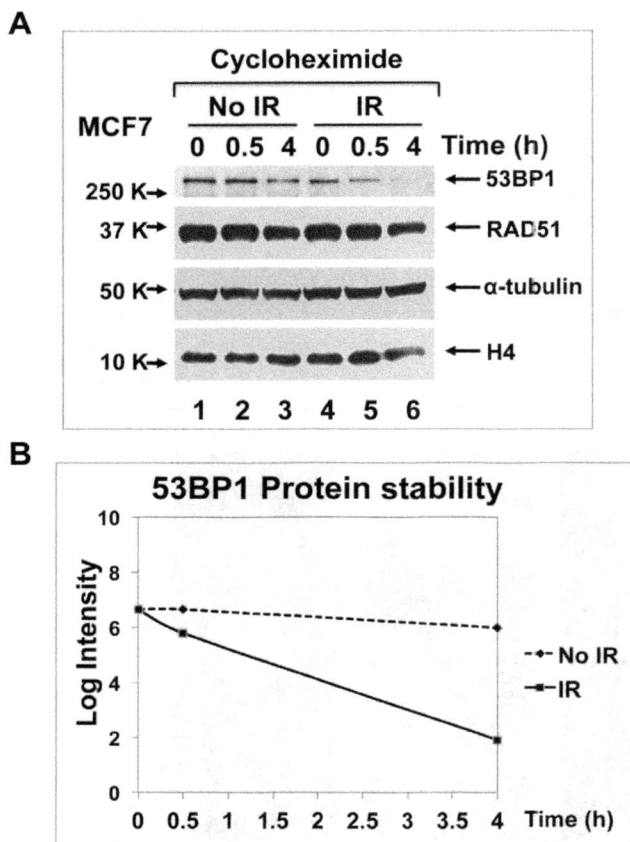

Figure 5. 53BP1 turnover is accelerated upon irradiation. A. Cycloheximide (100 µg/mL) was added to MCF7 cells with or without irradiation (10 Gy) and total cell lysates were prepared in SDS containing buffer according to the indicated time course (0, 0.5 and 4 h) and analyzed by immunoblotting as indicated. **B.** 53BP1 protein signal in immunoblot in A was measured by densitometry in each sample.

We observed chromatin-bound foci of 53BP1 that co-localized with γ-H2AX stain after irradiation and in the presence of MG132 (Figure 7C). The NP40 in the presence of 300 mM NaCl was sufficient to remove the loosely tethered bulk 53BP1 and reveal 53BP1 bound to the damage sites, consistent with a previous observation [47]. This result indicated that proteasome inhibition 30 minutes prior to DNA damage did not abrogate tight binding of 53BP1 to damage sites in chromatin. While prior studies had suggested that the RNF8-mediated degradation of KDM4A/JMJD2A was required to expose H4K20me2 for the recruitment of 53BP1 to DNA damage sites [16,23,24], our result revealed that the block due to KDM4A/JMJD2A binding to the chromatin was not absolute, and 53BP1 could still bind to the damage sites. We conclude from Figure 7C that proteasome inhibition by MG132 does not abolish 53BP1 association with IRIF, but blocks the degradation of the unbound bulk 53BP1, accounting for the failure to recruit RIF1 to the damage sites.

Taken together with data of RNF8/RNF168 regulation on 53BP1 function, stability and IRIF, we conclude that 53BP1 is initially recruited to damage sites by RNF8 and RNF168. The same regulators then degrade 53BP1 and thereby ensure proper 53BP1 protein concentration within the nucleus to recruit the downstream response factor RIF1 to damage sites for further

efficient repair, consistent with the results that 53BP1 functions in DSB repair (the C-NHEJ pathway) dependent on RNF8 and RNF168 (Figure 1A). If 53BP1 degradation fails to occur and the protein remains in high concentration throughout the nucleus, then RIF1 fails to bind to the DNA lesion. These results are consistent with there existing in the nucleus of a cell two pools of 53BP1, a small pool of 53BP1 bound to chromatin at the site of DNA damage, and a large pool of unbound or bulk 53BP1 in the nucleoplasm. Unbound 53BP1 is, in essence, a decoy that inhibits the signal from 53BP1 bound to the damage site.

Discussion

53BP1 is a DSB repair protein previously identified to influence the NHEJ process [14,16], though its specific function had not been clearly defined in this pathway. In this study, we found: 1) 53BP1 positively regulates the C-NHEJ pathway in a RNF8- and RNF168-dependent manner; 2) 53BP1 has no effect on the end resection dependent pathways of Alt-NHEJ pathway or on HDR, but it did suppress the end resection dependent SSA pathway; 3) the localization of 53BP1 at sites of DSBs is accompanied by the ensuing removal of bulk 53BP1 from the nucleus except at sites of DNA damage; 4) RNF8 and RNF168 are each required for the proteasome-mediated degradation of bulk 53BP1 after DNA damage; and 5) failure to degrade bulk 53BP1 in the nucleoplasm results in the failure for RIF1 to localize appropriately to DNA damage sites.

53BP1 binding to DSB sites

The immunoblots of total 53BP1 from four different cell lines (HeLa, 293T, RPE and MCF10A) clearly indicate that most of 53BP1 is degraded following ionizing radiation (Figure 2A, B). This observation of regulation of 53BP1 protein stability must be linked with how it binds to DSBs. 53BP1 localization to the sites of DNA damage requires the recognition of histone methylation, in particular H4K20me2 [28] by its tandem Tudor domain [29,30]. 53BP1 also binds to a second epitope, H2A ubiquitinated on Lys-15 (H2AK15ub), a product of RNF168 ubiquitination on chromatin, via its ubiquitin-dependent recruitment (UDR) motif. Initial recruitment of 53BP1 to DSBs also requires K63-linked ubiquitination of 53BP1 by RNF168 [32]. We suggest that bivalent binding of 53BP1 to epitopes on chromatin (H4K20me2 and H2AK15ub), or its K63-linked ubiquitination, may block the 53BP1 K48-linked ubiquitin targeted degradation, which ensues after 53BP1 localization to DSBs, thus distinguishing between bulk 53BP1 in the nucleoplasm and 53BP1 bound at the damage site. Since bulk 53BP1 is not bound to these chromatin epitopes, it would be susceptible to ubiquitination and degradation. Recruitment thus has a different mechanistic implication for the 53BP1 protein: rather than a movement of all of the 53BP1 in the nucleoplasm to the sites of DNA damage, following the early time points (4 min) when those molecules near the damage site move to the marked chromatin and stably bind [22,48], most 53BP1 is then degraded. The stabilization signal at sites of damage (Figure 7C) is in effect a signal for appearance of prominent focus formation. Within the first 4 minutes following ionizing radiation, the RNF8 and RNF168 proteins create the binding sites, and 53BP1 binds to the sites on the chromatin [22]. We suggest that some limited movement of 53BP1 is taking place at these early time points so that the protein localizes to the DNA lesion and degradation of the bulk 53BP1 in the nucleoplasm would occur soon after localization. Plus, 53BP1 protein continues to be synthesized, allowing more protein to accumulate at the IRIF while some fraction of the protein continues to be degraded. This observation does not

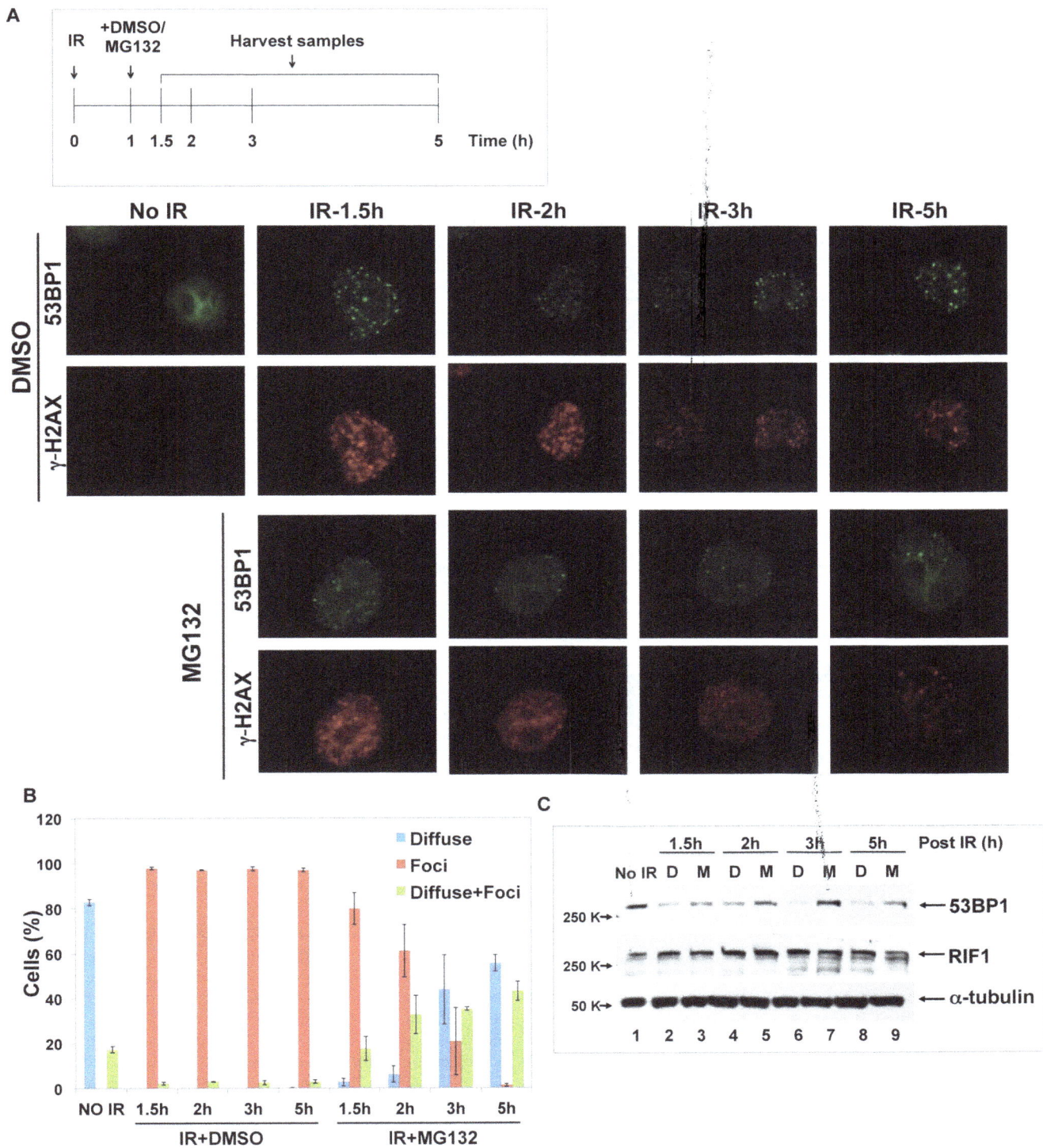

Figure 6. 53BP1 is degraded except when bound to a damage site. A. *upper* panel shows the workflow of the experiment. 1 h after 10 Gy X-ray irradiation was applied to the HeLa cells, DMSO or MG132 (20 μM) was added to the media. At time points 1.5 h, 2 h, 3 h, and 5 h post-IR, cells were either fixed for microscopy (A, *bottom*) or lysed for immunoblot analysis (C). **B**. in each sample, the percentage of the cells that have diffuse 53BP1 stain (blue), 53BP1 foci (orange) or diffuse 53BP1 stain with foci (green) was quantified (mean ±SEM; N = 3). **C**. cell lysates taken from panel A were subjected to immunoblot for 53BP1 and RIF1 stain. Samples were treated with DMSO vehicle (D, even lanes) or MG132 (M, lanes 3, 5, 7, 9). Tubulin was a loading control.

A

B

C

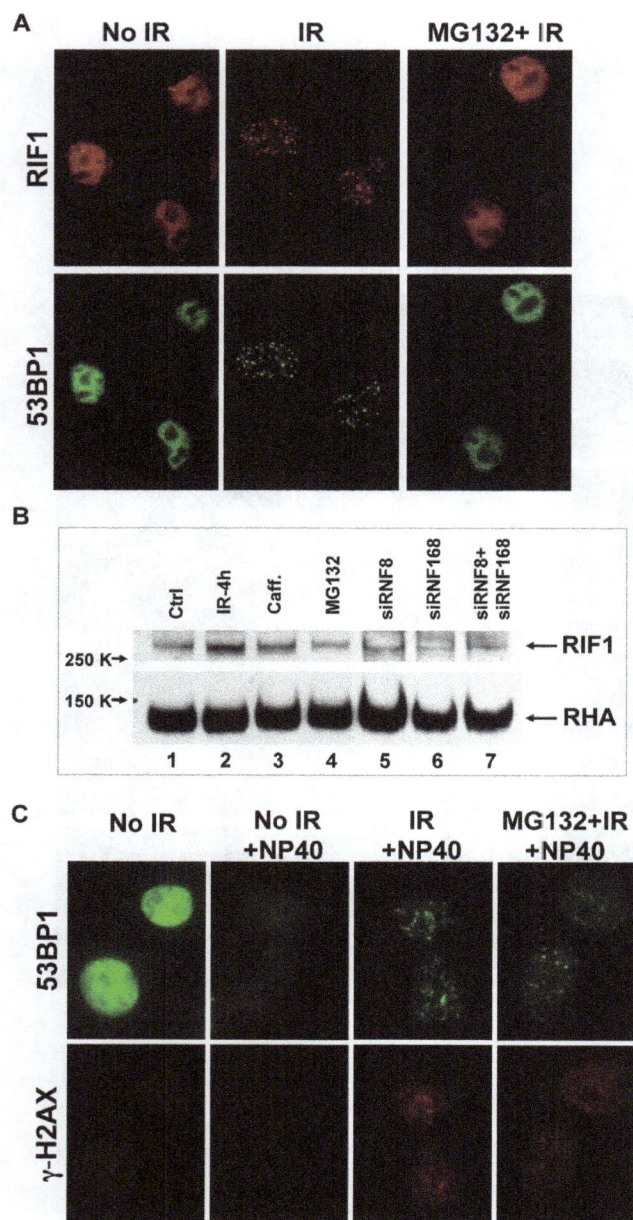

Figure 7. Inhibition of 53BP1 degradation causes failure to recruit RIF1 to the DSB sites. A. HeLa cells were treated with or without MG132 (20 µM) before exposure to 10 Gy X- irradiation. 4 h post-IR, cells were fixed for immunofluorescence microscopy as indicated. **B.** different treatments were applied to HeLa cells and immunoblot for RIF1 was done. Ctrl, no treatment (lane 1) or 4 h post 10 Gy-IR (lanes 2–7). Additional treatments included caffeine (10 mM; lane 3) MG132 (20 µM; lane 4), siRNA specific for RNF8 (lane 5), siRNA specific for RNF168 (lane 6), and mixed siRNA specific for both RNF8 and RNF168 (lane 7). **C.** MG132 (20 µM) was included in medium 30 minutes prior to exposure to 10 Gy X-ray irradiation. At 4 h post-IR, HeLa cells were extracted in situ with cell extraction buffer (Material and Methods) on ice for 15 min (+NP40) or not extracted. Cells were fixed and stained for immunofluorescence microscopy as above.

negate the prior observations that define the 53BP1 binding site on damaged chromatin, but rather amplifies our understanding of how the two mechanisms work in synergy to effect repairs.

53BP1 was shown to constitutively associate with chromatin in the absence of DNA damage independent of RNF8 [4,49] and likely this portion of chromatin-bound 53BP1 might quickly move to the damage sites at the early time points as some molecules are in the neighborhood of a DSB.

Mapping the region of 53BP1 that is required for its degradation upon IR and the identification of the site(s) within 53BP1 at which K48-linked ubiquitination occurs will have great interest. Mutation of the K48-linked ubiquitin acceptor site on 53BP1 will enable the testing the impact of 53BP1 degradation on IRIF and DSB repair pathways.

Previous studies have shown that 53BP1 was transiently immobilized at the sites of damage by Fluorescence Recovery After Photobleaching (FRAP) experiments [38,50]. It is possible that there is dynamic exchange between the IRIF-bound 53BP1 and newly synthesized protein consistent with prior observations [50].

Other prior studies have used laser micro-irradiation for generation of localized damage in cellular DNA, and 53BP1 localized to these DNA damage tracts [22,38,51]. We suggest that such a result does not contradict our current observations, but rather we posit that the damage induced by the laser micro-irradiation may have been local, but it induced a signal of the damaged state throughout the nucleus and causing bulk 53BP1 degradation. The implication for the mechanism of 53BP1 damage site binding is consistent with the observation that inhibition of the proteasome blocks 53BP1 degradation and blocks the appearance of 53BP1-containing IRIF in the nuclei (Figure 3B). When unbound bulk 53BP1 is removed from the nucleus by the detergent, tightly-bound 53BP1 reveals itself as IRIF in the presence of MG132, confirming that upon irradiation, 53BP1 is degraded everywhere with the exception of DNA damage sites where 53BP1 is bound and stabilized (Figure 7C). Excessive 53BP1 has been shown to be repressive to end-joining activity [16], and we suggest that this repressive activity may be due to the unbound pool of 53BP1 acting as a competitive inhibitor of the damage site bound 53BP1. Bulk 53BP1 prevents RIF1 from binding to the DNA in the undamaged state. After IR, bulk 53BP1 is degraded, and RIF1 is recruited to damage sites by bound 53BP1 to execute inhibition of end resection [15,46,52]. RIF1 is one of a few proteins identified to date that requires 53BP1 for its recruitment to DSBs [45] and is involved in C-NHEJ [46].

53BP1 versus BRCA1 function in DSB repair

53BP1 has been previously implicated as a competitor with BRCA1 in leading to opposite directions in the DSB repair process, NHEJ versus homologous recombination, respectively [13,14,46,53]. Consistent with this notion, depletion of 53BP1 partially suppressed the effects of depletion of BRCA1 on homologous recombination by partially rescuing defects caused by the loss of BRCA1 (Figure 1D, E). 53BP1 suppressed the SSA pathway, but had no effect on HDR, however in each case its depletion could partially correct the defect due to depletion of BRCA1. On the other hand, depletion of BRCA1 caused deficits of varying magnitude to all four DSB pathways, including C-NHEJ, which was stimulated independently by both 53BP1 and by BRCA1, suggesting that the opposing function of BRCA1 versus 53BP1 is actually complex.

The roles of RNF8 and RNF168

We found that 53BP1 specific function in the C-NHEJ process is RNF8- and RNF168-dependent since these factors regulated the degradation of 53BP1. This result is in agreement with a study that RIF1 accumulation at DSB sites is dependent on RNF8 and

RNF168 [46]. RNF8 and RNF168 are the E3 ligases that conjugate ubiquitin to histones H2A and H2AX. These provide the binding site for 53BP1 at DSBs [29]. The simplest interpretation of the results would be that one of RNF8 or RNF168 has the additional activity of ubiquitinating bulk 53BP1 via the K48-linked side chain, marking it for degradation by the ubiquitin-proteasome system (UPS). Since depletion of either factor stabilizes 53BP1 as effectively as simultaneously depleting both, since RNF8 is upstream of RNF168, and since RNF168 interacts directly with 53BP1 [32], we suggest that RNF168 is the E3 ligase that directly targets 53BP1 for proteasome mediated degradation. Such a notion would imply that RNF168 can change its specificity for different E2 factors from one E2 that targets ubiquitin Lys-63 linkages when modifying H2A to another E2 that targets ubiquitin Lys-48 linkages when modifying bulk 53BP1. Alternatively, it is possible that an as yet undetermined E3 ubiquitin ligase is downstream of RNF8 and RNF168 and which mediates the K48-linked ubiquitination of the bulk 53BP1.

Our results indicate that RNF8 and RNF168 function differently in each DSB repair pathway. For the C-NHEJ pathway, RNF8, RNF168, and 53BP1 have an epistatic relationship since pairwise depletion in any combination of these three factors equally affected the repair rate as single depletion (Figure 1A). By contrast, simultaneous depletion of RNF8 and RNF168 had a more severe effect on the Alt-NHEJ pathway. We interpret this finding to mean that these two ubiquitin ligases function independently in this NHEJ pathway. Similarly, in the HDR pathway depletion of the RNF8 and RNF168 were additive in their effect. These results dissect out distinct interrelationships between these two ubiquitin ligases and with 53BP1, and these give clues to the mechanisms by which they act in the DSB repair.

In summary, this study found that following ionizing radiation, there are effectively two populations of 53BP1: stable DSB-bound 53BP1 and bulk 53BP1 protein that is degraded by the proteasome pathway. Degradation of 53BP1 prevents competition between bulk 53BP1 and the DSB-bound 53BP1 for the recruitment of RIF1 to DNA damage sites. In addition, it was found that 53BP1 functions positively in the sequence conserving precise NHEJ process dependent on RNF8 and RNF168, but 53BP1 has no role in the error prone Alt-NHEJ process. Loss of 53BP1 protein would thus be highly mutagenic since the error-prone NHEJ repair would predominate. This study also dissected how the RNF8 and RNF168 protein function in the four DSB repair pathways and indicate that relationships between these factors are pathway-specific.

Supporting Information

File S1 Supporting Figures. Figure S1. 53BP1 protein is everywhere diminished except at DNA damage sites. RPE cells or MCF10A cells were subjected to immunofluorescence microscopy 4 h post-irradiation (10 Gy). Cells were stained for 53BP1 (green; *top*) and γ-H2AX (red; *bottom*). **Figure S2. In HeLa cells, 53BP1 turnover was accelerated upon irradiation.** Procedure in lane 3–8 was done as in Figure 5A except that HeLa cells were analyzed and two controls were included: no irradiation and irradiation (post 4 h IR) in lane 1 and 2.

Acknowledgments

We are grateful to Jeremy Stark (Beckman Research Institute of the City of Hope) and to Kuniyoshi Iwabuchi (Kanazawa Medical University) for the kind gift of plasmid reagents used in this study.

Author Contributions

Conceived and designed the experiments: YH JDP NM. Performed the experiments: YH NM. Analyzed the data: CW KH YH. Contributed reagents/materials/analysis tools: FX. Wrote the paper: YH JDP NM.

References

1. Fishman-Lobell J, Rudin N, Haber JE (1992) Two alternative pathways of double-strand break repair that are kinetically separable and independently modulated. Mol Cell Biol 12: 1292–1303.

2. Ma Y, Lu H, Schwarz K, Lieber MR (2005) Repair of double-strand DNA breaks by the human nonhomologous DNA end joining pathway: the iterative processing model. Cell Cycle 4: 1193–1200.

3. Lieber MR, Wilson TE (2010) SnapShot: Nonhomologous DNA end joining (NHEJ). Cell 142: 496–496 e491.

4. Bothmer A, Robbiani DF, Di Virgilio M, Bunting SF, Klein IA, et al. (2011) Regulation of DNA end joining, resection, and immunoglobulin class switch recombination by 53BP1. Mol Cell 42: 319–329.

5. Coleman KA, Greenberg RA (2011) The BRCA1-RAP80 complex regulates DNA repair mechanism utilization by restricting end resection. J Biol Chem 286: 13669–13680.

6. Difilippantonio S, Gapud E, Wong N, Huang CY, Mahowald G, et al. (2008) 53BP1 facilitates long-range DNA end-joining during V(D)J recombination. Nature 456: 529–533.

7. Dimitrova N, Chen YC, Spector DL, de Lange T (2008) 53BP1 promotes nonhomologous end joining of telomeres by increasing chromatin mobility. Nature 456: 524–528.

8. Noon AT, Shibata A, Rief N, Lobrich M, Stewart GS, et al. (2010) 53BP1-dependent robust localized KAP-1 phosphorylation is essential for heterochromatic DNA double-strand break repair. Nat Cell Biol 12: 177–184.

9. Xie A, Hartlerode A, Stucki M, Odate S, Puget N, et al. (2007) Distinct roles of chromatin-associated proteins MDC1 and 53BP1 in mammalian double-strand break repair. Mol Cell 28: 1045–1057.

10. Bunting SF, Callen E, Kozak ML, Kim JM, Wong N, et al. (2012) BRCA1 functions independently of homologous recombination in DNA interstrand crosslink repair. Mol Cell 46: 125–135.

11. Manis JP, Morales JC, Xia Z, Kutok JL, Alt FW, et al. (2004) 53BP1 links DNA damage-response pathways to immunoglobulin heavy chain class-switch recombination. Nat Immunol 5: 481–487.

12. Ward IM, Reina-San-Martin B, Olaru A, Minn K, Tamada K, et al. (2004) 53BP1 is required for class switch recombination. J Cell Biol 165: 459–464.

13. Bouwman P, Aly A, Escandell JM, Pieterse M, Bartkova J, et al. (2010) 53BP1 loss rescues BRCA1 deficiency and is associated with triple-negative and BRCA-mutated breast cancers. Nat Struct Mol Biol 17: 688–695.

14. Bunting SF, Callen E, Wong N, Chen HT, Polato F, et al. (2010) 53BP1 inhibits homologous recombination in Brca1-deficient cells by blocking resection of DNA breaks. Cell 141: 243–254.

15. Zimmermann M, Lottersberger F, Buonomo SB, Sfeir A, de Lange T (2013) 53BP1 regulates DSB repair using Rif1 to control 5′ end resection. Science 339: 700–704.

16. Butler LR, Densham RM, Jia J, Garvin AJ, Stone HR, et al. (2012) The proteasomal de-ubiquitinating enzyme POH1 promotes the double-strand DNA break response. EMBO J 31: 3918–3934.

17. Al-Hakim A, Escribano-Diaz C, Landry MC, O'Donnell L, Panier S, et al. (2010) The ubiquitous role of ubiquitin in the DNA damage response. DNA Repair (Amst) 9: 1229–1240.

18. Bekker-Jensen S, Mailand N (2011) The ubiquitin- and SUMO-dependent signaling response to DNA double-strand breaks. FEBS Lett 585: 2914–2919.

19. Doil C, Mailand N, Bekker-Jensen S, Menard P, Larsen DH, et al. (2009) RNF168 binds and amplifies ubiquitin conjugates on damaged chromosomes to allow accumulation of repair proteins. Cell 136: 435–446.

20. Huen MS, Grant R, Manke I, Minn K, Yu X, et al. (2007) RNF8 transduces the DNA-damage signal via histone ubiquitylation and checkpoint protein assembly. Cell 131: 901–914.

21. Kolas NK, Chapman JR, Nakada S, Ylanko J, Chahwan R, et al. (2007) Orchestration of the DNA-damage response by the RNF8 ubiquitin ligase. Science 318: 1637–1640.

22. Mailand N, Bekker-Jensen S, Faustrup H, Melander F, Bartek J, et al. (2007) RNF8 ubiquitylates histones at DNA double-strand breaks and promotes assembly of repair proteins. Cell 131: 887–900.

23. Mallette FA, Mattiroli F, Cui G, Young LC, Hendzel MJ, et al. (2012) RNF8-and RNF168-dependent degradation of KDM4A/JMJD2A triggers 53BP1 recruitment to DNA damage sites. EMBO J 31: 1865–1878.

24. Mallette FA, Richard S (2012) K48-linked ubiquitination and protein degradation regulate 53BP1 recruitment at DNA damage sites. Cell Res 22: 1221–1223.

25. Meerang M, Ritz D, Paliwal S, Garajova Z, Bosshard M, et al. (2011) The ubiquitin-selective segregase VCP/p97 orchestrates the response to DNA double-strand breaks. Nat Cell Biol 13: 1376–1382.

26. Stewart GS (2009) Solving the RIDDLE of 53BP1 recruitment to sites of damage. Cell Cycle 8: 1532–1538.

27. Wang B, Elledge SJ (2007) Ubc13/Rnf8 ubiquitin ligases control foci formation of the Rap80/Abraxas/Brca1/Brcc36 complex in response to DNA damage. Proc Natl Acad Sci U S A 104: 20759–20763.

28. Botuyan MV, Lee J, Ward IM, Kim JE, Thompson JR, et al. (2006) Structural basis for the methylation state-specific recognition of histone H4-K20 by 53BP1 and Crb2 in DNA repair. Cell 127: 1361–1373.

29. Fradet-Turcotte A, Canny MD, Escribano-Diaz C, Orthwein A, Leung CC, et al. (2013) 53BP1 is a reader of the DNA-damage-induced H2A Lys 15 ubiquitin mark. Nature 499: 50–54.

30. Huyen Y, Zgheib O, Ditullio RA Jr, Gorgoulis VG, Zacharatos P, et al. (2004) Methylated lysine 79 of histone H3 targets 53BP1 to DNA double-strand breaks. Nature 432: 406–411.

31. Sanders SL, Portoso M, Mata J, Bahler J, Allshire RC, et al. (2004) Methylation of histone H4 lysine 20 controls recruitment of Crb2 to sites of DNA damage. Cell 119: 603–614.

32. Bohgaki M, Bohgaki T, El Ghamrasni S, Srikumar T, Maire G, et al. (2013) RNF168 ubiquitylates 53BP1 and controls its response to DNA double-strand breaks. Proc Natl Acad Sci U S A 110: 20982–20987.

33. Bennardo N, Cheng A, Huang N, Stark JM (2008) Alternative-NHEJ is a mechanistically distinct pathway of mammalian chromosome break repair. PLoS Genet 4: e1000110.

34. Ransburgh DJ, Chiba N, Ishioka C, Toland AE, Parvin JD (2010) Identification of breast tumor mutations in BRCA1 that abolish its function in homologous DNA recombination. Cancer Res 70: 988–995.

35. Towler WI, Zhang J, Ransburgh DJ, Toland AE, Ishioka C, et al. (2013) Analysis of BRCA1 variants in double-strand break repair by homologous recombination and single-strand annealing. Hum Mutat 34: 439–445.

36. Hu Y, Parvin JD (2014) Small Ubiquitin-like Modifier (SUMO) Isoforms and Conjugation-independent Function in DNA Double-strand Break Repair Pathways. J Biol Chem 289: 21289–21295.

37. Zhuang J, Jiang G, Willers H, Xia F (2009) Exonuclease function of human Mre11 promotes deletional nonhomologous end joining. J Biol Chem 284: 30565–30573.

38. Galanty Y, Belotserkovskaya R, Coates J, Polo S, Miller KM, et al. (2009) Mammalian SUMO E3-ligases PIAS1 and PIAS4 promote responses to DNA double-strand breaks. Nature 462: 935–939.

39. Zhang F, Bick G, Park JY, Andreassen PR (2012) MDC1 and RNF8 function in a pathway that directs BRCA1-dependent localization of PALB2 required for homologous recombination. J Cell Sci 125: 6049–6057.

40. Lottersberger F, Bothmer A, Robbiani DF, Nussenzweig MC, de Lange T (2013) Role of 53BP1 oligomerization in regulating double-strand break repair. Proc Natl Acad Sci U S A 110: 2146–2151.

41. Zhong Q, Chen CF, Chen PL, Lee WH (2002) BRCA1 facilitates microhomology-mediated end joining of DNA double strand breaks. J Biol Chem 277: 28641–28647.

42. Nakada S, Yonamine RM, Matsuo K (2012) RNF8 regulates assembly of RAD51 at DNA double-strand breaks in the absence of BRCA1 and 53BP1. Cancer Res 72: 4974–4983.

43. Stark JM, Pierce AJ, Oh J, Pastink A, Jasin M (2004) Genetic steps of mammalian homologous repair with distinct mutagenic consequences. Mol Cell Biol 24: 9305–9316.

44. van Vugt MA, Gardino AK, Linding R, Ostheimer GJ, Reinhardt HC, et al. (2010) A mitotic phosphorylation feedback network connects Cdk1, Plk1, 53BP1, and Chk2 to inactivate the G(2)/M DNA damage checkpoint. PLoS Biol 8: e1000287.

45. Silverman J, Takai H, Buonomo SB, Eisenhaber F, de Lange T (2004) Human Rif1, ortholog of a yeast telomeric protein, is regulated by ATM and 53BP1 and functions in the S-phase checkpoint. Genes Dev 18: 2108–2119.

46. Escribano-Diaz C, Orthwein A, Fradet-Turcotte A, Xing M, Young JT, et al. (2013) A cell cycle-dependent regulatory circuit composed of 53BP1-RIF1 and BRCA1-CtIP controls DNA repair pathway choice. Mol Cell 49: 872–883.

47. Iwabuchi K, Basu BP, Kysela B, Kurihara T, Shibata M, et al. (2003) Potential role for 53BP1 in DNA end-joining repair through direct interaction with DNA. J Biol Chem 278: 36487–36495.

48. Ogiwara H, Ui A, Otsuka A, Satoh H, Yokomi I, et al. (2011) Histone acetylation by CBP and p300 at double-strand break sites facilitates SWI/SNF chromatin remodeling and the recruitment of non-homologous end joining factors. Oncogene 30: 2135–2146.

49. Santos MA, Huen MS, Jankovic M, Chen HT, Lopez-Contreras AJ, et al. (2010) Class switching and meiotic defects in mice lacking the E3 ubiquitin ligase RNF8. J Exp Med 207: 973–981.

50. Asaithamby A, Chen DJ (2009) Cellular responses to DNA double-strand breaks after low-dose gamma-irradiation. Nucleic Acids Res 37: 3912–3923.

51. Bekker-Jensen S, Lukas C, Melander F, Bartek J, Lukas J (2005) Dynamic assembly and sustained retention of 53BP1 at the sites of DNA damage are controlled by Mdc1/NFBD1. J Cell Biol 170: 201–211.

52. Chapman JR, Barral P, Vannier JB, Borel V, Steger M, et al. (2013) RIF1 is essential for 53BP1-dependent nonhomologous end joining and suppression of DNA double-strand break resection. Mol Cell 49: 858–871.

53. Tang J, Cho NW, Cui G, Manion EM, Shanbhag NM, et al. (2013) Acetylation limits 53BP1 association with damaged chromatin to promote homologous recombination. Nat Struct Mol Biol 20: 317–325.

Permissions

List of Contributors

Scott Cukras, Nicholas Morffy and Younghoon Kee
Department of Cell Biology, Microbiology, and Molecular Biology, College of Arts and Sciences, University of South Florida, Tampa, Florida, United States of America

Takbum Ohn
Department of Cellular & Molecular Medicine, College of Medicine, Chosun University, Gwangju, Republic of Korea

Yuri L. Bunimovich
Department of Molecular and Medical Pharmacology, University of California Los Angeles, Los Angeles, California, United States of America
Crump Institute for Molecular Imaging, University of California Los Angeles, Los Angeles, California, United States of America

Evan Nair-Gill, Mireille Riedinger and Melissa N. McCracken
Department of Molecular and Medical Pharmacology, University of California Los Angeles, Los Angeles, California, United States of America

Donghui Cheng
Howard Hughes Medical Institute, University of California Los Angeles, Los Angeles, California, United States of America

Jami McLaughlin
Department of Microbiology, Immunology, and Molecular Genetics, David Geffen School of Medicine, University of California Los Angeles, Los Angeles, California, United States of America

Caius G. Radu
Department of Molecular and Medical Pharmacology, University of California Los Angeles, Los Angeles, California, United States of America
Crump Institute for Molecular Imaging, University of California Los Angeles, Los Angeles, California, United States of America
Ahmanson Translational Imaging Division, David Geffen School of Medicine, University of California Los Angeles, Los Angeles, California, United States of America

Owen N. Witte
Department of Molecular and Medical Pharmacology, University of California Los Angeles, Los Angeles, California, United States of America

Howard Hughes Medical Institute, University of California Los Angeles, Los Angeles, California, United States of America
Eli and Edythe Broad Center for Regenerative Medicine and Stem Cell Research, University of California Los Angeles, Los Angeles, California, United States of America
Department of Microbiology, Immunology, and Molecular Genetics, David Geffen School of Medicine, University of California Los Angeles, Los Angeles, California, United States of America

Yu-Fen Lin, Hung-Ying Shih, David J. Chen and Benjamin P. C. Chen
Department of Radiation Oncology, University of Texas Southwestern Medical Center at Dallas, Dallas, Texas, United States of America

Hatsumi Nagasawa, Takamitsu A. Kato, Joel S. Bedford and John R. Brogan
Department of Environmental and Radiological Health Sciences, Colorado State University, Fort Collins, Colorado, United States of America

John B. Little
Department of Genetics and Complex Diseases, Harvard School of Public Health, Boston, Massachusetts, United States of America

Xian-Jin Xie
Department of Clinical Sciences, University of Texas Southwestern Medical Center at Dallas, Dallas, Texas, United States of America

Paul F. Wilson Jr.
Department of Biosciences, Brookhaven National Laboratory, Upton, New York, United States of America

Akihiro Kurimasa
Institute of Regenerative Medicine and Biofunction, Graduate School of Medical Science, Tottori University, Tottori, Japan

Chang-Rong Zhang, Shan Zhang, Jun Xia, Fang-Fang Li, Wen-Qiang Xia, Shu-Sheng Liu and Xiao-Wei Wang
Ministry of Agriculture Key Laboratory of Agricultural Entomology, Institute of Insect Sciences, Zhejiang University, Hangzhou, China

Xiangjing Gao, Guanglin Zhang, Linfeng Chi, Yihua Wu, Chunlan Yan
Collaborative Innovation Center for Diagnosis and Treatment of Infectious Diseases, The First Affiliated Hospital, Zhejiang University, Hangzhou, Zhejiang, China
Department of Toxicology, Zhejiang University School of Public Health, Hangzhou, Zhejiang, China

Liya Kong
Department of preventative medicine, Zhejiang Chinese Medical University, Hangzhou, China

Xianghong Lu
Lishui People's Hospital, Lishui, Zhejiang, China

Ying Jiang
Center Testing International Corporation, Shenzhen, Guangdong, China

Penelope Duerksen-Hughes
Department of Basic Science, Loma Linda University School of Medicine, Loma Linda, Californina, United States of America

Xinqiang Zhu
Department of Toxicology, Zhejiang University School of Public Health, Hangzhou, Zhejiang, China

Jun Yang
Collaborative Innovation Center for Diagnosis and Treatment of Infectious Diseases, The First Affiliated Hospital, Zhejiang University, Hangzhou, Zhejiang, China
Department of Toxicology, Hangzhou Normal University School of Public Health, Hangzhou, Zhejiang, China
Department of Biomedicine, College of Biotechnology, Zhejiang Agriculture and Forestry University, Hangzhou, China

Sazan Ali and Marie-Roberte Guichaoua
Institut Méditerranéen de Biodiversité et d'Ecologie marine et continentale (IMBE), Centre National de la Recherche Scientifique (CNRS) UMR 7263/ Institut de Recherche pour le Développement (IRD) 237, Faculté de Médecine, Aix-Marseille Université (AMU), Marseille, France

Gérard Steinmetz and Odette Prat
Institute of Environmental Biology and Biotechnology (IBEB), Life Science division, French Alternative Energy and Atomic Energy Commission (CEA), Marcoule, Bagnols-sur-Céze, France

Guillaume Montillet, Marie-Héléne Perrard and Philippe Durand
Institut de Génomique Fonctionnelle de Lyon (IGFL), Centre National de la Recherche Scientifique (CNRS) UMR 5242/ Institut National de la Recherche Agronomique (INRA), Ecole Normale Supérieure de Lyon (ENS), Lyon, France

Anderson Loundou
Unité d'Aide Méthodologiqueàla Recherche clinique, Faculté de Médecine, Aix-Marseille Université (AMU), Marseille, France

Andrea Plecenikova
Gregor Mendel Institute, Austrian Academy of Sciences, Vienna Biocenter (VBC), Vienna, Austria
Department of Genetics, Faculty of Natural Sciences, Comenius University in Bratislava, Bratislava, Slovakia

Miroslava Slaninova
Department of Genetics, Faculty of Natural Sciences, Comenius University in Bratislava, Bratislava, Slovakia

Karel Riha
Gregor Mendel Institute, Austrian Academy of Sciences, Vienna Biocenter (VBC), Vienna, Austria
Central European Institute of Technology (CEITEC), Masaryk University, Brno, Czech Republic

Federico M. Lauro
School of Biotechnology and Biomolecular Sciences, The University of New South Wales, Sydney, New South Wales, Australia
Singapore Centre on Environmental Life Sciences Engineering (SCELSE), Nanyang Technological University, Singapore

Emiley A. Eloe-Fadrosh, Taylor K. S. Richter and Douglas H. Bartlett
Marine Biology Research Division, Scripps Institution of Oceanography, University of
California San Diego, La Jolla, California, United States of America

Nicola Vitulo
CRIBI Biotechnology Centre, University of Padua, Padova, Italy

Steven Ferriera and Justin H. Johnson
J. Craig Venter Institute, Rockville, Maryland, United States of America

Ameena H. El-Bibany, Andrea G. Bodnar and Helena C. Reinardy
Molecular Discovery Laboratory, Bermuda Institute of Ocean Sciences, St. George's, Bermuda

Erwin Reiling
National Institute for Public Health and the Environment, Bilthoven, The Netherlands
Department of Cell Biology and Genetics, Center for Biomedical Genetics, Erasmus MC, Rotterdam, The Netherlands

Martijn E. T. Dollé, Bhawani Nagarajah, Marianne Roodbergen and Piet de With
National Institute for Public Health and the Environment, Bilthoven, The Netherlands

Sameh A. Youssef and Alain de Bruin
Faculty of Veterinary Medicine, Department of Pathobiology, Utrecht University, Utrecht, The Netherlands

Moonsook Lee and Jan Vijg
Department of Genetics, Albert Einstein College of Medicine, Bronx, New York, United States of America

Jan H. Hoeijmakers
Department of Cell Biology and Genetics, Center for Biomedical Genetics, Erasmus MC, Rotterdam, The Netherlands

Harry van Steeg
National Institute for Public Health and the Environment, Bilthoven, The Netherlands
Department of Toxicogenetics, Leiden University Medical Center, Leiden, The Netherlands

Paul Hasty
Department of Molecular Medicine and Institute of Biotechnology, Barshop Institute for Longevity and Aging Studies, Cancer Therapy and Research Center, University of Texas Health Science Center at San Antonio, San Antonio, Texas, United States of America

Erwin Reiling
National Institute for Public Health and the Environment, Bilthoven, The Netherlands
Department of Cell Biology and Genetics, Center for Biomedical Genetics, Erasmus MC, Rotterdam, The Netherlands

Martijn E. T. Dollé, Bhawani Nagarajah, Marianne Roodbergen and Piet de With
National Institute for Public Health and the Environment, Bilthoven, The Netherlands

Sameh A. Youssef and Alain de Bruin
Faculty of Veterinary Medicine, Department of Pathobiology, Utrecht University, Utrecht, The Netherlands

Moonsook Lee and Jan Vijg
Department of Genetics, Albert Einstein College of Medicine, Bronx, New York, United States of America

Jan H. Hoeijmakers
Department of Cell Biology and Genetics, Center for Biomedical Genetics, Erasmus MC, Rotterdam, The Netherlands

Harry van Steeg
National Institute for Public Health and the Environment, Bilthoven, The Netherlands
Department of Toxicogenetics, Leiden University Medical Center, Leiden, The Netherlands

Paul Hasty
Department of Molecular Medicine and Institute of Biotechnology, Barshop Institute for Longevity and Aging Studies, Cancer Therapy and Research Center, University of Texas Health Science Center at San Antonio, San Antonio, Texas, United States of America

Mohammad Bani Ismail, Miki Shinohara and Akira Shinohara
Institute for Protein Research, Graduate School of Science, Osaka University, Suita, Osaka, Japan

Ayako Ishii, Aya Kurosawa and Shinta Saito
Graduate School of Nanobioscience, Yokohama City University, Yokohama, Japan

Noritaka Adachi
Graduate School of Nanobioscience, Yokohama City University, Yokohama, Japan
Advanced Medical Research Center, Yokohama City University, Yokohama, Japan

Andreas Luch
Department of Product Safety, German Federal Institute for Risk Assessment (BfR), Berlin, Germany

Flurina C. Clement Frey, Regula Meier, Jia Fei and Hanspeter Naegeli
Institute of Pharmacology and Toxicology, University of Zürich-Vetsuisse, Zürich, Switzerland

Takao Oishi and Masahiko Takada
Systems Neuroscience Section, Primate Research Institute, Kyoto University, Inuyama, Japan

Hiroo Imai, Masanori Imamura and Hirohisa Hirai
Molecular Biology Section, Primate Research Institute, Kyoto University, Inuyama, Japan

Yasuhiro Go
Department of Brain Sciences, Center for Novel Science Initiatives, National Institutes of Natural Sciences, Tokyo, Japan

Department of Developmental Physiology, National Institute for Physiological Sciences, Okazaki, Japan

Balázs Szalkai
PIT Bioinformatics Group, Eötvös University, Budapest, Hungary

Ildikó Scheer and Kinga Nagy
Laboratory of Genome Metabolism, Institute of Enzymology, Research Center for Natural Sciences, Hungarian Academy of Sciences, Budapest, Hungary

Beáta G. Vértessy
Laboratory of Genome Metabolism, Institute of Enzymology, Research Center for Natural Sciences, Hungarian Academy of Sciences, Budapest, Hungary Department of Applied Biotechnology and Food Sciences, Budapest University of Technology and Economics, Budapest, Hungary

Vince Grolmusz
PIT Bioinformatics Group, Eötvös University, Budapest, Hungary Uratim Ltd., Budapest, Hungary

Anurag Kirti, Hema Rajaram and Shree Kumar Apte
Molecular Biology Division, Bhabha Atomic Research Centre, Trombay, Mumbai, India

Akihisa Osakabe, Kenji Otawa, Hiroaki Tachiwana and Hitoshi Kurumizaka
Laboratory of Structural Biology, Graduate School of Advanced Science and Engineering, Waseda University, Tokyo, Japan

Yuichiro Takahashi, Hirokazu Murakami, Yukako Oma and Masahiko Harata
Laboratory of Molecular Biology, Graduate School of Agricultural Science, Tohoku University, Sendai, Japan

Hitoshi Nishijima and Kei-ich Shibahara
Department of Integrated Genetics, National Institute of Genetics, Mishima, Japan

Dylan P. G. Short, Suraj Gurung, Patrik Inderbitzin and Krishna V. Subbarao
Department of Plant Pathology, University of California Davis, Salinas, CA, United States of America

Xiaoping Hu
State Key Laboratory of Crop Stress Biology for Arid Areas and College of Plant Protection, Northwest A&F University, Yangling, Shaanxi, China

Dylan P. G. Short, Suraj Gurung, Patrik Inderbitzin and Krishna V. Subbarao
Department of Plant Pathology, University of California Davis, Salinas, CA, United States of America

Xiaoping Hu
State Key Laboratory of Crop Stress Biology for Arid Areas and College of Plant Protection, Northwest A&F University, Yangling, Shaanxi, China

Yufeng Wang, Yasuhiro Kuramitsu, Kazuhiro Tokuda, Byron Baron, Takao Kitagawa and Junko Akada
Department of Biochemistry and Functional Proteomics, Yamaguchi University Graduate School of Medicine, Ube, Yamaguchi, Japan

Shin-ichiro Maehara and Yoshihiko Maehara
Department of Surgery and Science, Graduate School of Medical Science, Kyusyu University, Fukuokashi, Fukuoka, Japan

Kazuyuki Nakamura
Centre of Clinical Laboratories in Tokuyama Medical Association Hospital, Shunan, Japan

Yiheng Hu, Chao Wang, Kun Huang and Jeffrey D. Parvin
Department of Biomedical Informatics, The Ohio State University, Columbus, Ohio, United States of America

Neelima Mondal
School of Life Sciences, Jawaharlal Nehru University, New Delhi, India

Fen Xia
Department of Radiation Oncology, The Ohio State University, Columbus, Ohio, United States of America

Index

A

Acetivorans, 142
Acetylation, 97, 110, 193
Acidocaldarius, 142, 144-145
Aeruginosa, 26-28, 30-33
Agriculture, 39, 46, 48, 99, 186-187
Agronomique, 58
Anesthesia, 17, 107, 147
Anisokaryosis, 101-102, 104-105
Annotation, 40-41, 71, 83-84, 157, 173, 189-191, 199
Anthraquinone, 207
Antibodies, 3, 15-17, 49-50, 112-113, 136-137, 139, 141-143, 184-185, 201-207, 210
Antimicrobial, 39-40, 43, 46
Apoptotic, 55, 69-70, 73, 100, 161, 207
Apyrimidinic, 79, 104, 108, 140, 146, 150
Ascoymycota, 186-187
Autophagosome, 200-201, 203-204, 207

B

Bacteriocyte, 46
Bacterium, 40, 82, 90-91
Bernstein, 47, 173, 208
Bioindicator, 93, 96, 98
Biosphere, 82, 90
Biotechnology, 3, 48-50, 58, 82, 101, 154, 201

C

Camptothecin, 3, 5, 174, 178, 182, 184
Cannabinoids, 206, 208
Capsulatum, 194, 197, 199
Carcinogenicity, 71, 135, 144
Carotenoids, 40, 47
Catalytic, 5, 14, 19, 27, 29, 36, 38, 81, 101, 108-109, 162, 189
Catastrophe, 57
Chaetomium, 194, 197, 199
Chemotaxis, 85, 88-89
Chlamydomonas, 73-74, 77, 80-81
Chloroform, 16, 59, 74, 83, 108
Chlorophyll, 164, 170-173
Cholesterol, 65, 72
Crustaceum, 197, 199

D

Deoxyribonucleoside, 14, 17, 21, 23, 25-26

D

Diethylstilbestrol, 71
Diplotene, 58-59, 62-63, 66, 68, 70
Dithiothreelectrophoretic, 166
Diversity, 39, 82, 89-91, 157, 187
Dothideomycetes, 199

E

Echinoderms, 92-93, 95, 98-100
Ecotoxicological, 92-93, 99-100
Electrophoresis, 5, 37, 40, 84, 113, 117, 124, 183-184, 195
Endocrine, 58-59, 64, 66-67, 70-72
Endosymbiont, 40, 46
Enzymatic, 23, 27, 46, 161

F

Fibroblast, 1-2, 8, 13, 22, 123-129
Fontanillas, 186

G

Galactopyranoside, 83, 108
Glycosylase, 44, 140-141, 143, 155, 160
Gorgoulis, 57, 221
Graminearum, 197, 199

H

Haematococca, 188, 197, 199
Haloferax, 155, 162
Healthcare, 16, 166, 182-184
Hematopoiesis, 14-15, 25, 27, 47
Hematoxylin, 107, 147
Hemoglobin, 147, 153
Homothallic, 198-199
Hyaluronan, 145-148, 150, 152-153
Hybridization, 59, 74-75, 78, 90, 180, 198

I

Immunoblot, 50, 52, 210-212, 215-219
Immunoglobulin, 22, 59, 220
Immunostaining, 26, 59, 112, 114-115, 118-119, 121-122, 184
Inoculating, 74, 83, 182
Integration, 73, 75-77, 80-81, 126-134, 180, 207
Intracellular, 14-16, 19-21, 200, 204, 207
Intramolecular, 182
Invertebrates, 47, 93, 98, 100
Iresembryonic, 15
Isolation, 37, 40, 74, 81, 90-91, 144
Isopropyl, 108, 183

K
Keratoplasticum, 198

L
Lethality, 101, 103, 108, 208
Lipopolysaccharides, 43
Longevity, 98-102, 106-107, 198
Lymphomas, 101, 103-104, 106-107

M
Machinmolecular, 187
Melanogaster, 46-47, 80, 155
Mercaptoethanol, 133, 146, 183
Metagenomic, 88-89, 154-158, 160-162

N
Neddylation, 1, 5-8, 11-13
Neurodegeneration, 200

O
Oligonucleotide, 163, 166-167, 171, 184, 211
Oxidative, 14, 18, 24-25, 27, 47, 59, 61, 65, 92, 99-100, 135, 140, 142, 144, 173

P
Pathogenesis, 45, 47, 124, 198-199
Penicillin, 30, 136, 146, 184
Perivascular, 148, 152
Pharmacol, 26, 57, 144, 207-208
Phosphatidylinositol, 203
Phosphorylation, 14-15, 17-20, 24, 26-27, 29-30, 36, 48-49, 56-57, 117, 185, 209, 220-221
Phylogenetic, 47, 82, 98, 154, 187, 198

R
Radiodurans, 171, 173
Radiosensitive, 15, 21, 30, 32, 35-36
Receptors, 43, 58, 66, 187, 197
Reductase, 11, 14, 21, 27, 66, 89, 200, 207
Ribosomal, 42-43, 46, 84, 88

S
Sonication, 153, 170, 198, 199
Spermatocytes, 46, 50, 51, 54, 56, 57
Starvation, 187, 191, 192, 194

www.ingramcontent.com/pod-product-compliance
Lightning Source LLC
Chambersburg PA
CBHW080637200326
41458CB00013B/4658